Handbook of Plant Disease Identification and Management

Handbook of Plant Disease Identification and Management

Balaji Aglave

CRC Press
Taylor & Francis Group
Boca Raton London New York

CRC Press is an imprint of the
Taylor & Francis Group, an **Informa** business

CRC Press
Taylor & Francis Group
6000 Broken Sound Parkway NW, Suite 300
Boca Raton, FL 33487-2742

First issued in paperback 2021

ISBN-13: 978-1-138-58547-8 (hbk)
ISBN-13: 978-1-03-209466-3 (pbk)

Library of Congress Cataloging-in-Publication Data

Names: Aglave, Balaji, author.
Title: Handbook of plant disease identification and management / Balaji
Aglave.
Description: Boca Raton, Florida : CRC Press, [2019]
Identifiers: LCCN 2018017131| ISBN 9781138585478 (hardback : alk. paper) |
ISBN 9780429504907 (ebook)
Subjects: LCSH: Plant diseases--Handbooks, manuals, etc. | Phytopathogenic
microorganisms--Control.
Classification: LCC SB731 .A34 2018 | DDC 632/.3--dc23
LC record available at https://lccn.loc.gov/2018017131

**Visit the Taylor & Francis Web site at
http://www.taylorandfrancis.com**

**and the CRC Press Web site at
http://www.crcpress.com**

Contents

Preface

Plant diseases are becoming major constraints in agricultural production with increases in the number of hybrid and genetically modified cultivators who are focused on increasing yield. Climatic changes are also creating favorable conditions for most of the diseases. At the same time, there are many reports of new pathogens that can cause considerable damage to agricultural crop. Disease identification plays a key role in overall diseases management. One disease may show different symptoms; on the contrary, different diseases may show same kind of symptoms. Scope and depth of our knowledge of plant and crop physiology are rapidly expanding, and plant physiologists are continuously making new discoveries.

The *Handbook of Plant Disease Identification and Management* will be a plant pathology and diseases related work, which will center around the topic of crop diseases. This handbook will provide fundamental knowledge about how to identify the disease, how to track disease development, and IPM (Integrated pest management) by using diverse ways like chemical, biological, and physical methods. It is a unique, comprehensive, and complete collection of the topics in plant pathology to serve as an all-inclusive resource and up-to-date reference to effectively cover the information relevant to plant diseases.

The sociology of crop pathology is discussed in this handbook as well as a review of a great variety of techniques for the diagnosis of crop disease, losses due to crop diseases, and theory behind the disease management. It also explores topics on how society is constraining the possibilities for management; management of diseases through changing the environment; biological control of crop diseases; weed management through pathogens; and the epidemiologic and genetic concepts of managing host genes.

Subsequent chapters present the management of crop disease with chemicals and some examples of diseases that benefit man and even a few that benefit plants. This book also describes the organization and operation of society-supported disease management activities, as well as important advisory services provided by the industry.

This handbook will act as a complete guide for academic researchers, students, and growers to understand the basics of plant disease identification. In this handbook I have tried to explain in layman's terms the disease cycle with favorable conditions for disease development that will help growers to manage the diseases and help researchers in their future study. This work is intended for an individual conducting research in plant pathology to broaden his views, stimulate his thinking, and help to synthesize ideas. This book will be very beneficial for industrial professionals as well.

Balaji Aglave

About the Author

Balaji A. Aglave was born in Yermala, Maharashtra, India. He obtained his Bachelor of Science degree in Agricultural Sciences, 2002, and Master of Science degree in Agricultural Biotechnology, 2004 from Marathwada Agricultural University, Parbhani Maharashtra (India). His Ph.D. was in Biotechnology with specialization in Plant Pathology, and he obtained the degree from Government Institute of Science, Aurangabad, Maharashtra (India) in 2009. After completing his master's education, Dr. Aglave worked for National Research Center for Citrus, Nagpur M.S. India as a Senior Research Fellow. As a Senior Research Fellow, his areas of study were molecular and serological diagnosis of citrus viruses, diagnosis and characterization of major citrus viruses, development of non-radioactive probes, and diagnostic kits for plant virus detection. In 2005, he was hired as the Head of the Department of Biotechnology by H.P.T. Arts and R.Y.K. Science College, Nashik, M.S. India (University of Pune). He worked as Head of the Department for four years after which, in 2010, he was hired by Florida Ag Research as a Scientist and Lab Director for the Division of Plant Pathology and Nematology at East Coast research station of Pacific Ag Research, San Luis Obispo, California. In addition to his Ph.D., he mastered the area of plant nematology by completing *Plant Parasitic Nematode Identification* course at Clemson University. Dr. Aglave is an expert in plant diseases and common plant parasites, and his insights in these topics are invaluable to the field. Presently, his work as Scientist and Lab Director aims at developing innovative technologies to improve crop production and environmental quality, testing product effectiveness against nematodes and diseases, studying crop/fruit rot and decay, soil fumigation studies, etc.

Dr. Aglave has published two books, viz. *Biotechnology Review*, A&A Publisher, Tampa, Florida, in 2012 and *Techniques in Chemistry, biophysics and Instrumentation*, Shanti Prakashan, Delhi, India, in 2011. From 2012, he has been working as an active member of various well reputed committees like the Industry Committee, Nematology Committee, and Soil Microbiology and Root Disease committee of American Phytopathology Society. He has acted as the Chairman of Board of studies in Biotechnology subcommittee in Applied Biotechnology, University of Pune, India and Chairman of Board of studies in Microbiology subcommittee in Applied Wine Technology, University of Pune, India. He is currently working as an Editorial Board Member of Advances in Plants and Agriculture Research, African Journal of Food Science, and as a Reviewer for several journals in Agriculture sciences. He has 20 years of academic, research, and industry experience in the field of plant pathology and nematology.

From 2008 to present, Dr. Aglave has conducted multiple Conferences and Workshops covering diverse areas of biotechnology like techniques in molecular biology, techniques in genetic engineering, trends in biotechnology, etc. He delivered more than 50 seminars and plenary lectures in various scientific meetings and at various institutes in India and the United States. His research work is also published in over 100 journal research articles, review articles, and abstracts. Also, he holds numerous internationally published research abstracts and technical summaries completed on topics such as disease and insect management. Currently, Dr. Aglave is continuing his research work on biotechnology at Florida Ag Research, Thonotosassa, Florida.

1 Strawberry

Strawberry, *Fragaria × ananassa* (Weston) Duchesne ex Rozier is a small fruit, grown throughout the world. It has a peculiar red color with a unique shape and flavor. Different berry plant cultivars or cultivated varieties have been developed for different regions and climates throughout the world. The characteristics that distinguish varieties may include fruit ripening time frame, plant disease resistance, cold tolerance, and specific berry traits such as size, shape, firmness, and flavor. Strawberries are a high-value crop in the United States, which is the world's largest producer of strawberries, accounting for nearly 1/3 of the world's total production. Other major strawberry producing countries of the world are South Korea, Japan, Spain, Poland, and the Russian Federation.

In Brittany, France, during the late eighteenth century, the first garden strawberry was grown. Prior to this, wild strawberries and cultivated selections from wild strawberry species were the common sources of the fruit (Figure 1.1).

Strawberry plants can be affected by many diseases. For example, the leaves of strawberry may be infected by powdery mildew, leaf spot caused by the fungus *Sphaerella fragariae*, leaf blight caused by the fungus *Phomopsis obscurans*, and by a variety of slime molds. The crown and roots may fall victim to black root rot, red stele, verticillium wilt, and nematodes. The fruits are subject to damage from *Rhizopus* rot, gray mold, and leather rot. Thus, it is crucial to have different preventive methods and ways to manage strawberry diseases. The methods discussed in this chapter are very useful for strawberry cultivation and disease prevention.

1.1 ANTHRACNOSE OF STRAWBERRY

Anthracnose is the term used to identify strawberry diseases caused by the fungus *Colletotrichum*. Anthracnose is an important disease of strawberry that can affect foliage, runners, crowns, and fruit. The disease is caused by several species of fungi in the genus *Colletotrichum*: *C. acutatum*, *C. fragariae*, and *C. gloeosporioides*. They all cause similar or nearly identical symptoms on the strawberry plant. The 2 most destructive forms of the disease are crown rot, usually associated with *C. fragariae*, and fruit rot, usually associated with *C. acutatum*.

Historically, anthracnose has generally been restricted to the southern United States and was not common in the northern United States. It has generally been considered to be a "warm weather" or "southern disease" of strawberry. Epidemics of anthracnose fruit rot caused by *C. acutatum* have occurred in Ohio, but the crown rot phase has been observed only a few times in the mid-1980s.[1]

Over the past few years, the incidence of anthracnose fruit rot in northern production areas has increased, and there is a concern about the potential impact of this disease in northern, perennial-production systems. Although the disease occurs sporadically and is not common in most plantings in Ohio, when it does occur, it can be devastating, resulting in 100% loss of fruit.

1.1.1 Causal Organism

Anthracnose is caused by distinct species of *Colletotrichum*, mentioned in table below.

Species	Associated Disease Phase	Economic Importance
C. acutatum	Fruit rot	High
C. gloeosporioides	Crown rot	Low to moderate
C. fragariae	Crown rot	Low

FIGURE 1.1   Strawberry fruit.

1.1.2   Symptoms

1.1.2.1   Fruit Rot

Anthracnose fruit rot is caused by *C. acutatum*. It can infect green fruit but is found most often on ripe fruit. Round, firm, sunken spots develop on ripening fruit (Figure 1.2). Spots may range from tan to dark brown. Under rainy or humid conditions, masses of fungal spores develop around the center of spots in a cream to salmon-colored slimy matrix (Figure 1.3). Spots often enlarge until the entire fruit is affected. Diseased fruit frequently become mummified.

1.1.2.2   Crown Rot

Crown Rot is caused by *C. gloeosporioides* and *C. fragariae*. The first visible symptoms of crown rot are sudden leaf wilting and plant death. Crown infections are initiated by spores splashing or washing into central buds from leaves or petioles or by the fungus growing directly into crown tissue.

When crowns of wilted plants are split lengthwise, reddish brown streaking/marbling is visible (Figure 1.4). Although strawberry crowns show discoloration regardless of the cause of death, the reddish marbling pattern is most characteristic of anthracnose. Anthracnose crown rot is active during warm weather but becomes dormant during colder months. However, disease progression resumes when the soil warms in spring. Other symptoms like spots on petioles and stolons and leaf spot are also observed.

1.1.2.3   Petioles and Stolons

Small, dark lesions (dead spots) can appear on stolons and petioles anytime during warm weather. Lesions gradually become black, dry, and sunken. When a lesion girdles a stolon, the unrooted daughter plant beyond the lesion dies. Similarly, a lesion on the petiole often results in death of the attached leaf. Anthracnose symptoms on petioles and stolons are often confused with diseases caused by *Rhizoctonia* and various other leaf spot fungi.

FIGURE 1.2   Sunken spots on fruit due to anthracnose.

FIGURE 1.3   Masses of fungal spores develop in decayed areas on infected fruit.

1.1.2.4   Leaf Spot

Small, round, black to gray spots can appear on expanding leaflets even before petiole or stolon symptoms are noticed. Spores produced in these lesions can wash down into crowns and initiate crown rot.

1.1.3   CAUSE AND DISEASE DEVELOPMENT

Strawberry anthracnose is caused by at least two species of the fungus *Colletotrichum*. *C. fragariae* is primarily a crown rotting fungus, and *C. acutatum* primarily rots fruit. *C. acutatum* is the most common species causing fruit rot in Ohio. The disease is probably introduced into new plantings on infected plants. Recent research indicates that the fungus can grow and produce spores on the surface of apparently healthy leaves. These fungi can persist in infected plants as dormant spores or other fungal structures.

1.1.4   FAVORABLE CONDITIONS OF DISEASE DEVELOPMENT

During warm and rainy or humid weather, the fungi become active and rapidly initiate disease development. Once the disease is established in the field, the fungus can overwinter on infected plants and plant debris, such as old dead leaves and mummified fruit. Spore production, spore germination, and infection of strawberry fruits are favored by warm, humid weather and rainfall. In spring and early summer, spores are produced in abundance on previously infected plant debris.

FIGURE 1.4   Reddish-brown discoloration of strawberry crown.

The spores are spread by splashing rain, wind-driven rain, and by people or equipment moving through the field. They are not airborne, so they do not spread over long distances in the wind. Spores require free water on the plant surface to germinate and infect.

The optimum temperature for infection on both immature and mature fruit is between 25°C and 30°C. Under favorable conditions, the fungus produces secondary spores on infected fruit. These spores are spread by rain and result in new infections throughout the growing season. Disease development can occur very rapidly. Up to 90% of the fruit can be infected within a week or less. Both immature and mature fruit are susceptible to infection; however, the disease is most common on ripening or matured fruit.

Short distance disease spread can occur in the field via

- Rain splash
- Overhead irrigation water
- Movement of contaminated equipment

Long distance spread is accomplished by

- Movement of strawberry transplants from the nursery to the grower

1.1.5 DISEASE CYCLE

Infected transplants and soil from infected transplants appear to be the primary source of inoculum in most instances, especially in annual production systems. This may be especially true for *C. fragariae*, which has a limited host range and does not survive in soil over the summer. In perennial systems, the fungi may overseason in infected plants and debris, providing inoculum for the following fruiting season. Spores (conidia) may be dispersed in the field by wind-driven rain, splashing water, insects, movement of workers, equipment, or animals. Disease development and spread are minimal in most cases under cool, dry conditions. Crown infections often occur in the nursery but do not appear until after planting. The fungus continues to develop in newly planted nursery infected plants, which may suddenly die during warm weather in the fall or early spring of the following year (Figure 1.5).

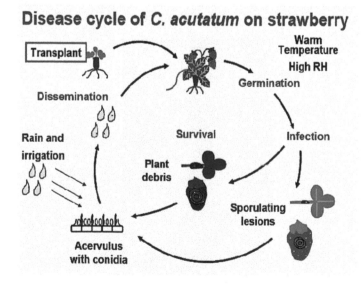

FIGURE 1.5   Disease cycle of *C. acutatum* on strawberry.

1.1.6 MANAGEMENT

Diseases can be managed using the following strategies:

- Chemical control
- Biological control
- Cultural control
- Integrated pest management

1.1.6.1 Chemical Control

Anthracnose fruit rot is a common problem in many areas, and its occurrence is increasing across the Midwest United States. The disease is very important in plasticulture systems. Once anthracnose fruit rot is established in a planting, it is difficult to control and can be very severe, resulting in complete loss of the crop. Captan and Thiram are protectant fungicides that have some activity against anthracnose. If used in a protectant program, they will provide some level of control. Abound, Cabrio, and Pristine are strobilurin fungicides and are labeled for control of anthracnose on strawberry. They have good activity against anthracnose on strawberry of all currently registered fungicides. For purposes of fungicide resistance management and increased efficacy, Abound, Cabrio, and Pristine should be used in rotation with or in combination with Captan or Thiram. Abound, Cabrio, and Pristine are the same class of chemistry so they should not be alternated with each other as a fungicide-resistance strategy. The label states that no more than two applications of one of these fungicides can be made without switching to a fungicide with a different mode of action. Switch has also been reported to have moderate to good activity against anthracnose fruit rot. Therefore, Switch may be used in alternation with Abound, Cabrio, or Pristine for anthracnose control and fungicide resistance management.[2]

1.1.6.2 Biological Control

Trichoderma isolates are known for their ability to control plant pathogens. It has been shown that various isolates of *Trichoderma*, including *T. harzianum* isolate T-39 from the commercial biological control product TRICHODEX, were effective in controlling anthracnose (*C. acutatum*) in strawberry, under controlled and greenhouse conditions. Three selected *Trichoderma* strains, namely T-39, T-161, and T-166, were evaluated in large-scale experiments using different timing application and dosage rates for the reduction of strawberry anthracnose. All possible combinations of single, double, or triple mixtures of *Trichoderma* strains, applied at 0.4% and 0.8% concentrations, and at seven- or ten-day intervals, resulted in reduction of anthracnose severity; the higher concentration (0.8%) was superior in control whether used with single isolates or because of combined application of two isolates, each at 0.4%. Isolates T-39 applied at 0.4% at two-day intervals, T-166 at 0.4%, or T-161 combined with T-39 at 0.4% were as effective as the chemical fungicide fenhexamid.

1.1.6.3 Cultural Control

Using drip irrigation and clean planting stock is important components of managing this disease. Thoroughly washing all soil from plants before planting will reduce disease in crowns and fruit. It may be worthwhile to dip trays of long-term cold storage (-2°C) transplants into a hot water bath for seven minutes right before planting to reduce occurrence of this disease. Prepare plants for this treatment by thoroughly washing them to remove all dirt; then place them in a circulating water bath that is held at a constant temperature of 49°C. Afterward, submerge them in very cold water and then plant them as soon as possible. (This treatment is not recommended for fresh-dug transplants that have only been stored at 0.5°C.)

Clean field equipment before using it to ensure that contaminated soil and plant parts are not transported into a field or from an infested part of the field to a non-infested section. Crop rotation with a non-host crop can also help in reducing levels of this pathogen in the soil. Also important is good weed management in and around the field to destroy any weeds that may harbor the pathogen.

Recent research has demonstrated the importance of removing the weeds from the fields after they are destroyed because the pathogen can still produce spores even though the weeds are dead.

The following strategies can be followed:

1.1.6.3.1 Use Disease-Free Planting Material

The disease is introduced to the field with infected plant material. The best way to avoid the disease is to begin with disease-free planting material. Although there are no nurseries that can certify plants to be free of fungal and bacterial plant pathogens, inspection of plants for the disease before planting is recommended.

1.1.6.3.2 Proper Irrigation

If the field was previously infected, or the disease is present in the field, minimize the amount of overhead irrigation used. The fungus is spread by splashing water. Avoid the use of overhead irrigation and use drip irrigation if possible.

1.1.6.3.3 Mulching

Plastic mulch increases the level of splash-dispersal of the pathogen. Mulching with straw is recommended in perennial matted row plantings to reduce water splash and disease spread.

1.1.6.3.4 Remove Infected Plant Parts

Infected plant parts serve as a source of inoculum for the disease. Remove as much old, infected plant debris as possible. Try to remove infected berries from the planting during harvest.

Table 1.1 lists materials in order of usefulness in an integrated pest management (IPM) Program, considering efficacy. Also, consider the general properties of the fungicide as well as information relating to environmental impact. Not all registered pesticides are listed. Always read the label of the product being used (Table 1.2).

1.2 POWDERY MILDEW OF STRAWBERRY

Powdery mildew is considered a moderate disease that can affect fruit, leaves, and flowers. This disease produces white patches of web-like growth that develop on both the lower and upper leaf surface. The edges of the leaves may curl upward. Immature fruit may fail to ripen, become hard, crack, and turn a reddish color with raised seeds. Powdery mildew is favored by warm, dry conditions followed by moisture on leaves from overnight dew or rainfall. Spores can be spread by wind and can overwinter in trash from the previous and current crops. The disease affects all cultivated strawberries worldwide. No variety is resistant, but each differs in susceptibility.

1.2.1 CAUSAL ORGANISM

Powdery mildew is mostly caused by the fungi mentioned as follows:

Species	Associated Disease Phase	Economic Importance
Podosphaera aphanis (Previously known as *Sphaerotheca macularis*)	Leaves, fruits, and flowers	High

1.2.2 SYMPTOMS

P. aphanis infects leaves, flowers, and fruit. Early foliar infections are characterized by small white patches of fungus growing on the lower leaf surface. On susceptible cultivars, dense mycelia growth

TABLE 1.1
Materials in IPM Program

Common Name (Trade Name)	Amount/Acre[a]	R.E.I.[b] (Hours)	P.H.I.[b] (Days)
METHYL BROMIDE/CHLOROPICRIN	300–400 lb	48	0
Sequential application of:	9–12 gal (shank)	5 days	0
1,3 - DICHLOROPROPENE/CHLOROPICRIN (Telone C5)			
.... OR............			
1,3 - DICHLOROPROPENE/CHLOROPICRIN (InLine)	28–33 gal (drip)	5 days	0
.... OR...............			
CHLOROPICRIN	15–30 gal (shank)	48	0
(MetaPicrin)	15–21.85 gal (drip)	48	0
(Tri-Clor)			
Followed 5–7 days later by:	37.5–75 gal	48	0
METAM SODIUM (Vapam HL, Sectagon 42)	30–60 gal	48	0
.... OR...............			
METAM POTASSIUM (K-Pam HL)			
AT PLANTING: AZOXYSTROBIN	5–8 fl oz/100 gal	4	0
(Abound)			
FOLIAR FUNGICIDES: CYPRODINIL/FLUDIOXONIL	11–14 oz	12	0
(Switch) 62.5WG			
CAPTAN 50WP	4 lb	24	0
AZOXYSTROBIN (Abound)	6.2–15.4 fl oz	4	0

[a] Apply all materials in 200-gal water/acre to ensure adequate coverage.

[b] Restricted entry interval (R.E.I.) is the number of hours (unless otherwise noted) from treatment until the treated area can be safely entered without protective clothing. Preharvest interval (P.H.I.) is the number of days from treatment to harvest. In some cases, the REI exceeds the PHI. The longer of two intervals is the minimum time that must elapse before harvest.

and numerous chains of conidia give these patches a powdery appearance (Figure 1.6). Under favorable conditions, the patches expand and coalesce until the entire lower surface of the leaf is covered (Figure 1.7). In some strawberry cultivars, relatively little mycelium is produced, making it difficult to see the white patches. Instead, irregular yellow or reddish-brown spots develop on colonized areas on the lower leaf surface, and eventually break through to the upper surface (Figure 1.8). The edges of heavily infected leaves curl upward (Figure 1.9). At times, dark round structures (cleistothecia) are produced in the mycelia on the undersides of leaves. Cleistothecia are initially white but turn black as they mature. The fungus also infects flowers, which may produce aborted or malformed fruit. In addition, *P. aphanis* colonizes older fruit producing a fuzzy mycelial growth on the seeds (Figure 1.10). Both types of infection may reduce fruit quality and marketable yields.

1.2.3 CAUSE AND DISEASE DEVELOPMENT

P. aphanis is an obligate parasite that only infects living tissue of wild or cultivated strawberry. The fungus readily infects living, green leaves in the nursery. Thus, infected transplants are normally the primary source of inoculum for fruiting fields in Florida. When conditions are favorable, conidia produced on infected plants are wind dispersed.

1.2.4 FAVORABLE CONDITIONS

Development and spread of powdery mildew are favored by moderate to high humidity and temperatures between 15.5°C and 27°C. Rain, dew, and overhead irrigation inhibit the fungus. Because

TABLE 1.2

Fungicides Registered for Control of Powdery Mildew of Strawberries in Florida

Product Name (Active Ingredient)	Fungicide Group	Maximum Rate Per Acre Per: Application	Season	Min. Days To Harvest	Remarks
Abound (azoxystrobin)	11	15.4 fl oz.	1.92 qt	0	Do not make more than 2 consecutive appl. and no more than 4 appl./crop year. See label for instructions on dipping transplants.
Bumper 41.8 EC (propiconazole)	3	4 fl oz.	16 fl. oz.	0	Do not make more than 2 consecutive applications.
Cabrio EG (pyraclostrobin)	11	14 fl. oz.	70 fl. oz.	0	Do not make more than 2 consecutive applications and no more than 5 appl./ crop year.
Nova 40 W (myclobutanil)	3	5 oz.	30 oz.	0	Do not plant rotational crops until 30 days after last application.
Orbit (propiconazole)	3	4 fl.oz.	16 fl. oz.	0	Do not make more than 2 consecutive applications.
(potassium bicarbonate) many brands[a]	NC	varies	varies	1	Do not mix with highly acidic products.
Pristine (pyraclostrobin + boscalid)	11+7	23 oz.	115 oz.	0	Do not make more than 2 consecutive appl. and no more than 5 appl./ crop.
Procure 50WS (triflumizole)	3	8 oz.	32 oz.	1	Do not plant leafy vegetables within 30 days or root vegetables within 60 days or rotational crops not on label for one year after application.
Quintec (quinoxyfen)	13	6 fl. oz.	24 fl. oz.	1	Do not make more than 2 consecutive applications or more than 4 applications per crop. Do not plant crops not on label for 30 days after application.
Rally 40W (myclobutanil)	3	5 oz.	30 oz.	0	Do not plant rotational crops until 30 days after last application.
Sonoma 40 WSP (myclobutanil)	3	5 oz.	30 oz.	0	Do not plant rotational crops until 30 days after last application.
(sulfur) many brands[b]	M1 or M9	varies	varies	1	Do not use when temperatures exceed 27°C to 30°C.
Switch 62.5 WG (cyprodinil + fludioxonil)	9+12	14 oz.	56 oz.	0	Do not make more than 2 consecutive applications. Do not plant crops not on the label for 30 days after last application.
T-Methyl 70 W WSB (thiophanate-methyl)	1	1 lb.	4 lb.	1	Fungicides from different chemical groups should be used in spray program for disease resistance management.
Topsin 4.5 FL (thiophanate-methyl)	1	20 fl. oz.	80 fl. oz.	1	Fungicides from different chemical groups should be used in spray program for disease resistance management.

(Continued)

TABLE 1.2 (CONTINUED)

Fungicides Registered for Control of Powdery Mildew of Strawberries in Florida

Product Name (Active Ingredient)	Fungicide Group	Maximum Rate Per Acre Per: Application	Season	Min. Days To Harvest	Remarks
Topsin M 70 WP[c] Topsin M WSB[c] (thiophanate-methyl)	1	1 lb.	4 lb.	1	Fungicides from different chemical groups should be used in spray program for disease resistance management.

[a] e.g. Kaligreen, Armicarb 100, Milstop

[b] e.g. Micro Sulf,, Sulfur 90W, Super-Six, Microthiol Disperss, Wettable Sulfur, Kumulus DF, Dusting Sulfur-IAP, Thioperse 80%, Yellow Jacket Dusting Sulfur, Yellow Jacket Wettable sulfur.

[c] Fungicide group (FRAC Code): Numbers (1–37) and letters (M) are used to distinguish the fungicide mode of action groups. All fungicides within the same group (with same number or letter) indicate same active ingredient or similar mode of action. This information must be considered for fungicide resistance management decisions. M = Multi site inhibitors, fungicide resistance risk is low; NC = not classified.

Source: http://www.frac.info/ (FRAC = Fungicide Resistance Action Committee).

dry conditions and high humidity are common in greenhouses and plastic tunnels, powdery mildew is typically more severe in protected culture. In open fields in central Florida, the disease is typically most severe in November and December and usually subsides in January and early February but may reappear in late February and March.

1.2.5 DISEASE CYCLE

Little is known about either the life cycle of the pathogen or the disease cycle in Powdery Mildew. Disease may occur in the fall, allowing the pathogen to overwinter on strawberry plants in a particular field. Mycelium from the previous fall infection apparently initiates new disease in the spring.

1.2.6 MANAGEMENT

The disease can be managed using the following strategies:

- Chemical control
- Biological control
- Cultural control
- Integrated pest management

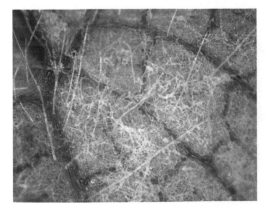

FIGURE 1.6 Mycelia of *S. macularis* on strawberry leaf surface.

FIGURE 1.7   Lower leaf surface of strawberry covered with powdery mildew.

FIGURE 1.8   Reddish-brown spot reaction caused by *S. macularis*.

FIGURE 1.9   Curling leaves on severely infected plants.

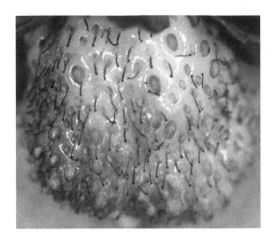

FIGURE 1.10 *S. macularis* on seeds.

1.2.6.1 Chemical Control

Fungicides should be applied at the first sign of disease to control powdery mildew on susceptible cultivars. This is especially important when using protectant fungicides such as elemental sulfur. Systemic fungicides have some limited curative action. These include Rally, whose active ingredient is myclobutanil, and which was formerly named Nova. Sonoma is a competing brand that also contains myclobutanil. Fungicides in the same chemical class as Rally and Sonoma include Procure, Bumper, and Orbit. These products are treated as a group since they belong to the same fungicide class and have similar properties. All share a common, single mode of action and, for this reason, should be rotated with other fungicides with different properties to avoid the development of resistance. Quintec is a recently introduced and effective fungicide with a different mode of action than other powdery mildew products. Other rotational options include the benzimidazole fungicide Topsin M and the strobilurin fungicides Abound, Cabrio, and Pristine, but caution should be taken to not exceed four applications of these products per season. In addition, powdery mildew was recently added to the label of Switch. Controlling foliar infections helps to prevent fruit infections.

1.2.6.2 Biological Control

- Serenade MAX (*Bacillus subtilis* strain QST 713) at 1–3 lb/A is registered for suppression only. As such it is not recommended for use in the Pacific Northwest. Four-hour re-entry.
- Sonata (*Bacillus pumilis* strain QST 2808) at 2–4 quarts/A is registered for suppression only. As such it is not recommended for use in the PNW. May be applied up to and including the day of harvest. Four-hour re-entry.

1.2.6.3 Cultural Control

Avoid overhead irrigation and excess use of nitrogen and use resistant cultivars where practical. Destroying old leaves by renovating plants after harvest may help reduce inoculums. Plant resistant cultivars.

The following materials are listed in order of usefulness in an IPM Program, considering efficacy. Also, consider the general properties of the fungicide as well as information relating to environmental impact. Not all registered pesticides are listed. Always read the label of the product being used (Table 1.3).

1.3 LEAF SCORCH OF STRAWBERRY

Leaf scorch is caused by the fungus *Diplocarpon earliana*. The leaf scorch fungus can infect leaves, petioles, runners, fruit stalks, and caps of strawberry plants. Leaf scorch is common on older leaves

TABLE 1.3

Fungicides Registered for Control of *Botrytis* Fruit Rot of Strawberries in Florida[a]

Product Name (Active Ingredient)	Fungicide Group	Maximum Rate Per Acre Per Application Season		Min. Days to Harvest	Remarks
Abound (azoxystrobin)	11	15.4 fl oz	1.92 qt	0	For suppression of *Botrytis* on the foliage. Do not make more than 2 sequential applications of Group 11 fungicides and no more than 4 applications of Group 11 fungicides per crop year.
Cabrio EG (pyraclostrobin)	11	14 fl oz	70 fl oz	0	For suppression of *Botrytis* on the foliage. Do not make more than 2 sequential applications of Group 11 fungicides and no more than 5 applications of Group 11 fungicides per crop year.
Captan 80 WDG (captan)	M4	3.75 lb	30 lb	1	Rate per treated acre.
Captec 4L (captan)	M4	3 qt	24 qt	1	Rate per treated acre.
Captevate 68 WDG (captan + fenhexamid)	M4 + 17	5.25 lb	21 lb	0	Do not make more than 2 consecutive applications of Group 17 fungicides and no more than 4 applications of Group 17 fungicides per crop year.
Elevate 50 WDG (fenhexamid)	17	1.5 lb	6 lb	0	Do not make more than 2 consecutive applications of Group 17 fungicides and no more than 4 applications of Group 17 fungicides per crop year.
Iprodione 4L AG (iprodione)	3	2 pt	2 pt	N/A	Do not make more than 1 application per season. Do not apply after first fruiting flower.
Pristine (pyraclostrobin + boscalid)	11 + 7	23 oz	115 oz	0	Do not make more than 2 sequential applications of Group 11 fungicides and no more than 5 applications of Group 11 fungicides per crop year.
Rovral 4 Flowable (iprodione)	2	2 pt	2 pt	N/A	Do not make more than 1 application per season. Do not apply after bloom initiation.
Scala SC (pyrimethanil)	9	18 fl. oz	54 fl. oz	1	Do not make more than 2 consecutive applications of Group 9 fungicides. Do not use more than 2 of 6 applications of Group 9 fungicides in any one season.
Serenade ASO	44	6 qt.	-	0	For improved performance, use in a tank mix or rotational program with other registered fungicide.
Serenade Max	44	3 lb.	-	0	For improved performance, use in a tank mix or rotational program with other registered fungicide.
Switch 62.5 WG (cyprodinil + fludioxonil)	9 + 12	14 oz	56 oz	0	Do not make more than 2 consecutive applications. Do not plant crops not on the label for 30 days after last application.

(Continued)

TABLE 1.3 (CONTINUED)

Fungicides Registered for Control of *Botrytis* Fruit Rot of Strawberries in Florida[a]

Product Name (Active Ingredient)	Fungicide Group	Maximum Rate Per Acre Per Application	Maximum Rate Per Season	Min. Days to Harvest	Remarks
Thiram 65 WSB (thiram)	M2	5 lb	25 lb	3	Do not rotate treated crops with other crops for which Thiram is not registered
Thiophanate-methyl 85 WDG (thiophanate-methyl)	1	0.8 lb	3.2 lb	1	Should always be tank-mixed or alternated with a product of a different fungicide group.
T-Methyl 70 W WSB (thiophanate-methyl)	1	1 lb	4 lb	1	Should always be tank-mixed or alternated with a product of a different fungicide group.
Topsin4.5FL(thiophanate-methyl)	1	20 fl. oz	80 fl. oz	1	Should always be tank-mixed or alternated with a product of a different fungicide group.
Topsin M 70 WP[b] Topsin M WSB[b] (thiophanate-methyl)	1	1 lb	4 lb	1	Should always be tank-mixed or alternated with a product of a different fungicide group.

Consult the product label for specific use requirements and restrictions.

[a] Recommendations given in this fact sheet are based on experimentation and statements from the manufacturer.

[b] Fungicide group (FRAC Code): Numbers (1–37) and letters (M) are used to distinguish the fungicide mode of action groups. All fungicides within the same group (with same number or letter) indicate same active ingredient or similar mode of action. This information must be considered for fungicide resistance management decisions. M = Multi site inhibitors, fungicide resistance risk is low.

Source: http://www.frac.info/ (FRAC = Fungicide Resistance Action Committee).

and at the end of the season, but can also affect leaf stalks, fruit stalks, flowers, and fruit. This disease produces small purple spots that first appear on older leaves and gradually enlarge, join other spots and finally produce large dead patches giving the leaves a scorched appearance.

1.3.1 CAUSAL ORGANISM

Species	Associated Disease Phase	Economic Importance
D. earliana	Leaf scorch	High

1.3.2 SYMPTOMS

1.3.2.1 Leaves

Leaf spots (lesions) may take two forms: pinpoint lesions in large or small numbers, and blotchy type lesions measuring 1/4 to 1/2 inches in diameter (Figure 1.11). Lesions are typically reddish to purple, coalescing to give a burned appearance to the plants. They often appear as numerous irregular, purplish to brownish blotches, 1–5 mm in diameter, developing on the leaf surface (laminae). The centers of these lesions do not become white or gray, as with leaf spot (*Mycosphaerella fragariae*). The blotches coalesce irregularly when numerous, and tissue between the blotches turns purplish to bright red. As the disease progresses, leaves turn brown,

FIGURE 1.11   Advanced scorch symptoms on strawberry leaves.

dry out, and turn up at the margins, assuming a burned or "scorched" appearance, as indicated in the name leaf scorch.

1.3.2.2   Leaf Stems (Petioles)

Lesions are typically elongated, sunken, purplish-brown or reddish-brown spots or streaks (Figure 1.12). Advanced lesions can girdle the petiole and kill the leaf.

1.3.2.3   Fruit

All parts of the flower truss and fruits may be infected. Peduncles and pedicels may develop elongated lesions and purplish streaks. In severe cases, tissues are girdled, resulting in the death of flowers and fruits. Infected petals wither and fall off. Irregular brown areas form on infected sepals, often on the margins or tips. These infections lead to fruit with dead calyxes ("dead cap", "dead burr") which are less attractive to consumers, resulting in lower market grades. Signs (visible presence of the pathogen)—using a hand lens, look for small dark spots or fungal fruiting bodies (acervuli) with glistening spore masses (Figure 1.13). As leaf lesions enlarge, they may gradually resemble drops of tar due to the production of large numbers of the minute black acervuli. Rarely, you might see apothecia develop on advanced lesions in leaves and strawberry leaf residues.

1.3.3   Cause and Disease Development

The fungus overwinters on infected leaves that survive the winter. In the spring, conidia are produced on both leaf surfaces in speck-sized black acervuli. The fungus also produces ascospores in the early spring, within disk-shaped apothecia (fungal fruiting structures) that appear as black dots

FIGURE 1.12   Strawberry leaf scorch on petioles.

FIGURE 1.13   Red blotches on the leaves and on the unripened fruit.

in old lesions on the lower surface of diseased leaves that died during winter. In the presence of moisture, ascospores germinate within 24 hours and infect the plant through the lower leaf surface. After symptom development, conidia are produced on the leaf spots in large numbers throughout the growing season. Therefore, repeated infections occur whenever weather conditions are favorable. Conidia are spread mainly by splashing water.

1.3.4   FAVORABLE CONDITIONS OF DISEASE DEVELOPMENT

The fungus overwinters on infected leaves. The fungus produces spore-forming structures in the spring on both surfaces of dead leaves. These structures produce spores abundantly in midsummer. The disease is favored most in warm conditions. The disease is most severe at temperatures from 20°C to 25°C. In the presence of free water, these spores can germinate and infect the plant within 24 hours. Older and middle-aged leaves are infected more easily than young ones.

1.3.5   DISEASE CYCLE

Leaf scorch can progress year-round in most climates but dry conditions, and temperatures above 35°C and below freezing markedly reduce the rate of disease. In North America, scorch can continue to develop in foliage beneath snow cover at temperatures around -4°C–3°C. Symptoms appear quickly on leaves of early growth in spring when scorch is more commonly severe, increasing in late spring and late summer to mid-fall. Infection by ascospores has received little attention. Acervuli can remain dormant for long periods in dry leaves but mature quickly during wet periods. The sticky conidia are dispersed from the acervuli by splashing rain, dew, sprinkler water, and probably by arthropods. Conidia directly penetrate the cuticle and develop into a subcuticular intercellular mycelium. Lesions begin to appear 6–15 days after infection at favorable temperatures (15°C–30°C) provided a post-infection wetness period of nine-hours occurs (18-hours for very young leaves). Mature acervuli form 1–25 days after infection when the microclimate is favorable (leaf age dependent). (Figure 1.14).

1.3.6   MANAGEMENT

The disease can be managed using the following strategies:

- Chemical control
- Cultural control

1.3.6.1   Chemical Control

Several fungicides are registered for control of strawberry leaf diseases. Topsin M, Captan, Thiram, Nova, and Syllit (previously marketed as Cyprex) are all registered for use on strawberries. The

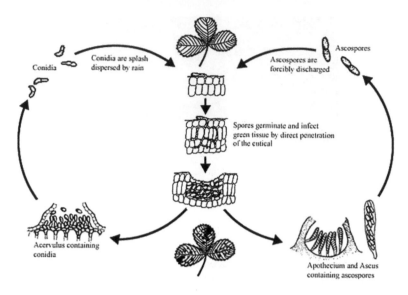

FIGURE 1.14 Strawberry leaf scorch (Red Spot) disease cycle.

label states that Topsin M cannot be applied before early bloom; thus, applications made very early in the season (as new growth starts) should use Syllit, Captan, Nova, or Thiram. The strobilurin fungicides (Cabrio, Abound, and Pristine) also have excellent activity against leaf diseases. If leaf diseases are a serious problem, post-harvest or post-renovation applications of these fungicides may be required. Nova and the strobilurin fungicides have the highest level of activity against leaf diseases. An alternating program of Nova and a strobilurin fungicide should provide excellent control of leaf diseases as well as fungicide resistance management.

1.3.6.2 Cultural Control

- Cultivars differ greatly in their resistance to leaf scorch. If it is a serious problem, use a more resistant cultivar.
- Don't use too much nitrogen fertilizer. It can cause soft, succulent foliage that is more susceptible to leaf scorch.
- Allow good air circulation for optimum drying by spacing plants appropriately and keeping weeds under control.
- Summer renovation will help reduce inoculums level. In Oregon, after every two to four weeks of each harvest it is recommended to renovate crop types.

1.4 CRINKLE VIRUS

Strawberry Crinkle Virus (SCV), also known as Strawberry Frizz Virus, is a viral disease first reported in *F. vesca*. Its normal vector is the aphid *Chaetosiphon fragaefolii* and *C. jacobi*. The virus is carried by mechanical inoculation.

1.4.1 Causal Organism

The causal virus is *Cytorhabdovirus*, of family *Rhabdoviridae*. SCV is transmitted in a persistent propagative manner by the principal natural aphid vector *C. fragaefolii*. Infectivity of aphids is retained lifelong. The length of a transmission cycle in nature depends on temperature conditions since lower temperatures extend the incubation period in strawberry and the latent period in the vector. The virus also multiplies in aphid species other than *C. fragaefolii* when injected.

1.4.2 SPECIES AFFECTED

SCV has a narrow natural host range among species of *Fragaria*. It occurs on the wild species *F. vesca*, *F. virginiana*, and *F. chiloensis*, as well as on cultivated strawberries, *F. ananassa*.

1.4.3 SYMPTOMS

Symptoms vary in relation to strain and strawberry cultivar. Mild strains are symptomless in all cultivars ("strawberry latent virus"). Severe strains, in susceptible cultivars, cause distortion and crinkling of the leaves, with leaflets unequal in size and small irregularly shaped chlorotic spots, often associated with the veins (Figure 1.15).

1.4.4 MEANS OF MOVEMENT AND TRANSMISSION

Under natural conditions, SCV is dispersed locally by the strawberry aphid *C. fragaefolii*. Movement also occurs with runners or with propagated material from tissue culture.

1.4.5 PREVENTION AND CONTROL

Propagation of virus-free plants and control of vectors are the essential measures.

1.5 LATENT C VIRUS (SLCV)

The pathogen responsible for strawberry latent C disease has not been isolated or described morphologically, and its affinities are not known. Its normal vector is the aphid *C. fraguefolii*, which is widespread in Europe. The disease is otherwise only graft-transmissible. The organism behaves as a latent virus, normally giving no obvious symptoms on cultivated strawberries except in combination with other virus diseases, such as crinkle, mottle, vein banding, or yellows. It then causes moderate to severe degeneration.

1.5.1 CAUSAL ORGANISM

Rhabdovirus, of *Rhabdoviridae*, the causal agent of the disease, has not been morphologically described, but cross inoculations or natural complexes indicate that it is distinct from known strawberry viruses. In addition, electron microscopy of *F. vesca* (wild Strawberry) showing symptoms of strawberry latent C disease indicated the presence of virus particles belonging to the rhabdovirus group, which were accumulated in the peri-nuclear space and in the nuclei, whereas strawberry crinkle rhabdovirus particles were in the cytoplasm.

FIGURE 1.15 Crinkling of the leaves and unequally sized leaves.

1.5.2   Species Affected by SLCV

F. chiloensis (Chilean strawberry)
F. vesca (wild strawberry)
F. virginiana (scarlet strawberry (United Kingdom))

1.5.3   Symptoms

The pathogen alone causes no obvious symptoms in commercial strawberry cultivars. In the presence of other viruses, it causes moderate to severe degeneration in the form of extreme stunting, curling, and twisting of the leaves or an intensification of symptoms attributable to the other viruses. The chronic dwarf symptoms range from severe to moderate, but still show an obvious reduction in leaf size.

1.5.4   Means of Movement and Transmission

In the field, the disease is probably transmitted by the aphid vectors. In international trade, infected propagating material, including tissue cultures, is liable to carry the disease; infected strawberry material from the United States has been intercepted in the United Kingdom.

1.5.5   Prevention and Control

As a control method, heat treatment and meristem tip culture, applied separately, are only partly successful in eliminating the pathogen. The main control procedure is based on the use of certified virus-free planting material. The frequency of detection of SLCV in the field appears to be directly related to the presence of nearby sources in the planting. The production of cultivar clones free of SLCV and moderate care in isolation of seedling, selection, and nursery blocks from known sources, followed by the continued replacement of certified fruiting-field stocks, and possibly the use of aphicides, should result in the disappearance of this disease.

1.6   MILD YELLOW EDGE VIRUS

Strawberry mild yellow edge (SMYE) disease is one of the major diseases of strawberries in most parts of the world; however, because of the interaction of cultivars, viruses and virus strains, crop management, and environment, it is difficult to assess the importance of the disease in terms of economic loss. Alone, it is not particularly damaging to most cultivars, but it seldom occurs alone. The complex of the disease with other pathogens, for example, strawberry mottle agent, strawberry crinkle rhabdovirus, strawberry vein banding *caulimovirus*, or strawberry pallidosis agent, can cause severe loss of plant vigor, yield, and fruit quality. It was first reported in *F. vesca*; from California, USA and England; by Horne (1922); Harris (1933).

1.6.1   Causal Organism

Recent investigations have shown that SMYE disease is probably caused by a virus complex consisting of a *Potexvirus* (SMYE-associated *Potexvirus*) as well as a virus originally designated SMYE *Luteovirus*, but which is now recognized as a strain or synonym of soybean dwarf *Luteovirus*.

1.6.2   Species Affected

In nature, both viruses have been found only in *Fragaria spp*. The wild species *F. virginiana, F. vesca* and some clones of *F. chiloensis* show symptoms; *F. ovalis* is a symptomless carrier. Most strawberry cultivars are symptomless carriers of the disease.

FIGURE 1.16   Symptoms of SMYE Virus in a strawberry plant.

1.6.3   SYMPTOMS

Cultivated strawberries usually remain symptomless.
   If symptoms appear, they generally are:

* Leaflets cupped
* Chlorotic margins
* Vigor reduced
* Chlorotic vein netting
* Necrosis of youngest leaves (Figure 1.16)

1.6.4   MEANS OF MOVEMENT AND TRANSMISSION

It is transmitted by a vector; an insect; *Chaetosiphon fragaraefolii, C. thomasi, C. thomasi jacobi*; Aphididae. The vector is transmitted in a persistent manner. Virus can help the vector transmission of another virus (SMYE-associated *Potexvirus*); transmitted by grafting; not transmitted by mechanical inoculation; not transmitted by contact between plants (of *F. vesca* clone); not transmitted by seed; not transmitted by pollen.

1.6.5   PREVENTION AND CONTROL

Control of the virus can be achieved by thermotherapy or meristem culture, combined with planting of certified virus-free material. Thermotherapy for SMYE was successful at approximately 50% when the central growing point was excised, and plants were almost completely defoliated during treatment for nine weeks at 38°C. The technique stimulated the development of side crowns, which could then be excised and rooted in sand at normal greenhouse temperatures.[3]

1.7   MOTTLE DISEASE

Strawberry mottle virus (SMoV), also known as Strawberry Mild Crinkle Virus, is a serious pathogen of strawberries (*Fragaria ananassa*) worldwide and is transmitted by aphids in a semi-persistent manner. Severe strains of SMoV may reduce yield by up to 30% and losses can be up to 80% in mixed infections with other viruses. SMoV occurs in many areas where strawberries are grown.

1.7.1   CAUSAL ORGANISM

The virus belongs to family *Secoviridae* and it has not been assigned with any genus name. It is commonly referred to as SMoV.

1.7.2   SPECIES AFFECTED

- *Fragaria × ananassa*
- *F. virginiana*
- *F. vesca*

1.7.3   SYMPTOMS

Dwarfing of leaves, mottle, vein clearing, and stunting are common symptoms caused by this virus.

1.7.4   MEANS OF MOVEMENT AND TRANSMISSION

It is transmitted by a vector, an insect, *C. fragaraefolii*, *C. thomasi*, *C. minor*, *C. jacobi*, *Aphis gossypii*; Aphididae. The virus is transmitted in a semi-persistent manner. The virus is lost by the vector when it molts; it does not multiply in the vector; is not transmitted congenitally to the progeny of the vector; is transmitted by mechanical inoculation; is transmitted by grafting; is not transmitted by contact between plants; and is not transmitted by seed.

1.7.5   PREVENTION AND CONTROL

- *Always buy plants which are certified as virus-free.* It is unwise to accept plants from old strawberry beds—these will almost certainly be infected with one or more viruses.
- *Destroy and replace plants as soon as yields start to fall,* usually after two or three years. Do not use runners from these plants, which will certainly be infected. Instead, buy new certified, virus-free stock
- If possible, *avoid replanting strawberries on the same site.* Since it is not practical to determine which virus is present on the basis of symptoms, because these are so variable, it is prudent to assume that some of the species spread by soil nematodes may be involved. These will persist in the soil and infect new plants.

1.8   NECROTIC SHOCK VIRUS

For many years, strawberry necrotic shock disease was thought to be caused by a strain of tobacco streak virus (TSV). Tzanetakis, et al. (2004) found that strawberry necrotic shock disease is caused by a different virus and not by a strain of TSV. It was then that the name Strawberry Necrotic Shock Virus (SNSV) was suggested for this virus instead of TSV.

1.8.1   CAUSAL ORGANISM

As mentioned above, the causal organism was once thought to be a strain of TSV, but the further research suggested that it was the SNSV and not TSV.

1.8.2   SPECIES AFFECTED

There are no symptoms seen in commercial cultivars of Strawberry caused by SNSV. Grafted susceptible indicator strawberry plants (*F. vesca*) may show a severe necrotic reaction in new leaves. These symptoms are however temporary, and the new growth appears to be normal and healthy.

1.8.3   SYMPTOMS

Severe necrotic reaction on the leaves is a major symptom observed. Symptoms may also include chlorosis, stunting, and leaf malformation. The commercial cultivars show no symptoms, but there is a visible reduction in the yield and runner production of these cultivars.

1.8.4 Means of Movement and Transmission

Transmission of this virus occurs through seed, pollen, or thrips. This virus has a wide host range, and host plant species near strawberry fields can serve as sources of inoculum.

1.8.5 Prevention and Control

As described, the commercial cultivars show no symptoms of the virus. It becomes very difficult to manage the disease as there are no visible symptoms. So, the most practical way to minimize the risk of infection on commercial fields is to use clean plant material (tissue-cultured, and virus-tested) and to follow best management practices for insect and weed control.

1.9 VEIN BANDING *CAULIMOVIRUS*

Strawberry Vein Banding Virus (SVBV) is a plant pathogenic virus and a member of the family *Caulimoviridae*. It was first described by Fraizer after a differential aphid transmission to susceptible wild strawberries. He identified suitable virus indicators and demonstrated virus transmission by various aphids, dodder, and grafting. He also established the inability of the virus to transmit via sap.

1.9.1 Causal Organism

The virus, *Caulimovirus*, belonging to family *Caulimoviridae*.

1.9.2 Species Affected

The virus, *Caulimovirus,* is known to occur only on *Fragaria* spp. The main host is *F. vesca* (wild strawberry). Commercial strawberries may also be infected, but diagnostic symptoms are usually only apparent when strawberry latent C "rhabdovirus" is present simultaneously.

1.9.3 Symptoms

1.9.3.1 On *F. vesca*

Symptoms initially appear on the youngest developing leaf; there is an epinasty of midribs and petioles, a tendency for opposite halves of leaflets to be appressed, irregularly wavy leaflet margins, and slight crinkling of the laminae. Usually, the above symptoms are mild and not all present simultaneously. It is not until the affected leaf expands that clearing, followed by yellowish banding of some or all of the veins, becomes visible. Often, the coloration occurs in scattered discontinuous streaks of varying lengths along the main and secondary veins. The second and third leaves formed after onset of symptoms are affected more severely than the first or any subsequent leaf; in older leaves, chlorotic streaks are reduced in number, scattered, and confined to portions of the leaflets. This may be followed by the appearance of a series of apparently healthy leaves and then reappearance of mild or severe symptoms.

1.9.3.2 On Commercial Strawberries

There are no very diagnostic symptoms but, if strawberry latent C disease is also present, the reaction to infection is intermediate to that on *F. vesca*. As affected leaves mature, the vein-banded areas may gradually disappear, or they may become brownish-red or necrotic. Especially on outdoor plants, the veins become discolored, without previous chlorosis. Affected leaflets characteristically exhibit epinasty, mild crinkling, and wavy margins.

1.9.4 Means of Movement and Transmission

In the field, the virus is transmitted by aphid vectors. Because of the ability of certain aphid species to undertake long, high-altitude flights, wide natural dissemination is possible. This is, however, limited by the relatively short persistence of the virus in the vector. In international trade, SVBV is liable to be carried on infected plants and propagating material of strawberries. The following aphids are cited as vectors: *Acyrthosiphon pelargonii, Amphorophora rubi, Aphis idaei, A rubifolii, Aulacorthum solani, C. fragaefolii, C. jacobi, C. tetrarhodum, C. thomasi, Macrosiphum rosae, Myzus ascalonicus, M. ornatus, M. persicae.*

Of these species, *Chaetosiphon* spp. are the most efficient vectors in glasshouse experiments, although other genera are probably important vectors when they occur in large numbers and frequently move from plant to plant. Aphids can acquire and transmit the virus in 30–120 minutes, but persistence in the vector is short, usually less than eight hours (semi-persistent type). There are differences in the efficiency of clonal lines of aphids, and evidence that some species will transmit only certain strains of SVBV. *A. gossypii, A. fabae, A. solani*, and *Macrosiphum euphorbiae* failed to transmit the virus in a limited number of trials.

The virus is transmissible by grafting and by means of *Cuscuta subinclusa*. Attempts to transmit SVBV mechanically have been unsuccessful. The incubation period in the indicator host varies from two to five weeks depending on the strain.

1.9.5 Prevention and Control

There are no specific control measures. SVBV is highly resistant to inactivation by heat therapy but it can be eliminated from plants by means of meristem tip culture. As a consequence, the use of certified planting material is the best control procedure, and certification schemes for the production of healthy planting material of strawberry are in operation in several countries. Control of aphids with insecticides could reduce the incidence of the disease.

1.10 *PHOMOPSIS* LEAF BLIGHT OF STRAWBERRY

Phomopsis leaf blight is a common disease of strawberry in the eastern United States. Although the fungus infects leaves early in the growing season, leaf blight symptoms are most apparent on older leaves near or after harvest in Ohio. The economic importance of leaf blight in Ohio appears to be relatively minor; however, incidence of the disease has been increasing. The disease can weaken strawberry plants through the destruction of older foliage. Weakened plants can result in reduced yields the following year. In years highly favorable for disease development, leaf blight can cause defoliation and, in some cases, the death of plants.

Especially in warmer climates, the fungus that causes leaf blight can also cause a fruit rot called soft rot. The first observation of *Phomopsis* fruit rot (soft rot) in Ohio was on plants growing under plastic culture in *1999*. Although not common in Ohio, *Phomopsis* fruit rot can result in serious losses.

1.10.1 Causal Organism

Phomopsis leaf blight is majorly caused by the following mentioned fungal organism:

Species	Associated Disease Phase	Economic Importance
Phomopsis obscurans	Leaf blight (Soft rot)	Serious

FIGURE 1.17 Leaf Spot like symptoms.

1.10.2 SYMPTOMS

Leaf blight is caused by the fungus *P. obscurans*. Leaf blight is found most commonly on plants after harvest. The disease is distinctively different from both leaf spot and leaf scorch. However, the young lesions resemble that of Strawberry Leaf Spot (Figure 1.17). The enlarging leaf spots of this disease are round to elliptical or angular and a quarter of an inch to an inch in diameter (Figure 1.18). Spots are initially reddish purple. Later, they develop a darker brown or reddish-brown center surrounded by a light-brown area with a purple border. Similar spots may sometimes develop on the fruit caps. Usually, only one to six lesions develop on a leaflet. Often the infected area becomes V-shaped with the widest part of the "V" at the leaf margin. New lesions appear throughout the summer and fall if weather conditions are favorable. Older leaves become blighted and may die in large numbers. This disease is usually more destructive on slow-growing or weak plants. The same fungus can cause an enlarging, soft, pale-pink rot at the stem end of the fruit. Black specks of pycnidia often develop within the central areas of the older lesions. Initial symptoms on fruit are round, light pink, and water-soaked lesions (Figure 1.19). Frequently, two or more lesions may coalesce into large soft dark brown lesions. Information on resistance to leaf blight in currently used varieties is limited. If growers encounter a high level of disease on certain varieties, these varieties should be avoided.

FIGURE 1.18 *Phomopsis* leaf blight on strawberry.

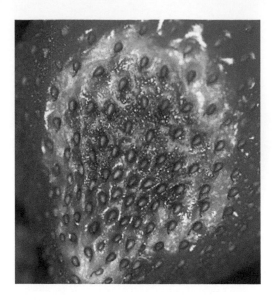

FIGURE 1.19   *Phomopsis* soft rot.

1.10.3   CAUSE AND DISEASE DEVELOPMENT

This fungus produces conidia in speck-sized, black pycnidia (fungal fruiting bodies) embedded in the centers of older leaf lesions. Conidia ooze out of pycnidia during damp weather when temperatures are high. Conidia are splashed to new leaf tissue where they germinate in the presence of free water to initiate new infections on leaves and fruit. The fungus overwinters on either infected leaves that survive the winter or in dead tissue on old infected leaves.

1.10.4   FAVORABLE CONDITIONS OF DISEASE DEVELOPMENT

The disease is spread largely by wind and splashing water from rain or overhead irrigation. The optimum conditions for disease development are temperatures ranging from 26°C–32°C and 72 hours of leaf wetness. Leaf Blight has worldwide distribution but tends to be more severe in cooler climates.

Extended wet periods, particularly in autumn, are the most favorable conditions for the development of disease.

1.10.5   DISEASE CYCLE

P. obscurans overwinters as mycelium and pycnidia on old leaves attached to the plant. Conidia from pycnidia are viable in early spring and are rain splashed onto new leaves. Primary infection may occur early in the growing season, but symptom expression and rapid disease progression do not occur until mid-season.

1.10.6   MANAGEMENT

The disease survives from year to year in infected leaf debris. Management practices include:

- Removal and burial of infected leaves during renovation,
- Use of tolerant varieties,
- Application of an effective and properly timed fungicide, such as Equal 65 WP. Unfortunately, Equal 65 WP applied in cold temperatures and cold weather can injure strawberry leaves.

Unfortunately, *Phomopsis* leaf blight occurs sporadically from year to year. Infection depends on weather conditions leading up to harvest. If we could predict when infection takes place, growers could target fungicide applications more precisely.

In a recent study at Ohio State University, researchers looked at the influence of temperature, leaf wetness duration and leaf age on the infection of strawberry leaves cv. "Honeye" and "Earliglow". Their objective was to develop a prediction model for leaf blight. Disease incidence and severity were most influenced by the age of the leaf. The younger the leaf at the time of infection, the higher the disease observed four weeks later. Leaf wetness duration also significantly influenced the disease. Longer leaf wetness periods resulted in higher levels of the disease. Surprisingly, relationship between infection and temperature on disease severity and incidence was minor.

The results of this study may help to explain the severity of leaf blight observed in 2003. Extended rainy periods in late May and early June were ideal for infection of the new leaves. In addition, overhead irrigation for frost protection resulted in more extended leaf wetness periods. Although August was very warm and dry, September and October were very wet, which resulted in more infections.

High levels of leaf blight may survive this winter and could result in higher disease pressure this coming spring. The disease prediction model developed in Ohio has not been tested in Ontario; however, results suggest that fungicides should be applied to protect new leaves, prior to a leaf wetness event greater than five hours, regardless of temperature. This will be particularly important during next spring and after renovation when plants are producing an abundance of new susceptible leaves.

1.11   GRAY MOLD

Botrytis fruit rot, also called Gray Mold, is a major disease of Strawberries throughout the world. The disease, caused by the fungus *Botrytis cinerea*, is responsible for fruit losses of 50% or more during cool, wet seasons.

In addition to Strawberries, *Botrytis* also causes economic losses for many other crop plants.

The disease affects fruit in the field, resulting in severe pre-harvesting losses. It also affects fruits after harvest, since infections that begin in the field continue to develop during storage and transit at refrigeration.

Fruit turns brown at the calyx end and the fungus produces a gray cotton-like growth on the surface.

1.11.1   CAUSAL ORGANISM

Botrytis fruit rot is majorly caused by a single species of *Botrytis* mentioned below:

Species	Associated Disease Phase	Economic Importance
B. cinerea	Fruit rot	High

1.11.2   SYMPTOMS

Strawberry flowers are highly susceptible to *B. cinerea* and may be blighted directly (Figure 1.20). However, symptoms usually are observed later, on green and ripening fruit. Lesions typically develop on the stem end of the fruit and are often associated with infected stamens or dead petals adhering to the fruit or trapped beneath the calyx (Figure 1.21). Lesions begin as small, firm, light brown spots that enlarge quickly (Figure 1.22). During periods of rainy weather, heavy dews, or high relative humidity, lesions become covered with masses of tan to gray spores (Figure 1.23). Large

FIGURE 1.20   Flower blighted by *B. cinerea*.

numbers of spores are released as visible gray puffs when infected fruit are disturbed. *Botrytis* may consume and mummify the entire fruit (Figure 1.24).

1.11.3 Cause and Disease Development

B. cinerea is a common colonizer of strawberry foliage in the nursery and is also present on dying vegetation around strawberry fields. After transplanting, spores produced on old dying leaves rapidly colonize new emerging leaves without causing visible symptoms. These conidia are dispersed by air, water, and harvesters to infect flowers during the main bloom period in January and February. Cool to mild temperatures and prolonged leaf wetness promote spore production, germination, and infection of stamens, petals, and other floral parts. Flower infections often progress slowly, with lesions becoming visible on green and ripening fruit two to four weeks after infection. Direct infection of fruit by spores is not considered important in the field or after harvest. However, the pathogen also spreads from diseased fruit to healthy fruit by direct contact (Figure 1.25). As the epidemic progresses, diseased fruit, mummified fruit, and decayed flowers and pedicles become important new sources of inoculum. *Botrytis* fruit rot is especially damaging in annual production systems characterized by prolonged flowering and fruiting cycles. In Florida, the second crop of

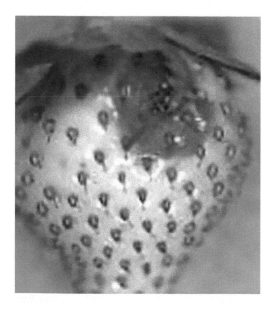

FIGURE 1.21   *Botrytis* lesion from colonized petal (arrow).

FIGURE 1.22   *Botrytis* lesion without spores.

fruit that ripens in February and March are more seriously affected than the first crop of fruit that ripen in December and January.

1.11.4   FAVORABLE CONDITIONS OF DISEASE DEVELOPMENT

Disease development is favored by wet conditions accompanied by temperatures between 5°C and 30°C. Conditions that impede drying of fruit wetted by rain or sprinkler irrigation will encourage *Botrytis* rot.

FIGURE 1.23   *Botrytis* lesion with spores.

FIGURE 1.24   *Botrytis*-mummified fruit.

The gray mold fungus is readily airborne and commonly encountered. Winter carryover is greatest in fields in which there is a large amount of dead plant material, on which the fungus develops. Mild, wet, humid weather is most favorable for infection. Most infections of the fruit result from blossom infections that remain latent in the developing berry, becoming active and causing a rot when the fruit ripens.

1.11.5   DISEASE CYCLE

B. cinerea may colonize and produce conidia on almost any plant debris. It overwinters in strawberry plantings on decayed foliage and fruit from the previous season. Increasing temperatures and moisture in the spring promotes fungal growth and the production of conidia, which are spread by wind and rain to the developing strawberry plants. *Botrytis* conidia are abundant throughout the growing season in most strawberry growing areas.

Strawberries are susceptible to *Botrytis* during bloom and again as fruits ripen. During the blossom blight phase of the disease, the fungus colonizes senescing flower parts, turning the blossoms brown. Blossom infections establish the fungus within the plant and produce inoculums that can spread the fungus to other plants. Cool, wet weather and particularly frost injury favor blossom infections. The fungus can then move into developing fruit and remain quiescent until the fruits start to mature, at which time the rot becomes noticeable.

FIGURE 1.25   Fruit-to-fruit spread of *B. cinerea*.

FIGURE 1.26   Leaf Blight.

Infections may be associated with senescent petals adhering to sepals at the stem end of green or ripe fruit. Infected senescent petals adhering to leaves may also result in leaf blight (Figure 1.26). Abundant gray-brown, fluffy, fungal growth on infected tissue is responsible for the disease's name "Gray Mold".

Fruit infections may be noticeable on green fruits; however, they are most apparent on ripe fruit where abundant sporulation may develop. Fruits touching the ground or in areas where poor air drainage does not allow for rapid drying are most likely to become rotted. When conditions are conducive to abundant blossom infection, the chance of a high level of fruit rot developing at harvest is increased. Figure 1.27 shows the cycle of the disease.

1.11.6   MANAGEMENT

The disease can be managed using the following strategies:

- Chemical control
- Biological control
- Cultural control
- Integrated pest management

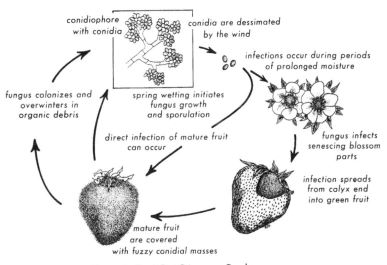

FIGURE 1.27   Disease cycle of *Botrytis* fruit rot on strawberry.

1.11.6.1 Chemical Control

Gray mold control can be aided by applying protective fungicides beginning at or before bloom and continuing until harvest. Where gray mold has been a significant problem before, applications should begin at the white bud stage of flower development. Also, where frost has damaged a planting and a marketable crop remains, great care should be taken to maintain a strict fungicide spray program. In commercial fields in Central Florida, fungicide applications are usually necessary to suppress sporulation and protect flowers from infection. A good disease management program is based on regular applications of a broad-spectrum protective fungicide such as Captan or Thiram. Applications at low rates should begin after overhead irrigation for plant establishment has ended and continued throughout the season. Strawberries bloom from November to March in Florida, but peak blooms occur in November and January/February. Disease incidence is usually low in the first bloom and the regular protectant applications are sufficient to prevent significant early-season losses. During the second peak bloom, fungicides with good activity against *Botrytis* fruit rot can be substituted for protective applications. Captevate®, Elevate®, Pristine®, Scala®, and Switch® are among the most effective fungicides for control of *Botrytis* fruit rot (Table 1.3). The first application should be made at 10% bloom (usually late January). Susceptible cultivars may require up to four applications at weekly intervals to protect flowers throughout the bloom period. Applications are especially critical during periods of mild temperatures and prolonged wetness caused by rains, fogs, or heavy dews. Once this critical period has ended, normal applications of Captan or Thiram can be resumed, usually at high label rates. Applications of protectant fungicides are usually sufficient to control *Botrytis* fruit rot in March when the disease is naturally suppressed by hot weather.

1.11.6.2 Biological Control

Trichoderma isolates are known for their ability to control plant pathogens. It has been shown that various isolates of *Trichoderma*, including *T. harzianum* which isolate T-39 from the commercial biological control product TRICHODEX, were effective in controlling gray mold (*B. cinerea*) in strawberry under controlled and greenhouse conditions. Three selected *Trichoderma* strains, namely T-39, T-161, and T-166, were evaluated in large-scale experiments using different timing application and dosage rates for the reduction of strawberry gray mold. All possible combinations of single, double, or triple mixtures of *Trichoderma* strains, applied at 0.4% and 0.8% concentrations, and at seven- or ten-day intervals were tried. Only a few treatments resulted in significant control of gray mold. Isolates T-39 applied at 0.4% at two-day intervals, T-166 at 0.4%, or T-161 combined with T-39 at 0.4% were as effective as the chemical fungicide fenhexamide. The biocontrol isolates were identified to the respective species *T. harzianum* (T-39), *T. hamatum* (T-105), *T. atroviride* (T-161), and *T. longibrachiatum* (T-166), according to internal transcribed spacer sequence analysis.

1.11.6.3 Cultural Control

Botrytis fruit rot can be controlled by both chemical and cultural measures. Cultural practices include the use of resistant cultivars and the physical removal of infected plant parts (plant sanitation). Although there are no commercial cultivars highly resistant to this disease, "Camarosa", "Carmine", and the newly released "FL Radiance" and "FL Elyana" are less susceptible to *Botrytis* fruit rot than "Strawberry Festival", "Treasure", and "Sweet Charlie". The Californian cultivar "Camino Real" has been proven highly susceptible under Florida conditions. Cultivars with large clasping calyces are generally more susceptible because moisture trapped between the calyx and the receptacle promotes the spread of the pathogen from stamens and petals to the developing fruit. Removal of senescing and dying leaves after establishment helps to eliminate a potential source of inoculum. However, studies have shown that leaf pruning modestly reduces disease incidence, but does not increase marketable yield, and is not practical due to the high cost of labor. Yields may even be reduced when pruning includes the removal of partially green leaves. However, the removal of diseased and culled fruit from the plant canopy during normal harvest operations is considered vital to successful management of *Botrytis* fruit rot.

Certain cultural practices help control gray mold by promoting faster drying of foliage and fruit while other practices reduce exposure to fungal inoculum.

- Select a planting site with good soil drainage and air circulation.
- Expose planting to full sun.
- Orient plant rows toward the prevailing wind.
- Apply appropriate nitrogen levels to prevent excessive foliage from developing.
- Mulch plants with straw to reduce fruit contact with the soil.
- Pick fruit frequently.
- Cull out and remove diseased berries from the planting.
- Handle berries with care to avoid bruising and refrigerate harvested fruit promptly at 0°C–10°C (32°F–50°F).

The following materials are listed in order of usefulness in an IPM Program, considering efficacy. Also, consider the general properties of the fungicide as well as information relating to environmental impact. Not all registered pesticides are listed. Always read the label of the product being used.

1.12 LEATHER ROT OF STRAWBERRY

Leather rot is caused by the soilborne pathogen *Phytophthora cactorum*. The leather-rot pathogen is a fungus-like organism called an oomycete and is not a true fungus. Leather rot has been reported in many regions throughout the United States. In many areas, it is considered a minor disease of little economic importance. However, excessive rainfall during May, June, and July can lead to severe fruit losses and quality reduction. In 1981, many commercial growers in Ohio lost up to 50% of their crop to leather rot. The leather rot pathogen primarily attacks the fruit but may also infect the blossoms. The key control methods are maintaining a good layer of straw mulch between the fruit and the soil, selecting well-drained planting sites, improving water drainage through tiling before planting, or using other methods to improve soil drainage. Avoiding soils that become saturated with water is critical for leather rot control.

1.12.1 CAUSAL ORGANISM

Leather Rot is majorly caused by single species of *Phytophthora* mentioned below:

Species	Associated Disease Phase	Economic Importance
P. cactorum	Fruit rot	Low

1.12.2 SYMPTOMS

The leather rot pathogen can infect berries at any stage of development. When the disease is serious, infection of green fruit is common. On green berries, diseased areas may be dark brown or natural green outlined by a brown margin (Figure 1.28). As the rot spreads, the entire berry becomes brown, maintains a rough texture, and is leathery in appearance. The disease is more difficult to detect on ripe fruit. On fully mature berries, symptoms may range from a little color change to discoloration that is brown to dark purple (Figure 1.29). Generally, infected mature fruit

FIGURE 1.28 Leather rot symptoms on an immature strawberry fruit.

are dull in color and are not shiny or glossy. Infected ripe fruit are usually softer to the touch than healthy fruit. When diseased berries are cut across, a marked darkening of the water-conducting system to each seed can be observed. In later stages of decay, mature fruits also become tough and leathery. Occasionally, a white moldy growth can be observed on the surface of infected fruit. In time, infected fruit dry up to form stiff, shriveled mummies. Berries that are affected by leather rot have a distinctive and very unpleasant odor and taste. Even healthy tissue on a slightly rotted berry is bitter. This presents a special problem to growers in pick-your-own operations. An infected mature berry with little color change may appear normal and be picked and processed with healthy berries. Consumers have complained of bitter-tasting jam or jelly made with berries from fields where leather rot was a problem. Leather rot is most commonly observed in poorly drained areas where there is or has been free-standing water or on berries in direct contact with the soil.

1.12.3 Cause and Disease Development

The pathogen survives the winter as thick-walled resting spores, called oospores, that form within infected fruit as they mummify. These oospores can remain viable in soil for long periods of time. In the spring, oospores germinate in the presence of free water and produce a second type of spore called a sporangium. A third type of spore called a zoospore is produced inside the sporangium. Up to 50 zoospores may be produced inside one sporangium. The zoospores have tails (flagella) and can swim in a film of water. In the presence of free water on the fruit surface, the zoospores germinate and infect the fruit. In later stages of disease development, sporangia are produced on the surface of infected fruit under moist conditions. The disease is spread by splashing or windblown water from rain or overhead irrigation. Sporangia and/or zoospores are carried in water from the surface of the infected fruit to healthy fruit where new infections occur.

FIGURE 1.29 Leather rot symptoms on a mature strawberry fruit. Note the light, off-color area on the fruit.

1.12.4 FAVORABLE CONDITIONS OF DISEASE DEVELOPMENT

Under the proper environmental conditions, the disease can spread very quickly. A wet period (free water on fruit surface) of two hours is sufficient for infection. The optimum temperatures for infection are between 16°C and 25°C. As the length of the wet period increases, the temperature range at which infection can occur becomes much broader. As infected fruit dry up and mummify, they fall to the ground and lie at or slightly below the soil surface. Oospores formed within the mummified fruit enable the fungus to survive the winter and cause new infections the following year, thus completing the disease cycle.

1.12.5 DISEASE CYCLE

Leather rot is caused by *P. cactorum*, the same organism that causes crown and collar rots of apple and other deciduous fruit trees. This fungus is present in many soils throughout New York.

P. cactorum persists in the soil as thick-walled resting spores (oospores), which can survive in a dormant state for many years. When the soil is moist or wet, some of the oospores in the soil germinate and form structures called sporangia, which are filled with the infection spores of the fungus (zoospores). These microscopic zoospores are released into the soil when it is flooded or puddled and swim to the surface using the tail-like structures that they possess.

A leather rot epidemic can begin when strawberry fruit becomes infected after lying in puddled water containing zoospores or when the puddled water is splashed onto them by rain or sprinklers.

Following initial infection, the leather rot fungus forms additional sporangia on the fruit surface during periods of plentiful rainfall and high relative humidity. The sporangia are spread through the air by wind and rain and cause new infections. Still more sporangia may be produced on newly infected fruit and continue to spread the disease as long as weather conditions remain favorable. The leather rot fungus eventually forms its resting spores (oospores) within the infected fruit, and these spores are returned to the soil when the fruit falls to the ground and decays (Figure 1.30).

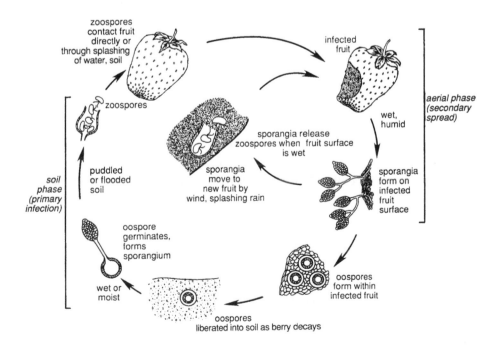

FIGURE 1.30 Leather Rot disease cycle.

1.12.6   Management

Leather rot is not common on annual plantings of strawberries in California because it is usually controlled by preplant fumigation and plastic mulches. Cultural practices play an important role in disease prevention; soil solarization may also provide control. Plantings held for two or three years, however, could be infected by the leather rot pathogen.

1.12.6.1   Chemical Control

Most fungicides currently available for use on strawberries are generally ineffective for controlling leather rot. Although Captan and Thiram are beneficial in suppressing leather rot, they will not provide adequate control if an epidemic develops. Furthermore, the use of these fungicides is severely restricted or prohibited during harvest due to re-entry restrictions or preharvest intervals. Ridomil is registered for use on strawberries for control of red stele and leather rot. Ridomil is very effective for control of leather rot and may be applied in the spring after the ground thaws and before first growth. This early application is recommended primarily for control of red stele but may be beneficial in providing some control of leather rot. A second application is recommended specifically for leather rot and can be made during the growing season at fruit set. Aliette 80% WDG is also registered for use on strawberries and should provide good control of both red stele and leather rot. It can be applied from the initiation of bloom through harvest on a seven- to fourteen-day schedule and has no preharvest restriction.

1.12.6.2   Cultural Control

Ensure that fields are prepared so that they have adequate water drainage. Remove diseased fruit and use plastic mulches. Avoid overhead irrigation; use drip irrigation. Straw mulch has been effective in controlling this disease in the eastern United States.

1.12.6.3   Soil Solarization

This is a unique management strategy to prevent leather rot. In warmer areas of the state, solarization has been shown to be effective for the control of soilborne pathogens and weeds. Solarization is carried out after the beds are formed and can be effective if weather conditions are ideal (30–45 days of hot weather that promotes soil temperatures of at least 50°C). The effectiveness of solarization can be increased by solarizing after incorporating the residue of a cruciferous crop, in particular broccoli or mustards, into the soil or following an application of metam sodium (40 gal/A).

Table 1.4 outlines the percentages of fungicide effectivness at disease control: (Table 1.4)

TABLE 1.4

Effect of Fungicides on Control of Strawberry Leather Rot

Treatment and Rate (a.i/ha)	Leather Rot (%)[a]	Marketable Fruit (%)[b]	Total No. of Fruits	Total Yield (kg)[c]	Percent Disease Control
Pyraclostrobin (0.20 kg)	0.5 a[d]	96.8 a	1080 a	10.8 a	99
Azoxystrobin (0.28 kg)	0.4 a	97.8 a	1054 a	10.9 a	99
Phosphorous acid (2.35 kg)	0.8 a	96.8 a	1065 a	10.4 a	98
Mefenoxam (0.56 kg)	0.3 a	97.9 a	1080 a	9.6 a	99
Untreated control	58.1 b	35.9 b	1144 a	7.7 b	–

[a] Mean percentage of *Phytophthora cactorum*-infected fruit from three harvest dates (June 3, 7, and 10).
[b] Mean percentage of marketable fruit from the above three harvest dates.
[c] Total yield from the above three harvest dates for 3 m of crop row per replication.
[d] For percentages, the analysis was based on the angular transformation. Numbers followed by the same letter within columns do not differ significantly according to Duncan's modified (Bayesian) LSD test ($P = 0.05$).

1.13 *RHIZOPUS* FRUIT ROT

Rhizopus fruit rot, or leak, is primarily a postharvest or storage rot, but it may also occur in the field on ripe fruit. The disease is caused by the fungus *Rhizopus* spp.

1.13.1 CAUSAL ORGANISM

Species	Associated Disease Phase	Economic Importance
Rhizopus stolonifer, Rhizopus nigricans and Other spp.	Fruit rot	N/A

1.13.2 SYMPTOMS

Initial infections of *Rhizopus* fruit rot appear as discolored, water-soaked spots on fruit. These lesions enlarge rapidly, releasing enzymes that leave the berry limp, brown, and leaky (Figure 1.31). Under conditions of high relative humidity, the berry rapidly becomes covered with a coat of white mycelium and sporangiophores. The sporangiophores develop black, spherical sporangia, each containing thousands of spores. When disrupted, these sporulating berries release a cloud containing millions of spores. *Rhizopus* and mucor fruit rots closely resemble each other and may be difficult to differentiate in the field.

1.13.3 CAUSE AND DISEASE DEVELOPMENT

The fungus is an excellent saprophyte that lives on and helps break down decaying organic matter. It invades strawberries through wounds and secretes enzymes that degrade and kill the tissue ahead of the actual fungal growth. The fungus is active most of the year in California and survives cold periods as mycelium or spores on organic debris. Spores are airborne. The pathogen has a large host range and is prevalent worldwide.

1.13.4 FAVORABLE CONDITIONS OF DISEASE DEVELOPMENT

Damaged and overripe fruit exposed to warm temperature and high humidity are the most favored conditions for the disease development.

1.13.5 DISEASE CYCLE

The fungus survives on crop debris and in the soil between seasons. *Rhizopus* can only infect through wounds. Under favorable conditions of high temperature and moisture, sporulation is rapid and abundant. Spores are disseminated by air and by insects.

FIGURE 1.31   Fruit break down caused by *Rhizopus* rot.

1.13.6   MANAGEMENT

Rhizopus stops growing at temperatures below 8°C–10°C (46°F–50°F), so rapid postharvest cooling of fruit is essential for disease control. Field sanitation also is extremely important: do not leave discarded plant refuse or berries in the furrows and be sure to remove all ripe fruit from the field. There are some benefits to the use of protective fungicides, but unless the disease is widespread throughout the field, this pathogen should not cause excessive damage.

1.13.6.1   Cultural Control

Field sanitation is extremely important. Handle fruit with care always. Remove all ripe fruit from the field at harvest. Be sure when fruit is being picked that the entire fruit is removed from the stem, and the fleshy receptacle of the fruit is not left behind as it can serve as a site for invasion by fungus. Cultivars with thick cuticles are less susceptible to *Rhizopus* fruit rot because they are better able to resist infection.

1.13.6.2   Organically Acceptable Methods

Sanitation, cultivar selection, and rapid postharvest cooling are acceptable for use in an organically certified crop.

1.13.6.3   Treatment Decisions

Fungicide treatment is not recommended.

1.14   RED STELE ROOT ROT OF STRAWBERRY

Red stele root rot is a destructive disease in most strawberry producing regions of the world where soils tend to be cool and wet. It is very common in poorly drained soils, particularly during wet spring seasons or those following a rainy autumn. The disease is most destructive in heavy clay soils that are saturated with water during cool weather. Once established in the soil, the fungus remains alive for up to 13 years and possibly longer, regardless of the crop rotation used. Most infections are passed directly from strawberry plant to strawberry plant. Red stele usually does not appear in a new planting until the spring of the first bearing year, from about full bloom to harvest. Minor symptoms of root infection may appear, however, in late fall of the first growing season. Damage increases each year that susceptible cultivars are grown in the infested soil.

Red stele may appear to be well distributed over an entire strawberry field or patch during a cool, wet spring. Normally, however, the disease is most prevalent in lower or poorly drained areas.

1.14.1   CAUSAL ORGANISM

Species	Associated Disease Phase	Economic Importance
Phytophthora fragariae	Root rot	Major

1.14.2   SYMPTOMS

1.14.2.1   Aboveground

Symptoms of red stele rarely occur in the first year of strawberry growth unless plants were severely diseased before planting or if soil conditions were suitable for rapid fungal growth. Usually, red stele is first noticed during bloom of the second year. The symptoms will be most noticeable in low or soil compacted areas of a field where water drainage is poor. Strawberry plants infected with *P. fragariae* will show a general lack of vigor with poor runner growth and small berries. New

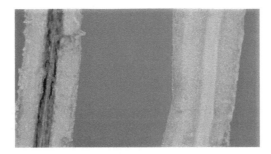

FIGURE 1.32   Longitudinal section of a healthy (left) and red stele-infected (right) strawberry root.

leaves may appear bluish-green, while older leaves sometimes turn red, orange, or yellow. The leaves tend to wilt during warm weather or drought stress. Severely diseased plants may collapse prior to fruiting. Although these aboveground symptoms are typical for red stele, they may resemble symptoms caused by other types of root disorders; therefore, roots also need to be examined.

1.14.2.2   Below Ground

To correctly diagnose red stele, strawberry roots should be sampled during early spring and summer up until the time of harvest. Samples taken after harvest are not reliable because infected roots may have already begun to decay. When taking a plant sample, dig rather than pull the plant from the ground. Examine the roots of plants which are just beginning to show signs of wilting. If red stele is present, the roots will appear unbranched and will be lacking feeder roots. This "rat-tail" appearance of the root is a diagnostic trait of red stele. Select a white root with a rotted tip and make a lengthwise cut at the point where diseased root tissue meets healthy tissue. Red stele infected roots will have a reddish-brown core, but the outer tissue will be white. The discoloration will begin at the root tip and move upward, but usually will not move into the crowd (Figures 1.32 and 1.33).

1.14.3   CAUSE AND DISEASE DEVELOPMENT

Often poor drainage leads to general root rot and contributes to red stele root rot. These plants need well-drained soil and will not do well in other soils. Poor drainage stresses plants and roots die from insufficient oxygen. Poor drainage may result from heavy (clay-like) soil, high water table, etc. It is usually a winter problem and not noticed at the time. Such soil may appear well-drained in the summer.

FIGURE 1.33   Red stele infected roots on the left and healthy roots on the right. Note the absence of numerous roots.

Some fungi attack roots that are stressed because of wet soil conditions and cause further root damage. They may be involved in general root rot. Red stele root rot disease is caused by *P. fragariae* and is favored by wet soil.

1.14.4 FAVORABLE CONDITIONS OF DISEASE DEVELOPMENT

Normally, the disease is prevalent only in the lower or poorly drained areas of the planting; however, it may become fairly well distributed over the entire field, especially during a cool, wet spring. The red stele fungus may become active at a soil temperature of 4°C. However, the optimum soil temperature for growth and disease development is between 13°C and 15.5°C. Under favorable conditions of high soil moisture and cool temperatures, plants will show typical disease symptoms within ten days after infection.

Soil types do not affect the presence or absence of the red stele fungus. It grows in any soil with a pH of 4.0–7.6 but will not grow in an alkaline soil (a pH of 8.0 or above). Heavy clay soils, which retain moisture for long periods of time, provide a conducive environment for the development of the red stele disease because the zoospores can spread greater distances and produce more infection sites.

1.14.5 DISEASE CYCLE

The red stele fungus is spread from one field, or area, to another primarily by the distribution of nursery infected plants. Infection is then spread within the field by moving water, and by soil carried on implements and shoes. Once in the field, thick-walled resting spores (oospores) in infected roots produce large numbers of motile spores (zoospores) that swim about when soil moisture is high, infecting the tips of the young, fleshy roots and destroying their water- and food-conducting tissues. Infection and growth of the fungus in roots reduces the flow of water and nutrients to the developing leaves and fruit causing drought-like symptoms in the plant (Figure 1.34).

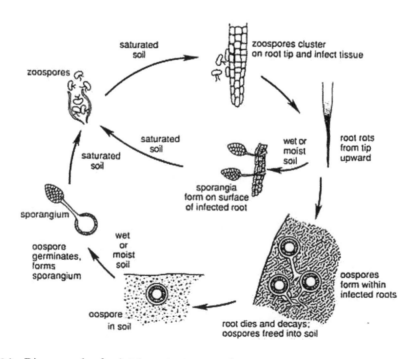

FIGURE 1.34   Disease cycle of red stele root rot on strawberry.

1.14.6 MANAGEMENT

The disease can be managed using the following strategies:

- Chemical control
- Biological control
- Cultural control

1.14.6.1 Chemical Control

Aliette WDG (Fosetyl-Al) and Ridomil Gold 480 EC (metalaxyl-M) are two very effective fungicides registered in Ontario for diseases caused by *Pythium* spp. and *Phytophthora* spp. such as red stele in strawberries. Although both fungicides are effective against root diseases caused by *Phyphthora* spp., they are very different in the way they control these pathogenic fungi and the way they move in plants.

Ridomil was originally targeted to protect crops from foliar diseases; however, it is now widely used for controlling many soilborne diseases as well. Ridomil Gold 480 EC acts on susceptible fungi by inhibiting RNA synthesis. The end result is that Ridomil Gold 480 EC interferes with the development and germination of *Phytophthora* spp.

Ridomil Gold 480 EC is very soluble in water and moves systemically up from roots into stems and then leaves with the transpiration stream of the plant. There is very little downward movement in plants, and therefore, it is important to apply this fungicide as a soil drench for best results against *Phytophthora* root diseases. Ridomil Gold 480 EC can be applied as a soil drench in the fall and the spring for strawberries. In fact, fall is the best time to apply Ridomil Gold 480 EC to control red stele in strawberries. Pay close attention to application timing and always read and follow the label. Ridomil Gold 480 EC should not be applied in the spring to plants bearing strawberries.

Aliette WDG, a phosphonate type of fungicide, on the other hand, is one of the first fungicides developed that can move both up and down in plants. On berries, Aliette is only registered for foliar applications. Aliette can be used as a drench to prevent *Phytophthora* root. Once inside the plant, the active ingredient fosetyl-al is broken down rapidly into phosphorous acid, which is extremely soluble in water and toxic to many *Phytophthora* species. Aliette works in two ways. It acts directly on the invading pathogens to stop their growth and sporangia or spore sack production. It also acts indirectly by stimulating the plant to activate its own defense system, thus helping to prevent future infections from taking place. Plants that have their defense system already activated prior to the invasion by a pathogen can defend much more effectively than plants that do not have their defense system pre-activated.

Regardless of the way these two effective fungicides work, these fungicides should never be used exclusively to control either red stele in strawberries or *Phytophthora* root rot in raspberries, blueberries, or apples. Ridomil Gold 480 EC and Aliette should be alternated with each other and be included as part of an integrated disease management system to reduce the potential of resistance developing.

1.14.6.2 Cultural Control

Since significant production and movement of infective zoospores occurs only during periods when the soil is completely saturated, the key to control is drainage. Strawberries should not be planted in low-lying or heavy soils where water accumulates or is slow to drain. On marginal soils, planting strawberries on beds raised at least ten inches high will bring much of the root system above the zone of greatest pathogen activity and the severity of red stele root rot should be significantly reduced.

The only practical method of controlling red stele is to grow certified, disease-free plants of resistant cultivars. Only resistant varieties should be planted in a field where red stele is known to have caused losses within the last five to ten years. Resistant cultivars include Darrow, Delite, Earliglow, Guardian, Midway, Pathfinder, Redchief, Redglow, Sparkel (Paymaster), Stelemaster, Sunrise, and

Surecrop. All of these cultivars are adapted to conditions in Illinois. However, not all are resistant in all infested soils because different races or strains of the fungus occur. These races vary in their ability to infect the different cultivars. A cultivar that is resistant to red stele in one area may be susceptible in another. Several races of *P. fragariae* have been found in Illinois. Earliglow, Guardian, Redchief, Sunrise, and Surecrop are the only resistant cultivars presently suggested for use in Illinois. They are resistant to three or more races of the fungus. Even these cultivars should be rotated with other crops to reduce the chance that a new, more virulent race of the fungus may appear) one that could attack resistant cultivars. Always plant small "trial" plots of new varieties to test them for resistance to red stele on your farm and to evaluate their performance before you make extensive plantings.

Whenever possible, select a planting site that has never had red stele, has good to excellent drainage, and is located where water from nearby land will not drain through it. Avoid low, wet spots.

If possible, use your own tools and machinery for setting out a strawberry field and carrying out general cultural practices. If you borrow equipment, be sure to clean off the soil and plant debris thoroughly before using it.

Soil fumigation with soil sterilants and/or pesticide applications may be helpful in situations where resistant varieties are not available or are not adapted. Extreme care should be taken not to reinfest a fumigated field by using contaminated equipment or plants. Soil fumigation should be the last resort in controlling the red stele disease. The first step is to use resistant varieties and a selection of well-drained planting sites.

It is important to minimize the chance of introducing the red stele fungus into a field where it does not already exist. Buy nursery stock only from a reputable supplier, and take care not to transfer soil on farm implements from an infested field into a clean one. New fungicides active against red stele also help in controlling this disease but are most effective when used in combination with good soil—water management practices.

1.14.6.3 Biological Control

After *in vitro* screening of more than 100 bacterial isolates from the rhizosphere on their antagonistic effect against *P. fragariae* var. *fragariae*, the causal agent of red stele disease of strawberry, three bacteria out of different genera *Raoultella terrigena* (G-584), *Bacillus amyloliquefaciens* (G-V1), and *Pseudomonas fluorescens* (2R1-7) were found with the highest inhibitory effect on the mycelial growth of both *Phytophthora* spp. For the management of the fungal disease, the antagonistic bacteria were further evaluated under greenhouse and field conditions. In the greenhouse all three bacteria were significantly effective in reducing red core, exhibited a similar level of control as the chemical fungicide Aliette of up to 59%. In field trials conducted at different locations in Germany under artificially and naturally infested soil conditions in two seasons, 2003–2005, different levels of biocontrol were performed by the tested bacteria. In trial during the first season under artificial conditions, the three rhizobacteria showed a significant control of up to 45% against the disease and in the next season, only *B. amyloliquefaciens* was effective against red stele. Under natural conditions, a significant effect of 37.5% was observed from a mixture of *R. terrigena* and *B. amyloliquefaciens* in the first season, and in the second season *R. terrigena* showed a significant effect of 45.1% in the northern part of Germany. In the south, *R. terrigena* and *B. amyloliquefaciens* were significantly efficient up to 51.5% and the overall effects were similar to Aliette.

1.15 *PHYTOPHTHORA* CROWN ROT

Phytophthora crown and root rot caused by *P. cactorum* is a disease of long-standing importance in strawberry. It is responsible for sporadic but serious production losses. Pre-plant soil fumigation, improved cultural practices, and systemic oomycete fungicides have helped to minimize the losses, but the pathogen's ability to survive indefinitely in soil and its capacity for rapid reproduction have prevented its eradication from strawberry production systems. The pathogen causes loss primarily by killing plants, but it also can reduce growth and yield through sub-lethal infections.

1.15.1 CAUSAL ORGANISM

Species	Associated Disease Phase	Economic Importance
P. cactorum	Crown rot and root rot	N/A

1.15.2 SYMPTOMS

Symptoms of disease caused by *P. cactorum* vary with the stage in the production system and the time of year. Early in the season, either at nurseries or fruiting fields, infected plants may exhibit stunting. As weather warms, the most notable symptom of infection, at least on susceptible cultivars, is plant collapse (Figure 1.35) associated with crown rot (Figure 1.36). However, it is difficult to reliably distinguish crown necrosis caused by *P. cactorum* from that induced by *C. acutatum* or other pathogens, especially in the later stages of disease. Furthermore, in the early stages after infection, crown rot caused by *P. cactorum* may be limited to outer regions or sectors of the plant crown. Diagnostic tests are required to determine with certainty which pathogen or pathogens are associated with the problem. At nurseries, *P. cactorum* causes runner lesions in addition to crown and root rot. Many of the roots of daughter plants infected by *P. cactorum* exhibit regions of dark necrosis, which may be limited to the outer (cortex) or extend into the inner (stele) portions of the root. The pathogen also can be carried on nursery stock, lacking clear symptoms of disease.

1.15.3 CAUSE AND DISEASE DEVELOPMENT

Of the *Phytophthora* species involved, *P. cactorum* is the most common; the others are much less prevalent on strawberry. *Phytophthora* is soilborne. When the soil becomes saturated with water, the pathogen can produce and release zoospores, which swim through water-filled pores to infect plant tissue. *Phytophthora* species also produce resilient spores (chlamydospores, oospores) that enable them to survive in soil for long periods without a host or under adverse conditions.

1.15.4 FAVORABLE CONDITIONS OF DISEASE DEVELOPMENT

Infections can occur during cool to moderate temperatures, which are typical throughout coastal fruit-production cycles.

FIGURE 1.35   Symptoms of *Phytophthora* crown and root rot caused by *Phytophthora cactourm*. Typical "plant collapse" in a commercial fruiting field.

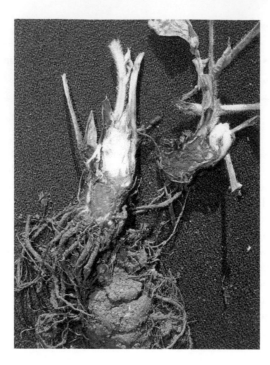

FIGURE 1.36   Symptoms of *Phytophthora* crown and root rot caused by *P. cactourm*. Magnified crown rot.

1.15.5   DISEASE CYCLE

The most important propagule for this pathogen is zoospores, which originate from hyphae or germinating oospores and sporangia. In many cases, this pathogen may enter a field through infected transplants. Infection by *P. cactorum* usually occurs during warm periods with prolonged wetness.

Motile zoospores are released from sporangia during saturated soil conditions and enter through wounds. Once the zoospore reaches a host, it infects, and developing hyphae of the fungus colonize the host (Figure 1.37).

Disease expression is influenced by time of planting and environmental conditions. Plantings established in fall may have wilted plants soon after planting but it is possible the disease will not be expressed until the following spring after the pathogen has resumed activity.

1.15.6   MANAGEMENT

The key to effective management of disease caused by *P. cactorum* is integrated prevention. No single disease control measure is completely effective for management of *Phytophthora* crown rot in strawberry cultivars highly susceptible to the pathogen, but combined approaches can be very effective. Effective pre-plant soil fumigation helps to insure production of clean nursery stock and minimize the risk of infection in fruiting fields. Pre-plant fumigation with mixtures of methyl bromide and chloropicrin (MB: CP) typically kill all or most inoculum of the pathogen at soil depths in the surface 2–3 ft of soil. Trials indicated that CP or mixtures of 1,3-dichloropropene (1,3-D) and CP can approach the effectiveness of MB: CP, but only at rates of at least 300 lb/A. With drip applications, drip-line placement and water application amount during the fumigation can critically influence effectiveness of CP and 1,3-D: CP for control of *P. cactorum* in soil. Because these fumigants have less diffusion potential than MB, uniform wetting of the bed is essential for effective control of the inoculum. Virtually impermeable film (VIF) properly applied over plant beds can significantly reduce fumigant emissions to the atmosphere and improve control of weed seeds

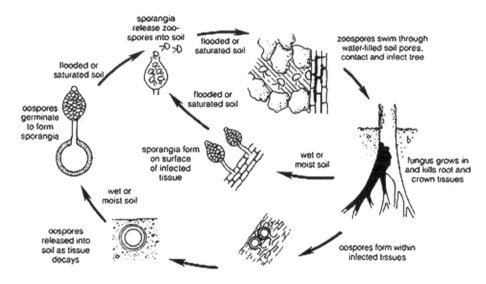

**Disease cycle of Phytophthora
root and crown rots.**

FIGURE 1.37   Disease cycle of *P. cactorum*.

and pathogen inoculum near the soil surface; however, control of *P. cactorum* and other soilborne inoculum at soil depths of 1 ft or more was not significantly improved by VIF (Table 1.5).

1.16   *RHIZOCTONIA* ROOT ROT

Rhizoctonia Root Rot, also known as black root rot of strawberries is a very complex and serious disease that has been reported in strawberry fields around the world. Several pathogenic organisms have been associated with the disease; however, the soilborne fungus *Rhizoctonia fragariae* is probably the most frequently isolated pathogen from strawberry roots exhibiting symptoms of black root rot.

1.16.1   CAUSAL ORGANISM

Species	Associated Disease Phase	Economic Importance
R. fragariae	Black root rot	Major

1.16.2   SYMPTOMS

Affected plants may be scattered throughout a strawberry planting or grouped in one or more parts of it. Plants with black root rot are less vigorous than normal plants and produce fewer runners. Individual or groups of leaves may wilt, discolor, and die. Entire plants may die when black root rot is severe. Affected plants should be carefully dug up (not pulled) and their root systems examined. Plants with black root rot will exhibit 1 or more of the following root symptoms:

* Root system smaller than in normal plants.
* Main root with lesions—these are darker than the rest of the root.

TABLE 1.5

Effects of Treatments with Aliette or Ridomil on Productivity of Two Strawberry Cultivars in Non-Infested Soil and Soil Infested with *Phytophthora cactoruma*[a]

Strawberry Cultivar	Soil Treatment[b]	Chemical Treatment Program[c]	Marketable Yield (Total Grams per Plant)
Diamante	Infestation w/ *P. cactorum*	• Aliette plant dip and spray	1031
		• Water control plant dip and spray	572
		• Ridomil soil drench	1163
		• Water control soil drench	659
	Non-infested control	• Aliette plant dip and spray	1113
		• Water control plant dip and spray	1097
		• Ridomil soil drench	1172
		• Water control soil drench	1128
Aromas	Infestation w/ *P. cactorum*	• Aliette plant dip and spray	1388
		• Water control plant dip and spray	938
		• Ridomil soil drench	1400
		• Water control soil drench	891
	Non-infested control	• Aliette plant dip and spray	1481
		• Water control plant dip and spray	1250
		• Ridomil soil drench	1463
		• Water control soil drench	1384

[a] From a field trial at Monterey Bay Academy in 2001/02.

[b] After pre-plant fumigation with methyl bromide-chloropicrin mixture, the soil treatments were applied to each planting hole in 100 ml of V8 juiceoat- vermiculite medium that was either permeated with *P. cactorum* (the infestion treatment) or sterile (the non-infested control).

[c] The pre-plant dip and spray treatments with Aliette were applied at maximum label rates; one pre-plant dip and five foliar sprays were applied over the growing season. The drench program with Ridomil simulated drip chemigation with the material; the maximum label rate was used, with one treatment applied at planting and two more applied during the growing season.

- Feeder roots are lacking.
- Feeder roots with dark zones or lesions.
- All or part (usually the tip) of main roots killed. A cross-section of a dead root shows it is blackened throughout (Figures 1.38 and 1.39).

1.16.3   CAUSE AND DISEASE DEVELOPMENT

One or more of the following factors may be involved in black root rot problems: soil fungi (e.g., *Rhizoctonia, Fusarium, Pythium*) nematodes (microscopic round worms), winter injury, fertilizer burn, drought, excess salts, herbicide injury, wet soils, or pH imbalance. In some cases, environmental factors may predispose plants to an attack by the root rotting fungi.

1.16.4   FAVORABLE CONDITIONS OF DISEASE DEVELOPMENT

Wet soils, excess salt conditions of the soil, pH imbalance, and winter injury favor the development of the disease.

FIGURE 1.38   Strawberry plant dying from black root rot.

1.16.5   MANAGEMENT

There are no cures or guaranteed controls for black root rot. Control measures center around proper planting and care of strawberry plants. The following strategies can be followed:

- Use only healthy, white-rooted strawberries when planting.
- Plant in well-drained soils.
- Maintain plant vigor with adequate fertilization and cultivation.
- Irrigate strawberries during dry periods.
- Fumigation may be practical for commercial plantings. It is not recommended for home plantings.

FIGURE 1.39   Black root rot developing on primary and feeder roots of strawberry.

1.17 CHARCOAL ROT OF STRAWBERRY

Charcoal rot, caused by *Macrophomina phaseolina*, is a relatively new disease in Florida. This disease was first observed in December 2001, when collapsed and dying strawberry plants from a commercial field were submitted to our diagnostic clinic. During the 2003–2004 season, *M. phaseolina* was isolated from dying strawberry plants from the original field and two additional farms. Since then, a few additional samples are received in our diagnostic clinic every season. Affected plants are often found along field margins or other areas that were inadequately fumigated with methyl bromide. Charcoal rot has also been reported on strawberry in France, India, and Illinois.

1.17.1 CAUSAL ORGANISM

Species	Associated Disease Phase	Economic Importance
M. phaseolina	Wilt and Crown Rot	Minor

1.17.2 SYMPTOMS

Symptoms caused by *M. phaseolina* are similar to those caused by other crown-rot pathogens such as *Colletotrichum* and *Phytophthora* species. Plants initially show signs of water stress and subsequently collapse (Figure 1.40). Cutting the crowns of affected plants reveals reddish-brown necrotic areas on the margins and along the woody vascular ring (Figure 1.41). To confirm a diagnosis, a sample must be submitted to a Diagnostic Clinic and the pathogen must be isolated from the diseased crowns and identified.

1.17.3 CAUSE AND DISEASE DEVELOPMENT

Very little is known regarding this disease on strawberries. *M. phaseolina* is a common soilborne pathogen in many warm areas of the world and has a very broad host range. Many vegetable crops planted as second crops after strawberry such as squash, cantaloupe, and peppers, legumes, and others are susceptible. Those infections may increase inoculum levels of *M. phaseolina* in the soil in the off-season for strawberries.

1.17.4 FAVORABLE CONDITIONS OF DISEASE DEVELOPMENT

In general, high temperatures and low soil moisture favor infection and disease development.

FIGURE 1.40   Plant wilt symptom of charcoal rot.

FIGURE 1.41   Internal crown symptoms of charcoal rot.

1.17.5   MANAGEMENT

No fungicides are labeled for control of charcoal rot on strawberries. Topsin M® is labeled for control of charcoal rot on other crops. Preliminary results with Topsin M® have shown that application of this product may delay the onset of symptoms. Studies are currently being conducted to determine if cultivars differ in susceptibility to charcoal rot. This disease may be an emerging threat as the Florida strawberry industry makes the transition from methyl bromide to other fumigants.

1.18   GNOMONIA FRUIT ROT AND LEAF BLOTCH

Gnomonia fruit rot, leaf blotch, and stem-end rot caused by *Gnomonia comari* P. Karst. (anamorph, *Zythia fragariae* Laibach) were observed in a strawberry fruit production field at Watsonville, CA, in 1996. *Z. fragariae* has been known for years to attack leaves and cause leaf blotch but this is the first time that the perfect stage, *G. comari*, was identified and documented to infect fruits and cause stem-end rot in California.

1.18.1   CAUSAL ORGANISM

Species	Associated Disease Phase	Economic Importance
G. comari	Fruit Rot, Leaf Blotch, Stem-End Rot	Serious

1.18.2   SYMPTOMS

Plants become infected between flowering and harvesting. Figure 1.42 shows how the fruiting bodies of the fungus develop on trash and from there the spores are produced that infect the next crop. Figure 1.43 shows very early symptoms of the disease on leaves. It is important to start controlling the disease at this stage. The fungus first infects the calyx (see Figure 1.44), and disease spreads into the fruit as a rot. Both green and ripe fruit may be infected. Infected fruit ripens early and turns pale red to brown. They remain firm but are often invaded by other fruit rots such as gray mold (Figure 1.45).

FIGURE 1.42   Black "dots" are the fruiting bodies of the fungus on strawberry residue. Spores infect the next crop.

FIGURE 1.43   Early symptoms of infection on a leaf (small lesions on the underside).

FIGURE 1.44   The start of stem-end rot in fruit caused by *Gnomoniopsis* (on the tips of the calyx).

1.18.3   Cause and Disease Development

Trash from previous and current strawberry crops left in the soil is the source of infection. Planting material can carry the spores systemically, but levels are usually quite low, and it is unlikely that runners are a significant source of infection (Figure 1.46).

FIGURE 1.45   Leaf symptoms of *Gnomoniopsis*.

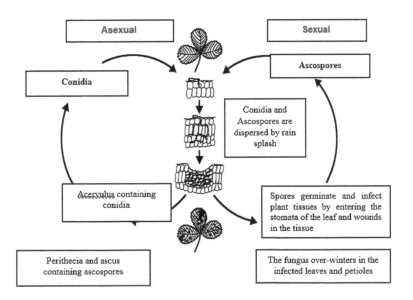

FIGURE 1.46   Life cycle of *G. comari*.

1.19   LEAF SPOT OF STRAWBERRY

Leaf spot is one of the most common and widespread diseases of strawberry. *M. fragariae* (asexual stage *Ramularia tulasnei* Sacc.) is also the cause of black seed disease on strawberry fruit, which occurs occasionally in North America where *Mycosphaerella* leaf spot is present. Prior to the development of resistant cultivars and improved control programs, leaf spot was the most economically important strawberry disease.

1.19.1 CAUSAL ORGANISM

Species	Associated Disease Phase	Economic Importance
M. fragariae	Leaf Spot. Also Affects the Petioles, Stolons, and Fruits	Low

1.19.2 SYMPTOMS

1.19.2.1 Leaves

Leaf symptoms vary with strawberry cultivar, the strain of the fungus causing disease, and environmental conditions. Leaf lesions or "spots" are small and round (3–8 mm diameter), dark purple to reddish in color, and are found on the upper leaf surfaces (Figure 1.47). The center of the spots becomes tan to gray to almost white over time, while the broad margins remain dark purple. Lesion centers on younger leaves stay light brown, with a definite reddish purple to rusty brown margin. Numerous spots may coalesce and cause the death of the leaf. Large, spreading lesions that involve large portions of the leaflet are formed on some highly susceptible cultivars; the centers of which remain light brown. In warm humid weather, typical solid rusty brown lesions without purple borders or light-colored centers may form on young leaves. Lesions are evident on the undersurface of the leaf but are less intense in color, appearing as indistinct tan or bluish areas.

1.19.2.2 Leaf Stems (Petioles), Runners, Fruit Stalks (Pedicels), Berry Caps (Calyxes)

Symptoms are almost identical to those on leaves, except for fruit. Only young tender plant parts are infected by this pathogen (Figure 1.48).

1.19.2.3 Fruit

Superficial black spots (6 mm in diameter) form on ripe berries under moist conditions. These spots surround groups of seeds (achenes) on the fruit surface. The surrounding tissue becomes brownish black, hard, and leathery. The pulp beneath the infected area also becomes discolored; however, no general decay of the infected berry occurs. Usually, only one or two spots occur on a berry, but some may have as many as eight to ten "black-seeds" (Figure 1.49). Symptoms are most conspicuous on white, unripe fruit and on ripe fruit of light-colored cultivars. Economic losses in this case are due to the unattractiveness of "black seed" spots on fruit, rather than fruit rot.

FIGURE 1.47   Typical foliar symptoms of leaf spot on strawberry leaves.

FIGURE 1.48   *M. fragariae*, pathogen of strawberry, leaf spot on sepals.

1.19.3   CAUSE AND DISEASE DEVELOPMENT

M. fragariae overwinters in lesions in old leaves and produces its first spores in about mid-May. The spores fall on other leaves and germinate when it rains. After an incubation period ranging from 15 to 30 days, new spots appear and produce new spores that infect other young leaflets. This cycle can be repeated several times during a single growing season. Strawberry leaf spot is spread by water. During rainfall events or spray irrigation, the water droplets that make contact with the leaves tear spores away from the lesions and project them onto new leaves. Unlike several other species of fungi, the spores are not transported by the wind, limiting their propagation. Heavy, frequent downpours can, however, result in epidemic outbreaks of the disease.

1.19.4   FAVORABLE CONDITIONS OF DISEASE DEVELOPMENT

Leaf spot may reach economic threshold levels, provided young leaves and inoculum are present, under conditions of high temperatures and long periods of leaf wetness. Research results show most severe infection of young leaves to occur during periods of leaf wetness from 12 to 96 hours when temperatures fall in the range of 15°C–20°C. This data suggests fungicide treatments should be applied in early spring and after renovation of plantings if inoculum was present.

1.19.5   DISEASE CYCLE

In the south, perithecia and sclerotia are absent. Conidia are produced in small dark fruiting bodies (pseudothecia) within leaf lesions and serve as inoculum. In this instance, infection is a continuous

FIGURE 1.49   Black seed symptoms on strawberry.

process with older lesions producing conidia to infect young leaves during each season. Conidia landing on leaf surfaces produce germ tubes which penetrate through natural leaf openings (stomata) on the upper and lower surfaces of leaves. New conidia are produced on clusters (fascicles) of conidiophores which grow out through stomata. These are carried to new leaves by rain splash, and the disease cycle begins again (Figure 1.50).

In northern growing regions, the life cycle is somewhat different. Three sources of primary inoculum may be present: conidia overwintering on living leaves, conidia from overwintering sclerotia, and ascospores. Abundant conidia, produced in early summer on lesions on both upper and lower leaf surfaces and lesions on other plant parts, are spread primarily by water splash. High rainfall can lead to disease of epidemic proportions. Sclerotia are produced profusely during the winter on dead infected leaves. These may also produce abundant conidia in the spring. Conidia also develop on occasion from the bases (apices) of perithecia. Perithecia are produced primarily on upper surfaces of overwintered leaves. From these, perithecia are wind disseminated. It is not known if these serve as an important source of primary inoculum, but they are most probably a means by which genetically different strains of the fungus may travel long distances. *M. fragariae* establishes in the stigma at the time of flowering and then grows to the achene. From there it infects surrounding berry (receptacle) tissue. Conidia produced in leaf infections are probably the primary inoculum source for fruit infections.

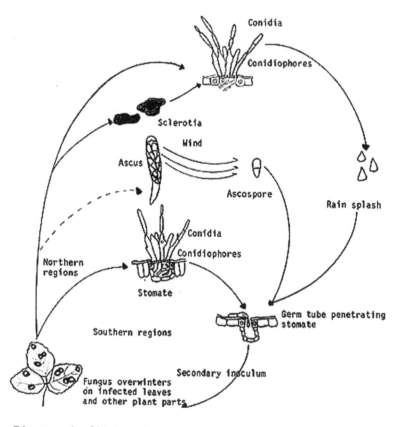

FIGURE 1.50   Disease cycle of *M. fragariae*.

1.19.6   MANAGEMENT

1.19.6.1   Chemical Control

Keep varying the fungicides used to prevent or promote the development of resistance. Bear in mind that too many applications of any pesticide can lead to the development of resistance. Captan, Folpet, Dodine, and copper are a few of the fungicides recommended by Réseau d'avertissements phytosanitaires (RAP). It is recommended that copper be applied (in the form of tribasic copper sulfate) during the year of harvest. It is important not to abuse this metal since it accumulates in the soil and could reach toxic concentrations.

1.19.6.2   Cultural Control

1.19.6.2.1   Scouting for the Disease

Despite all precautions, it is impossible never to have problems with leaf spot. The provisional scouting method for leaf spot approved by the RAP consists in observing 100 leaflets at random (one leaflet per plant) twice—once at the end of September of the planting year and again during flowering in the first growing season. In the spring, treatment is not recommended if less than 25% of the old leaves have symptoms the previous fall, and, during the growing season, if less than 10% of the new leaflets are infected at flowering.

1.19.6.2.2   Umbrella Effect

If the scouting results indicate that control measures are required, preventive measures should be taken. The threshold limit value for leaf spot is higher than that for gray mold rot, which attacks the fruit directly. However, recent Quebec studies reveal that owing to the method of infection of *M. fragariae*, the best strategy is to obtain the "umbrella effect". The "umbrella" strategy is based on the fact that only young strawberry leaflets are susceptible to the disease. It is important, therefore, to protect the young foliage through fungicide applications during the rapid growth period of strawberries. The foliage must then be protected until the plant has formed a few leaves and these leaves have passed the stage at which they are susceptible to the disease. The leaves at the very top of the plant, which are free of disease, protect the new leaves, hence the name "umbrella". This prevents spores on infected leaves from contaminating the smaller leaves located below by dripping rainwater. When the crop is mowed, the same approach must be taken, i.e., to treat until the umbrella effect is obtained.

1.19.6.3   Preventive Treatment

M. fragariae requires water to produce infection. When treatment is necessary, it is highly recommended that preventive treatment be carried out when weather forecasts call for rain within 24 hours.

The following materials are listed in order of usefulness in an IPM Program, considering efficacy. Also, consider the general properties of the fungicide as well as information relating to environmental impact. Not all registered pesticides are listed. Always read the label of the product being used (Table 1.6).

1.20   BACTERIAL LEAF SPOT OF STRAWBERRY

Bacterial leaf spot, also known as Angular Leaf Spot is a bacterial disease caused by *Xanthomonas fragariae*, a pathogen highly specific to both the wild and the cultivated strawberry, *F. ananassa*. Bacterial angular leaf spot disease on strawberries has been increasing in importance to strawberry producers in recent years because it is spread by infected but asymptomatic strawberry plantlets used in annual row culture systems. Until now there have been no control methods for the disease

TABLE 1.6

Preventive Treatment

Common Name (Trade Name)	Mode of Action Group Name (Number1)	Amount/ Acre[a]	R.E.I.[b] (Hours)	P.H.I.[b] (Days)
Chlorothalonil (BravoWeatherStik)	Chloronitrile (M5)	1.5 pt	12	N.A.
Myclobutanil (Rally) 40W	Demethylation inhibitor (3)	2.5–5 oz	24	0
Triflumizole (Procure) 50WS	Demethylation inhibitor (3)	4–8 oz	12	1

N.A. Not applicable.

Group numbers are assigned by the Fungicide Resistance Action Committee (FRAC) according to different modes of actions (for more information, see http://www.frac.info/). Fungicides with a different group number are suitable to alternate in a resistance management program. In California, make no more than one application of fungicides with mode of action Group numbers 1, 4, 9, 11, or 17 before rotating to a fungicide with a different mode of action Group number; for fungicides with other Group numbers, make no more than two consecutive applications before rotating to fungicide with a different mode of action Group number.

[a] Apply all materials in 200-gal water/acre to ensure adequate coverage.

[b] Restricted entry interval (R.E.I.) is the number of hours (unless otherwise noted) from treatment until the treated area can be safely entered without protective clothing. Preharvest interval (P.H.I.) is the number of days from treatment to harvest. In some cases, the REI exceeds the PHI. The longer of two intervals is the minimum time that must elapse before harvest.

and no resistant varieties. Four genetically distinct strains of the pathogen, *X. fragariae*, have been identified and used to screen a collection of 81 strawberry accessions for resistance to the pathogen.

1.20.1 CAUSAL ORGANISM

Species	Associated Disease Phase	Economic Importance
X. fragariae	Leaf Spot	High

1.20.2 SYMPTOMS

Infection first appears as minute, water-soaked spots on the lower surface of leaves. The lesions enlarge to form translucent, angular spots that are delineated by small veins and often exude a viscous ooze of bacteria and bacterial exudates, which appear as a whitish and scaly film after drying. As the disease progresses, lesions coalesce and reddish-brown spots, which later become necrotic, appear on the upper surface of the leaves. A chlorotic halo usually surrounds the infected area. *X. fragariae*, the causal agent of ALS, is a slow-growing, gram-negative bacterium that produces water-soaked lesions on the lower leaf surfaces (Figure 1.51). The bacteria enter the leaf through the stomata (tiny spores that are most abundant on the lower surface of the leaf). Lesions begin as small and irregular spots on the undersurface of the leaflets. When moisture is high on the leaves, lesions ooze sticky droplets of bacteria. As the disease develops, lesions enlarge and coalesce to form reddish-brown spots, which later become necrotic (Figure 1.52).

A practical way to recognize the disease is to place the leaves against a source of background light where the translucent spots can be seen (Figure 1.53). The tissue with older damage eventually dies and dries up, giving leaves a ragged appearance.

During severe epidemics, the pathogen can also cause lesions on the calyx of fruit that are identical to foliar lesions (Figure 1.54). When severe, these calyxes can dry up and make the fruit unmarketable.

FIGURE 1.51   Water-soaked lesions of angular leaf spot.

FIGURE 1.52   Reddish-brown spots of angular leaf spot.

FIGURE 1.53   Translucent spots of angular leaf spot.

FIGURE 1.54   Water-soaked lesions of angular leaf spot on the calyx.

1.20.3   CAUSE AND DISEASE DEVELOPMENT

The primary source of inoculum in a new field is contaminated transplants. Secondary inoculum comes from bacteria that exude from lesions under high moisture conditions. Bacteria can survive on dry infested leaves and tissue buried in the soil for up to 1 year. The pathogen can be spread easily by harvesting operations when wet and cool conditions favor the production of bacterial exudate. The pathogen also can be dispersed by rain and overhead sprinkler irrigation. If the disease invades the vascular system of the plant, the disease will be difficult to control. Affected plants may wilt and die.

1.20.4   FAVORABLE CONDITIONS OF DISEASE DEVELOPMENT

Not much is known in this respect, and more research is underway to determine which conditions are most favorable for disease development and spread. Some report moderate to low daytime temperatures and nighttime temperatures below be freezing are needed. Most researchers agree high humidity is also a key factor. The development of the disease seems favor by warm days (20°C) and cold nights (-2°C–4°C).

The primary source of inoculum in a new field is contaminated transplants. Secondary inoculum comes from bacteria that exude from lesions under high moisture conditions. Bacteria can survive on dry infested leaves and tissue buried in the soil for up to one year.

The pathogen can be spread easily by harvesting operations when wet and cool conditions favor the production of bacterial exudate. The pathogen also can be dispersed by rain and overhead sprinkler irrigation.

1.20.5   DISEASE CYCLE

Inoculum for primary spring infections in new growth comes primarily from infected transplants or systemically infected overwintered plants and dead leaves. This bacterium is resistant to adverse conditions such as desiccation and can survive for long periods in dry leaf debris or buried leaves in soil.

Bacteria exuded from the undersides of leaves under high moisture conditions serve as the secondary source of inoculum in plantings. Angular leaf spot (ALS) bacteria are carried from plant to plant by splashing water from rain or overhead irrigation, as well as harvesting operations. The motile bacterial cells may enter the plant through drops of dew, guttation droplets, and rain or irrigation water.

1.20.6 MANAGEMENT

The best way to control ALS is to use pathogen-free transplants. Since this is not always possible, growers should avoid harvesting and moving equipment through infested fields when the plants are wet. Minimizing the use of overhead sprinklers during plant establishment and for freeze protection also reduces the spread of the disease. The use of surfactant-type spray adjuvants should also be avoided when ALS is a threat since these products often help bacteria penetrate through the stomata and may enhance disease development.

Copper-based products can provide effective control of the disease in some instances, but low rates of copper should be used since phytotoxicity (reddening of older leaves, slow plant growth, and yield decrease) has been documented with repeated sprays. Many copper products are labeled for ALS control on strawberry, such as copper hydroxide, copper oxychloride, basic copper sulfate, cuprous oxide, and various other copper compounds. These active ingredients suppress ALS, but it is important to apply the correct amount. Trial results have shown that preventive, weekly applications of copper fungicides at 0.3 lb of *metallic copper* per acre were effective in reducing disease symptoms without causing phytotoxicity on the plants. However, trial results have also shown that when disease pressure is low to moderate, the use of copper sprays did not significantly increase yield. Copper products can increase yield and decrease the possibility of fruit rejection only when environmental conditions arc highly favorable for infection and spread.

Many other products have been tested over the years in the search for an alternative to copper. Actigard®, a plant-resistant activator manufactured by Syngenta, has been shown to suppress ALS. Actigard® is used to control bacterial spot disease on tomatoes in Florida, but it is not currently approved for use on strawberry.

1.21 THE DAGGER NEMATODE OF STRAWBERRY

The Dagger Nematode, also known as American Dagger Nematode, is one of many species of the genus *Xiphinema*. The common name "Dagger Nematode" applies to all species of the genus. *X. americanum* was first described in 1913 by N. A. Cobb, who had recovered it from the roots of corn, grass and citrus found growing on both the "Atlantic and Pacific slopes of the United States". Found in both agricultural and forest soils, *X. americanum* has been referred to as the most destructive plant-parasitic nematode in America. It has been reported that many nematodes identified as *X. americanum* from various parts of the world are probably a number of different closely related species.

1.21.1 DISTRIBUTION OF THE CAUSAL ORGANISM

X. americanum has a worldwide distribution. It has been found in Canada, the United States, Mexico, Central and South America, and in the Caribbean Islands; has also been recorded from Africa, Japan, India, parts of Europe, Australia, U.S.S.R., and Pakistan.

1.21.2 SYMPTOMS

The symptoms that plants exhibit in response to the pathogenicity of *X. americanum* are similar to those of other migratory ectoparasitic nematodes of roots. It is common to see poor growth and or stunting of the plant, yellowing or wilting of the foliage, and reduced root systems, which can include root necrosis, lack of feeder or secondary roots, and occasional tufts of stubby rootlets.

The dagger nematode causes the devitalization of root tips and overall root death when they feed at the root tips and root sides of strawberry plants. Reddish-brown lesions that turn black and necrotic with time result at the sites of feeding, and result in reduced root systems and stunted tops.

X. americanum is listed as a C-rated pest in California due to its wide host range of California crops. C-rated pests are widespread, and are of known economic or environmental detriment, according to The California Department of Food and Agriculture. Due to *X. americanum's* difficulty in maintaining high populations in frequently tilled soils, the dagger nematode is mainly an economic problem on biennial and perennial crops rather than annual crops (except for damage to emerging seedlings).

1.21.3 Disease Cycle

Relatively little critical biological and ecological work has been done on this important genus— some species are difficult to maintain in greenhouse cultures, e.g. *X. americanum*, others apparently have a very long life-cycle, e.g., *X. diversicaudatum*. All stages occur in the soil.

Eggs, from which first-stage juveniles emerge, are deposited singly in water films around soil particles and are not enclosed in an egg mass. There are three or four juvenile stages and sexually mature adults. Males are rare in most species and are apparently unnecessary for reproduction. *X. americanum* can live as long as three to five years.

Reproduction by fertilization from a male is rare if not nonexistent due to the lack of male *X. americanum* individuals, and therefore females reproduce parthogenetically. All of the stages of *X. americanum* occur in the soil, with no particular stage as an important survival stage. In places with low winter temperatures, however, the egg is the primary survival structure. (Figure 1.21.1).

1.21.4 Management

1.21.4.1 Chemical Control

Now there are no products with proven post-plant efficacy registered for use on blueberry. Preplant assessment of nematode levels is necessary to determine if preplant fumigation is required (Figure 1.55).

- Basamid G. Avoid application when soil is over 32°C. Do not apply within 3 to 4 ft of growing plants or closer than the drip line of larger plants. Do not harvest within one year of application. 24-hour re-entry plus ventilation. **Restricted-use pesticide**.
- Paladin at 35 to 51.3 gal/A. Buffer zone from 35 to 690 ft depending on the rate used and acreage treated. Two- to five-day entry restriction period. See the label for details. Unknown efficacy in the PNW. **Restricted-use pesticide**.
- Use Telone II the fall before spring planting. Allow two to three weeks between treating and planting, or wait until odor has left the soil. Do not treat extremely heavy soils. Five-day re-entry. **Restricted-use pesticide**.

1.21.4.2 Cultural Control

- Locate plantings on soil that have been tested and found free of dagger nematodes.
- Control broadleaf weeds to eliminate virus hosts.
- Plant virus-tested (and found to be free of all known viruses) certified stock.
- Crop rotation is another form of control for *X. americanum*. It has been shown that certain non-host plants may deny the nematode population an adequate food source for reproduction, and thus greatly reduce its population in the soil. This is termed passive suppression.

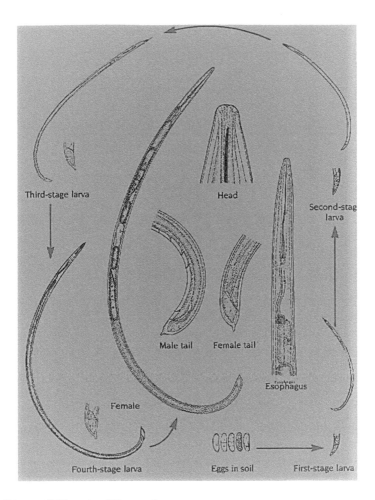

FIGURE 1.55   Disease (life) cycle of *X. americanum.*

- *X. americanum* can only travel via run-off and in damp soil; therefore, if soils are kept dry enough the nematodes can be localized and quarantined.
- Additionally, if the soil is tilled frequently, *X. americanum* will likely not be in high enough in population density to cause any noticeable symptoms in its hosts. There is also evidence of *X. americanum* resistance and "tolerance" seen in certain species of grapes that appeared to be better adapted to the parasite.

1.22   THE FOLIAR NEMATODE

The most important plant-parasitic nematode of strawberry is the foliar (leaf) nematode called *Aphelenchoides fragariae*. *A. fragariae* has a very extensive host range including vegetables, ornamentals, fruits, broad acre crops, and weeds. *A. fragariae* is a common and widespread migratory ecto-parasite (outside) and endo-parasite (inside) of leaves, buds, and stems.

1.22.1   THE CAUSAL ORGANISM

A. fragariae is a foliar nematode. which has an extensive host range and is widely distributed through the tropical and temperate zones around the world. It is a frequently encountered and economically damaging pest in the foliage plant and nursery industries. As its name suggests, *A.*

FIGURE 1.56   Tight aggregation of strawberry crown with malformed leaves.

fragariae is also a major pest of strawberries worldwide. It has recently been detected in Australian strawberry crops, causing some major losses, whereas historically *A. besseyi* has caused problems in the Australian industry. This nematode should not be confused with *A. ritzemabosi*, another bud and leaf nematode that occurs mainly in chrysanthemum, or *A. besseyi* which causes "crimp disease" in strawberries.

1.22.2   SYMPTOMS

On strawberries, abnormal plant growth with stunting and deformation of buds, leaves, and flowers is the first symptom. The malformations include twisting and puckering of leaves (Figure 1.56), undersized leaves with crinkled edges, tight aggregation of crowns, reddened and stunted petioles, and flower stalks with aborted or partly aborted flowers. *A. fragariae* can severely impact the yield of this high-value crop, as heavily infected plants do not produce fruit (Figure 1.57).

1.22.3   FAVORABLE CONDITIONS

A. fragariae can become a serious pathogen in nursery grown plants where environmental conditions such as warm temperatures and high humidity favor a rapid buildup of the population.

1.22.4   DISEASE CYCLE

After the growing season, *Aphalenchoides fragariae* adults and juveniles may remain in soils for up to three months, while eggs may stay dormant for years until favorable conditions arise.

FIGURE 1.57   Heavily infested strawberry plant (left) adjacent to asymptomatic plant.

Overwintering only occurs in dead plant tissues, and nematodes may successfully remain dormant in temperatures as low as 3°C. Once moist conditions return in the spring, nematodes become active and feed ecto-parasitically on crowns, runners, and new buds of their host strawberry plants, only occasionally being found in leaf tissue. Nematodes reproduce sexually with females laying up to 30 eggs in ideal fertile conditions of approximately 18°C. A typical life cycle lasts between 10 and 13 days of which juveniles undergo three molting stages. Multiple life cycles will occur in one growing season as long as conditions are favorable. However, if conditions become unfavorable, adult and juvenile nematodes may become dormant either in strawberry or nearby weed tissue until ideal conditions arise. Nematodes can be dispersed within an area through irrigation, direct contact of healthy plants with infected, or poor sanitation techniques.

1.22.5 MANAGEMENT

Foliar nematode diseases can be minimized with good cultural practices. Growers should maintain strict sanitation and inspection of plant material to minimize losses. This includes the constant removal and destruction of infected plants and the use of nematode-free planting material for propagation. Excessive humidity and splashing of water on stems and leaves and contact between plants should be avoided. Hot-water treatments to ensure clean planting material have long been recommended. An emergency use permit for the application of liquid Nemacur® into the crown of strawberry plants has been granted for use at strawberry runner farms in Queensland (QLD), Australia. This is not an option for fruit growers as a six-week withholding period is required.

The following is a list of activities designed to prevent or contain incidence of foliar nematodes:

- Thoroughly rouge and burn infested plant material.
- Avoid contact between plants (in screen houses).
- Avoid the formation of water film on the leaf surface, use drip irrigation.
- Select planting material from healthy stocks or use certified planting materials.
- Produce runners in sterilized soil, or by tissue culture using nematode-free materials.
- Disinfect benches and tools before use.
- Treat planting material with hot water at 50°C for 20 minutes.
- Rotate crop with grain crops such as barley and rye.
- Test planting material, and soil of the block to be planted, for nematodes before planting the crop.

1.23   THE LESION NEMATODE OF STRAWBERRY

Lesion or root lesion nematode disease is caused by members of the genus *Pratylenchus*. The common name of these nematodes is derived from the often-conspicuous necrotic lesions they cause on host roots. In the past, the name meadow nematodes were used occasionally because of their abundance in this habitat from which the first species was described. Lesion nematodes are migratory endoparasites that enter the host root for feeding and reproduction and move freely through or out of the root tissue. They do not become sedentary in the roots, as do the cyst or root knot nematodes. Feeding is restricted almost entirely to the cortex of the root.[3]

1.23.1 THE CAUSAL ORGANISM

Lesion nematodes are essentially worldwide in distribution. Five of the more than 40 species of *Pratylenchus* have been reported to occur most commonly. These are *Pratylenchus penetrans*, *P. alleni*, *P. hexincisus*, *P. neglectus*, and *P. scribneri*. Lesion nematodes may exist as a single species at a given site, or as a complex of two or more species.

1.23.2 SYMPTOMS

Infected plants are dwarfed, off-color, and grow poorly. Damage is frequently seen as spots in the field. Roots have brown lesions. There is a reduction in leaf size and the number of leaves produced on heavily infected plants. Yields can be substantially reduced. Very high nematode populations may interact with other soilborne pathogens to kill initially weakened plants, resulting in barren areas in a field. The affected areas may increase in size as the season progresses. If growing conditions are otherwise favorable, however, moderately infected plants often appear to outgrow early damage, since the root systems are able to extend deep into the soil. Flowers or fruit may be reduced substantially in number and quality. The vigor of the host is reduced, and the plant may be predisposed to winter injury or other infectious diseases. Replacing the affected plant without soil treatment frequently results in poor growth and sometimes in the death of the new plant. Symptoms of lesion-nematode damage resemble those of other soilborne diseases, nutrient deficiencies, insect damage, or cultural and/or environmentally induced stress. Thus, a soil test is necessary to determine whether a nematode problem exists. The symptoms on infected roots initially are small, light-to-dark brown lesions (Figure 1.58). These lesions tend to expand and to merge as the growing season progresses, giving the roots a discolored appearance overall.

1.23.3 FAVORABLE CONDITIONS

The degree of damage caused by the nematode depends on a number of environmental conditions. Injury is usually most severe in light-textured soils that are low in nutrients (e.g., nitrogen, potassium, or calcium) and in organic matter. Plants under moisture and high-temperature stress are most likely to suffer damage. In a year of high stress, damage can be substantial. When growth conditions are optimum, there may be no noticeable injury, even with high populations of lesion nematodes in the roots. The disease symptoms are more pronounced and yield reduction is greater.

When abnormally high temperatures occur early in the growing season, rainfall is inadequate, soil fertility is low or imbalanced, or root-rot organisms attack the plants.

1.23.4 DISEASE CYCLE

The disease (life) cycle of a lesion nematode is rather simple. Root lesion nematodes are migratory endo-parasites. This means that they usually feed on the roots internally versus ecto-parasites that feed on the roots externally. The migratory endo-parasite enters the root by puncturing a hole in an outside cell with its stylet. Once the nematode has access to the inside of the cell, it continues to migrate from one cell to the next feeding along its way. As it moves through the cells and takes nutrients from them, the cells are left with tiny lesions that eventually become necrotic as the root begins to decay. As the nematode feeds on the cells in order get nutrients, the metacarpal pump

FIGURE 1.58   Note the necrotic areas on these strawberry roots caused by the lesion nematode.

(a part of a nematode) begins to "pump" nutrients which are ingested into the nematode. Over the course of feeding on a specific root, female nematodes lay single eggs that can hatch in the root (or if the egg was laid in the soil, it hatches in the soil), and the process begins again with the juvenile nematode. *Pratylenchus* does not need both male and female to reproduce; in fact, the females can and do lay eggs without the presence of males, i.e., eggs develop parthogenetically. For some species, the number of females drastically outnumbers the number of males. Sexual reproduction occurs in *P. penetrans* and *P. alleni*. After mating, the female lays its eggs singly or in small groups in the host root or in the soil near the root surface.

Lesion nematodes overwinter as eggs, larvae, or adults in host roots or soil. The length of the life cycle depends on the species and the soil temperature. Favorable temperature for *P. penetrans* is 25°C (77°F).

Lesion nematodes remain inactive when soil temperatures are below 15°C (59°F); except for *P. penetrans*, there is little activity until temperatures rise above 20°C (68°F). *P. penetrans* completes its life cycle in 30 days at 30°C (86°F), 35 days at 24°C (75°F), and 86 days at 15°C (59°F). Although the other species have not been thoroughly studied, their developmental biology is probably similar to that of *P. penetrans* (Figure 1.59).

1.23.5 Management

1.23.5.1 Chemical Control

1. Preplant soil fumigation.
   - Basamid G. Avoid application when soil is over 32°C. Do not apply within 3 to 4 feet of growing plants. Do not harvest within one year of application. Twenty-four-hour re-entry plus ventilation. **Restricted-use pesticide.**
   - Methyl bromide + chloropicrin (33 to 55%) under polyethylene film. See the label for rates. **Restricted-use pesticide.**
   - Midas 50:50 at 200 to 350 lb/A for broadcast application with standard tarps. Can only treat 40 acres per day. Do not plant within 10 days of application. Five-day re-entry plus ventilation. **Restricted-use pesticide**.

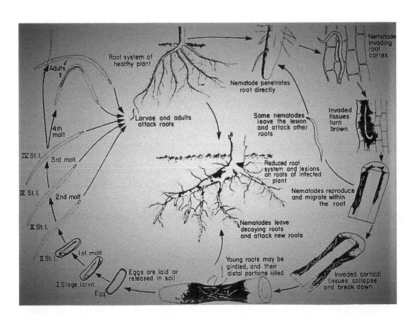

FIGURE 1.59 Disease (life) cycle of *P. penetrans*.

- Paladin at 35 to 51.3 gal/A. Buffer zone from 35 to 690 ft depending on the rate used and acreage treated. Two- to five-day entry restriction period. See the label for details. **Restricted-use pesticide**.
- Telone II at 24 to 36 gal/A broadcast on mineral soils. Wait two to three weeks after treating to plant, or until odor leaves the soil. Do not treat extremely heavy soils. Five-day re-entry. **Restricted-use pesticide.**
- Telone C-17 at 27 to 41 gal/A broadcast on mineral soils. Wait two to three weeks after treating to plant, or until odor leaves the soil. Do not treat extremely heavy soils. Five-day re-entry. **Restricted-use pesticide.**
- Vapam (32.7% metam sodium) at 50 to 100 gal/A. Immediately roll the soil and follow up with tarps or a light watering. May use through an irrigation system. Fourty-eight-hour re-entry and/or while tarps are being removed. **Restricted-use pesticide.**

2. Ecozin Plus at 25 to 56 oz/A through a drip irrigation system. Label suggests using additives to aid penetration into the soil and to make applications in the morning. Efficacy in the Pacific Northwest is unknown. Four-hour re-entry.
3. Nema-Q at 1.5 to 3 gal/A for the first application followed by four to six weekly applications at 1 qt/A. Efficacy in the Pacific Northwest is unknown. Two-day re-entry.

1.23.5.2   Cultural Control

- **Maintain optimum growing conditions**. The greatest damage by lesion nematodes occurs on plants that are under stress. The amount of injury to plants can be reduced, or in some cases eliminated by maintaining optimum growing conditions. Plants should be provided with adequate moisture, nutrients, and soil aeration at all times. Controlling other diseases and insects also reduce plant stress.
- **Rotate crops**. Despite the wide range of hosts for lesion nematodes, crop rotation can provide control in some instances. Keeping the crop free of weeds and volunteer plants that may serve as hosts for the species of lesion nematode involved can also aid in disease control.
- **Treat propagation material with heat**. A hot-water treatment is an effective method of eradicating lesion nematodes from the roots of transplants. The time and temperature to use will depend on the plant and variety and **must** be controlled closely. Temperatures of 45°C to 55°C (113°F to 131°F) sustained for 10 to 30 minutes are commonly used. Before making a large-scale treatment, treat several plants of each variety to make sure that heat damage will not occur.
- **Treat the soil with dry or moist heat**. This method is commonly used to control nematodes and other soilborne pathogens in the greenhouse and in the home. The method is economical and is highly effective if performed correctly. Nematodes are killed by exposure to temperatures of 40°C to 52°C (105°F to 126°F), depending on the species. Aerated steam is the most efficient method but baking small quantities of soil in an oven at 82°C (180°F) for 30 minutes or 71°C (160°F) for 60 minutes is also effective, especially for the homeowner where a small quantity of soil is needed.
- **Apply nematicides**. The use of chemical fumigants to control lesion nematodes can be effective and economical, especially where high-value crops are involved. Preplant fumigation with nematicides may be necessary in order to control replant and other lesion nematode diseases in orchards, nurseries, strawberry beds, and other areas.
- **Use certified plants**.

1.24   THE STING NEMATODE

The sting nematode, *Belonolaimus longicaudatus*, is a pest of major importance to commercial strawberry production. Although the disease was first noticed in strawberry in 1946 in Florida, it was not until 1950 that the problem was correctly identified as that of the sting nematode.

1.24.1 The Causal Organism

B. longicaudatus is found primarily in the sandy coastal plains of the Atlantic and Gulf coasts but also occurs naturally in sandy areas of some Midwestern plains states such as Kansas and Nebraska. Sting nematodes can be introduced to new areas on infested turf sod and have been introduced by this means to some golf courses in California and internationally to some of the Caribbean islands, Puerto Rico, Bermuda, and Australia. Sting nematodes require at least 80% sand content in soil to survive, so they are typically only found in sandy soil environments.

1.24.2 Symptoms

Strawberry production problems caused by sting nematode tend to occur in definite areas where transplants fail to grow-off normally. Infested areas consist of spots that vary in size and shape, but the boundary between diseased and healthy plants usually is fairly well defined. Initially, a field may have only a few such areas, which may then increase in size and number until the entire field becomes involved. The effect on strawberries is to cause both stunting and decline, the intensity of which is related to initial population levels and the rates to which populations increase during the course of strawberry crop growth. Affected plants become semi-dormant, with little or no new growth. Leaf edges turn brown, progressing or expanding from the edges to midrib to include the entire leaf. Leaves seldom become chlorotic, although cases have been reported in which leaf yellowing occurs when essential nutrients are present in limited supply. Since the outer older leaves die first the plant gradually decreases in size and eventually may be killed (Figure 1.60).

Sting nematode can be very damaging to nursery seedlings and transplants. As a rule, most other crop plants are not killed unless subjected to other adverse conditions but affected strawberry plants undergo progressive decline and may eventually die. Older plants that have already developed an extensive root system can still be severely affected. In this soil zone, plants can develop a dense root system, but no roots are able to penetrate below this upper layer (Figure 1.61). Such plants can be easily lifted or pulled from soil and are much more susceptible to droughty conditions and injury from fertilizer salt accumulations. Root tips are killed, forcing the development of new lateral roots, whose root tips in turn are killed (Figure 1.62).

1.24.3 Disease Cycle

The figure shows the process of embryogenesis of *B. longicaudatus* from the one-celled stage to a first-stage juvenile (J1) that molts inside the egg into a second-stage juvenile (J2) ready to hatch. After hatching, the J2 move through soil to the root system of a host plant where they congregate around root hairs. The J2 feed on the root hairs until they molt into third-stage juveniles (J3). The J3 move immediately to the meristems of either major or lateral roots to feed, as do all subsequent life-stages. The J3 molt to become fourth-stage juveniles (J4), which molt to become adults.

FIGURE 1.60 Strawberry plant stunting caused by the sting (*B. Longicaudatus*). Note irregular or patchy field distribution of stunted plants rather than throughout the entire field.

FIGURE 1.61   Sting nematode, *B. longicaudatus*, induced symptoms on strawberry roots. Note short, dark and discolored abbreviated roots with swollen root tips.

1.24.4   MANAGEMENT

1.24.4.1   Chemical Control

Sting nematodes can be effectively managed with nematicides. Unlike many of the endoparasitic nematodes that spend the majority of their life within roots, contact nematicides often work well on sting nematode. Both carbamate (aldicarb, carbofuran) and organophosphate (fenamiphos, ethoprop, turbufos) nematicides and fumigants (methyl bromide, 1,3-dichloropropene, metam sodium) are currently registered and can be effectively used to reduce sting nematode populations. On annual crops, nematicides applied either before or at planting usually provide acceptable levels of control

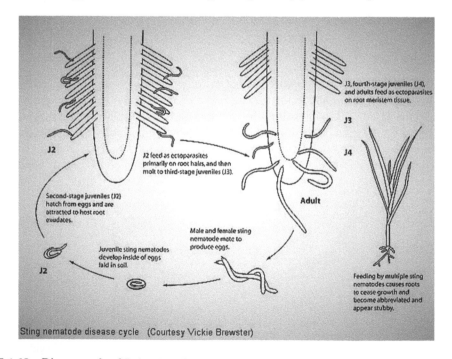

FIGURE 1.62   Disease cycle of *B. longicaudatus*.

by protecting newly developing root systems. On perennial crops such as turf grasses, seasonal application of post-plant nematicides during times of root growth may be required.

1.24.4.2   Cultural Control

When possible, avoid use of infested planting material. Most warm-season turf grasses are planted as sod or sprigs, so sting nematodes and other pathogens can be moved in the soil adhering to the sod. It is believed that this is the primary way that the sting nematode has become established in new areas, especially outside of its native geographical range. On turfgrasses, relieving additional stresses by raising mowing height, increasing irrigation frequency, improving aeration to roots, and reducing traffic can improve tolerance to sting nematodes. The addition of organic, and some inorganic, amendments to soil also can improve tolerance to sting nematodes by improving the water and nutrient-holding capacity of the soil. Organic amendments have also been shown to reduce population densities of sting nematodes in some studies. This may be due to direct effects of these additives on the nematodes or due to stimulation of antagonistic microorganisms in the soil.

1.24.4.3   Biological Control

Pasteuria usgae, an endospore-forming bacterium, is an obligate parasite of *B. longicaudatus*. This bacterium is found in soils throughout Florida and presumably other areas where sting nematodes occur. *P. usgae* was successfully introduced into a previously non-infested putting green resulting in the suppression of sting nematodes. Presently, the only method for infesting a field site with *P. usgae* is by adding soil from a site that already has sting nematodes infected with the bacterium. Unfortunately, this method is not economically feasible for commercial use. However, *in vitro* production of *P. usgae* is being attempted at this time. If these efforts are successful, *P. usgae* may become a viable inoculative biological control agent for sting nematodes in the future.

1.25   VERTICILLIUM WILT OF STRAWBERRY

Verticillium wilt of strawberry can be a major factor that limits the production severely. When a plant is severely infected by the verticillium wilt fungus, the probability of it surviving to produce a crop is greatly reduced. The *Verticillium* fungus can infect about 300 different host plants, including many fruits, vegetables, trees, shrubs, and flowers, as well as numerous weeds and some field crops. The fungus can survive in soil, and, once it becomes established in a field or garden, it may remain alive for 25 years or longer.

1.25.1   Causal Organism

Species	Associated Disease Phase	Economic Importance
Verticillium albo-atrum, V. dahliae	Roots, Wilting of Leaves	Major

1.25.2   Symptoms

Symptoms resemble drought stress. The first symptoms of verticillium wilt in new strawberry plantings often appear about the time runners begin to form. In older plantings, symptoms usually appear just before picking time. Symptoms on aboveground plant parts may differ with the susceptibility of the cultivar affected. In addition, aboveground symptoms are difficult to differentiate from those caused by other root-infecting fungi. Isolation from diseased tissue and culturing the fungus in the laboratory are necessary for positive disease identification.

On infected strawberry plants, the outer and older leaves droop, wilt, turn dry, and become reddish-yellow or dark brown at the margins and between the veins (Figure 1.63). The infected plants wilt

FIGURE 1.63 Verticillium wilt, causing browning of the outer leaves.

rapidly under stress. Few, if any, new leaves develop and those that do, tend to be stunted and may wilt and curl up along the mid-vein. Infected plants are often stunted and flattened with small yellowish leaves and appear to be suffering from a lack of water. Brownish to bluish black streaks or blotches may appear on the runners and leaf petioles. New roots that grow from the crown are often dwarfed with blackened tips. Brownish streaks may occur within the decaying crown and roots (Figure 1.64).

If the disease is serious, large numbers of plants may wilt and die rapidly. When the disease is not so serious, an occasional plant or several plants scattered over the entire planting may wilt and die.

1.25.3 CAUSE AND DISEASE DEVELOPMENT

The fungus is not host-specific and infects many weed species and crops worldwide. It is especially destructive in semiarid areas where soils are irrigated. Inoculum densities may be high following the planting of susceptible crops. Disease severity is greater when high levels of nitrogen are used.

1.25.4 FAVORABLE CONDITIONS OF DISEASE DEVELOPMENT

Cool and overcast days interspersed with warm and bright days are most conducive to the development of verticillium wilt disease. Infection and disease development occur when the soil temperatures are between 12°C and 30°C (53°F and 86°F), with an optimum of 21°C to 24°C (70°F to 75°F). Verticillium wilt is found mostly in the temperate climate zone. In Illinois, the disease is particularly severe during cool seasons.[6]

1.25.5 DISEASE CYCLE

The *Verticillium* fungus overwinters in the soil and plant debris as dormant mycelium or black, speck-sized bodies (microsclerotia). Those bodies remain viable for many years. When suitable

FIGURE 1.64 Effect of verticillium wilt on crown.

conditions occur, these microsclerotia germinate by putting forth one or more threadlike hyphae. These hyphae may penetrate the root hairs directly, but more infection is aided by breaks or wounds in rootlets caused by insects, cultivating or transplanting equipment, frost injury, or root-feeding nematodes.

Once inside the root, the fungus invades the water-conducting tissue (xylem). The spread of the fungus into the aerial parts of the plant may be hastened by the movement of conidia in the tran-spiration stream. These conidia become lodged in the vascular tissue where they germinate and produce small, mycelial mats. These mats, in turn, produce more conidia which are then carried upward. Runner plants may become infected by the movement of the fungus into the stolons from the diseased mother plant. Older mycelia produce micro-sclerotia in host tissues, completing the disease cycle.

1.25.6 MANAGEMENT

Preplant fumigation is an important component of managing verticillium wilt in strawberry fields. If fumigation is not desirable, select fields isolated from established growing areas, avoiding any fields with detectable levels of the pathogen or with a history of susceptible crops. Crop rotation with broccoli has been shown as an effective way to reduce *Verticillium* in the soil. Solarization of formed beds may be used to reduce pathogen levels in areas that get adequate amounts of sunshine and warm weather during the summer months, although the usefulness of this technique for reduc-ing verticillium wilt in strawberries is unknown.

1.25.6.1 Chemical Control

If economically feasible, and if available, use soil fumigation as a preplant treatment. When prop-erly done, fumigation kills soil insects and weed seeds as well as disease-causing fungi and nema-todes. Fumigation is usually done by commercial applicators who are licensed to handle restricted (dangerous) chemicals and not by the grower. The soil fumigants that are most effective against *Verticillium* include chloropicrin, chloropicrin-methyl bromide mixtures, chloropicrin chlorinated C hydrocarbon (DD) mixtures, Vapam, and Vorlex. These broad-spectrum soil fumigants are costly to apply. This cost is offset by larger yields of better quality fruit, control of soil pests (primarily weeds), and extended life of the planting. Soil fumigation should permit the growing of verticillium-susceptible cultivars. Many soil fumigants require treated soil to be covered with gas-proof sheeting (polyethylene or vinyl) for at least 24–48 hours after treatment. Planting cannot take place for an additional two to three weeks. When using a soil fumigant, follow all of the manufacturer's direc-tions and precautions carefully.

1.25.6.2 Cultural Control

If infested fields cannot be avoided and fumigation is not feasible, either solarize the soil or imple-ment a crop rotation program. Cover crops of cereal rye or ryegrass can help to reduce soil levels of *Verticillium*. Use relatively tolerant strawberry cultivars when practical. Also, use drip irrigation and avoid excess amounts of nitrogen fertilizer.

Soil solarization: In warmer areas of the state, solarization has been shown to be effective for the control of soilborne pathogens and weeds. Solarization is carried out after the beds are formed and can be effective if weather conditions are ideal (30–45 days of hot weather that promotes soil temperatures of at least 50°C). The effectiveness of solarization can be increased by solarizing after incorporating the residue of a cruciferous crop, in particular broccoli or mustards, into the soil or following an application of metam sodium (40 gal/A).

Crop rotation: Rotating strawberries with broccoli can significantly reduce levels of the *Verticillium* pathogen in the soil and has been shown to be an economically viable option under moderate levels of verticillium wilt disease pressure.

TABLE 1.7

Cultural Control

Common Name (Trade Name)	Amount/Acre[a]	R.E.I.[b] (Hours)	P.H.I.[b] (Days)
PREPLANT FUMIGATION: A. METHYL BROMIDE[c]/CHLOROPICRIN[c]	300–400 lb	48	0
B. Sequential application of: 1,3- DICHLOROPROPENE[c] /CHLOROPICRIN[c] (Telone C35) ....OR.................	9–12 gal (shank)	5 days	0
1,3 - DICHLOROPROPENE[c] /CHLOROPICRIN[c] (InLine) ....OR..............	28–33 gal (drip) 15–30 gal (shank)	5 days	0 0
CHLOROPICRIN[c] (MetaPicrin) (Tri-Clor)	15–21.85 gal (drip)	48 48	0
C. Followed 5-7 days later by: METAM SODIUM[c] (Vapam HL, Sectagon 42) ....OR................	37.5–75 gal	48	0
METAM POTASSIUM[c] (K-Pam HL)	30–60 gal	48	0

[a] Rates are per treated acre; for bed applications, the rate per acre may be lower.

[b] Restricted entry interval (R.E.I.) is the number of hours (unless otherwise noted) from treatment until the treated area can be safely entered without protective clothing. Preharvest interval (P.H.I.) is the number of days from treatment to harvest. In some cases, the REI exceeds the PHI. The longer of two intervals is the minimum time that must elapse before harvest.

[c] Permit required from county agricultural commissioner for purchase or use.

The following materials are listed in order of usefulness in an IPM Program, considering efficacy. Also, consider the general properties of the fungicide as well as information relating to environmental impact. Not all registered pesticides are listed. Always read the label of the product being used (Table 1.7).

REFERENCES

1. Ellis, M. A., and Erincik, O., 2016. *Anthracnose of Strawberry*, Ohio: Department of Plant Pathology, The Ohio State University. https://ohioline.osu.edu/factsheet/plpath-fru-16 (Apr 15, 2016).
2. Ellis, M., 2012. *Fungicides for Strawberry Disease Control*, Wooster OH: Department of Plant Pathology, The Ohio State University OARDC.
3. Converse, R. H., Martin, R. R., and Spiegel, S., 1987. Strawberry mild yellow-edge. In: *Virus Diseases of Small Fruits*, Ed. by Converse, R. H., Agriculture Handbook No. 631, Washington DC, USA: US Department of Agriculture, pp. 25-9.
4. Tzanetakis I. E., Mackey I. C., Martin R. R., 2004. Strawberry necrotic shock virus is a distinct virus and not a strain of Tobacco streak virus. *Arch Virol*. 2004 Oct, 149(10): 2001–11.
5. Ries, S. M., 1996. *Reports on Plant Diseases: Verticillium Wilt of Strawberry*, IL: University of Illinois. http://ipm.illinois.edu/diseases/series700/rpd707, June 1996.
6. University of Illinois Extension, 1999. College of Agricultural, Consumer and Environmental Sciences, *Reports on Plant Disease*, RPD No. 1103, May 1999.

2 Tomato

The tomato is one of the most important "protective foods" which has special nutritive values. Tomato (scientific name: *Lycopersicon esculentum*) belongs to the genus *Lycopersicon* under *Solanaceae* family. Tomato is an herbaceous sprawling plant which grows up to 3 m in height with a weak woody stem. Tomato is a native plant to Peruvian and Mexican regions. Tomato is the world's largest vegetable crop after potato and sweet potato, but it tops the list of canned vegetables. In 2014, world production of tomatoes was 170.8 million tonnes, with China accounting for 31% of the total, followed by India, the United States and Turkey as the major producers. There are around 7,500 tomato varieties grown for various purposes. Tomato diseases are rarely fatal if the proper management is employed. It is important to diagnose any tomato disease early, before it spreads to the entire tomato plant. Also, there is a possibility that other plants in the same family, such as potatoes, eggplants, and peppers may be affected if planted near an infected tomato. Here in this chapter, some common tomato diseases are discussed (Figure 2.1).

2.1 ANTHRACNOSE OF TOMATO

Anthracnose is a frequent problem in the latter part of the growing season on ripening tomato fruit. The disease results in a fruit rot which reduces the quality and yield of tomatoes. Anthracnose is a serious fungal disease causing circular, sunken lesions on ripe tomato fruit. These lesions reduce the marketability of the fruit and allow other fruit rot organisms to invade. Although the symptoms only show up when fruit ripen, control must begin much earlier in the season.

The anthracnose organism is spread primarily by splashing water. It can infect tomato foliage, providing a source of inoculum close to the fruit. Foliar symptoms, however, are inconspicuous. Anthracnose can infect both ripe and green fruit, but symptoms will not show up until the fruit begin to mature.

2.1.1 CAUSAL ORGANISM

Species	Associated Disease Phase	Economic Importance
Colletotrichum coccodes	Sunken Lesions on the Fruit Causing Complete Rotting of the Fruit	Severe

2.1.2 SYMPTOMS

Anthracnose is a common and widespread rot of ripe or overripe tomato fruit. Symptoms are rare on green fruit. Symptoms on ripe fruit are small, sunken, circular spots that may increase in size up to 1/2 inch in diameter (Figure 2.2). The center of older spots later becomes blackish. Spots may become numerous in severe cases and secondary rotting organisms may invade anthracnose lesions to completely rot infected fruit. The fungus forms small, dark survival structures called sclerotia in the centers of fruit spots (Figure 2.3). These sclerotia survive in soil for up to three years and cause infections either directly or by producing secondary spores. Green fruit are infected but do not show symptoms until ripening. The fungus then spreads from infected to healthy fruit as spores are splashed by rain or overhead irrigation, or by pickers working wet plants. Anthracnose is favored by warm rainy weather, overhead irrigation, and heavy defoliation caused by early blight.

FIGURE 2.1   Tomato.

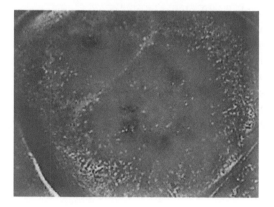

FIGURE 2.2   Sunken lesions on the fruit.

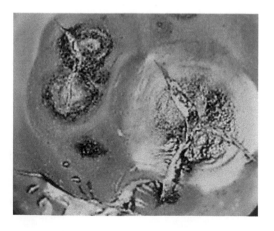

FIGURE 2.3   Secondary rot showing dark survival structures called sclerotia.

2.1.3 CAUSE AND DISEASE DEVELOPMENT

C. coccodes survives between crops on infested plant debris in the soil. Early in the growing season, spores from the soil splash on lower leaves of the tomato plant. Few symptoms develop on infected leaves, but the spores produced on foliage can be carried by splashing rain onto developing green fruit. Infected green fruit will not develop symptoms of anthracnose until they begin to ripen. Ripe fruit is very susceptible to this fungus.

2.1.4 FAVORABLE CONDITIONS OF DISEASE DEVELOPMENT

The anthracnose organism can be present in crop debris, weeds, and tomato foliage. It prefers wet conditions and temperatures of 10°C–30°C. At 20°C–25°C, it only takes 12 hours of leaf wetness (on unprotected leaves or fruit) for infection to occur. Within five or six days of a spore landing on a ripe fruit, a lesion can be seen.

2.1.5 DISEASE CYCLE

The fungus survives the winter as seed-like structures called sclerotia and as threadlike strands called hyphae in infested tomato debris. In late spring the lower leaves and fruit may become infected by germinating sclerotia and spores in the soil debris. Infections of the lower leaves of tomato plants are important sources of spores for secondary infections throughout the growing season. Senescent leaves with early blight infections and leaves with flea beetle injury are especially important spore sources because the fungus can colonize and produce new spores in these wounded areas.[1]

Growth of *C. coccodes* is most rapid at 27°C, although the fungus can cause infections over a wide range of temperatures (13°C–35°C). Wet weather promotes disease development, and splashing water in the form of rain or overhead irrigation favors the spread of the disease (Figure 2.4).

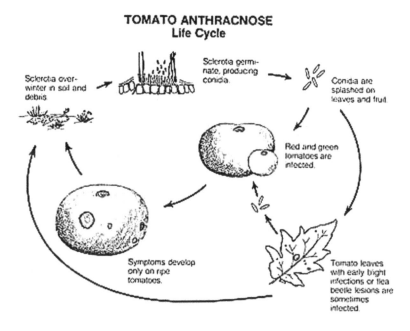

FIGURE 2.4   Disease cycle of *C. coccodes*.[1]

2.1.6 Management

- Harvest fruit as soon as possible after ripening.
- Avoid excessive overhead irrigation or use drip irrigation to reduce moisture levels on fruit and humidity in the plant canopy.
- Fungicide sprays used to control leaf diseases reduce losses from anthracnose when applied on a regular schedule and in a manner to achieve thorough fruit coverage.
- A three-year rotation may also reduce chances of infection.
- Organic fungicides recommended for anthracnose are copper formulations. However—and this is a big however—copper fungicides are minimally effective against anthracnose. (There are many copper containing fungicides. Some are organic, and some are not depending on formulation and additives. Make sure the product is labeled for tomatoes.)
- Synthetic fungicides (chlorothalonil) that are labeled for control of septoria leaf spot are recommended for anthracnose on tomato. Organic and synthetic fungicides are preventative. Begin fungicide applications from bloom and repeat every seven to ten days (follow label directions) until harvest. Check the label for the number of days between application and harvest (Table 2.1).

2.2   BACTERIAL CANKER OF TOMATO

Bacterial canker is caused by *Corynebacterium michiganense pv. Michiganense* (Cmm). Although usually sporadic in its occurrence, it is so destructive in nature that vigilance must be exercised in the selection and handling of seed stocks, the preparation and management of greenhouse soil beds or bags, and the selection and preparation of the ground for field production. Bacterial canker is a vascular (systemic) and parenchymatal (superficial) disease with a wide array of symptoms resulting in loss of photosynthetic area, wilting and premature death, and the production of unmarketable fruit. Early recognition of the disease, especially in greenhouse crops, is essential if the disease is to be contained. The organism is seedborne and can survive for short periods in soil, greenhouse structures, and equipment and for longer periods in plant debris.

2.2.1   Causal Organism

Species	Associated Disease Phase	Economic Importance
C. michiganense pv. michiganense	Necrotic Lesions on Older Leaves, Raised White Blisters on Young Green Fruit, known as "Bird's-Eye Spot"	Severe

TABLE 2.1
For Recommended Fungicides to Treat Anthracnose on Tomato

Fungicides for Tomato Anthracnose Control

Fungicide	Typical Application Interval	Examples of Trade Names
Azoxystrobin	7–14 days	Quadris
Chlorothalonil	7–14 days	Daconil, Terranil, Bravo, Echo, others Homeowner: Daconil, FungiGard, Liquid Fungicide, Encore, Monterey Bravo.
Copper products	7–14 days	Bordeaux, Kocide, Tenn-Cop Others Homeowner: Copper Fungicide, Bordeaux
Mancozeb and maneb	7–14 days	Dithane, Penncozeb, Manex, others Homeowner: Mancozeb, Maneb
Ziram	7–14 days	Ziram

2.2.2 Symptoms

2.2.2.1 Seedlings

Marginal necrosis, tan to dark necrotic patches on the leaves and stems, and small white raised blisters on infected leaves may be symptoms of bacterial canker infection on young plants. Stunting, wilting, and stem splitting can also occur, especially in grafted seedlings. However, symptoms can take several to many weeks to develop following infection and therefore may not be visible at the seedling or transplant stage.

2.2.2.2 Leaf and Plant

Leaf yellowing and necrosis around leaf margins called "firing" or "marginal necrosis" can indicate a foliar and/or systemic infection (Figure 2.5). When the stems or petioles are cut open, discoloration of the vascular tissues may be seen. In greenhouse-grown plants, symptoms appear as interveinal chlorotic to pale green patches that quickly become necrotic, giving a scorched appearance. Infected plants wilt beginning with the lower, older leaves, or leaves above the point of infection. Wilting may be asymmetric, appearing more on one side of the plant than the other. Infected leaves die, and light brown streaks or cankers, which may darken with age, develop on infected stems. Typical cankers can be common in the field but are rarely seen in the greenhouse. The vascular tissues become light brown to reddish brown and the pith appears mealy, brown and dry. Older plants tend to be less susceptible to Cmm than younger ones and the disease tends to be more severe on plants infected early vs. late in their growth cycle.

2.2.2.3 Fruit

Small dark spots on the fruit surrounded by a white halo or "bird's-eye" spots are characteristic of bacterial canker on field-grown fruit (Figure 2.6). Spots become raised and the centers turn brown with age. Infections, and the resulting spots occur when Cmm bacteria are deposited on fruit by splashing water from rain or overhead irrigation, or mechanically during handling of the plants. When internally infected fruit are opened, yellowing or browning caused by the decay of the tissues may be seen. In the greenhouse, bird's-eye spots are typically not observed, but fruit may appear netted or marbled, or they may remain symptomless. It is important to have an accurate diagnosis of any disease problem in tomatoes so that appropriate control measures can be taken. Diagnostic kits have been developed for rapid, on-site identification of Cmm. However, it is advisable to submit a tissue sample to a reputable laboratory for confirmation of the diagnosis.

FIGURE 2.5 "Firing" or "marginal necrosis" of leaves.

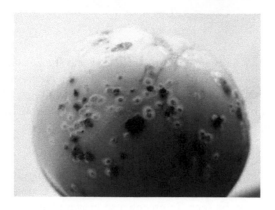

FIGURE 2.6   Bird's-eye spots on fruit.

2.2.3 CAUSE AND DISEASE DEVELOPMENT

The pathogen can survive in many environments including free living in infested soil for short periods, in over-seasoned plant debris in the soil, on weed hosts and volunteer plants, on contaminated stakes and in association with seed. Cmm is a seedborne pathogen, although rates of seedborne infestation may be very low. Volunteer tomato plants from an earlier infected crop may harbor the pathogen, as can cull piles of diseased tomato plants. Cmm can infect or survive on some weeds such as nightshade, and several wild *Lycopersicon* species, and these can act as reservoirs of Cmm for new infections. Overhead irrigation during seedling production, movement through foliage by production workers and rainfall on open fields favor the spread of Cmm, especially if plants have recently been staked or pruned. Once the disease appears in a field or greenhouse, the pathogen may spread to adjacent plants and infect them through pruning wounds and injury, or through naturally occurring pores along the leaf surface or leaf margins (hydathodes). The pathogen can also be moved quite easily by equipment during cultivation, especially with open field processor tomatoes.[2]

2.2.4 FAVORABLE CONDITIONS OF DISEASE DEVELOPMENT

Development of bacterial canker is favored by warm—24°C–32°C (75°F–90°F)—moist conditions. In greenhouses, the disease tends to be more severe in the summer during long, hot days when plants are stressed. Bacterial canker is more likely to be found in wetter areas of the greenhouse (e.g. where water condenses and drips on plants) than in drier areas.

2.2.5 MANAGEMENT

Bacterial canker is one of the most difficult tomato diseases to control. First, there is the problem of detecting infected plants, due to the wide variability of symptom expression. Second, the highly infectious nature of the disease, the number of sources of inoculum, and the absence of effective chemicals for treatment mean that sanitation and preventive measures must be enforced.

- Use only certified disease-free seed from canker-free plants. Never save seed from a source known to have had bacterial canker. If it is necessary to use non-certified seed or seed of unknown origin, make sure the seed has been extracted by the standard acetic acid extraction method or by the fermentation process. Make sure seed has not been prepared by centrifuge extraction, which can lead to high levels of seed contamination. Although the acid or fermentation treatment will eliminate seed coat contamination, it does not completely control embryonic infection.

- Plant only certified disease-free transplants that have been produced under a vigorous inspection program. It is usually not possible to distinguish between infected or healthy seedlings at the time of transplanting.
- Once the disease is suspected or confirmed in a greenhouse crop, aids to pollination and high-volume, high-pressure pesticide spraying should stop. These restrictions will decrease the risk of spread, especially when superficial symptoms are present. Remove diseased plants as soon as they are detected by cutting the plants off at the ground line and placing them in a plastic bag for disposal. At least several "healthy" plants on either side of the infected plants should also be removed. If the area of diseased plants is limited, there is a good chance that the disease can be contained. Every effort must be made to isolate affected areas from the rest of the crop. Hands, shoes, tools, and crop support wires should be disinfected. Hypochlorite, or laundry bleach, is not satisfactory as a greenhouse disinfectant. Quaternary ammonium compounds as used to disinfect potato storages are recommended. In the field, if bacterial canker becomes severe early in the season, fields should be plowed down to prevent spread to nearby healthy fields. If affected plants are found throughout the crop, not more than 100 plants per acre should be removed in an attempt to restrict spread. Pulling out more is rarely of benefit.
- Greenhouse seedbeds and soils must be sterilized to destroy the bacteria. Steam sterilization is preferred, but *methyl bromide is satisfactory if attention is paid to removing all debris. (Restricted-use pesticides are identified by an asterisk, *.) In the field, all plant debris must be plowed under, and affected areas should be rotated out of tomatoes for at least three years. Weeds belonging to the Solanaceae family should be destroyed.
- Fixed copper sprays may help in protecting healthy plants, particularly if only superficial symptoms are present.

2.3 BACTERIAL SPECK OF TOMATO

Bacterial speck is caused by *Pseudomonas syringae pv. tomato*. The bacterium is believed to be widespread and is often isolated from plant roots and soil particles. The bacterium is seedborne and probably overwinters within infected tomato plant debris. It has been associated with different plants near tomato fields, where it survives as a saprophyte. Recently it was shown that these bacteria may multiply at the base of leaf hair on healthy tomato leaves and later cause disease.

2.3.1 CAUSAL ORGANISM

Species	Associated Disease Phase	Economic Importance
P. syringae pv. tomato	Dark Brown to Black Lesions on Leaves, Fruits, and Stem	Severe

2.3.2 SYMPTOMS

The foliar symptoms of speck consist of small (1/8–1/4 inches) black lesions, often with a discrete yellow halo (Figure 2.7). The lesions of bacterial spot are similar, but tend to have a greasy appearance, whereas those of speck do not. Speck seems to curl the leaves more severely than spot. Both diseases affect flowers. Lesions on stems and petioles cannot be distinguished. Bacterial speck and spot are more clearly differentiated by symptom development on the fruit. Bacterial speck lesions are slightly raised but are generally much smaller (1/16 inch) than those of bacterial spot (Figure 2.8). Bacterial speck lesions are very superficial and do not crack or become scaly as in bacterial spot.

FIGURE 2.7   Black lesions with a discrete yellow halo on tomato leaf.

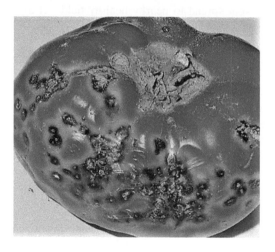

FIGURE 2.8   Bacterial speck symptoms on tomato fruit.

2.3.3   CAUSE AND DISEASE DEVELOPMENT

The sources of bacterial speck inoculum and the methods of spread are the same as those for bacterial spot. Studies have shown that the speck organism can survive in the crevices and cavities of the tomato seed coat for up to 20 years.

2.3.4   FAVORABLE CONDITIONS OF DISEASE DEVELOPMENT

Bacterial speck is favored by cool, moist environmental conditions. The virulent bacteria are spread mechanically and by wind-driven rain. The disease will develop rapidly at 24°C. However, disease development is readily apparent at 17°C. At 32°C pathogen populations are so severely depleted that typical symptoms are not evident.

2.3.5   MANAGEMENT

- Use disease-free, hot water—treated seed.
- Strive to obtain disease-free transplants that have been produced with a good protective spray program (mancozeb plus fixed copper, with streptomycin as a replacement bactericide for copper in later sprayings if weather conditions favor speck development). Note: Streptomycin can only be used on tomato plants before transplanting.

- Practice crop rotation because of the carryover of inoculum in plant debris and weeds.
- Follow good weed control and sanitation programs before establishing the current season crop.
- Practice a preventive copper + mancozeb spray program from anthesis until the first-formed fruit are 1/3 of their final size. After that point, the greatest risk of bacterial speck has passed; copper can be dropped from the program, and the full labeled rate of fungicide should be used to control foliar blights, especially early blight.
- Resistance to bacterial speck has been identified in three tomato species and will be added to commercial varieties.

2.4 BACTERIAL SPOT OF TOMATO

Bacterial spot is caused by *Xanthomonas campesiris pv. vesicatoria*. It is periodically a severe disease of tomatoes. Because bacterial spot and speck produce similar symptoms, they are often misdiagnosed.

2.4.1 CAUSAL ORGANISM

Species	Associated Disease Phase	Economic Importance
X. campesiris pv. vesicatoria	Lesions on Leaves and Water-Soaked Spots on Green Fruits	Severe

2.4.2 SYMPTOMS

Infected leaves show small, irregular, dark lesions, which can coalesce and cause the leaves to develop a general yellowing (Figure 2.9). Both spot and speck can occur on stems and petioles where they are indistinguishable. Flower infection with bacterial spot can be quite serious with pedicel infection causing early blossom to drop. The two diseases are most readily distinguished based on fruit symptoms. In the case of bacterial spot on green fruit, small water-soaked spots are first noticed. These spots become slightly raised and enlarge up to 1/8 to 1/4 inches in diameter. The center becomes irregular, brown, slightly sunken, with a rough, scabby surface (Figure 2.10). Although ripe fruit are not susceptible, lesions are very obvious if fruit is infected when green.

2.4.3 CAUSE AND DISEASE DEVELOPMENT

The bacterial spot pathogen may be carried as a contaminant on tomato seed. It is also capable of overwintering on plant debris in the soil and on volunteer host plants in abandoned fields. Because the bacteria have a limited survival period of days to weeks in the soil, contaminated seed is a common source of primary infection in nurseries and home gardens. In commercial fields, volunteer host plants are the main source of initial inoculum because the bacterial pathogen survives in lesions on those plants.

FIGURE 2.9 Small, irregular, dark lesions, which can coalesce and cause the leaves to develop a general yellowing.

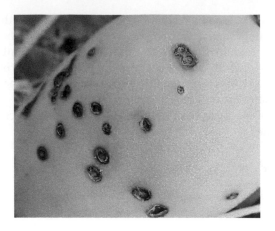

FIGURE 2.10   Symptoms of bacterial spot on tomato fruit.

2.4.4   FAVORABLE CONDITIONS OF DISEASE DEVELOPMENT

The bacterium becomes active when the temperature reaches above 24°C–30°C. Moist weather and splattering rain are essential for dissemination of the bacteria. Transplant production environments are favorable for the bacterium because overhead sprinklers are commonly used. The wet and crowded plants are at high risk of infection.

2.4.5   DISEASE CYCLE

Bacteria enter through stomata on the leaf surfaces and through wounds on the leaves and fruit caused by abrasion from sand particles and/or wind. Prolonged periods of high relative humidity favor infection and disease development. In a 24-hour period, the bacteria can multiply rapidly and produce millions of cells. Symptom development is delayed when relative humidity remains low for several days after infection. Although bacterial spot is a disease of warm, humid regions, it can develop in arid but well irrigated regions.

2.4.6   MANAGEMENT

Control is based on preventive steps taken during the entire season. Once the disease has started in a field, control is very difficult, especially during wet weather.

- Obtain seed that has been grown in regions without overhead irrigation and is certified free from the disease-causing bacteria. This is by far the most important step. Seed may be treated by washing for 40 minutes with continuous agitation in 2 parts Clorox Liquid Bleach (5.25% sodium hypochlorite) plus 8 parts water (e.g. 2 pints Clorox plus 8 pints water). Use 1 gallon of this solution for each pound of seed. Prepare fresh solution for each batch of seed treated. Rinse seed in clean water immediately after removal from the Clorox solution and promptly allow to dry prior to storing or treating with other chemicals. This treatment will likely reduce seed germination. Thus, before attempting to treat an entire seed lot perform a test using 50–100 seed and check for the effect on germination.
- Produce plants in sterilized soil or commercially prepared mixes.
- Avoid fields that have been planted with peppers or tomatoes within 1 year, especially if they had bacterial spot.
- Do not plant diseased plants. Inspect plants very carefully and reject infected lots—including your own! Use certified plants.

- Prevent bacterial leaf spot in the plant beds: keep the greenhouse as dry as possible and avoid splashing water; spray with a fixed copper (Tribasic Copper Sulfate 4 lb, Copper-Count N, or Citcop 4E 2 to 3 qt, or Kocide 101 1.0 to 1.5 lb per 100-gal water). The addition of 200 ppm streptomycin (Agri-mycin 17-1.0 lb in 100 gal of the copper spray with a spreader-sticker) will improve the effectiveness of the spray program. Make applications on a seven- to ten-day schedule if spots appear, and 1 day before pulling plants.
- In the field start spray schedule when the disease first appears. However, do not use streptomycin in the field.
  - Mix fixed copper (see E above for amount) and 1.5 lbs of mancozeb (Manzate 200DF or Dithane M-45) or maneb (Maneb 80WP, Maneb 75DF, Manex) in 100-gal of water. Do not use mancozeb on tomatoes within five days of harvest. Mancozeb is not registered for use on peppers; use maneb.
  - Adjust sprayer and speed of tractor to obtain complete coverage of all plant surfaces. Spray pressure of 200–400 psi is recommended and the use of at least three nozzles per row for peppers and five drop nozzles per row for tomatoes. Depending on plant size, use 50–150 gal/A of finished spray.
  - Adjust spray schedules according to the weather and presence of disease: (a) Spray one week after plants are set; (b) spray every five to seven days during rainy periods; spray on 10-day intervals during drier weather; (c) spray before rain is forecast but allow time for the spray to dry.

2.5 SEPTORIA LEAF SPOT ON TOMATO

Septoria leaf spot of tomato caused by the fungus *Septoria lycopersici* occurs on tomatoes worldwide. The fungus infects only solanaceous plants, of which the tomato is the most important. Tomatoes may often be infected with leaf spot and early blight (*Altemaria solani*) simultaneously, but the two diseases can be distinguished readily, and the control measures are similar.

2.5.1 CAUSAL ORGANISM

Species	Associated Disease Phase	Economic Importance
S. lycopersici	Leaf Spot	Severe

2.5.2 SYMPTOMS

Septoria leaf spot can occur at any stage of plant development. Symptoms may appear on young greenhouse seedlings ready for transplanting or be first observed on the lower, older leaves and stems when fruits are setting. The timing of symptom appearance can be correlated with the sources of inoculum and environmental factors and will be discussed later.

Small, water-soaked circular spots 1/16 to 1/8 inches (1.6 to 3.2 mm) in diameter first appear on the undersides of older leaves. The centers of the spots are gray, or tan and spots have a dark brown margin. As the spots mature, they enlarge to about 1/4 inch in diameter (6.4 mm) and may coalesce (Figure 2.11). In the center of the spots are many dark brown, pimple-like structures called pycnidia-fruiting bodies of the fungus (Figure 2.12). The structures are large enough to be seen with the unaided eye or with the aid of a hand lens. Pycnidia are absent from early blight lesions and from lesions produced by the gray leaf fungus *Stemphylium solani*, which is common in areas with consistently warm and humid conditions. Septoria leaf spot also lacks the target-like lesions so typical for *Altemaria* blight. Spots may also appear on stems, calyxes, and blossoms, but rarely on fruit. Heavily infected leaves will turn yellow, dry up, and drop off. This defoliation will result in sun-scalding of the fruit.

FIGURE 2.11   Enlarged matured leaf spots.

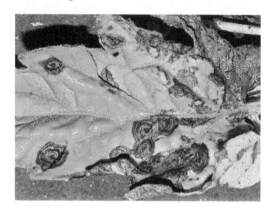

FIGURE 2.12   Mature lesions of septoria leaf spot showing black, speck-like pycnidial fruit bodies.

2.5.3   CAUSE AND DISEASE DEVELOPMENT

Septoria survives the winter on infected plant debris including tomato and related plants. The fungus may also be transmitted by infected seed, and spores can be present around growing facilities such as greenhouses, cold frames, flats, etc. Where spores have survived the winter, initial infections may begin early in the year. Otherwise, the fungus will not sporulate below 15°C, which delays the onset of infections. The spores are splashed by rain, blown by the wind, or carried by insects and other animals (including man), and once the initial infections have started the fungus can produce new spores, which rapidly increases the rate at which the disease spreads.

2.5.4   FAVORABLE CONDITIONS FOR DISEASE DEVELOPMENT

The disease is usually first seen in early to mid-August when the foliage has become sufficiently dense to restrict air movement within the canopy. After canopy closure the humidity remains high and any free water on leaf surfaces tends to dry more slowly. Infection can occur when the relative humidity has been at 100% for more than 48 hours. These conditions are cumulative, however, and can be spread over several days. The optimal temperature range for Septoria is between 20°C and 25°C. Under wet conditions, numerous spores (conidia) are produced in the pycnidia and are exuded when the fruiting structures are mature. The disease usually starts on the lowest leaves where the humidity tends to be the highest and where the fungal spores are most likely to land.

2.5.5   DISEASE CYCLE

S. lycopersici overwinters in old tomato debris and on wild Solanaceous plants, such as ground cherry, nightshade, and jimsonweed. Seeds and transplants may also carry the fungus. The disease is favored by moderate temperatures and abundant rainfall. Spore production is abundant when temperatures are 15.5°C–28°C (60°F–80°F). Spores are easily spread by wind and rain. Infection occurs on the lower leaves after the plants begin to set fruit.

2.5.6   Management

2.5.6.1   Cultural Control
- Dispose of crop refuse by plowing under or composting.
- Control weeds in and around the edge of the garden.
- Rotate tomatoes with cereals, corn, or legumes. A four-year rotation is recommended where disease has been severe.

2.5.6.2   Chemical Control
- Apply fungicides on a preventative schedule before the disease first appears on the lower leaves. Begin sprays when the first fruits of the first cluster are visible after blossom fruits have dropped. Apply fungicides every seven to ten days or more often when the weather is warm and wet. In home gardens the fungicides, chlorothalonil (e.g. Daconil 2787) or maneb (e.g. Maneb), can be used.
- In commercial plantings, chlorothalonil (e.g. Bravo, Terranil) and maneb (e.g. Dithane Rainshield NT, Penncozeb) can be rotated with the systemic fungicide, azoxystrobin (e.g. Quadris), every seven days. Alternating sprays is important in order to delay the development of resistant strains of the fungus to azoxystrobin. Refer to the current Virginia Pest Management Guide for Home Grounds and Animals (VCE Publication 456–018) or Commercial Vegetable Production Recommendations (VCE Publication 456–420) for details on fungicide application rates and timing.

2.5.6.3   Resistance
- No resistant cultivars are available.

2.5.6.3.1   Integrated Pest Management Strategies
- **Remove diseased leaves.** If caught early, the lower infected leaves can be removed and burned or destroyed. However, removing leaves above where fruit has formed will weaken the plant and expose fruit to sunscald. At the end of the season, collect all foliage from infected plants and dispose of or bury. Do not compost diseased plants.
- **Improve air circulation around the plants.** If the plants can still be handled without breaking them, stake or cage the plants to raise them off the ground and promote faster drying of the foliage.
- **Mulch around the base of the plants.** Mulching will reduce splashing soil, which may contain fungal spores associated with debris. Apply mulch after the soil has warmed.
- **Do not use overhead watering.** Overhead watering facilitates infection and spreads the disease. Use a soaker hose at the base of the plant to keep the foliage dry. Water early in the day.
- **Control weeds.** Nightshade and horsenettle are frequently hosts of Septoria leaf spot and should be eradicated around the garden site.
- **Use crop rotation.** Next year do not plant tomatoes back in the same location where diseased tomatoes grew. Wait one or two years before replanting tomatoes in these areas.
- **Use fungicidal sprays.** If the above measures do not control the disease, you may want to use fungicidal sprays. Fungicides will not cure infected leaves, but they will protect new leaves from becoming infected. Apply at seven- to ten-day intervals throughout the season.
- Apply chlorothalonil, maneb, macozeb, or a copper-based fungicide, such as Bordeaux mixture, copper hydroxide, copper sulfate, or copper oxychloride sulfate. Follow harvest restrictions listed on the pesticide label.

2.6 CUCUMBER MOSAIC VIRUS ON TOMATO

Cucumber mosaic virus (CMV) can occur wherever tomatoes are grown. The host range of the virus consists of more than 750 plant species including many vegetables (such as tomato, pepper, cucurbits, and legumes), weeds, and ornamentals. Strains of CMV have been reported which are specific to the tomato.

2.6.1 CAUSAL ORGANISM

This virus is a member of the

> Family: Bromoviridae
> Genus: *Cucumovirus*
> Species: CMV (*tomato fern leaf virus*).

This virus has a worldwide distribution and a very wide host range, the tomato being one of them.

2.6.2 SYMPTOMS

Tomatoes infected with CMV often are stunted and bushy (shortened internodes) and may have distorted and malformed leaves (Figure 2.13). Leaves may appear mottled (intermingling of dark green, light green, and yellow tissue), a similar symptom to those caused by other viruses. The most characteristic symptom of CMV is extreme filiformity, or shoe-stringing, of leaf blades (Figure 2.14). CMV symptoms can be transitory, that is, the lower or upper leaves can show symptoms while those in the midsection of the plant appear normal. Effect of CMV on yield depends on a number of factors, including plant age when infected and environmental conditions. Severely affected plants produce few fruit, which are usually small.

2.6.3 MEANS OF MOVEMENT AND TRANSMISSION

CMV is the second most important virus disease of tomato. CMV has an extensive host range and is transmitted by aphids in a non-persistent manner. Unlike TMV (Tobacco mosaic virus), CMV

FIGURE 2.13   Distorted and malformed leaves of tomato.

FIGURE 2.14   Shoe-stringing symptom.

is not seedborne in tomato and does not persist in plant debris in the soil or on workers' hands. CMV has been found in greenhouse planting. Seedlings grown outdoors and left unprotected by isolation before moving indoors are 1 likely source of infection. Other sources of inoculum are the spread of CMV by aphids from infected plants in adjoining greenhouses (weeds under benches, ornamentals, or other vegetables) and by viruliferous aphids entering through non-insect-proof vent windows.

2.6.4   POTATO VIRUS Y

Potato virus Y (PVY) occurs worldwide but has a narrow host range, affecting plants in the Solanaceae family (that is, tomatoes, potatoes, and peppers). It is transmitted by aphids. Near total crop failures have been reported when PVY was detected early in the season and high aphid populations were present.

2.6.5   SYMPTOMS

Symptoms on tomato vary according to the PVY strain, plant age, varieties infected, and environmental conditions. General symptoms on tomato are faint mottling and slight distortion of the leaves (Figure 2.6.3). Severe symptoms include deep brown, dead areas in the blade of nearly mature leaflets. Leaflets at the terminal end of a leaf usually are the most adversely affected, often showing severe necrosis. In many cases, all leaflets are affected. Leaves formed after the onset of PVY exhibit mild wrinkling, slight distortion, and mild mottling. Leaflets of plants infected for some time are rolled downward with curved petioles, giving the plant a drooping appearance. Stems often show a purplish streaking but no symptoms are produced on the fruit. Mature plants are stunted and unthrifty and yield is reduced.[3]

FIGURE 2.15   Faint mottling and slight distortion of the leaves of tomato.

2.6.6   MEANS OF MOVEMENT AND TRANSMISSION

PVY is transmitted in a non-persistent manner by many aphid species. Aphids can acquire the virus in less than 60 seconds from an infected plant and transmit it to a healthy plant in less than 60 seconds. The virus may be retained by the aphid for longer than 24 hours if feeding does not occur. PVY can also be transmitted mechanically. Potato is an important source of the virus for tomato and other solanaceous crops. The virus does not appear to be seed-transmitted (Figure 2.15).

2.6.7   TOBACCO ETCH VIRUS

Tobacco etch virus (TEV) infects tomatoes and peppers along with other plants in the Solanaceae family. The occurrence of TEV in tomato fields is closely associated with other infected solanaceous crops, especially pepper, and natural weed hosts, which serve as virus reservoirs.

2.6.8   SYMPTOMS

Leaves of infected plants are severely mottled, puckered, and wrinkled. Plants infected at an early age are severely stunted. Fruit from infected plants are mottled and never achieve marketable size. The younger the plants are when infected, the greater the reduction in yield.

2.6.9   MEANS OF MOVEMENT AND TRANSMISSION

The virus can be transmitted by at least 10 species of aphid in a non-persistent manner. There are no reports of seed transmission in any host plant.

2.6.10   PREVENTION AND CONTROL OF CMV, PVY, AND TEV

There are no good sources of resistance in tomato for CMV, PVY, or TEV, so other control strategies must be used. These include the following:

- Eradicate all biennial and perennial weeds and wild reservoir hosts in and around fields. Maintain a distance of at least 30 ft between susceptible crops, weeds, or other susceptible plants, including those in ditch banks, hedge or fence rows, and other locations.
- Plant earlier to avoid high aphid populations that occur later in the season.
- Plant late settings as far as possible from fields used to produce early tomatoes and peppers. These areas can act as sources of viruses and aphids for subsequent crops.
- Scout fields for the first occurrence of virus disease. Where feasible, pull up and destroy infected plants, but only after spraying them thoroughly with an insecticide to kill any insects they may be harboring.

- Use reflective mulches to repel aphids, thereby reducing the rate of spread of aphid-borne viruses.
- Monitor aphid populations early in the season and apply insecticide treatments when needed.
- Minimize plant handling to reduce the amount of virus spread mechanically.
- Avoid planting tomatoes near potato fields to control PVY.

2.7   COMMON MOSAIC OF TOMATO (TOBACCO/TOMATO MOSAIC)

TMV is distributed worldwide and may cause significant losses in the field and greenhouse. TMV is 1 of the most stable viruses known, able to survive in dried plant debris as long as 100 years. Many strains of TMV have been reported and characterized. TMV can be seedborne in tomato, is readily transmitted mechanically by human activities, and may be present in tobacco products. The virus is not spread by insects but occurs commonly in greenhouses or in the field.

2.7.1   CAUSAL ORGANISM

The virus belongs to

Genus: *Tobamovirus*
Species: Tomato Mosaic Virus (ToMV)

Several strains are known.

2.7.2   SYMPTOMS

The symptoms in tomato vary greatly in intensity depending upon the variety, virus strain, time of infection, light intensity, and temperature. High temperatures, for example, may mask foliar symptoms. The most characteristic symptom of the disease on leaves is a light- and dark-green mosaic pattern (Figure 2.16). Some strains (referred to as the acuba strains) may cause a striking yellow mosaic, whereas other strains may cause leaf malformation and "fern-leafing", suggestive of CMV infection. With the use of ToMV resistant or tolerant varieties, plants may be infected by some strains whose symptoms are latent. Ordinarily, the fruit from infected plants do not show mosaic symptoms but may be reduced in size and number. Occasionally the fruit will show disease symptoms which vary from an uneven ripening to an internal browning of the fruit wall (Figure 2.17). Brown wall typically occurs on the fruit of the first two clusters and appears several days prior to foliar symptoms. Under certain environmental conditions, some varieties with resistance

FIGURE 2.16   Symptoms of ToMV on tomato leaves.

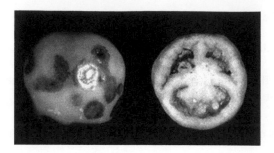

FIGURE 2.17   Symptoms of ToMV on tomato fruit.

(heterozygous) to ToMV will show necrotic streaks or spots on the stem, petiole, and foliage as well as on the fruit. This disorder is particularly evident in well-developed but unripe fruits of the first cluster, involving a collapse of cells in the fleshy parenchyma. The cause is attributed to a "shock reaction" following ToMV infection, with a number of other factors contributing to the severity of symptoms (high soil moisture, low nitrogen and boron, and sensitivity of the variety).

2.7.3 MEANS OF MOVEMENT AND TRANSMISSION

The virus may be introduced on infected seed. Only a small number of seedlings need to be infected for the virus to spread rapidly. It can also be spread by contaminated tools and the clothing and hands of workers during routine activities. It is readily transmitted by machinery or workers from infected to healthy plants during handling. Infested debris from a previous crop can lead to infection when the roots of the new tomato plants come in contact with the debris. Chewing insects can transmit the virus but are not considered a major source of infection. Tomato seed can carry the virus, but actual infection is thought to occur when plants are thinned or transplanted.

2.7.4 PREVENTION AND CONTROL

ToMV is spread readily by touch. The virus can survive on clothing in bits of plant debris for about two years, and can easily enter a new plant from a brief contact with a worker's contaminated hands or clothing. Tobacco products can carry the virus, and it can survive on the hands for hours after touching the tobacco product. Ensure that workers do not carry or use tobacco products near the plants and wash hands well (with soap to kill the virus) after using tobacco products. Ensure that workers wear clothing not contaminated with tomato, tobacco, or another host-plant material. Exclude non-essential people from greenhouses and growing areas.

Choose resistant varieties. Use disease-free seed and transplants, preferably certified ones. Avoid the use of freshly harvested seed (two years old is best if non-certified seed is used). Seed treatment with heat (two to four days at 70°C using dry seed) or trisodium phosphate (10% solution for 15 minutes) has been shown to kill the virus on the outside of the seed and, often, most of the virus inside the seed as well. Care must be taken to not kill the seeds, though. Use a two-year rotation away from susceptible species. In greenhouses, it is best to use fresh soil, as the steaming soil is not 100% effective in killing the virus. If soil is to be steamed, remove all parts of the plant from the soil, including the roots. Carefully clean all plant growing equipment and all greenhouse structures that come into contact with plants.

When working with plants, especially when picking out seedlings or transplanting, spray larger plants with a skim milk solution or a solution made of reconstituted powdered or condensed milk. Frequently dip hands, but not seedlings, into the milk. Wash hands frequently with soap while working with plants, using special care to clean out under nails. Rinse well after washing. Tools should be washed thoroughly, soaked for 30 minutes in 3% trisodium phosphate and not rinsed.

Steam sterilizing the potting soil and containers as well as all equipment after each crop can reduce disease incidence.

Sterilizing pruning utensils or snapping off suckers without touching the plant instead of knife pruning helps reduce disease incidence. Direct seeding in the field will reduce the spread of ToMV.

Another method for control of this disease is to artificially inoculate plants with a weak strain of the virus. This will not cause symptoms on the plants but protect them against disease-causing strains of the virus.

2.8 TOMATO SPOTTED WILT VIRUS

Tomato spotted wilt virus (TSWV) causes serious diseases of many economically important plants representing 35 plant families, including dicots and monocots. A unique feature is that TSWV is the only virus transmitted in a persistent manner by certain thrips species. At least six strains of TSWV have been reported; the symptoms produced and the range of plants infected vary among strains. Although previously it was a threat only to crops produced in tropical and subtropical regions, today the disease occurs worldwide, largely because of wider distribution of the western flower thrips and movement of virus-infected plant material. Early and accurate detection of infected plants and measures to reduce the vector population are discussed as critical steps for disease control.

2.8.1 CAUSAL ORGANISM

TSWV is the only member of an RNA-containing virus group that has membrane-bound spherical particles 70–90 mm in diameter. Tomato spotted wilt, first described in Australia in 1919, was later identified as a virus disease. It is now common in temperate, subtropical, and tropical regions around the world. Heavy crop losses of tomato in the field were reported in the 1980s in Louisiana and in tomato and lettuce in Hawaii. Other southern states reporting losses in tomato in recent years include Mississippi, Arkansas, Florida, Alabama, Georgia, and Tennessee.

2.8.2 SYMPTOMS

Tomato plants infected with TSWV are stunted and often die. Initially, leaves in the terminal portion of the plant stop growing, become distorted, and turn pale green. In young leaves, veins thicken and turn purple, causing the leaves to appear bronze (Figure 2.18). Necrotic spots or ring spots frequently occur on infected leaves. Stems of infected plants often have purplish-brown streaks. Infected fruit may exhibit numerous ring spots and blotches and may become distorted if infected when immature (Figure 2.19).

FIGURE 2.18 Symptoms of TSWV on tomato leaves.

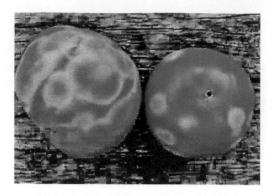

FIGURE 2.19   Symptoms of TSWV on tomato fruit.

2.8.3   MEANS OF MOVEMENT AND TRANSMISSION

The virus is transmitted by insects called thrips. It is not easily mechanically transmitted by rubbing. Juvenile thrips acquire the virus from infected weed hosts. They disperse it after becoming winged adults that feed on host tomatoes and other target plants. Consequently, this disease is more of a problem in spring crops and in certain favorable years.

Weed hosts identified as potential virus carriers include spiny amaranthus, wild lettuce (*Lactuca* sp.), pasture buttercup (*Ranunculus* sp.), *Solanum* sp., and sowthistle (*Sonchus* sp.).

2.8.4   PREVENTION AND CONTROL

Currently, there is no effective way to control TSWV. To reduce the source of infection

- Control weeds adjacent to the field. (TSWV can overwinter in weeds.)
- Apply systemic insecticides to the soil at planting to slow the initial spread of the virus into the field.
- Spray bordering weeds and the tomato crop with insecticides to suppress thrip populations and the spread of TSWV.
- Remove and destroy infected plants as soon as symptoms appear, to further reduce virus spread.

2.9   DISORDERS OF TOMATO

2.9.1   BLOSSOM-END ROT

Blossom-end rot (BER) is a physiological disorder, not a disease. The disorder is often prevalent in commercial as well as home garden tomatoes, and severe losses may occur if preventive control measures are not undertaken. BER is caused by calcium deficiency, usually induced by fluctuations in the plant's water supply. Because calcium is not a highly "mobile" element in the plant, even brief changes in the water supply can cause BER. Droughty soil or damage to the roots from excessive or improper cultivation (severe root pruning) can restrict water intake preventing the plants from getting the calcium that they need. Also, if plants are growing in highly acidic soil or are getting too much water from heavy rain, over-irrigation, or high relative humidity, they can develop calcium deficiency and BER.

2.9.1.1   Symptoms

Symptoms may occur at any stage in the development of the fruit, but, most commonly are first seen when the fruit is 1/3 to 1/2 of its full size. As the name of the disease implies, symptoms appear only at the blossom end of the fruit. Initially a small, water-soaked spot appears, which enlarges and

FIGURE 2.20   BER of tomato.

darkens rapidly as the fruits develop (Figure 2.20). The spot may enlarge until it covers as much as 1/3 to 1/2 of the entire fruit surface, or the spot may remain small and superficial. Large lesions soon dry out and become flattened, black, and leathery in appearance and texture.

2.9.1.2   Management

Control of BER is dependent upon maintaining adequate supplies of moisture and calcium to the developing fruits. Tomatoes should not be excessively hardened nor too succulent when set in the field. They should be planted in well-drained, adequately aerated soils. Tomatoes planted early in cold soil are likely to develop BER on the first fruits, with the severity of the disease often subsiding on fruits set later. Thus, planting tomatoes in warmer soils helps to alleviate the problem. Irrigation must be sufficient to maintain a steady even growth rate of the plants. Mulching of the soil is often helpful in maintaining adequate supplies of soil water in times of moisture stress. When cultivation is necessary, it should not be too near the plants nor too deep, so that valuable feeder roots remain uninjured and viable. In home gardens, shading the plants is often helpful when hot, dry winds are blowing, and soil moisture is low. Use of fertilizers low in nitrogen but high in superphosphate, such as 4-12-4 or 5-20-5, will do much to alleviate the problem of BER. In emergency situations, foliage can be sprayed with calcium chloride solutions. However, extreme caution must be exercised since calcium chloride can be phytotoxic if applied too frequently or in excessive amounts. Foliar treatment is not a substitute for proper treatment of the soil to maintain adequate supplies of water and calcium.

2.9.2   CATFACING

Insect damage, poor pollination, and environmental factors all cause catfacing, a term that describes the puckering, scarring, and deformation of strawberries, stone fruits, and tomatoes. You can recognize catfacing in tomatoes by the scarred indentations found on the blossom end of the fruit. Sometimes this scarring extends deep into the fruit cavity, making much of the fruit inedible.

2.9.2.1   Cause

Anything causing abnormal pistillate development will precipitate catfacing at the blossom end of the fruit. Growth regulators such as the herbicide 2,4-D have been shown to cause catfacing. Unfavorable weather such as a prolonged cold period during blossoming also appears related to a high incidence of catface. Catface is more prevalent on large-fruited, fresh market tomato varieties.

2.9.2.2   Symptoms

In concentric cracking, the fruit develop circular, concentric cracks around the stem end of the fruit. In radial cracking, the fruit cracks radiate from the stem end. Catface is expressed as a malformation and cracking of fruit at the blossom end, often exposing the locules (Figure 2.21).

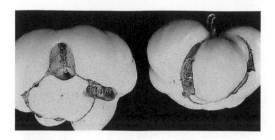

FIGURE 2.21   Catface expressed as malformation and cracking of fruit at the blossom end.

2.9.2.3   Management

Good growing practices, especially temperature control, should be followed in greenhouse production of field transplants. Excess nitrogen, aggressive pruning, and accidental exposure to hormonal herbicides should be avoided.

2.9.3   BLOTCHY RIPENING

Blotchy ripening of tomatoes is characterized by areas of the fruit that fail to ripen properly. White or yellow blotches appear on the surface of the ripening fruit while the tissue inside remains hard (Figure 2.9.3). It has been linked to potassium or boron deficiency and to high nitrogen levels. Blotchy ripening is caused by inadequate fertilizer application or low availability of nutrients.

Climatic, nutritional, and cultural problems may contribute to blotchy ripening. Low levels of potassium in plants and prolonged cloudy periods or inadequate light intensity have been associated with the disorder. Other possible contributing factors are high soil moisture, high humidity, low temperature, soil compaction, and excessive fertilization. These environmental factors can contribute to nutrient deficiencies or other imbalances that impede development of the red pigment in the fruit.

2.9.3.1   Management

Growers should provide balanced fertilization, and in greenhouses should avoid excessively high temperatures, if possible. Cultivars vary in susceptibility to this disorder (Figure 2.22).

2.9.4   SUNSCALD

Sunscald commonly affects tomatoes. It's generally the result of exposure to sunlight during extreme heat, though it may be caused by other factors as well. While this condition is not technically dangerous to plants, it can damage fruits and lead to other issues that could become a problem.

FIGURE 2.22   Blotchy ripening of tomato.

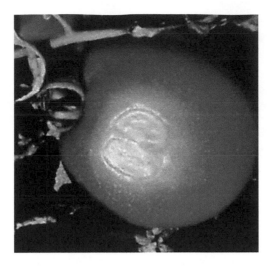

FIGURE 2.23   Sunscalded tomato.

2.9.4.1   Symptoms

The initial symptom is a whitish, shiny area that appears blistered. The killed, bleached tissues gradually collapse, forming a slightly sunken area that may become pale yellowish and wrinkled as the fruit ripens (Figure 2.23). The dead tissue is quickly invaded by secondary organisms and the fruit decays.

Sunscald occurs, quite simply, with overexposure of tomato plants to direct sunlight during hot weather. Tomato sunscald occurs most often on tomato varieties which have sparse foliage. Tomato foliage may be sparse due to the variety of tomato plant, or due to a variety of tomato leaf diseases such as septoria leaf spot, fusarium wilt, or verticullum wilt. Additionally, both early and late blight are factors in sparse foliage on tomato plants.

2.9.4.2   Management

To prevent Sunscald, practice good management techniques that include crop rotation, sanitation, and mulching to reduce the likelihood of defoliation by Septoria leaf spot and early blight. Also, if staking, don't remove too much of the foliage. Also, to prevent sunscald on tomato fruit, control foliar diseases and avoid heavy pruning or shoot removal.

2.9.5   FRUIT CRACKING

Concentric fruit cracking and radial fruit cracking are the two types. This is a physiological disorder that occurs as the fruit is sizing and results from variations in soil moisture and temperature. Growth cracks can occur during periods of rapid fruit growth when relative humidity and air temperatures are high or when water becomes abundantly available after a drought period. These cracks are easily invaded by secondary organisms that promote fruit rot. Varieties vary in their susceptibility to fruit cracking.

Similar to BER, cracking is associated with rapid fruit development and wide fluctuations in water availability to the plant. Fruit that has reached the ripening stage during dry weather may show considerable cracking if the dry period is followed by heavy rains and high temperatures. Tomato varieties differ considerably in the amount and severity of cracking under climatic conditions (Figure 2.24).

2.9.5.1   Radial Cracking

Radial cracking splits the tomato skin into long lines from the stem scar to the bottom end. Several factors cause this type of cracking, including uneven soil moisture levels, inappropriate pruning,

FIGURE 2.24   Tomato fruit cracking.

plants that go through periods of fast growth and slowed growth, and fertilizer too high in nitrogen and too low in potassium. To reduce radial cracking, buy plants that are not susceptible to cracking, water properly, mulch the garden, prune correctly, and fertilize as needed.

2.9.5.2   Concentric Cracking

Concentric cracking splits the tomato skin in a circular shape around the stem scar. As fruit ages, the skin becomes less flexible and cracking from uneven moisture is more likely. To prevent this type of cracking, pick all ripe fruits in a timely fashion.

2.9.5.3   Management

- As with BER, mulching and avoiding heavy applications of nitrogen fertilizer should help reduce fruit cracking.
- Supersonic and Jetstar are two varieties that show relatively low incidence of cracking.
- In the greenhouse or conservatory, control temperature and sunlight levels carefully to avoid extremes, using combinations of heating, ventilation, and white greenhouse paint as appropriate. A good maximum—minimum temperature thermometer is essential. Poor quality greenhouses often lack adequate ventilation arrangements and the temporary removal of glass panes during summer can help remedy this lack. Polythene tunnels have few options for decreasing the temperature except using extra shading or installing wind-up sides to increase ventilation.
- Managing temperature is almost impossible for tomatoes grown outdoors.
- Feed regularly to maintain high soil fertility. Special tomato fertilizers have high levels of potassium to encourage good fruit development.
- Water to maintain a constant level of soil moisture. This is especially important when growing in growing bags, pots, or other containers. Outdoors, water to maintain as constant a level of soil moisture as possible. Plants grown in border soil, indoors or out, usually have a more extensive root system, which helps protect them from fluctuations in water supply.

2.9.6   Physiological Leaf Roll

Physiological leaf roll may be associated with environmental stresses such as excess moisture, excess nitrogen, and transplant shock. Leaf roll may also be related to moisture conservation during periods of extreme heat and drought. Improper cultural practices such as severe pruning and root damage during cultivation can also cause leaf roll symptoms. Physiological leaf roll involves an initial upward cupping of the leaves, followed by an inward roll. In severe cases, the leaves roll up until the leaflets overlap.

2.9.6.1   Cause

The severity of leaf roll appears to be cultivar dependent. Cultivars selected for high yield tend to be most susceptible. Indeterminate cultivars of tomato are reported to be more sensitive to this disorder than determinate cultivars. Determinate varieties of tomatoes, also called "bush" tomatoes, stop growing when their first fruit sets, whereas indeterminate varieties, also called "vining" tomatoes, grow and flower until killed by frost or other harsh environmental conditions. In some cases, the condition is believed to occur most commonly when plants are pruned during dry soil conditions. In other cases, causes listed include growing high-yielding cultivars under high nitrogen fertility programs, phosphate deficiency, or extended dry periods. The disorder is also attributed in some areas to excess soil moisture coupled with prolonged high air temperatures.

2.9.6.2   Symptoms

Plants affected by tomato leaf roll suffer a greatly reduced growth rate and so become stunted or dwarfed. Leaflets are rolled upward and inward, while the leaves are often bent downward (epinasty) but are stiff rather than limp as in wilted plants. Leaves are thicker than normal and of a leathery texture and often have a purple tinge to the venation on the undersurface. The newly produced young leaves are paler in color than those on healthy plants. Fruit, if produced at all on affected plants, is smaller than normal, dry in texture, and un-saleable. Plants with an advanced infection will not produce fruit (Figure 2.25).

2.9.6.3   Management

- Planting determinate cultivars;
- Planting in well-drained soils and maintaining uniform, adequate soil moisture (~1 inch per week during the growing season depending on the area of production);
- Being careful not to over-fertilize—especially with nitrogen fertilizers—and providing appropriate phosphorus fertilizer (refer to your soil test results for specific fertilizer recommendations);
- Avoiding severe pruning; and
- Maintaining temperatures below 35°C by using shading or evaporative cooling.

FIGURE 2.25   Physiological Leaf Roll.

2.9.7　Puffiness

Fruit suffering from puffiness appear somewhat bloated and angular. Cavities inside the fruit may lack the normal gel and the fruit is dense.

2.9.7.1　Cause

Puffiness results from incomplete pollination, fertilization, and seed development. Often, this is due to cool temperatures during bloom that reduce insect activity and pollination. High nitrogen and low potassium can also lead to puffiness.

2.9.7.2　Symptoms

When this problem is slight, it may be impossible to detect puffiness until fruit are cut (Figure 2.26). Severe puffy fruit will appear to be flat-sided or angular in nature. When fruit are cut, open cavities are observed between the seed gel area and the outer wall. Fruits are also very light in relation to size.

2.9.7.3　Management

Use of "hot set" varieties can reduce the problem, but even these have limitations when night temperatures remain above about 24°C. Other factors such as high Nitrogen levels, low light, or rainy conditions can also cause seed set problems.

2.10　EARLY BLIGHT OF TOMATO

Early blight, caused by the fungus *Alternaria solani*, is one of the most common diseases of tomatoes worldwide. It occurs to some extent every year wherever tomatoes are grown. In spite of its name, the disease may occur at any time during the growing season. The fungus attacks leaves, stems, and fruit.

2.10.1　Causal Organism

Species	Associated Disease Phase	Economic Importance
A. solani	Rot on Leaves, Stem, and Fruits	Severe

2.10.2　Symptoms

The leaves, stems, and fruit on the vine may be affected. Symptoms typically appear soon after fruit set; starting on the lower leaves as tiny dark brown spots. Symptoms on leaves are most likely

FIGURE 2.26　Puffiness. Photo credit: Timothy Coolong, University of Kentucky.

FIGURE 2.27   Lesions on leaves have concentric rings surrounded by yellowing tissue.

to appear on the older foliage. Small dark spots enlarge into circular lesions consisting of concentric rings. The spots enlarge to over 1/2 inch in diameter and develop a grayish-white center with a darker border. The tissue surrounding the lesions becomes yellow and the spots later become irregular in shape. The leaf becomes yellow as greater parts of the tissue are affected (Figure 2.27). The lesions turn brown and eventually drop from the plant. Black pycnidia (fungal fruiting bodies that appear as pinhole sized black dots) form in the center of the spot as they mature. Defoliation occurs under prolonged periods of leaf wetness and high temperatures; exposed fruit become susceptible to sunburn damage.

Leaf lesions arising from *S. lycopersici*, another foliar pathogen of tomato, may be confused with early blight symptoms. However, *Septoria* lesions are lighter than in color with a small pepper-like fruiting body in the center of the lesion, and the disease generally arises under cooler temperatures than with the case of early blight. Stems and petioles affected by early blight have elliptical concentric lesions, which drastically weaken the plant (Figure 2.28). Lesions at the base of emerging seedlings can cause a collar rot. If this arises simultaneously on many seedlings, it may indicate contamination of tomato seeds or soil used for planting. Mature and ripe fruit may be affected (Figure 2.29). Lesions occur on the stem end or the upper shoulder and may be quite large. Fruit wounds or cracks may also have large, dark, leathery, sunken areas with concentric rings.

2.10.3   CAUSE AND DISEASE DEVELOPMENT

The pathogen survives primarily on infected crop debris and in the soil for years. It can also overwinter on volunteer hosts and weeds. Conidia are spread by wind and splashing water. Infection occurs directly or through stomata primarily on older leaf tissue and through wounds or through moist induced swelling of lenticels on tubers. Symptoms appear within a week of infection. Multiple

FIGURE 2.28   Dark concentric rings develop on the stem end of the fruit.

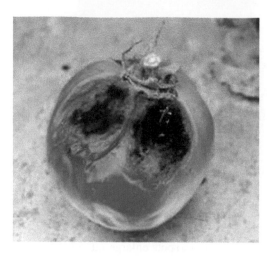

FIGURE 2.29   Alternaria lesions on tomato fruit covered by masses of black conidia.

cycles can occur in one crop season under favorable conditions. Susceptibility to A. solani increases with the age of the plant tissue and the plant, particularly, after fruit and tuber initiation.

2.10.4   FAVORABLE CONDITIONS OF DISEASE DEVELOPMENT

Alternaria spores germinate within two hours over a wide range of temperatures, but at 27°C to 29°C it may only take 1/2 hour. Another three to twelve hours are required for the fungus to penetrate the plant depending on temperature. After penetration, lesions may form within two to three days or the infection can remain dormant awaiting proper conditions (15.5°C with extended periods of wetness). *Alternaria* sporulate best at about 26.5°C when abundant moisture (as provided by rain, mist, fog, dew, irrigation) is present. Infections are most prevalent on poorly nourished or otherwise stressed plants.

The disease is favored by warm temperatures and extended periods of leaf wetness from frequent rain, overhead irrigation, or dews. The disease cycle is about five to seven days, so numerous repeating cycles can occur during the long growing season. Plants under periods of stress are more susceptible, for example, during fruiting, under attack from nematodes, when inadequately fertilized, or on older plants. Early blight may be more prevalent on old transplants or transplants lacking vigor or appear stressed by wilting.

2.10.5   DISEASE CYCLE

A. solani can survive from year to year in old, diseased vines left in the field. Splashing rain, running water, and moving machinery can spread the fungus in the field. Symptoms are usually visible about ten days after the plants are infected. The disease is greatly influenced by the degree of thriftiness of the plants. Infection takes place slowly unless the plants have been weakened or wounded (Figure 2.30).

2.10.6   MANAGEMENT

2.10.6.1   Cultural Control
- Obtain the best-certified seed or transplants. Prevention of seedling infection is very important.
- Practice crop rotation. Tomatoes should not be planted in areas where susceptible vegetables, such as tomato, potato, pepper, or eggplant, have been grown during the previous three or four years.
- Destroy solanaceous weeds, such as black nightshade or Jerusalem cherry, which can serve as hosts for the fungus.

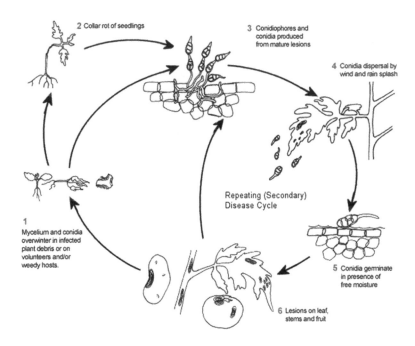

FIGURE 2.30   Disease cycle of *A. solani*.

- Space transplants to allow good air circulation, thereby permitting plants to dry off rapidly after rain and dews. This will reduce the risk of disease development.
- Plow under or remove old vines as soon as harvest is completed.

2.10.6.2   Chemical Control
- Apply a fungicide, such as chlorothalonil (e.g. Daconil 2787) or maneb (e.g. Maneb), on a preventative basis. Follow the label rates or consult the current Virginia Pest Management Guide for Home Grounds and Animals (VCE Publication 456–018) or Commercial Vegetable Production Recommendations (VCE Publication 456–420) for details on fungicide control (refer to Table 2.2).

2.10.6.3   Resistance
- Tomato cultivars vary in their resistance to the disease. The cultivars, Mountain Fresh, Mountain Supreme, and Plum Dandy, have resistance to early blight.

2.11   LATE BLIGHT OF TOMATO

Late blight of potatoes and tomatoes, the disease that was responsible for the Irish potato famine in the mid-nineteenth century, is caused by the fungus-like oomycete pathogen *Phytophthora infestans*. It can infect and destroy the leaves, stems, fruits, and tubers of potato and tomato plants. Before the disease appeared in Ireland, it caused a devastating epidemic in the early 1840s in the northeastern United States.

2.11.1   CAUSAL ORGANISM

Species	Associated Disease Phase	Economic Importance
P. infestans	Lesions on Leaves, Petiole, Stem, and Fruits	Severe

TABLE 2.2
Fungicide Control

Fungicides for Early Blight Control

Fungicide	Typical Application Interval	Examples of Trade Names
Azoxystrobin, Pyraclostrobin	7–14 days	Quadris, Amistar, Cabrio EG
Bacillus subtilis	5–7 days	Seranade
Chlorothalonil	7–14 days	Daconil, Bravo, Echo, Fungonil and others
Copper products	7–14 days	Bordeaux Mixture, Kocide, Tenn-Cop,Liqui-cop, Basicop, Camelot
Hydrogen dioxide	Commercial only, see label	Oxidate
Mancozeb and Maneb	7–14 days	Dithane, Penncozeb, Manex, Mancozeb, Maneb
Potassium bicarbonate	5–14 days as needed	Armicarb 100, Firststep
Ziram	7–14 days	Ziram

2.11.2 Symptoms

2.11.2.1 On Tomato Leaves

Lesions begin as indefinite, water-soaked spots that enlarge rapidly into pale green to brownish-black lesions and can cover large areas of the leaf. During wet weather, lesions on the abaxial surface of the leaf may be covered with a gray to white moldy growth (not to be confused with powdery mildew disease). On the undersides of larger lesions, a ring of moldy growth of the pathogen is often visible during humid weather. As the disease progresses, the foliage turns yellow and then brown, curls, shrivels, and dies. The late blight symptoms are distinct from and should not be confused with symptoms of powdery mildew disease, the spores of which appear usually on the upper leaf surface of tomato[4] (Figures 2.31 and 2.32).

2.11.2.2 On Tomato Petioles and Stems

Lesions begin as indefinite, water-soaked spots that enlarge rapidly into brown to black lesions that cover large areas of the petioles and stems. During wet weather, lesions may be covered with a gray to white moldy growth of the pathogen. Affected stems and petioles may eventually collapse at the point of infection, leading to the death of all distal parts of the plant (Figure 2.33).

FIGURE 2.31   Symptoms of leaf blight of tomato on tomato leaves.

FIGURE 2.32 A sign of the pathogen *P. infestans* can be visible as powdery whitish rings around the margins of the blighted areas.

2.11.2.3 On Tomato Fruits

Dark, olivaceous greasy spots develop on green fruit; a thin layer of white mycelium may be present during wet weather (Figure 2.34).

2.11.3 Cause and Disease Development

Severe late blight epidemics occur when *P. infestans* grows and reproduces rapidly on the host crop. Reproduction occurs via sporangia that are produced from infected plant tissues and is most rapid during conditions of high moisture and moderate temperatures (15.5°C–26.5°C). Sporangia disperse to healthy tissues via rain splash or on wind currents. Reproduction is asexual; each sporangium is an exact copy of the one that initiated the parent lesion, and each can initiate a new lesion.

   P. infestans (Mont.) de Bary is not a true fungus, but rather is regarded as a fungus-like organism. This pathogen is currently classified as an oomycete, which are members of the

FIGURE 2.33 Elongated blackened lesions on stem and petiole.

FIGURE 2.34   Dark olivaceous greasy spot on fruit.

kingdom *Chromista* (*Stramenopiles* or *Straminopiles*). Oomycetes belong to one of two orders, *Saprolegniales* and *Peronosporales*. The order *Personosporales* contains *Phytophthora* species and a number of other very important plant-pathogenic genera, including the genus *Pythium*. *P. infestans* which has worldwide distribution, but most severe epidemics occur in areas with frequent cool, moist weather.

2.11.4   Favorable Conditions of Disease Development

Daytime temperatures between 15.5°C and 21°C, night temperatures between 10°C and 15.5°C, and relative humidity near 100% are the ideal conditions for infection and the spread of late blight disease.

2.11.5   Disease Cycle

2.11.5.1   Dissemination

Sporangia or mycelial fragments are dispersed from infected plant organs by winds and/or by splashing raindrops or wind-driven rain.

2.11.5.2   Inoculation

Sporangia or mycelial fragments land on susceptible host organ(s).

2.11.5.3   Infection and Pathogen Development

Sporangia germinate directly via germ tubes and penetrate a plant organ, or sporangia release motile zoospores which in turn encyst on host organs and penetrate the tissues via a penetration peg.

2.11.5.4   Symptom and Disease Development

Mycelium of the pathogen penetrates cell walls directly and ramifies inter-cellularly throughout host tissues, rapidly destroying them and leading to the development of the characteristic necrotic late blight symptoms.

Sexual reproduction is rare in nature; more commonly, asexual reproduction occurs. Sporagiophores bearing asexually produced zoosporangia form on diseased tissues at a relative humidity of 91–100% and a temperature range of approximately 3.3°C–26.1°C with an optimum temperature between 18°C and 22°C.

P. infestans survives in plant debris or on volunteer tomato plants and on perennial weeds such as nightshade. Where both mating types are present (A1 and A2), *P. infestans* generates the thick-walled oospores that are longer-term survival propagules.

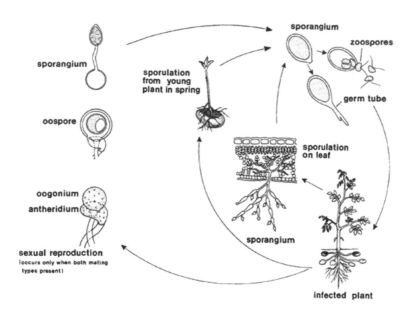

FIGURE 2.35   Disease cycle of *P. infestans.*

Dispersal of the pathogen is by wind, rain, or human-assisted via movement of infested or infected materials such as seed or tools.

Predisposing factors include cool, wet weather and high relative humidity, and large, densely planted crops of tomato (Figure 2.35).

2.11.6   MANAGEMENT

- The riskiest range of months to cultivate tomatoes in blight-prone areas in Hawaii is from November to April. Avoid winter plantings of tomato in areas where prolonged periods of cool, wet, humid weather tend to occur. The pathogen becomes inactive during dry periods.
- Select a tomato variety that reaches maturity quickly (i.e., early-bearing or short-season types). This will allow the grower to get the crop harvested as soon as possible. "Early season" tomato varieties require about 55–67 days to reach maturity; whereas "mid-season" and "late-season" varieties require from about 68–78 days and about 79–85+ days, respectively.
- Grow tomatoes in glasshouses or in spaces where there is humidity control and the plants are covered and protected from rainfall.
- Ensure quick and vigorous tomato seedling growth though adequate and supplemental plant nutrition. Silicate-containing fertilizers may increase pest resistance, especially in the seedling stages.
- Harvest early. If your tomatoes are nearly full size (yet green) and there is a big storm coming and you can use green tomatoes, you might as well harvest them.
- Stake up tomato plants, especially indeterminate types. Keep tomato stems and branches away from the ground.
- Plant blight-resistant tomato varieties when they become available. Late blight-resistant tomatoes are difficult to breed and to find in seed catalogs, and they may not necessarily perform well at your location and may be very susceptible to other diseases.
- Intercrop tomato with non-susceptible host plants, preferably non-solanaceous plants.
- Disperse the tomato plants around the property and avoid planting them all together in one place. Use wide plant spacing to allow air ventilation of the tomato canopy, which allows wet plant surfaces to dry off most rapidly.

- Practice good crop sanitation; inspect the plants regularly for late blight disease symptoms and promptly remove diseased material from the plot or garden (carefully detach diseased leaves, stems of fruits and destroy them).
- Eliminate cull piles in the vicinity of tomato plantings.
- Use disease-free tomato transplants; inspect seedlings for symptoms and destroy diseased plants promptly.
- Protect seedlings with fungicide sprays before transplanting.
- Destroy volunteer tomato plants.
- Avoid moving through the tomato garden or field when plant foliage is wet.
- Rotate crops; avoid successive crops of tomatoes in the same location; avoid planting a new crop of tomatoes beside a diseased crop.
- Do not plant potatoes near tomatoes, or if potatoes are planted, use a blight-resistant variety.
- Orient plant rows parallel to the prevailing winds to allow breezes to move between plant rows, which allow more rapid drying of foliage and reduced relative humidity in the plant canopy.
- Time irrigation to water plants early in the day, rather than late in the day. This allows foliage and soil to dry out before evening, lessening the duration of leaf wetness and lowering relative humidity. Do not spray foliage with water; keep water off leaves and stems (irrigate plants at ground level rather than overhead, except for foliar feeding).
- Control solanaceous weeds around the tomato garden.
- Where possible, apply fungicides on the basis of a weather-based disease forecasting system rather than on a calendar basis.
- Large-scale growers may decide to use a preventive fungicide spray regime, applying products on a calendar basis. It is best to rotate different products to avoid fungicide resistance.
- Other growers may apply fungicides as soon as symptoms are observed.
- Do not purchase diseased tomato seedlings—inspect them for symptoms first.
- Protect seedlings from nutritional stress and other pests. Potassium silicate fertilizer can help to grow seedlings that are more resistant to fungal diseases and some insect pests.

2.12 SOUTHERN BLIGHT ON TOMATO

Southern blight, also known as "southern wilt" and "southern stem rot" is a serious and frequent disease. It is caused by the soilborne fungus *Sclerotium rolfsii* and attacks a number of vegetable crops. The disease usually appears in "hot spots" in fields in early to mid-summer and continues until cooler, dryer weather prevails. Losses may vary from light and sporadic to almost total destruction of the crop.

2.12.1 Causal Organism

Species	Associated Disease Phase	Economic Importance
S. rolfsii	Lesions on Stem Causing Wilting of the Plant	Minor

2.12.2 Symptoms

Southern blight causes a sudden wilting of the foliage, followed by yellowing of the leaves and browning of stems and branches. Wilting and plant death result from a decay of the stem or crown at the soil line. Infected tissues are frequently covered with a white, fan-like fungal mat of mycelium. As the disease progresses, numerous small, round fungal bodies (sclerotia) appear embedded in the fungal mat (Figure 2.36). Initially the sclerotia are white; later becoming light brown, reddish

FIGURE 2.36   Fungal mat on the stem.

brown, or golden brown in color. Each is about the size of a mustard seed. Fruits can also be infected when in contact with soil or infected tissue, leading to a watery or mushy fruit rot with slightly sunken lesions. Sclerotia can also develop on fruit (Figure 2.37).

2.12.3   Favorable Conditions of Disease Development

Southern blight is not a common disease of tomatoes. The disease is favored by high humidity and soil moisture and warm to hot temperatures (29°C–35°C).

2.12.4   Disease Cycle

The fungus overwinters as sclerotia and in host debris in the soil. A characteristic of the fungus is that it is generally restricted to the upper two or three inches of soil and will not survive at deeper depths. In most soils, the fungus does not survive in significant numbers when a host is absent for two years or more. The fungus is more active in warm, wet weather, and it requires the presence of un-decomposed crop residue to initiate infection.

2.12.5   Management

2.12.5.1   Integrated Management

Southern blight is difficult to control because the fungus has a broad host range that includes over 500 plant species, and sclerotia can survive for several years in soil (Aycock, 1966). However, some

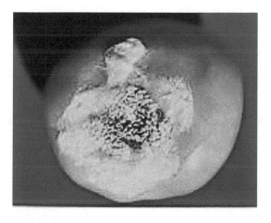

FIGURE 2.37   Formation of sclerotia on infected tomato fruit.

integration of management methods may help to reduce the impact of southern blight during vegetable production.

- Use of pathogen-free transplants and resistant cultivars: Transplant nurseries should be located far from vegetable production fields and avoid excess watering and high temperatures. Vegetable transplants should always be inspected for overall health prior to planting. The use of resistant cultivars is always a preferred method of disease management. Unfortunately, resistance to *S. rolfsii* has not been identified or is limited for many host plant species, and it is currently not a viable option for most vegetables. Resistance has been identified for some hosts. Six tomato breeding lines—5635M, 5707M, 5719M, 5737M, 5876M, and 5913M—were released jointly from Texas A&M University Research Center, Coastal Plain Experiment Station and the University of Georgia (Leeper et al. 1992). Recent attempts to reassess resistance to *S. rolfsii* in peppers confirmed some useful levels of resistance in several pepper species, including in the bell-type cultivar "Golden California Wonder", which is conferred by a single recessive gene (Dukes and Fery, 1984; Fery and Dukes, 2005).
- Crop rotation: Although crop rotation is a traditional and preferred method to control disease, it is not very effective in controlling southern blight because of the broad host range of *S. rolfsii* and the survivability of sclerotia in the soil. Yet, rotating with non-susceptible crops, such as corn or wheat, may help decrease disease incidence in following years by lowering initial inoculum (Mullen, 2001).
- Soil solarization: Sclerotia can be killed in four to six hours at 50°C (122°F) and in three hours at 55°C (131°F) (Ferreira and Boley, 1992). Covering moistened soil with clear polyethylene sheets during the summer season can reduce the number of viable sclerotia if the soil temperature under the sheet remains high enough for an appropriate length of time. While effective for smaller areas, solarization is generally impractical for larger commercial operations. Refer to http://edis.ifas.ufl.edu/in824 for more information about solarization.
- Deep plowing: Deep plowing in the fall or in the spring before bed preparation is another effective method. The ability of sclerotia to germinate is reduced with soil depths greater than 2.5 cm (1 inch). Soil depths of 8 cm (3.1 inch) or greater prevent germination completely due to the mechanical stress created by the soil over the sclerotia (Punja, 1985). However, growers need to be aware that studies of other soilborne pathogens have shown that deep plowing can spread the pathogen, thus changing the distribution of future disease outbreaks (Subbarao, Koike, and Hubbard, 1996).
- Soil amendments: Amending soils with organic fertilizers, biological control agents, and organic amendments—such as compost, oat, corn straw, and cotton gin trash—may help control southern blight. For example, the use of organic amendments, cotton gin trash, and swine manure was found to control southern blight through the improved colonization of soil by antagonistic *Trichoderma* spp. (Bulluck and Ristaino, 2002). Deep plowing the soil combined with applications of certain inorganic fertilizers—like calcium nitrate, urea, or ammonium bicarbonate—was also shown to control southern blight on processing carrots (Punja, 1986). Studies have proposed that the increased nitrogen inhibits sclerotia germination, whereas the increased calcium might alter host susceptibility. However, these approaches have not been tested in the sandy soils of Florida, and earlier studies indicated that the addition of inorganic fertilizers may be less effective in soils prone to leaching.

2.12.5.2 Chemical Control

The use of soil fumigants, such as methyl bromide, chloropicrin, and metam sodium, are the most practical means to treat seed beds and fields for a number of soilborne pathogens, including *S. rolfsii* (Mullen, 2001). They must be applied days to weeks before planting. However, the availability of methyl bromide is limited due to its status as an ozone-depleting material (EPA, 2009).

Preplant fungicides, such as Captan and pentachloronitrobenzene (PCNB), are effective in reducing disease severity. PCNB can effectively limit disease incidence when applied prior to infection and is registered for use on a limited number of vegetable crops. Some commercially available strobilurin fungicides (azoxystrobin, pyraclostrobin, and fluoxastrobin) are also labeled for the control of southern blight on certain vegetables and were found to provide some control of southern blight in peanut production (Culbreath, Brenneman, and Kemerait, 2009; Woodward, Brenneman, and Kemerait, 2007).

2.12.5.3 Biological Control

Some biological agents, such as *Trichoderma harzianum*, *Gliocladium virens*, *Trichoderma viride*, *Bacillus subtilis*, and *Penicillium* spp., were found to antagonize *S. rolfsii* and could suppress the disease. *G. virens* was found to reduce the number of sclerotia in soil to a depth of 30 cm, resulting in a decreased incidence of southern blight on tomato (Ristaino, Perry, and Lumsden, 1991). *Trichoderma koningii* also reduced the number of sclerotia and the plant-to-plant spread of southern blight in tomato fields (Latunde-Dada 1993). However, there is evidence from a greenhouse study that *G. virens* has better biocontrol capability against *S. rolfsii* than *Trichoderma* spp. (Papavizas and Lewis, 1989).

2.13 GRAY MOLD ON TOMATO

Botrytis blight, or gray mold, as it is commonly known, has an exceptionally wide host range with well over 200 reported hosts. The fungus can occur as both a parasite and a saprophyte on the same wide range of hosts. This fungus disease is intriguing in that it can cause a variety of plant diseases including damping off and blights of flowers, fruits, stems, and foliage of many vegetables and ornamentals. It is a major cause of postharvest rot of perishable plant produce, including tomatoes at harvest and in storage. The disease can occur both in the greenhouse and in the field.

2.13.1 CAUSAL ORGANISM

Species	Associated Disease Phase	Economic Importance
Botrytis cinerea	Lesions on Leaves and Stem and Ghost Spot on Fruit	Severe

2.13.2 SYMPTOMS

Stem lesions on seedling tomatoes can occur at, or just below, the soil level. Entry to the stem may occur through senescent cotyledons or damaged tissue. Stem lesions can also occur later during the growth of the crop. Stems can become infected through leaf scars, dead leaves, or any form of stem damage. Stem lesions often partially girdle the stem, but sometimes the whole stem is affected, and the plant is killed. Petiole lesions appear very similar to those on the stem and often result from infection and colonization of a leaflet. Leaflet lesions often start from senescent tissue or any physical or chemical damage. The pathogen can grow along a petiole to the main stem and can eventually form a lesion there. Flower parts that have fallen onto leaves are a common starting point for leaflet colonization. Pollen from flowers and the flower parts can act as a stimulant to *B. cinerea* spores, not only stimulating germination, but also increasing the virulence of the isolate. In the field the fungus appears as a gray, velvety covering of spores on dying flowers and on the calyx of fruit. Senescent flowers are frequently colonized by *Penicillium* spp. (blue mold), and this fungus may be confused with gray mold. Infections spread from flowers and fruit back toward the stem; the stem turns beige to white and develops a canker, which can girdle the entire flower head (Figure 2.38). Immature green fruit turn light brown or white, starting at the point where they touch

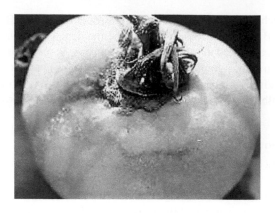

FIGURE 2.38   Velvety covering of spores on dying flowers.

other infected plant parts. A soft rot may develop with the fruit skin remaining intact, but the inner tissue becomes mushy and watery (Figure 2.39). Later, a gray fuzzy mold develops, and sclerotia may appear (Figure 2.37, note under dead calyx or sepals). If this stage occurs in the greenhouse, the floor of the house will be littered with fruit that have fallen off the plant. In the field, the alleyways will be filled with discarded fruit. Green fruit can also become infected directly by airborne spores instead of by contact with other infections. White circular (halo) spots appear on the fruit and have been termed "ghost spots". These spots will persist and can appear on green, breaker, and mature fruit (Figure 2.40). As fruit ripens, the color of the halos changes from white to yellow. The "ghost-spot" symptom results from spore germination and penetration of the fruit, which is only susceptible to attack up to cherry size. As soon as the surface of the fruit is shiny, it is no longer susceptible. Penetration of the mycelium of *Botrytis* into the fruit produces a host reaction preventing any further mycelial growth and results in localization of the pathogen. The halo forms around the point of entry.

2.13.3   CAUSE AND DISEASE DEVELOPMENT

Gray mold is one of the main causes of postharvest rot of fresh market tomatoes, and it occasionally affects processing tomatoes when there is a rain, heavy dew, or fog before harvest. Careful management of irrigation water keeps the disease to a minimum. Gray mold can be a problem in greenhouse-grown tomatoes. It can develop where supporting wires and strings rub against stems, causing a wound.

FIGURE 2.39   Soft rot of the fruit.

FIGURE 2.40   Ghost spots on fruit.

2.13.4   Favorable Conditions of Disease Development

The disease typically begins in cooler weather. Disease development is greatest under moderate temperatures of 18°C–24°C. The disease is favored by sufficient humidity in the canopy and, on tomato, is most severe on plants in acidic, sandy soils with high soil moisture. Dying flowers are the most favorable sites for infection. Infection may also result from direct contact with moist, infested soil or plant debris.

2.13.5   Disease Cycle

The fungus has a very wide host range and spores can be blown from other hosts. The fungus survives saprophytically on leftover plant debris. Small, black resting bodies (called sclerotia) may be produced, particularly in rotten fruit, which allows survival of the fungus during adverse conditions. Conidia from infected tissue are dispersed by wind and by splashing rain. The fungus is considered a weak pathogen that typically enters the plant through wounds or aging tissue.

The fungus overwinters as sclerotia or as mycelium in plant debris and may be seedborne as spores or mycelium in a few crops. Other crops may also serve as sources of the pathogen and are likely to cross-infect. Conidia are airborne and may also be carried on the surface of splashing raindrops. High relative humidities are necessary for prolific spore production. In the field, spores landing on tomato plants germinate and produce an infection when free water from rain, dew, fog, or irrigation occurs on the plant surface.

Optimum temperatures for infection are between 18°C and 24°C (65°F and 75°F), and infection can occur within five hours. High temperatures, above 28°C (82°F), suppress growth and spore production. Dying flowers are a favorable site for infection, but infections can also result from direct contact with moist infested soil or plant debris. In the greenhouse, stem lesions develop either by direct colonization of wounds or through infected leaves. The presence of external nutrients, such as pollen grains in the infection droplet, can markedly increase infection. The type of wound is said to influence stem lesion development; breaking off leaves is reported to give a lower incidence of stem lesions than cutting off leaves with a knife, leaving a stub (Figure 2.41).

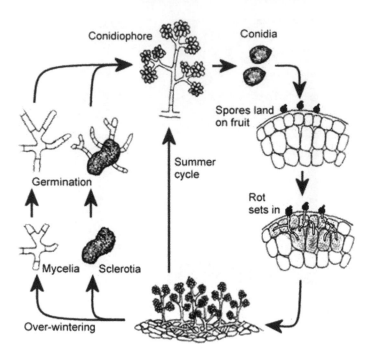

FIGURE 2.41   Disease cycle of *B. cinerea*.

2.13.6   Management

2.13.6.1   Cultural Control
- Since the signs of this fungus may be confused with other saprophytic fungi which colonize dead tissue, the presence of *Botrytis* by microscopic examination should be confirmed.
- Plants should be supplied adequate calcium by liming acidic soils and maintaining uniform soil moisture.
- A calcium-to-phosphorus ratio of 2 or higher in leaf petiole tissue decreases plant susceptibility.

2.13.6.2   Chemical Control
- Chlorothalonil, chlorothalonil plus mefenoxam, pyraclostrobin (suppression only), and boscalid are fungicides labeled for field application of gray mold on tomato.
- Also labeled for use on tomato are Pyrimethanil and *B. subtilis* strain QST 713. Both of these compounds should be applied with an appropriately labeled fungicide.

2.13.6.3   Resistance Management
- Resistant management strategies on fungicide labels (e.g., boscalid and pyraclostrobin) such as tank mixing with another fungicide, rotation of applications, and maximum rate use per application and per season should be followed.

2.14   ROOT KNOT NEMATODE

Root knot nematode (RKN) is a serious malady in tomato. The functional root system is modified into galls and it impairs uptake of water and nutrients. Poor development of root system makes the plant highly susceptible to drought. In addition, RKN in association with pseudomonas leads to bacterial wilt, which in association with soilborne *rhizoctonia* reduce seed germination and increases root rot problem.

RKNs are found worldwide and are known to affect over 2000 species of plants including various vegetable and crop species. There are several different species of RKN including *Meloidogyne javonica*, *M. arenaria*, *M. incognita* (southern RKN), *M. chitwoodi* (the Columbia RKN), and *M. hapla* (the northern RKN). Quarantines have been placed upon some countries and states known to have root knot to protect vegetable seed and crop production. Root knot is associated with other diseases such as crown gall and Fusarium diseases because the RKNs provide entry ways for *Agrobacterium* sp. and *Fusarium* sp. to infect plants as well.

2.14.1 SYMPTOMS AND DAMAGE

The northern RKN produces small, discrete galls while the southern RKNs produce large galls and massive root swellings. Infected plants are stunted, appear yellow or pale green in color, and wilt easily, even when soil moisture is adequate. Severe infestations can dramatically reduce yields and eventually kill plants. Damage from RKN feeding may also increase the incidence of other soilborne diseases such as Fusarium wilt and cause Fusarium wilt-resistant varieties to become susceptible (Figure 2.42).

2.14.2 LIFE CYCLE

The juveniles hatch from eggs, move through the soil and invade roots near the root tip. Occasionally they develop into males but usually become spherical-shaped females.

The presence of developing nematodes in the root stimulates the surrounding tissues to enlarge and produce the galls typical of infection by this nematode. Mature female nematodes then lay 100s of eggs on the root surface, which hatch in warm, moist soil to continue the life cycle.

Continued infection of galled tissue by second and later generations of nematodes causes the massive galls sometimes seen on plants such as tomatoes at the end of the growing season. The length of the life cycle depends on temperature and varies from four to six weeks in summer to ten to fifteen weeks in winter. Consequently, nematode multiplication and the degree of damage are greatest on crops grown from September to May.

Nematodes are basically aquatic animals and require a water film around soil particles before they can move. Also, nematode eggs will not hatch unless there is sufficient moisture in the soil. Thus, soil moisture conditions that are optimum for plant growth are also ideal for the development of RKN (Figure 2.43).

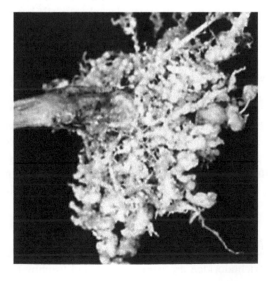

FIGURE 2.42   RKN on tomato roots.

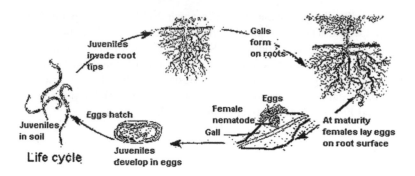

FIGURE 2.43   Life cycle of RKN.

2.14.3 MANAGEMENT

Management of root knot should focus on sanitation measures for preventing contamination of soils, reducing populations below damaging levels where infestations already exist, and variety selection. Sanitation measures include planting nematode-free tomato transplants and avoiding the introduction of nematodes on any other type of transplant stock or with soil. This is difficult in reality because soil clinging to plant roots may contain nematodes without obvious plant symptoms. Equipment and boots should be washed free of soil before working clean ground when moving from areas suspected of harboring nematodes. Strategies for reducing nematode populations include starving nematodes by two-year crop rotations with resistant crops like corn, milo, and nematode-resistant soybean varieties; or with clean (weed-free) fallow. Soil solarization may be effective in some situations, but soil fumigation provides more consistent control of nematode populations. Soil fumigants are restricted-use pesticides that can only be applied by certified applicators. Incorporation of cruciferous green manures such as cabbage, mustard, and rape into soil may also help reduce populations, particularly when combined with solarization. Many root knot resistant tomato varieties are available.

Fertilizers like calcium cyanide, sodium cyanide, and urea cyanamide can release ammonia and NH4, which is poisonous to nematodes. Seed treatment with Aldicarb or Carbofuran (2 gm/Kg) or nursery treatment (2 gm/m²) or seedling root dip (1000 mg/L) for 30 minutes or main field application of chemicals (2 Kg/ha) controls the nematode efficiently.

VAM fungus is proved to reduce the number of nematodes that develop into adults. Fungus *Paecilomyces lilacinus* can attack the eggs of nematodes when applied at 8 gm/plant.

Management of RKN disease infecting tomato, by the use of fungal bioagents *Acremonium strictum* and *T. harzianum* isolated from egg masses of *M. incognita* infecting tomato has been carried out. The rhizosphere and rhizoplane of RKN infested tomato revealed consistent association of *A. strictum*. In the present study *A. strictum* and other fungal bioagents viz. *Aspergillus niger*, *P. lilacinus*, *Rhizoctonia solani*, and *T. harzianum* isolated earlier from egg masses of *M. incognita*, identified and maintained have been investigated through in-vitro and in-vivo trials for their potentiality against *M. incognita*. Out of the above, isolated mycoflora *A. niger* was identified to be toxic against *M. incognita* while *A. strictum* and *T. harzianum* were found to possess both egg parasitic or opportunistic and toxic properties. A field trial with all the above fungal bioagents both alone and together showed significant promising performance by the dual treatment of *A. strictum* and *T. harzianum* in improving the health of the tomato plant with a remarkable reduction in *M. incognita* population.

2.15   STUBBY-ROOT NEMATODE

Nematodes in the family Trichodoridae are commonly called "stubby-root" nematodes, because feeding by these nematodes can cause a stunted or "stubby" appearing root system. *Paratrichodorus minor* is the most common species of stubby-root nematode in Florida, and in tropical and subtropical regions worldwide. *P. minor* is important because of the direct damage it causes to plant roots, and also because it can transmit certain plant viruses.

2.15.1 SYMPTOMS AND DAMAGE

Damage caused by *P. minor* usually occurs in irregularly shaped patches within a given field. Symptoms are usually more severe in sandy than in heavier soils. Aboveground symptoms include stunting, poor stand, wilting, nutrient deficiency, and lodging. Roots may appear abbreviated or "stubby" looking. However, all these symptoms can be caused by other factors, so the only way to verify if *P. minor* is a problem is to have a nematode assay conducted by a credible nematode diagnostic lab.

2.15.2 MANAGEMENT

P. minor is known to occur deeper in the soil than many other plant-parasitic nematodes. In experiments, a large percentage of *P. minor* populations occurred between 8 to 16 inches deep, below the typical treatment zone of soil fumigants. This allows many *P. minor* to escape being killed by fumigant treatments. Population numbers of *P. minor* are known to rebound following soil fumigation to numbers higher than if no treatment were used. Therefore, soil fumigants, while effective treatments for other plant-parasitic nematodes, often are not recommended for management of *P. minor*. Continuous cultivation of highly susceptible crops such as corn or sorghum can build up populations of *P. minor* to damaging numbers and may require the use of nematicides on subsequent crops. Summer legumes such as velvet bean or cowpea tend to keep populations of *P. minor* low and may reduce reliance on nematicides. Systemic nematicides have shown greater effectiveness for management of *P. minor* because the active ingredient is taken up into the plant roots. This protects the plant from *P. minor* while its roots are getting established.

2.15.3 STING NEMATODE ON TOMATO

Sting nematodes are among the most destructive plant-parasitic nematodes on a wide range of plants. Adults can reach lengths greater than 3 mm, making them 1 of the largest plant-parasitic nematodes. While there are several species of sting nematodes described, only *Belonolaimus longicaudatus* Rau is known to cause widespread crop damage.

2.15.4 SYMPTOMS

Plants damaged by sting nematodes often wilt, may be stunted, and may show symptoms of nutrient deficiency. Seedlings may sprout from the soil and then cease growing altogether. Plant death may occur with high population densities of sting nematodes. All of these symptoms may be caused by a number of plant diseases and disorders. Therefore, the only way to be certain whether sting nematodes are a problem is to have a soil nematode assay conducted by a credible diagnostic facility.

2.16 FUSARIUM CROWN AND ROOT ROT

Fusarium crown and root rot (FCRR) of tomato are caused by the fungus *Fusarium oxysporum f.* sp. *radicis-lycopersici* (FORL). It is a sporadic problem. Infected plants have a water-soaked appearance at the base, and roots are usually discolored or rotting. Crown and root rot is favored by cool temperatures (10°C to 20°C/50°F to 68°F). Low soil pH, ammoniacal nitrogen, and waterlogged soil also exacerbate the disease. FCRR has occurred in Canada, Mexico, Israel, Japan, many countries in Europe, and states in United States including Florida, California, New Jersey, New York, New

Hampshire, Ohio, Pennsylvania, and Texas. It is a serious problem for seedling and greenhouse fruit production and can cause significant yield decreases in field-grown, staked tomatoes.

2.16.1 CAUSAL ORGANISM

Species	Associated Disease Phase	Economic Importance
F. oxysporum f. sp. *radicis-lycopersici*	Crown Rot and Root Rot	Severe

2.16.2 SYMPTOMS

Wide ranges of symptoms are associated with FORL. The fungus invades susceptible plants through wounds and natural openings created by newly emerging roots. Early symptoms caused by FORL in tomato seedlings include stunting, yellowing, and premature abscission of cotyledons and lower leaves. A pronounced brown lesion that girdles the hypocotyls, root rot, wilting, and seedling death are advanced symptoms. Typically, the first symptom in the mature plants is a yellowing along the margins of the oldest leaves when the first fruit is at or near maturity. Yellowing is soon followed by necrosis and collapse of the leaf petiole. Symptom development progresses slowly upward on successively younger leaves. Some plants may be stunted and wilt quickly and wither. Older plants may wilt slowly and still be alive at the end of the harvest. Wilting first occurs during the warmest part of the day and plants appear to recover at night. Infected plants may be stunted, totally wilt and die, or persist in a weakened state, producing reduced numbers of inferior fruit.

As the disease progresses dry, brown lesions develop in the cortex of the tap or main lateral roots and taproot often rots away. Chocolate brown lesions develop at or near the soil line and extend into the vascular system. When diseased plants are sectioned lengthwise, extensive brown discoloration and rot are evident in the cortex of the crown and roots. This brown vascular discoloration typically does not extend more than 25–30 cm above the soil line, which helps to distinguish this disease from Fusarium wilt, where discoloration may extend 1 m high (Figure 2.44). Stem cankers may develop at or above the soil line. Following rains and during fogs, the pink sporulation of the pathogen can be profuse on exposed necrotic lesions. The fungus produces masses of white mycelium and yellow to orange spores in necrotic stem lesions on dead and dying plants.

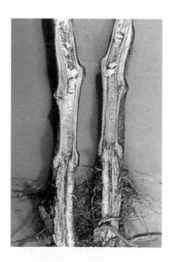

FIGURE 2.44   Fusarium crown rot: Vascular discoloration.

2.16.3 CAUSE AND DISEASE DEVELOPMENT

The fungus apparently enters plants soon after seeding or transplanting. Generally, the first indication that a plant is affected by the disease occurs when the first fruits are nearing maturity. At this time some plants lose turgor and die quickly. Rapidly wilting plants generally occur only when first fruit are maturing. When the basal vascular region is exposed, necrosis of the vascular tissue is observed. Immediately before the onset of symptoms, affected plants are as large as healthy plants, and have as much fruit load. Other plants go into a "slow wilt" at about the same time as the first plants with the rapid wilt are seen. "Slow wilt" is characterized by slow general yellowing and eventual browning of the margins of the older lower leaves, followed by loss of some of these leaves. The yellowing and browning of leaves progresses upward to younger leaves. Some plants with slow wilt eventually die, but in most cases these plants survive and have a new flush of growth when fruits have been harvested.

2.16.4 FAVORABLE CONDITIONS OF DISEASE DEVELOPMENT

Crown and root rot is favored by cool temperatures (10°C to 20°C/50°F to 68°F). Low soil pH, ammoniacal nitrogen, and waterlogged soil also exacerbate the disease.

2.16.5 DISEASE CYCLE

F. oxysporum is an abundant and active saprophyte in soil and organic matter, with some specific forms that are plant pathogenic.[6] Its saprophytic ability enables it to survive in the soil between crop cycles in infected plant debris. The fungus can survive either as mycelium or as any of its three different spore types.[7]

Healthy plants can become infected by *F. oxysporum* if the soil in which they are growing is contaminated with the fungus. The fungus can invade the plant either with its sporangial germ tube or mycelium by invading the plant's roots. The roots can be infected directly through the root tips, through wounds in the roots, or at the formation point of lateral roots (Agrios, 1988). Once inside the plant, the mycelium grows through the root cortex intercellularly. When the mycelium reaches the xylem, it invades the vessels through the xylem's pits. At this point, the mycelium remains in the vessels, where it usually advances upward toward the stem and crown of the plant. As it grows, the mycelium branches and produces microconidia, which are carried upward within the vessel by way of the plant's sap stream. When the microconidia germinate, the mycelium can penetrate the upper wall of the xylem vessel, enabling more microconidia to be produced in the next vessel. The fungus can also advance laterally as the mycelium penetrates the adjacent xylem vessels through the xylem pits (Agrios, 1988).

Due to the growth of the fungus within the plant's vascular tissue, the plant's water supply is greatly affected. This lack of water induces the leaves' stomata to close, the leaves wilt, and the plant eventually dies. It is at this point that the fungus invades the plant's parenchymatous tissue until it finally reaches the surface of the dead tissue, where it sporulates abundantly (Agrios, 1988). The resulting spores can then be used as new inoculum for further spread of the fungus.

2.16.6 MANAGEMENT

2.16.6.1 Cultural control

- Aggressive sanitation programs are very important, starting with a very clean house by cleaning or disinfecting. The worse cases of this disease have been associated with attempts to reuse items without sanitizing them, especially items that come in direct contact with the soil mix. Crown rot is likely to occur with a higher frequency where direct seeding is used, instead of healthy transplants and where the soil contains high levels of chloride salts. The utmost sanitation production scheme for transplants in greenhouses should be used so that individuals or equipment used within or around the transplant site do not become contaminated with disease-causing organisms from the field. Finally, transplants should be transported, pulled, and set without tissue damage, as damaged tissues are likely to be sites for infection. Transplanting should be done when soil or media is 20°C or above. The selection and application of fertilizers can significantly influence disease development. For example, increasing soil pH and minimizing the use of ammoniacal nitrogen help in controlling FCRR. Dead tomato plants need to be completely removed.

- One of the most important components in an integrated disease control program is the selection and planting of cultivars that are resistant to pathogens. The term resistance usually describes the plant host's ability to suppress or retard the activity and progress of a pathogenic agent, which results in the absence or reduction of symptoms. Some resistant varieties are available for FCRR for mostly greenhouse production. The use of crown rot resistant cultivars is increasing but is currently not widely accepted due to horticultural characteristics that make these varieties less competitive than standard varieties. The following are some cultivars with resistance: Trend, Trust, Medallion, Match, Switch, and Blitz for greenhouse production; Charleston and Conquest for field production. FCRR resistance is conferred by a single dominant gene and already has been incorporated into commercially available cultivars.

- Crop rotation is a historical method of crop production that reduces soil pest problems by removing susceptible plants from an infested area for a period of time long enough to reduce pest populations to tolerable levels. Rotation away from tomatoes may be necessary in fields with a recurring crown rot problem. Avoid rotation with susceptible crops, such as eggplant or peppers; use non-hosts, such as lettuce instead. Crop rotation as a control strategy may be limited in controlling FCRR because the fungus can survive in soil for many years. Capital field improvements such as irrigation systems and the availability of suitable land also limit adoption of crop rotation as a pest control strategy. Once a grower has invested in an irrigation system for a piece of land, the grower is less likely to rotate to a lower value crop.

- Another cultural method in controlling FCRR is soil solarization. Solarization is a non-chemical soil disinfestation method, first developed for soilborne disease control in Israel and California during the 1970s. Soil solarization is defined as heating the soil by solar energy resulting in both physical and biological processes to control pathogens and other soil pests. Solarization depends on solar energy to heat the soil to temperatures that are lethal to pathogenic organisms. This is accomplished by covering moist soil with a clear plastic film or mulch during a two- to eight-week period with plentiful solar radiation. Most soilborne pests and plant pathogens are mesophilic and are killed at temperatures between 40°C and 60°C. At these elevated temperatures, dysfunction of membranes and increased respiration are responsible for death. However, death depends on the thermal dose, a product of temperature and exposure time. Exposure to long periods of sub-lethal temperatures may effectively control diseases by reducing the ability of propagules to germinate, thus increasing the susceptibility to biological control organisms and decreasing the ability to infect the host. Soil solarization has been demonstrated to control FCRR. In studies with different solarization methods, soil solarization reduced populations of FCRR down to a depth of 5 cm.

- Crown rot incidence was significantly reduced by Metam Sodium (29%), solarization + Metam Sodium (51%) and by Methyl bromide+chloropicrin (50%), while disease severity was significantly reduced (74%) by both the latter two treatments. No significant differences in marketable yield were observed among the treatments. Preliminary studies carried out in the open field showed that 12 days of soil solarization reduced survival of FORL propagules significantly. The effectiveness of pathogen control was improved by combing solarization with manure or extending the solarization treatment to 27 days. In a closed greenhouse, solarization and biofumigation with bovine manure proved effective in reducing the viability of FORL chlamydospores, reducing disease incidence, and in increasing commercial yield.

- Grafting also is used to control of crown and root rot of tomato. Resistant rootstocks provide excellent control of many tomato soilborne pathogens and particularly *Fusarium oxysporum f.* sp. *lycopersici.*, FORL, *P. lycopersici*, and *Meloidogyne* spp. This technique, which initially was considered too expensive, is now widely used at a commercial level in many Mediterranean countries and North America. In general, without grafting, the tomato plant density per hectare is about 18,000 plants. When grafted plants are used, the same yield could be obtained with 1/2 the plant population (9,000 plants ha-1). In addition to controlling some soilborne pathogens, tomato grafting promotes growth, increases yield, increases plant tolerance to low temperature, extends the growth period and improves fruit quality. Susceptible tomato plants grafted onto FCRR-tolerant hybrid rootstock (He-man), even cropped in a severe FORL infested soil, remained healthy during the growing season and gave a profitable yield.

2.16.6.2   Chemical Control

- Chemical control of FCRR in steam-sterilized soil by using a captafol drench proved effective in preventing re-infestation by airborne FORL conidia. Mihuta-Grimm reported that the application of benomyl at 0.090 g a.i./L on a 21-day schedule to plants growing on rockwool productions slabs resulted in optimum FCRR control. Yield from infected transplants treated with benomyl, however, was not significantly different from that of control plants. Other candidate fungicides proved to be phytotoxic at levels needed to control FCRR. Although fungicides such as benomyl or captafol have been demonstrated to be effective, they have some drawbacks. They are expensive, cause environmental pollution, and may induce pathogen resistance. Fungicides added to seeds can also cause stunting and chlorosis of young seedlings, and results may vary as fungicides are absorbed or inactivated by components of the soil or planting medium. Captafol is no longer labeled for usage, thus it cannot be used commercially.

- Fumigation with methyl bromide (MBr) + chloropicrin formulations have been the most commonly used preplant practice for control of FCRR in tomatoes. Application of MBr + chloropicrin significantly reduces the incidence and severity of the disease. Mbr: chloropicrin (67:33, by volume) reduced populations of FORL to a depth of 35 cm. However, FCRR incidence is still very high. Even with the use of MBr as a preplant fumigant, epidemics of FCRR have occurred in commercial production fields. MBr is a powerful soil fumigant providing effective control of a wide range of soilborne pathogens and pests, including fungi, bacteria, nematodes, insects, mites, weeds, and parasitic plants. Despite these major advantages, the use of MBr has been associated with major problems, including the depletion of the ozone layer. Because of this, its production and use will be phased out on a worldwide scale, by 2005 in the United States and European Union and other developed countries and by 2015 in the developing countries. An estimated 22.2 million Kgs of MBr are applied annually for preplant soil fumigation in the United States. Many strawberry and tomato growers have depended on MBr for control of nematodes, weeds, and fungal pathogens. The ban on MBr production and use has prompted the study of new chemical alternatives for the control of soilborne pests. However, these materials tend to provide a narrower spectrum of control than MBr, have less predictable efficacy, and may have their own problems with environmental pollution and safety. Five soil fumigants (1,3-dichloropropene, chloropicrin, dozamet, fosthiazate, and sodium methyldithiocarbamat, a contact nematicide and several combinations with pebulate herbicide were compared to MBr/chloropicrin (98 and 2%, respectively) for control of nutsedge, Fusarium wilt and crown rot and nematodes in tomato. Fusarium crown rot was reduced by MBr and 1,3 dichloro-propene + chloropicrin in the spring, but in the fall all chemical treatments, except those containing SMDC, provided better crown rot control than MBr. It had been reported that crown rot incidence (defined as a percentage number of plants infected by the pathogen) was significantly reduced by Metam Sodium (29%), solarization + Metam Sodium (51%), and by MBr + chloropicrin (50%), while disease severity (defined as the percentage of crown discoloration) was significantly reduced (74%) by both of the latter two treatments.[8] No significant differences in marketable yield were observed among the treatments.

- A fresh market tomato study comparing metam sodium and MBr fumigation to an untreated control reported that yields and fruit quality obtained with metam sodium were equivalent to those achieved with MBr fumigation. Metam sodium has been demonstrated to significantly reduce crown rot incidence and when combined with solarization, control was equivalent to MBr + chloropicrin. Metam sodium could reduce Fusarium crown rot only when thoroughly incorporated in the planting bed, such as through application to the soil prior to bed formation.

- Plantpro45TM, a new "low risk" iodine-based compound, was investigated as a potential alternative in controlling FCRR. Plantpro45TM provided significant control of Fusarium crown rot of tomato in naturally infested fields. Under greenhouse conditions, soil drench with Plantpro 45TM at 80 ppm a.i. followed by planting 21 days later and a foliar application at 80 ppm one week after planting increased root and shoot weight and improved root condition of tomato when grown in field soil naturally infested with FORL. Final disease incidence ratings revealed that plots pretreated with Plantpro45 were comparable to MBr for control of FCRR.

2.16.6.3 Biological Control

- Research has demonstrated that biological control of FCRR has been successful in some instances under greenhouse and field conditions. The fungus *Trichoderma*, a natural soil-inhabiting genus, has been used successfully to control Fusarium crown rot and root rot of tomatoes. The potential of *T. harzianum*, *Aspergillus ochraceus Wilhelm*, and *Penicillium funiculosum Thom* in controlling FCRR of field-grown tomatoes was shown. Sivan and Chet used *T. harzianum* in combination with soil sterilization and reduced rates of MBr to obtained significant control of tomato crown and root rot in the field with transplants colonized by *T. harzianum* during greenhouse propagation.

- Datnoff et al. conducted field experiments in Florida to evaluate commercial formulation of two fungi, *T. harzianum* and *Glomus intraradices* Schenck & Smith, for control of FCRR of tomato[8]. Compared to untreated controls, significant reduction in disease incidence was obtained with treatment by biocontrol agents. The interaction between *G. intradices* and FORL and effect of *G. intradices* on tomato plants were investigated. Caron et al. reported that tomato crown and root rot was decreased with *G. intradices*. However, there was no growth response of tomato plants to inoculation with the biocontrol agent. *T. harzianum* applied, as a peat-bran preparation to the rooting medium at the time tomatoes were transplanted. Such an application resulted in significant decrease in Fusarium crown rot through the growing season in field conditions. Yield increased as much as 26.2% over the controls in response to the treatment. Nemec et al. evaluated some biocontrol agents; *T. harzianum*, *G. intraradices*, and *Streptomyces griseverdis*, for controlling root diseases of vegetable crops and citrus. At the end of the study, they found that all biological control agents reduced FCRR in tomato in the field. *T. harzianum* and *B. subtilis* were the most effective biocontrol agents. *Paenibacillus macerans* and *T. harzianum* were evaluated for promoting plant growth and suppressing FCRR under fumigated and non-fumigated field conditions. *T. harzianum* and *P. macerans* significantly reduced the severity of FCRR. *T. harzianum* reduced the severity of FCRR by 12% and *P. macerans* by 9% in comparison to the untreated control in the non-fumigated treatments. No differences were observed between the biologicals and the untreated control in the MBr treated plots. Datnoff and Pernezny also reported that *T. harzianum* and *P. macerans* alone or in combination, significantly increased the growth of tomato transplants in the greenhouse and after outplanting into the field 30 days later.

- The potential of *T. harzianum* as a biocontrol agent in a soilless culture system was investigated with tomato plants infected with FORL. The application of *Trichoderma* reduced the incidence and spread of FCRR in tomatoes on an artificial growing medium. Marois and Mitchell reported that in greenhouse and growth chamber experiments, the fungal biocontrol amendment significantly reduced the mean lesion length and the incidence of FCRR. Under greenhouse conditions, the incidence of crown rot of tomato was reduced by up to 80% 75 days after sowing when *T. harzianum* T35 was applied as either a seed coating or a wheat bran-peat preparation applied to soil.

- Much research has been done on the potential of nonpathogenic *F. oxysporum* for control of FCRR. Louter and Edgington and Brammall and Higgens used isolates of avirulent *F. oxysporum* and isolates of *F. solani*, respectively, to reduce the effects of FORL on tomato plants. It was suggested that the fungi acted through either cross protection (Louter and Edmington, 1985) or competition for infection court sites.

- Alabouvette and Couteadier studied the efficacy of nonpathogenic *F. oxysporum* strain Fo47 and the fungicide himexazol for control of the root diseases of tomato in the greenhouse. Both treatments gave a good control of FCRR; the yield was slightly higher with the biological treatment and the cost of the biocontrol treatment was less than the cost of the chemical product. Under greenhouse conditions, control of FCRR of tomato can be achieved by introduction of either strain Fo47, or fluorescent *Pseudomonas* strain C7, or by the association of both into the growing medium. four nonpathogenic isolates of *F. oxysporum* (26B, 43A, 43AN1, and 43AN2) and an isolate of *T. harzianum* (Th2) were found to be effective in protecting tomato seedlings from FCRR. However, the *T. harzianum* isolate was less effective than the *F. oxysporum* isolate at reducing disease.

- There are also some bacteria, especially *Pseudomonas* spp., that have been shown to be effective in controlling FCRR. *Pseudomonas fluorescens* strain CHA0 suppressed crown and root rot of tomato. M'Piga et al. reported that *P. fluorescens* colonizes and grows in the outer root tissues of whole tomato plants and sensitizes them to respond rapidly and efficiently to FORL attack in addition to exhibiting an antimicrobial activity in the plant.

- *Streptomyces griseoviridis* strain K61 (MycostopTM) has been tested against *F. oxysporum*-induced crown rot in Israel and in the United Kingdom. A clear reduction of the disease was observed, but complete control was not achieved by using MycostopTM. MycostopTM is a live formulated strain of the bacterium *S. griseoviridis* that was discovered in Finnish peat. It is labeled for use on greenhouse tomato and is available from at least two suppliers in the United States. *Streptomyces* sp. Di-944, a rhizobacterium from tomato, suppressed *Rhizoctonia* damping off and Fusarium root rot in plug transplants when applied to seeds or added to the potting medium. Antibiosis was suspected as a key mechanism of biocontrol.

- Among the most promising bioactive oligosaccharides is chitosan (poly-N-glucosamine), a mostly deacetylated derivative of chitin occurring in the cell walls of several fungi, which is readily extracted from the chitin of crustacean shell wastes. Oligomers of chitosan, which are likely to be released by the action of plant encoded-chitinase from walls of invading fungi, can protect tomato roots against FORL when applied to the seed or roots. Chitosan, derived from crab-shell chitin, was applied as seed coating and substrate amendment prior to infection with the fungus FORL. Experiments were performed either on a mixture of peat, perlite, and vermiculite, or on bacto-agar in petri dishes. In both cases, a combination of seed coating and substrate amendment was found to significantly reduce disease incidence.

- The potential of *Bacillus pumilus* strain SE 34, either alone or in combination with chitosan, for inducing defense reactions in tomato plants inoculated with the vascular fungus FORL, was studied by light and transmission electron microscopy. Treatment of the roots with *B. pumilus* alone or in combination with chitosan prior to inoculation with FORL, substantially reduced symptom severity of FCRR as compared with untreated controls. Although some small, brownish lesions could be occasionally seen on the lateral roots, their frequency and severity never reached levels similar to those observed in control plants.
- Chitosan, oligandrin, and crude glucan, isolated from the mycoparasite *P. oligandrum*, were applied to decapitated tomato plants and evaluated for their potential to induce defense mechanisms in root tissues infected by FORL. A significant decrease in disease incidence was monitored in oligandrin- and chitosan-treated plants as compared to water-treated plants, whereas glucans from *P. oligandrum* cell walls failed to induce a resistance response. In root tissues from oligandrin-treated plants, restriction of fungal growth to the outer root tissues, decrease in pathogen viability, and formation of aggregated deposits, which often accumulated at the surface of invading hyphae, were the most striking features of the reaction. In chitosan-treated plants, the main response was the formation of enlarged wall appositions at sites of attempted penetration of the reaction.
- Lettuce residue soil amendments and lettuce intercropping were considered for biological control. Co-planted lettuce and *T. harzianum* strain Th2 provided protection from naturally occurring FCRR in a commercial tomato crop.

2.16.6.4 Integrated Management

At present, FCRR is difficult to control in field-grown tomatoes because the pathogen rapidly colonizes sterilized soil and persists for long periods. However, an integration of the following management procedures may help to reduce the impact of crown and root rot:

- Use disease-free transplants. Transplant houses should not be located near tomato production fields. Avoid overwatering, which makes the transplants more susceptible to crown and root rot. Disinfect transplant trays by steaming before reuse.
- Use a preplant fumigant. The soil should be of good tilth and adequately moist for at least two weeks prior to fumigation. Use an appropriate chisel spacing and depth and immediately cover the bed with plastic mulch following fumigation.
- Optimize cultural practices in the field. Avoid injuring transplants when they are set in the field. Physical damage and injury from excessive soluble salts may make young plants more susceptible to crown and root rot. The use of water drawn from wells rather than ditches for watering of transplants may help to prevent recontamination of fumigated soil. Avoid ammoniacal nitrogen and maintain the soil pH at 6–7. Rapidly plow in crop debris following final harvest. Disinfest tomato stakes before reuse or use new stakes.
- Rotate with a non-susceptible crop. Incomplete knowledge of the host range of FORL makes precise recommendations in this area difficult. Current research data suggests that leguminous crops should be avoided in favor of corn and similar crops. Rotation and intercropping with lettuce had reduced FORL in greenhouse-grown tomatoes.
- Significant progress has been made in breeding for resistance to FCRR in field-grown tomatoes. Although the commonly used commercial cultivars do not have resistance, some resistant cultivars, such as Conquest, are available for field use and Trust for greenhouse use.
- Additional management strategies under investigation include the use of biological control, cover crops, and soil solarization alone or in combination with fumigants.

2.17   FUSARIUM WILT OF TOMATO

If your tomato plants yellow and wilt on one side of the plant or one side of a leaf, they may have Fusarium wilt. Fusarium wilt on tomatoes is caused by *Fusarium oxysporum* sp. *lycopersici*. It is a soil-born fungus that is found throughout the world, especially in warm regions. The organism is specific for tomato and is very long lived in all regions. The disease develops more quickly in soils that are high in nitrogen and low in potassium. In addition, plants grown in sandy soils tend to contract this disease more often. The fungus works its way up through the plant's roots, clogging water-conducting tissue in the stem. This prevents water from reaching branches and leaves, starving the plant. Affected plants produce very few tomatoes. Often, the entire plant dies.

2.17.1   CAUSAL ORGANISM

Species	Associated Disease Phase	Economic Importance
F. oxysporum sp. *lycopersici*	NA	NA

2.17.2   SYMPTOMS

The first symptom of the disease is a slight wilting of the plants. Fusarium wilt symptoms also include strong downward bending of petioles, yellowing, wilting, and dying of the lower leaves, often on 1 side of the plant. These symptoms may appear on successively younger leaves with one or more branches being affected and others remaining healthy. Root necrosis is often extensive. After a few weeks, browning of the vascular system may be seen by cutting the stem open with a knife. This brown discoloration inside the stem may extend from the roots of the plant to the top. Plant growth is stunted, and under warm conditions the plant may die (Figure 2.45).

2.17.3   CAUSE AND DISEASE DEVELOPMENT

Fusarium may be introduced to soils in several ways: old crop residues, transplants, wind, water, implement-borne soils, or mulches. This fungus becomes established readily in most soils and can remain in the soil for years. When susceptible tomatoes are planted in infested soil, their roots are also subject to attack by these fungi. The disease is much more serious when accompanied by RKN.

2.17.4   FAVORABLE CONDITIONS OF DISEASE DEVELOPMENT

Disease development is favored by warm soil temperatures, and symptoms are most prevalent when temperatures range from 27°C–32°C.

FIGURE 2.45   Tomato plants killed by fusarium wilt.

2.17.5 DISEASE CYCLE

Fusarium fungi survive in the soil or associated with plant debris for up to ten years. Disease development is favored by warm soil temperatures, and symptoms are most prevalent when temperatures range from 27°C–32°C. The fungi enter the plants through their roots and are then spread throughout the plant by the plant's water-conducting vessels. (Refer to Disease Cycle of FCRR.)

2.17.6 MANAGEMENT

2.17.6.1 Integrated Pest Management Strategies

- Plant resistant varieties. These varieties are labeled VF and include cultivars such as "Spring Giant", "Burpee VF", "Supersonic", "Celebrity", "Manalucie", "Better Boy", and "Small Fry".
- Remove infected plants from the garden. Removal of infected plants will help limit the disease's spread. Soil sterilization or fumigation will eliminate wilt fungi from the soil but are impractical for home gardeners. Soil replacement should be considered.
- Avoid over-application of high nitrogen fertilizers. High soil nitrogen levels accompanied by low potassium levels can increase susceptibility to the fungus. Use a soil test to determine potassium levels and other nutrient deficiencies.
- Avoid activity in wet plantings. Movement of wet soil from place to place via shoes or tools will spread the disease.
- Sanitize stakes and tomato cages at the end of the season. Avoid using soil-encrusted tools and supports season after season. A thorough cleaning with water will reduce most risks of transmitting the disease.
- For four years, do not plant solanaceous plants in the area where infection occurred. Tomato, potato, pepper, and eggplant are all susceptible to the disease and may allow its survival year after year in the same planting area.

2.18 VERTICILLIUM WILT OF TOMATO

Verticillium wilt is a soilborne fungal disease that results in the yellowing, and eventual browning and death of foliage, particularly in branches closest to the soil. The wilt starts as yellow, V-shaped areas that narrow at the leaf margins. These yellow areas grow over time, turn brown, and then the leaf dies. Often, entire branches are infected.[9]

2.18.1 CAUSAL ORGANISM

Species	Associated Disease Phase	Economic Importance
Verticillium albo-atrum Or *Verticillium dahliae*	Wilting of Leaves and Abnormal Fruits	Severe

2.18.2 SYMPTOMS

In spite of the name verticillium wilt, a true wilt seldom occurs in tomato, at least not until late in the season. Rather, under good conditions of moisture and nutrition, yellow blotches on the lower leaves may be the first symptoms, then brown veins appear (Figure 2.46), and finally chocolate brown dead spots. Often, no symptoms are seen until the plant is bearing heavily or a dry period occurs. The bottom leaves become pale, then the tips and edges die and leaves finally die and drop off. V-shaped lesions at leaf tips are typical of verticillium wilt of tomato (Figure 2.47). The spots

FIGURE 2.46   Yellow blotches on the lower leaves.

FIGURE 2.47   Typical V-shaped lesions on tomato leaves associated with verticillium wilt.

may be confused with *Alternaria* early blight, but they are not definite, nor do they develop concentric bulls-eye rings.

The leaves may wilt, die, and drop off. The disease symptoms progress up the stem, and the plant becomes stunted. Only the top leaves stay green. Fruits remain small, develop yellow shoulders, and may sunburn because of loss of leaves.

Infection takes place directly when the fungus threads enter the root hair. It is aided in its entrance if rootlets are broken or nematodes have fed on the root system. The fungus grows rapidly up the xylem, or sap-conducting channels. Its activity there results in interference with the normal upward movement of water and nutrients. The fungus produces a toxin that contributes to the wilting and spotting of the leaves. Diagnosis involves making a vertical slice of the main stem just above the soil line and observing a brown color in the conducting tissues under the bark. This discoloration can be traced upward as well as downward into the roots. In contrast to Fusarium wilt, verticillium wilt discoloration seldom extends more than ten to twelve inches above the soil, even though its toxins may progress farther.

2.18.3   FAVORABLE CONDITIONS OF DISEASE DEVELOPMENT

The pathogen is sensitive to soil moisture and temperature. Tomatoes must have at least a day of saturated soil before infection occurs. Soil temperatures must be moderate or cool for infection to take place: 24°C (75°F) is optimum with 13°C (55°F) minimum and 30°C (86°F) maximum.

2.18.4   DISEASE CYCLE

Verticillium wilt is the result of a fungus called *V. albo-atrum* and is present in most soils which tend to be cool for long periods of time. It stays alive season after season by living on the dying underground parts of infected plants. It often attacks and multiplies on the roots of common weeds such as ragweed and cocklebur.

2.18.5 MANAGEMENT

Management of this disease is difficult since the pathogen survives in the soil and can infect many species of plants. As with many diseases, no single management strategy will solve the problem. Rather, a combination of methods should be used to decrease its effects. When a positive diagnosis has been made, the following recommendations may be followed:

- Whenever possible, plant resistant varieties. There are many verticillium-resistant varieties of tomatoes. These are labeled "V" for verticillium-resistance. Note: Verticillium-resistant plants may still develop verticillium wilt if there is a high population of nematodes in the soil.
- Remove and destroy any infested plant material to prevent the fungi from overwintering in the debris and creating new infections.
- Keep plants healthy by watering and fertilizing as needed.
- Fields should be kept weed-free since many weeds are hosts for the pathogen.
- Susceptible crops can be rotated with non-hosts such as cereals and grasses, although four to six years may be required since the fungi can survive for long periods in the soil.
- Soil fumigants are effective in reducing disease severity but are not recommended for use in gardening.

By far the most feasible and economic control is the use of verticillium-tolerant tomato cultivars of which there are many with varying maturities and excellent horticultural qualities. These include the following:

New Yorker (V)	Earlirouge	Basket Vee
Springset	Supersteak	Campbell 17
Pic Red	Campbell 1327	Big Set
Jet Star	Fireball (V)	Setmore
Supersonic	Beefmaster	Small Fry
Heinz 1350	Better Boy	Terrific
Heinz 1439	Bonus	Big Girl
Westover	Gardener (V)	Mainpak
Royal Flush	Monte Carlo	Early Cascade
Floramerica	Nova (Paste)	Jumbo
Veebrite	Crimson Vee (Paste)	Wonder Boy
Veemore	Veeroma (Paste)	Rutgers 39
Veegan	Veepick (Paste)	Ultra-Boy
Veeset	Ramapo	Ultra-Girl
Burpee VF Hyb.	Moreton Hyb.	Rushmore
Starshot	Spring Giant	Jetfire

2.19 POWDERY MILDEW OF TOMATO

Tomato powdery mildew may be caused by 3 pathogens worldwide. *Leveillula taurica* (*Oidiopsis taurica*) is a pathogen of a wide range of host species in warm arid to semiarid climates in Asia, the Mediterranean, Africa, and more recently, the southwest United States. *Erysiphe orontii* (*E. cichoracearum* and *E. polyphaga*) is another species common to many host plants in both temperate and tropical regions. However, the recent disease outbreak reported in eastern North America was due to a distinct fungus *Oidium lycopersicum*, based on characteristics including appressoria shape, conidia, and conidiophore morphology.

2.19.1 CAUSAL ORGANISM

Species	Associated Disease Phase	Economic Importance
L. taurica or E. orontii or O. lycopersicum	Leaf Chlorosis, Premature Senescence, Reduction in Fruit Size and Quality	Severe

2.19.2 SYMPTOMS

Powdery mildew of tomato is caused by the fungus *O. lycopersicum or E. orontii or L. taurica*. Symptoms first appear as light green to bright yellow spots on the upper surface of the leaf (Figure 2.48). These spots usually don't have very distinct margins and gradually become more noticeable as they develop the white, powdery appearance typical of powdery mildews (Figure 2.49). However, this is where this disease differs from most other powdery mildews that we encounter. The powdery mildew of tomato is apparently much more aggressive than other mildews. Once leaves are infected, they quickly brown and shrivel on the plant. This rapid death of infected leaves and defoliation of plants is not typical of most mildews. The fungus is readily spread to nearby leaves or plants since abundant, powdery spores are produced and are easily carried by air currents or production activities.

FIGURE 2.48   Light green to bright yellow spots on the upper surface of the leaf.

FIGURE 2.49   White, powdery appearance typical of powdery mildew.

2.19.3 FAVORABLE CONDITIONS OF DISEASE DEVELOPMENT

The conditions that favor the development of disease include relative humidity levels greater than 50% (optimum RH > 90%) and temperatures ranging from 10°C–35°C (50°F–95°F) (best below 86°F/30°C). Unlike the situation with most other fungal diseases, free water on leaf surfaces is not necessary for infection. Moderate temperatures (15.5°C–27°C) and shady conditions generally are the most favorable for powdery mildew development. Spores and fungal growth are sensitive to extreme heat (above 32°C) and direct sunlight.

2.19.4 DISEASE CYCLE

The fungus does not overwinter outdoors on tomato transplants grown in Nevada or other areas of the Southwest where the fungus overwinters. Locally-grown transplants, if grown in a greenhouse free of the disease, are the best assurance of a healthy crop. Since powdery mildew spores are airborne and are blown north with prevailing wind currents, most plantings have low levels of disease by late summer. These late-season infections do not result in significant losses.

Most powdery mildew fungi grow as thin layers of mycelium (fungal tissue) on the surface of the affected plant part. Spores, which are the primary means of dispersal, make up the bulk of the white, powdery growth visible on the plant's surface and are produced in chains that can be seen with a hand lens; in contrast, spores of downy mildew grow on branched stalks that look like tiny trees.

Powdery mildew spores are carried by wind to new hosts. Although humidity requirements for germination vary, all powdery mildew species can germinate and infect in the absence of free water. In fact, spores of some powdery mildew fungi are killed, and germination is inhibited by water on plant surfaces for extended periods.

2.19.5 MANAGEMENT

The use of healthy, locally grown plants may eliminate the need for applying a fungicide later in the year. Sulfur sprays or dusts are effective if complete coverage applications start early and are repeated every seven to ten days. Plant damage by ground equipment can be reduced by leaving space at planting time for wheel tracks. Flowable sulfur may provide better protection than wettable sulfur because of the highly adhesive property and because it stays in suspension longer, thus preventing obstruction of spray nozzles.

When using transplants from areas where the fungus overwinters, applications of a fungicide should begin before any symptoms are apparent, usually in the first ten days of July. Visual inspection of the lower leaves should be made on a weekly basis beginning in late June. When symptoms are seen, the fungicide program should be closely followed. Thorough coverage is critical for effective control, and repeated applications are necessary. The only fungicide currently registered for use on tomatoes is sulfur. Fungicides are usually not necessary when using disease-free transplants.

Sulfur may cause phytotoxicity (burning) if applied when temperatures exceed 35°C. Therefore, make applications during cool weather and in the evening. Some sulfur labels indicate that applications should not be made within 40 days of harvest if the tomatoes are to be canned in metal containers. Check with your buyer to determine if this precaution needs to be followed.

The following list of fungicides are not specifically labeled for control of powdery mildew of tomato. However, it is legal to follow the directions suggested for control of tomato russet mite.

2.19.5.1 Sulfur

Wettable powder—Wilbur Ellis Red Top Spray Sulfur 97%, 6–10 lbs/A; FMC-Kolo spray 81.25%, 7 lbs/A, full coverage;

Home use: Lilly Miller Sulfur or Ortho Flotox 90%. Dust—FMC-Kolo dust or Wilbur Ellis Redtop Ben-sul 60%, 40–55 lbs/A, full coverage;

Home use: Lilly Miller Sulfur or Ortho Flotox 90%, dust to cover.
Flowable—1/2–1 gal/A full coverage.
Stauffer Magnetic Six; Stoller-That or That Big 8 (64%)
Dry Flowable 80%—Sandoz Thiolux, 3–10 lbs/A.

REFERENCES

1. Miller S., and Huang R., *Bacterial Canker of Tomato.* https://www.lincolnu.edu/c/document_library/get_file?uuid=585aefcc-b271-46d3-878e-0ab895232094&groupId=145912.
2. Sikora E. J., 2011. *Virus Diseases of Tomato,* Extension Plant Pathologist, Associate Professor, Plant Pathology, Auburn University: Auburn, AL, June 2011.
3. Nelson S. C., 2008. *Late Blight of Tomato (Phytophthora infestans),* Honolulu, Hawai'i: Department of Plant and Environmental Protection Sciences, CTAHR University of Hawai'i at Mänoa, August 2008.
4. Reddy P. P., 2014. *Biointensive Integrated Pest Management in Horticultural Ecosystems,* Springer Publications: New York, NY, pp. 83.
5. Smith, I. M., Dunez J., Phillips D. H., Lelliott R. A., and Archer S. A., eds. 1988. *European Handbook of Plant Diseases*, Oxford: Blackwell Scientific Publications, pp. 583.
6. Agrios, G. N., 1988. *Plant Pathology*, 3rd ed. New York: Academic Press, Inc., pp. 803.
7. McGovern, R. J., and Datnoff L. E., 1992. Fusarium crown and root rot: Reevaluation of management strategies. Proceedings of the Florida Tomato Institute, (FTI'1992), Gainesville, FL: University of Florida-IFAS, pp. 75–86.
8. Ozbay N., and Newman S. E., 2004. Fusarium crown and root rot of tomato and control methods. *Plant Pathology Journal*, 3: 9–18.
9. Aycoc k R., 1966. Stem rot and other diseases caused by *Sclerotium rolfsii. Tech. Bull.*, 174. Raleigh, NC: North Carolina State University Agricultural Experiment Station.
10. Leeper, P. W., Phatak, S. C., and George, B. F., 1992. Southern blight-resistant tomato breeding lines: 5635M, 5707M, 5719M, 5737M, 5876M, and 5913M. *Hortscience* 7: 475–478.
11. Smith, I. M., Dunez J, Phillips D. H., Lelliott R. A., and Archer S. A., eds. European handbook of plant diseases, Oxford, United Kingdom: Blackwell Scientific Publications. pp., 583 1988.
12. Louter, J. H., and Edgington, L. V., 1985. Cross protection of greenhouse tomato against fusarium crown and root rot. *J. Plant Pathol.*, 7: 445–446.

3 Citrus

The large citrus fruits of today evolved originally from small, edible berries over millions of years. Citrus plants are native to subtropical and tropical regions of Asia and the Malay Archipelago, and they were initially domesticated in these areas. The generic name "citrus", originated from Latin and it is referred to the plant now known as Citron (*C. medica*). These plants are large shrubs or small to moderate-sized trees, reaching 5–15 m (16–49 ft) tall. They have spiny shoots and alternately arranged evergreen leaves with an entire margin. The world's largest citrus-producing countries are Brazil, China, the United States, Mexico, India, and Spain.

Citrus plants are prone to infestation by aphids, whitefly, and scale insects. Also, the viral infections of citrus are a major concern for which some of these ectoparasites serve as vectors. The newest threat to citrus groves in the United States is the Asian citrus psyllid. In this chapter, detailed information about citrus diseases and their preventive measures is discussed.

3.1 *ALTERNARIA* BROWN SPOT OF CITRUS

Alternaria alternata is an opportunistic pathogen on over 380 hosts. In citrus, it especially affects mandarins. Alternaria brown spot first appeared in Florida 30 years ago. Currently, brown spot is also known to occur in South Africa, Turkey, Israel, Spain, and Colombia. It has become a severe problem on some varieties in recent years. Grapefruit can also be affected, although it is not a common commercial problem in that species.

3.1.1 CAUSAL ORGANISM

Species	Associated Disease Phase	Economic Importance
A. alternata	Rot of Leaves, Twigs, and Fruits	Severe

3.1.2 SYMPTOMS

Alternaria brown spot attacks leaves, twigs, and young fruit. On young leaves, lesions first appear as small brown to black spots, which soon become surrounded by yellow halos (Figure 3.1). A fungal-specific toxin that is responsible for much of the necrosis produces the halo. The spots on young leaves can appear as early as 36–48 hours after infection. The toxin is sometimes translocated in the vascular system, producing chlorosis and necrosis that extend out along the veins from the lesions. Spots enlarge as the leaves mature and if the disease is severe the leaves may drop, or the entire shoot may die. Young shoots are also attacked, usually producing lesions 1–10 mm in diameter. Shoot infection and abscission of infected, young leaves produce dieback of twigs.[1]

Fruitlets may be infected soon after petal fall, and even a small lesion causes immediate abscission. The lesion is then surrounded by yellow halo (Figure 3.2).

On more mature fruit, symptoms vary from small, dark specks to large, black lesions on the peel. Fruits are susceptible for at least three months after petal fall. Even after that time, some fruits may fall as the result of earlier infections. Symptoms of this disease are sometimes confused with those of anthracnose.

3.1.3 FAVORABLE CONDITIONS FOR DISEASE DEVELOPMENT

Spore production is greatest at relative humidity of above 85%. Spores are airborne and release into the air is triggered by rainfall or by a sharp change in relative humidity. Once the spores

FIGURE 3.1   Small brown to black spot surrounded by yellow halo.

are released, they are moved by the wind to susceptible tissue where they can infect. When temperatures are favorable (20°C–29°C (68°F–83°F)), the length of the wetting period required for infection is about eight to ten hours. When temperatures drop below 17°C or above 32°C (63°F and 90°F, respectively), the fungus requires extended leaf wetness duration (>24 hours) to cause significant infections.

FIGURE 3.2   Lesions on fruit surrounded by yellow halo.

3.1.4 DISEASE CYCLE

Spores of the fungus are thick-walled, multicellular, and pigmented and thus tolerate adverse conditions well.

Spores are produced primarily on old lesions on mature leaves that remain on the tree as well as those which have fallen to the ground, but they are not produced on fruit. Spores are airborne and carried by winds. Rain events or sudden changes in relative humidity trigger spore release.

The length of the wetting period required for infection is about eight to ten hours when the temperature is favorable (20°C–29°C). At temperatures of less than 17°C, extended periods of leaf wetness (greater than 24 hours) are needed before much infection occurs. Most of the infection probably follows rains, but dew is often sufficient for infection (Figure 3.3).

3.1.5 MANAGEMENT

There are many management practices that are helpful in reducing disease severity.

- In new planting of susceptible varieties, use disease-free nursery stock.
- Trees grown in greenhouses without overhead irrigation are usually free of Alternaria. If foliage remains dry, the disease never develops.
- Selection of appropriate planting sites: Choose a location with good air circulation and wind movement. Avoid foggy areas.
- Increase the spacing between trees and pruning tree skirts.
- Avoid excessive vegetative growth: Control over-fertilization and over-watering (Table 3.1).

3.2 CITRUS BLACK SPOT

Citrus black spot is one of the most important diseases in major citrus production areas of the world, such as Asia, South America, South Africa, and Australia. This disease is primarily important as a pre-harvest disease and causes severe lesions on the rind which significantly decreases the fruit quality and its marketability.

Late maturing acid lime, mandarins, and grapefruit are the most susceptible varieties.

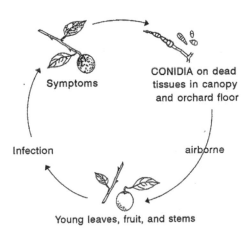

FIGURE 3.3 Disease cycle of *A. alternata* causing alternaria brown spot.

3.2.1 CAUSAL ORGANISM

Species	Associated Disease Phase	Economic Importance
Phyllosticta citricarpa (Asexual Stage)	Lesions on Rind	Severe
Guignardia citricarpa (Sexual Stage)		

3.2.2 SYMPTOMS

There are four main fruit symptom types:

3.2.2.1 Hard Spot

- This is the most common diagnostic symptom (Figure 3.4).
- The lesions are small, round, sunken with tan centers and with a brick-red to chocolate brown margin.
- Green halos are often seen around the hard spot lesions.
- Fungal structures appear as slightly elevated black dots.
- Lesions are most prevalent on the side of fruit that receives the most exposure to sunlight.

3.2.2.2 False Melanose

- Also known as Speckled Blotch.
- These are observed as numerous small, slightly raised lesions that can be tan to dark brown (Figure 3.5).
- The slightly raised lesions are 1–3 mm in diameter.

TABLE 3.1
Recommended Chemical Controls

Pesticide	FRAC MOA[a]	Mature Trees Rate/Acre[b]
Copper fungicide	M1	Use label rate.
Ferbam Granuflo	M3	5–7.5 lb.
Abound 2.08 F[c]	11	12.0–15.5 fl oz. Do not apply more than 92.3 fl oz/acre/season for all uses.
Gem 25 WG[c]	11	4.0–8.0 oz. Do not apply more than 32 oz/acre/season for all uses.
Gem 500 SC[c]	11	1.9–3.8 fl oz. Do not apply more than 15.2 fl oz/acre/season for all uses.
Headline[c]	11	12–15 fl oz. Do not apply more than 54 fl oz/acre/season for all uses.
Pristine[d]	7/11	16–18.5 oz Do not apply more than 74 oz/acre/season for all uses.
Quadris Top[d]	11/3	15.4 fl oz. Do not apply more than 61.5 fl oz/acre/season for all uses. Do not apply more than 0.5 lb ai/acre/season of difenconazole. Do not apply more than 1.5 lb ai/acre/season of azoxystrobin.

Note: This table is from PP-147, one of a series of the Plant Pathology Department, Florida Cooperative Extension Service, Institute of Food and Agricultural Sciences, University of Florida. Original publication date December 1995. Revised February 2012.

[a] Mode of action class for citrus pesticides from the Fungicide Resistance Action Committee (FRAC) 2011. Refer to ENY624, Pesticide Resistance and Resistance Management, in the 2012 Florida Citrus Pest Management Guide for more details.

[b] Lower rates can be used on smaller trees. Do not use less than minimum label rate.

[c] Do not use more than 4 applications of strobilurin fungicides/season. Do not make more than 2 sequential applications of strobilurin fungicides.

[d] Do not make more than 4 applications of Pristine or Quadris Top/season. Do not make more than 2 sequential applications of Pristine or Quadris Top before alternating to a non-strobilurin, SDHI or DMI fungicide.

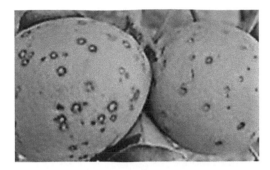

FIGURE 3.4   Hard spots.

- It may occur on green fruit and does not have pycnidia (fungal structures).
- False melanose may become hard spot later in the season.

3.2.2.3   Cracked Spot

- Cracked spot has large, flat, dark brown lesions with raised cracks on their surface.
- It is thought to be caused by an interaction between the pathogen and rust mites.
- It occurs on green as well as mature fruit and can become hard spot later in the season.

3.2.2.4   Virulent Spot

- Lesions are sunken and irregular in shape and occur on heavily infected, mature fruit toward the end of the season.
- In high humidity, large numbers of pycnidia may develop.
- The lesions can turn brown to black with a leathery texture that eventually covers the entire fruit (Figure 3.6).
- Virulent spot may cause premature fruit to drop and serious post-harvest losses since the symptoms may extend into the fleshy part of the fruit.

3.2.2.5   Symptoms on Leaves and Stem

- Leaf and stem symptoms although not as common as fruit symptoms can occur when there is insufficient disease control on any cultivar.

FIGURE 3.5   False melanose (speckle spot).

FIGURE 3.6   Early virulent form of citrus black spot.

- They are most commonly found on acid lime, a very susceptible species.
- Lesions begin as small reddish-brown lesions that are slightly raised.
- With age, they become round sunken necrotic spots with gray centers and prominent margins that are brick-red to chocolate brown.

3.2.3   CAUSE AND DISEASE DEVELOPMENT

Wind-borne ascospores are forcibly ejected from fungal fruiting bodies embedded in leaves in the leaf litter under trees and are carried by air currents, approximately 75 ft (25 m) from the leaf litter. Rain splash may also move spores from fruit infected with conidia and/or leaf litter (conidia and ascospores), but only moves the spores a few inches. Live leaves that have latent infections (infections that are not visible) are common means of long-distance spread. These often are moved as trash in loads of fruit. Infected nursery stock is another potential means of spread. This can occur very easily since these latent infections cannot be seen in otherwise healthy looking trees. Leaf litter movement may be either by wind or human activities. Humans are the main form of long-distance movement.[2]

3.2.4   FAVORABLE CONDITIONS OF DISEASE DEVELOPMENT

Moist climate favors the development of the disease. Spores are released only when the leaf litter is wetted by heavy dew, rainfall, or irrigation.

3.2.5   DISEASE CYCLE

As with many diseases, timing is important for black spot to occur. Inoculum in the leaf litter needs to be available during the period when the host is susceptible, and the environment is favorable for infection. Fruits are susceptible from fruit set until five to six months later when they become age resistant. Both the ascospores (sexual spores) and the conidia (asexual spores) of *G. citricarpa* are able to infect susceptible tissues. Ascospores are found in microscopic fungal structures embedded in the leaf litter. They are the most important source of inoculum, in some regions causing nearly all infections. Ascospores have never been found in fruit lesions or lesions on attached leaves. Spores are released when the leaf litter is wetted by heavy dew, rainfall, or irrigation and can be carried by air currents over long distances. Dark brown or black pycnidia, structures that produce conidia, are formed on fruit, fruit pedicles, and leaf lesions. They are also abundant on dead leaves. Conidia are not wind-borne but may reach susceptible fruit by rain splash. These spores are not considered a significant source of inoculum in climates with dry summers; however, in climates with frequent summer rains, conidia play a larger role in the epidemic when there are multiple fruit ages present on trees simultaneously. Often late hanging fruit with lesions remain on the tree and spores can be washed onto young susceptible fruit.

Infections are latent until the fruit becomes fully grown or mature. At this point the fungus may grow further into the rind producing black spot symptoms months after infection, often near or after harvest. Symptom development is increased in high light intensity, intensifying temperatures, drought, and low tree vigor (Figure 3.7).

3.2.6 MANAGEMENT

- Always plant clean, certified nursery stock. Keeping nursery stock clean is much easier with the new covered nursery regulations but black spot is still a threat. This will help prevent movement of black spot and other diseases into newly established grove plantings.[2]
- Increase airflow in the grove to reduce leaf wetness where possible. *G. citricarpa* needs 24–48 hours of leaf wetness for spore germination and infection as do many other fungal diseases.[2]
- Reduce leaf litter on grove floor to decrease ascospore load through enhanced micro-sprinkler irrigation.[2]
- Fungicides registered for citrus in Florida that have been found effective in other countries: Copper products (all formulations have been found to be equivalent).
- The best fungicide application method is with an airblast sprayer. Aerial applications are not likely to get adequate canopy penetration for control. It is important that the leaves and fruit are covered with fungicide.[2]
- For enhanced coverage, increase the gallons used to 250 gal/A for applications to ensure full coverage (Table 3.2).[2]

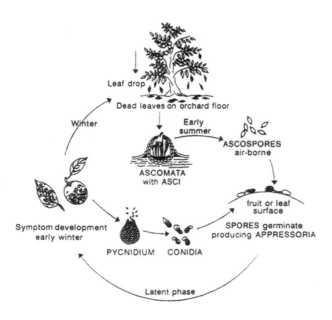

FIGURE 3.7 Disease cycle of *G. citricarpa*.

TABLE 3.2

Recommended Chemical Controls for Citrus Black Spot

Pesticide	FRAC MOA[a]	Mature Trees Rate/Acre[b]
Copper fungicide	M1	Use label rate.
Abound 2.08F[c]	11	12.4–15.4 fl oz. Do not apply more than 92.3 fl oz/acre/season for all uses. Best applied with petroleum oil.
Gem 25WG[c]	11	4.0–8.0 oz. Do not apply more than 32 oz/acre/season for all uses.
Gem 500 SC[c]	11	1.9–3.8 fl oz. Do not apply more than 15.2 fl oz/acre/season for all uses. Best applied with petroleum oil.
Headline[c]	11	9–12 fl oz. Do not apply more than 54 fl oz/acre/season for all uses. Best applied with petroleum oil.

[a] Mode of action class for citrus pesticides from the Fungicide Resistance Action Committee (FRAC) 20111. Refer to ENY-624, "Pesticide Resistance and Resistance Management," in the *2012 Florida Citrus Pest Management Guide* for more details.

[b] Lower rates can be used on smaller trees. Do not use less than minimum label rate.

[c] Do not use more than 4 applications of strobilurin fungicides/season. Do not make more than 2 sequential applications of strobilurin fungicides.

3.3 GREASY SPOT OF CITRUS

Greasy spot is a major foliar and fruit disease on citrus. It causes premature leaf drop beginning in the fall and continues through winter and spring. As a result, yields of the following crop are reduced. Furthermore, cold damage has been observed to be more severe on severely defoliated trees. Rind blemish from this disease causes a downgrading of fruit intended for the fresh fruit market and this can be particularly severe on grapefruit.

3.3.1 CAUSAL ORGANISM

Species	Associated Disease Phase	Economic Importance
Mycosphaerella citri	Greasy Spot on Leaves and Rind Blotch on Fruit	Severe

3.3.2 SYMPTOMS

The first symptom to appear is a small yellowish blister on the underside of the leaf. This is matched by a yellow mottle on the upper surface of the leaf. Later, infected areas of the leaf turn dark brown and become greasy in appearance (Figure 3.8). Leaf drop of infected leaves is common.

On the fruit, symptoms may take several months to appear after infection. They take the form of brown specks in the skin, where cells have died (Figure 3.9).

On most cultivars, the specks are too small to cause a significant blemish, but coloring may be delayed in areas around them. This results in unsightly patches of green on the ripe fruit. In grapefruit, the lesions are larger and often form large, speckled patches. The lesions are pink at first, and later turn brown.

3.3.3 CAUSE AND DISEASE DEVELOPMENT

Spores are produced in decomposing fallen leaves and are released when the leaves become wet. Germination of the spores requires high temperatures and high humidity.

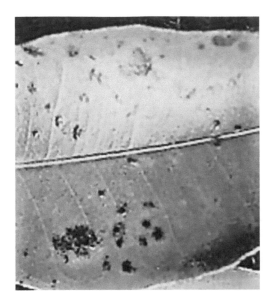

FIGURE 3.8   Typical leaf symptoms of greasy spot infection.

3.3.4   FAVORABLE CONDITIONS OF DISEASE DEVELOPMENT

High relative humidity and high temperatures are required for spore germination and the subsequent fungal growth on the leaf surface. Optimum germination and growth occur between 25°C (77°F) to 30°C (86°F) in the presence of free water or near 100% relative humidity.

3.3.5   DISEASE CYCLE

No sexual fruiting structures are produced in greasy spot lesions on living leaves. Pseudothecia on decomposing leaves are aggregated, have papillate ostioles, and measure up to 90 μm in diameter. Ascospores are slightly fusiform, have a single septum, and often contain two oil globules in each cell. They are hyaline and measure 2 to 3×6 to 12 μm. Ascospores of *M. citri* are produced in pseudothecia in decomposing leaf litter on the grove floor (Figure 3.10). Once mature, ascospores are forcibly ejected following wetting of the litter and subsequently dispersed by air currents. Ascospores

FIGURE 3.9   Greasy spot: Rind blotch on grapefruit.

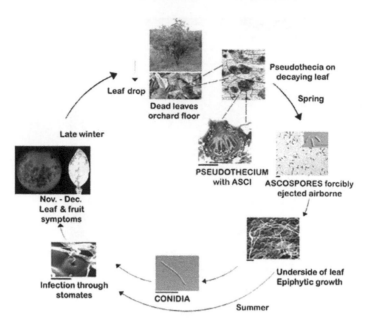

FIGURE 3.10   Disease cycle of greasy spot caused by *M. citri*.

deposited on the underside of the leaf germinate and form epiphytic mycelia. Development of the epiphytic growth requires high temperatures and extended periods of high humidity or free moisture. Appressoria form over stomata, and the fungus penetrates into the mesophyll of the leaf. Nearly all infections occur through the lower leaf surface since citrus leaves have stomata only on the abaxial side. Numerous penetrations are required for the development of macroscopic symptoms. Colonization of the leaf is very slow, and symptoms appear only after 45–60 days, even on highly susceptible species under optimal conditions. Undergrove conditions, infection of leaves occurs mostly in the summer rainy season and symptoms develop in late fall or winter. Symptoms develop more rapidly when winter temperatures are warm. Symptomatic leaves abscise prematurely, and most of the greasy spot—induced leaf drop—occurs in late winter and early spring. The conidial stage *Stenella citri-grisea*, is found in nature only on epiphytic mycelium in late summer. Conidia are not believed to play an important role in disease development.

3.3.6   MANAGEMENT

Greasy spot is relatively easy to control with well-timed sprays of many products. The epiphytic growth on the leaf surface is exposed and readily killed by many materials.

The products that have been traditionally used for control of greasy spot are copper fungicides and petroleum oils.

- Copper fungicides directly kill germinating ascospores and epiphytic mycelium and prevent infections. The activity of oil for control of greasy spot has been investigated but is still not well understood. Oil does not appear to inhibit ascospore germination or germ tube growth but does prevent leaf penetration. It also slows fungal development in the mesophyll and symptom development.
- Petroleum oils are widely used for control of diseases caused by *Mycosphaerella* spp., such as yellow Sigatoka disease of banana, but their activity is not well understood in those cases either. Petroleum oil controls foliar infection but has not been highly effective for rind blotch control. However, we have found that the higher viscosity oils used in more recent years also effectively control rind blotch.

- Dithiocarbamate fungicides have also been used in the past, but they have a relatively short residual activity and are not highly effective.
- When benomyl was introduced, it was widely used and very effective for foliar and fruit symptoms. However, resistance developed quickly, and none of the Benzimidazole fungicides are currently recommended for greasy spot control.
- Fenbuconazole and the strobilurin fungicides are quite effective for control of foliar disease and rind blotch and are currently recommended in Florida, along with petroleum oils and/or copper fungicides.
- Foliar fertilizers, especially those containing heavy metals such as zinc, manganese, and iron, are quite effective for greasy spot control if applied at sufficiently high rates. Other nutritional products, fish oils, and biocontrol agents have all shown at least some activity against greasy spot.
- Acaricides are also active on greasy spot. It appears that they act directly rather than indirectly through action on mites that aggravate disease since acaricides are effective in the absence of mites. It appears that any product that is toxic to epiphytic mycelium will reduce greasy spot severity if applied at the proper time.
- Most fungicides have no effect, but benzimidazoles delay the formation of pseudothecia. Benomyl substantially reduces inoculum production, but only for four to six weeks (Table 3.3).

3.4 ALTERNARIA ROT OF CITRUS

Alternaria rot, also called black rot or navel rot, is caused by the fungus *Alternaria* spp. It is most common in navel oranges, Minneola and Orlando tangelos, and occasionally in lemons and limes. The fungus grows on dead citrus tissue during wet weather. It produces airborne spores which can land and grow on the blossom end of the fruit. Premature fruit coloring and fruit drop are commonly associated with infection.

3.4.1 CAUSAL ORGANISM

Species	Associated Disease Phase	Economic Importance
Alternaria citri	Stem-End Rot on Fruit	Minor

3.4.2 SYMPTOMS

Alternaria rot occurs primarily as a stem-end rot on fruit stored for a long time, but sometimes the decay develops at the stylar end of fruit in the orchard where it may cause premature fruit drop. Fruits infected with *A. citri* change color prematurely and may develop a light brown to black firm spot on the rind at or near the stylar end (Figure 3.11). Some fruits, however, show no external evidence of infection and must be cut to reveal center rot.

3.4.3 FAVORABLE CONDITIONS FOR DISEASE DEVELOPMENT

Alternaria rot is often associated with cold damage. Frost damaged tissue is more likely to be infected (Figure 3.12). The rot inside the fruit may not be evident from the outside.

3.4.4 DISEASE CYCLE

A. citri grows saprophytically on dead citrus tissue and produces airborne conidia. Initially, it establishes a quiescent infection in the button or stylar end of the fruit. Entrance to the fruit is facilitated

TABLE 3.3

Recommended Chemical Controls for Greasy Spot

Pesticide	FRAC MOA[a]	Mature Trees Rate/Acre[b]
Petroleum Oil 97+% (FC 435-66, FC 455-88, or 470 oil)	NR[c]	5–10 gal. Do not apply when temperatures exceed 34°C. 470 weight oil has not been evaluated for effects on fruit coloring or ripening. These oils are more likely to be phytotoxic than lighter oils.
Copper Fungicide	M1	Use label rate.
Copper Fungicide + Petroleum Oil 97+% (FC 435-66, FC 455-88, or 470 oil)	M1 and NR	Use label rate + 5 gal. Do not apply when temperatures exceed 34°C. 470 weight oil has not been evaluated for effects on fruit coloring or ripening. These oils are more likely to be phytotoxic than lighter oils.
Abound 2.08F[d]	11	12.40–15.45 fl oz. Do not apply more than 92.3 fl oz/acre/season for all uses. Best applied with petroleum oil.
Enable 2F	3	8 fl oz. Do not apply more than 3 times per year; no more than 24 fl oz./acre. Minimum retreatment interval is 21 days.
Gem 25WG[d]	11	4.0–8.0 oz. Do not apply more than 32 oz/acre/season for all uses.
Gem 500 SC[d]	11	1.9–3.8 fl oz. Do not apply more than 15.2 fl oz/acre/season for all uses. Best applied with petroleum oil.
Headline[d]	11	9–12 fl oz. Do not apply more than 54 fl oz/acre/season for all uses. Best applied with petroleum oil.
Quadris Top[d]	11/3	10–15.4 fl oz. Do not apply more than 61.5 fl oz/acre/season for all uses. Do not apply more than 0.5 lb ai/acre/season difenconazole. Do not apply more than 1.5 lb ai/acre/season azoxystrobin.

Note: This document is PP-144, one of a series of the Plant Pathology Department, Florida Cooperative Extension Service, Institute of Food and Agricultural Sciences, University of Florida. Original publication date December 1995. Revised February 2012.

[a] Mode of action class for citrus pesticides from the Fungicide Resistance Action Committee (FRAC) 2011. Refer to ENY624, Pesticide Resistance and Resistance Management, in the *2012 Florida Citrus Pest Management Guide* for more details.

[b] Lower rates can be used on smaller trees. Do not use less than minimum label rate.

[c] No resistance potential exists for these products.

[d] Do not use more than 4 applications of strobilurin-containing fungicides/season. Do not make more than 2 sequential applications of strobilurin fungicides.

FIGURE 3.11   Alternaria infection causes rotten spots that may eventually cover as much as 1/4 of the fruit.

FIGURE 3.12   Alternaria rot inside the fruit after freeze damage.

if growth cracks form at the stylar end. The fungus does not grow from the button into the fruit until the button becomes senescent.

Alternaria rot is more likely to occur if the fruit has been weakened by adverse conditions in the field, during storage, or is overmature.

3.4.5   MANAGEMENT

Healthy, excellent quality fruit are more resistant to Alternaria rot than stressed or damaged fruits, especially oranges with split navels. Preventing stress can reduce the incidence of splitting and Alternaria rot.

Postharvest treatments with imazalil, 2,4-D, or both have provided some control. The growth regulator 2,4-D delays senescence and thereby restricts colonization of the host.[3]

3.5   CITRUS BROWN ROT

Citrus brown rot (CBR) is one of the most common diseases of Citrus. Brown rot is the most common fruit rot observed in the orchard.

3.5.1   CAUSAL ORGANISM

Species	Associated Disease Phase	Economic Importance
Phytophthora spp. • *P. citrophthora* • *P. nicotianae* • *P. syringae* • *P. hibernalis*	Decay of the Fruit	Severe

3.5.2   SYMPTOMS

Symptoms appear primarily on mature or nearly mature fruit. Initially, the firm, leathery lesions have a water-soaked appearance, but they soon turn soft and have a tan to olive-brown color (Figure 3.13) and a pungent odor. Infected fruit eventually drops. Occasionally, twigs, leaves, and blossoms are

FIGURE 3.13   Tan to olive-brown lesion on citrus.

infected, turning brown and dying. At a high humidity, fruit becomes covered by a delicate white growth of the fungus (Figure 3.14).

The most serious aspect: Fruit infected before harvest may not show symptoms. If infected fruit gets mixed with healthy fruit, the disease may spread quickly from fruit to fruit in storage and during transit.

3.5.3   CAUSE AND DISEASE DEVELOPMENT

Brown rot is caused multiple species of *Phytophthora* when conditions are cool and wet. Brown rot develops mainly on fruit growing near the ground when *Phytophthora* spores from the soil are splashed onto the tree skirts during rainstorms. Infections develop under continued wet conditions. Fruit in the early stage of the disease may go unnoticed at harvest and infect other fruit during storage.

FIGURE 3.14   Fruit covered with delicate white fungus.

3.5.4 Favorable Conditions of Disease Development

Phytophthora brown rot is a problem usually associated with restricted air and/or water drainage. It commonly appears from mid-August through October following periods of extended high rainfall. It can be confused with fruit drop from other causes at that time of the year. If caused by *P. nicotianae*, brown rot is limited to the lower third of the canopy because the fungus is splashed onto fruit from the soil. *P. palmivora* produces airborne sporangia and can affect fruit throughout the canopy.

3.5.5 Disease Cycle

For the disease to develop, 18 hours of wetness is required for sporangia production and zoospore release. Three hours of wetness is required for infection. The length of the continuous rainy period is the most important predictor of brown rot epidemics. Zoospores produced in sporangia on the ground may be splashed up onto low-hanging fruit. Thus, brown rot mainly develops on fruit growing near the ground (Figure 3.15).[4]

3.5.6 Management

3.5.6.1 Cultural Practices

- Pruning tree skirts can significantly reduce brown rot.
- Sprinkler irrigation: Water should be directed away from the tree canopy.

3.5.6.2 Fungicidal Protection

Rates for pesticides are given as the maximum amount required to treat mature citrus trees unless otherwise noted. To treat smaller trees with commercial application equipment including handguns, mix the per acre rate for mature trees in 250 gallons of water. Calibrate and arrange nozzles to deliver thorough distribution and treat as many acres as this volume of spray allows (Table 3.4).

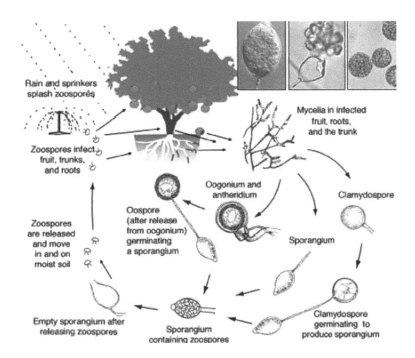

FIGURE 3.15 Disease cycle of *Phytophthora* spp. On citrus causing citrus brown rot.

TABLE 3.4

Recommended Chemical Controls for Brown Rot of Fruit

Pesticide	FRAC MOA[a]	Mature Trees Rate/Acre[b]
Aliette WDG	33	5 lb
Phostrol	33	4.5 pints
ProPhyt	33	4 pints
copper fungicide	M1	Use label rate.

Note: This table is from PP-148, one of a series of the Plant Pathology Department, Florida Cooperative Extension Service, Institute of Food and Agricultural Sciences, University of Florida. Original publication date December 1995. Revised February 2012.

[a] Mode of action class for citrus pesticides from the Fungicide Resistance Action Committee (FRAC) 2011. Refer to ENY-624, Pesticide Resistance and Resistance Management, in the *2012 Florida Citrus Pest Management Guide* for more details.

[b] Lower rates may be used on smaller trees. Do not use less than minimum label rate.

3.6 PHYTOPHTHORA FOOT ROT AND ROOT ROT OF CITRUS

Foot rot results from scion infection near the ground level, producing bark lesions that extend down to the bud union on resistant rootstocks. Crown rot results from bark infection below the soil line when susceptible rootstocks are used. Root rot occurs when the cortex of fibrous roots is infected, turns soft, and appears water-soaked. Phytophthora foot rot is also known as gummosis.

3.6.1 CAUSAL ORGANISM

Species	Associated Disease Phase	Economic Importance
P. nicotianae or *P. palmivora*	Foot Rot and Root Rot	Severe

3.6.2 SYMPTOMS

3.6.2.1 Foot Rot/Gummosis

An early symptom of phytophthora foot rot or phytophthora gummosis is sap oozing from small cracks in the infected bark, giving the tree a bleeding appearance. The gumming may be washed off during heavy rain. The bark stays firm, dries, and eventually cracks and sloughs off. Lesions spread around the circumference of the trunk, slowly girdling the tree. Decline may occur rapidly within a year, especially under conditions favorable for disease development, or may occur over several years (Figure 3.16).

3.6.2.2 Root Rot

Phytophthora root rot causes a slow decline of the tree. The leaves turn light green or yellow and may drop, depending on the amount of infection. The disease destroys the feeder roots of susceptible rootstocks. The pathogen infects the root cortex, which turns soft and separates from the stele, becomes somewhat discolored, and appears water-soaked. If the destruction of feeder roots occurs faster than their regeneration, the uptake of water and nutrients will be severely limited (Figure 3.17).

FIGURE 3.16   Foot rot gummosis on citrus tree.

The fibrous roots slough their cortex leaving only the white threadlike stele, which gives the root system a stringy appearance. The tree will grow poorly, stored energy reserves will be depleted, and production will decline. In advanced stages of decline, the production of new fibrous roots cannot keep pace with root death. The tree is unable to maintain adequate water and mineral uptake, and nutrient reserves in the root are depleted by the repeated fungal attacks. This results in the reduction of fruit size and production, loss of leaves, and twig dieback of the canopy.

Disease symptoms are often difficult to distinguish from nematode, salt, or flooding damage; only a laboratory analysis can provide positive identification (Figure 3.18).

FIGURE 3.17   *Phytophthora* foot rot of sweet orange showing light gumming. (Picture courtesy: J. H. Graham, CREC.)

FIGURE 3.18   Phytophthora root rot.

3.6.3   Cause and Disease Development

Phytophthora fungi are present in almost all citrus orchards. Under moist conditions, the fungi produce large numbers of motile zoospores, which are splashed onto the tree trunks. The *Phytophthora* species cause gummosis to develop rapidly under moist, cool conditions. Hot summer weather slows disease spread and helps drying and healing of the lesions.[3]

Secondary infections often occur through lesions created by *Phytophthora*. *Phytophthora citrophthora* is a winter root rot that also causes brown fruit rot and gummosis. *P. citrophthora* is active during cool seasons when citrus roots are inactive and their resistance to infection is low. *Phytophthora parasitica* is active during warm weather when roots are growing.

3.6.4   Disease Cycle

The disease cycle of *P. nicotianae* and *P. citrophthora* begins with the production of sporangia which release large numbers of zoospores. *P. nicotianae* produces chlamydospores in abundance while most isolates of *P. citrophthora* do not. *P. citrophthora* rarely produces oospores, whereas *P. nicotianae* commonly produces oospores. With time and appropriate conditions zoospores encyst and germinate to form mycelia. The optimum temperature for mycelial growth is 30°C–32°C for *P. nicotianae* and 24°C–28°C for *P. citrophthora*.

Sporangial production by *P. nicotianae* and *P. citrophthora* is favored by small deficits in matric water potential, but not by saturated conditions unless sporangia are produced on citrus root pieces. The optimal for sporangium formation probably represents a compromise between requirements for free water and aeration. Nutrition depletion and light also stimulate sporangial production from mycelium.

Indirect germination of sporangia to produce zoospores requires free water and is stimulated by a drop in temperature. Under moist conditions sporangia may also germinate directly by the growth of germ tubes, but the correlation between soil saturation and severity of phytophthora root rot suggests that indirect germination is more important in the root disease cycle.

Chlamydospore production by *P. nicotianae* occurs under unfavorable conditions for fungal growth, i.e., nutrient depletion, and low oxygen levels and temperatures (15°C–18°C). Water requirements for germination of chlamydospores are similar to those for sporangia. Chlamydospores of *P. nicotianae* appear to become dormant below 15°C, so exposure to temperatures of 28°C–32°C is used to stimulate germination. Nutrients that are acquired from soil extracts and excised citrus roots are known to stimulate chlamydospore germination.

The requirements for oospore germination are thought to be nearly identical to those of chlamydospores. Oospore maturation appears to be an important factor in germinability of *Phytophthora* spp. Periods of alternating high and low temperatures may also be a prerequisite for uniform germination (Tables 3.5 and 3.6).

3.6.5 MANAGEMENT

3.7 ANTHRACNOSE OF CITRUS

Anthracnose is a decay that develops primarily on fruit subjected to ethylene during commercial de-greening. The fungus is a common symptomless inhabitant of citrus rind and only manifests itself when the rind is weakened. In some instances, ethylene treatment causes sufficient weakening to induce the disease.

3.7.1 CAUSAL ORGANISM

Species	Associated Disease Phase	Economic Importance
Colletotrichum gloeosporioides (Asexual Stage) *Glomerella cingulate* (Sexual Stage)	Bruised and Injured Rind with Lesions	Severe

3.7.2 SYMPTOMS

3.7.2.1 Leaf

Common symptoms are a more or less circular, flat area, light tan in color with a prominent purple margin that at a later phase of infection will show the fruiting bodies of the fungus (tiny dispersed black flecks). Tissues injured by various environmental factors (such as mesophyll collapse or heavy infestations of spider mites) are more susceptible to anthracnose colonization (Figure 3.19).

3.7.2.2 Fruit

Anthracnose usually only occurs on fruit that have been injured by other agents, such as sunburn, chemical burn, pest damage, bruising, or extended storage periods. The lesions are brown to black spots of 1.5 mm or greater diameter. The decay is usually firm and dry but if deep enough, can soften the fruit. If kept under humid conditions, the spore masses are pink to salmon, but if kept dry, the spores appear brown to black. On ethylene de-greened fruit, lesions are flat and silver in color with a leathery texture. On de-greened fruit, much of the rind is affected (Figure 3.20). The lesions will eventually become brown to gray-black leading to soft rot (Figure 3.21).

It should be noted that leaves and fruit infected with other diseases (*Alternaria*, citrus canker) may also be colonized by the fruiting bodies of *C. gloeosporioides*. The fruiting bodies (black flecks) can be seen over the disease of concern.

3.7.3 CAUSE AND DISEASE DEVELOPMENT

The fungus, *C. gloeosporioides*, which causes anthracnose, is very common in citrus orchards. It grows in the deadwood of the tree canopy and produces spores that are carried in water to

TABLE 3.5

Recommended Chemical Controls for Phytophthora Foot Rot and Root Rot—Fosetyl-AL and Phosphite Salts Products

Pesticide	FRAC MOA[a]	Mature Trees Rate/Acre[b]	Method of Application	Comments
Aliette WDG[c,d]	33			Protectant and curative systemic. Buffering to pH 6 or higher is recommended to avoid phytotoxicity when copper has been used prior to, with, or following Aliette.
Nonbearing		5 lb/100 gal.	Foliar spray	
		2.5–5 lb/5 gals	Trunk paint or spray[e]	
		Up to 5 lb/acre	Microsprinkler	Adjust rate according to tree size.
Bearing		5 lb/acre or 1 lb/100 gal	Foliar spray in 100–250 gal/acre. Do not exceed 500 gal/acre.	Apply up to 4 times/year (e.g., March, May, July, and September) for fibrous root rot control.
		5 lb/10 gal/acre	Aerial	Fly every middle. Do not apply in less than 10 gal/acre.
		5 lb/acre	Surface spray on weed-free area followed by 0.5 inch irrigation or by microsprinkler in 0.1–0.3 inch of water.	Apply up to 4 times/year (e.g., March, May, July, and September) for fibrous root rot control.
Phostrol	33			Protectant and curative systemic. Do not apply when trees are under water stress or high temperature conditions.
Bearing or Nonbearing		4.5 pt/acre	Foliar spray	Apply up to 4 times/year (e.g. March, May, July, and September).
Bearing or Nonbearing		2–5 pt/5 gal	Trunk paint or spray.	Trunk paint or spray.
ProPhyt	33			Protectant and curative systemic. Do not apply when trees are under water stress or high temperature conditions.
Nonbearing		2 gal/100 gal	Drench	1/2 pt solution per seedling in 2 gallon pot; can be applied through microsprinkler.
Bearing		4 pt/acre	Foliar spray	Apply up to 4 times/year (e.g., March, May, July, and September) for fibrous root rot control.

* This information is from SL127, one of a series of the Soil and Water Science Department, Florida Cooperative Extension Service, Institute of Food and Agricultural Sciences, University of Florida.

a Mode of action class for citrus pesticides from the Fungicide Resistance Action Committee (FRAC) 2011. Refer to ENY624, Pesticide Resistance and Resistance Management, in the 2012 Florida Citrus Pest Management Guide for more details.

b Lower rates may be used on smaller trees. Do not use less than the minimum label rate.

c For combinations of application methods, do not exceed 4 applications or 20 lb/acre/year.

d Fungicide treatments control fibrous root rot on highly susceptible sweet orange rootstock, but are not effective against structural root rot and will not reverse tree decline.

e Apply in May prior to summer rains and/or in the fall prior to wrapping trees for freeze protection.

TABLE 3.6

Recommended Chemical Controls for Phytophthora Foot Rot and Root Rot—Mefenoxam and Copper Products

Pesticide	FRACMOA[a]	Mature Trees Rate/Acre[b]	Method of Application	Comments
Ridomil Gold SL[c,d]	4			Protectant and curative systemic. Do not apply tank mixes of Ridomil and residual herbicides to trees less than 3 years old. Apply herbicide first, then wait 3–4 weeks to apply Ridomil. Do not apply to bare roots. Do not apply rates higher than 1 qt./A to citrus resets or new plantings (less than 5 years old) to prevent potential phytotoxicity. Do not make trunk gummosis sprays and soil applications to the same tree in the same cropping season. Time applications to coincide with root flushes.
Nonbearing		1 qt/acre of treated soil surface	Surface spray on weed-free area, followed immediately by 0.5 inch irrigation or by microsprinkler in 0.1–0.3 inch of water.	Make the 1st application at time of planting. Make up to 2 additional applications per year at 3-month intervals for maximum control; in most cases a late spring and late summer application should be sufficient
			1/2 pt/grove acre	Through irrigation injection
		1.0–1.5 fl. oz/20 trees	Individual Tree Treatment for Resets/New Plantings: Mix desired amount of Ridomil Gold SL in a water solution. Apply as a directed spray to individual trees (generally 8–12 fl.oz./tree) around the base of the tree and outward to cover the fibrous root system. Follow with sprinkler irrigation to move product into root zone.	Make 1st application at time of planting. Make up to 2 additional applications per year at 3-month intervals for maximum control; in most cases a late spring and late summer application should be sufficient.
Bearing		1 pt/acre of treated soil surface if propagule counts are 10–20 propagules/cm³ soil	Surface spray on weed-free area, followed immediately by 0.5 inch irrigation or microsprinkler in 0.1–0.3 inch of water.	Begin applications during the spring root flush period. Apply up to 3 times/year on 3-month intervals (late spring, summer, early fall).
		1 qt/acre of treated soil surface if propagule counts are >20 propagules/cm³ soil		

(*Continued*)

TABLE 3.6 (CONTINUED)

Recommended Chemical Controls for Phytophthora Foot Rot and Root Rot—Mefenoxam and Copper Products

Pesticide	FRACMOA[a]	Mature Trees Rate/Acre[b]	Method of Application	Comments
Bearing		1/2 pt/grove acre if propagule counts are 10–20 propagules/cm³soil 1 pt/grove acre if propagule counts are >20 propagules/cm³soil	Through irrigation injection	May be applied up to 3 times/yr.
Ridomil Gold GR[b]	4		Trunk Spray for Gummosis: Spray the trunks to thoroughly wet the cankers.	Do not apply Ridomil Gold GR and residual herbicides to trees less than 3 years old simultaneously. Apply herbicide first, then wait 3–4 weeks to apply Ridomil. Do not apply more than 240 lb of Ridomil Gold GR/acre/year. Time applications to coincide with root flushes.
Nonbearing		40–80 lb/acre of treated soil surface	Apply as banded application under the canopy. For banded applications, use a band wide enough to cover the root system. If rain is not expected for 3 days, follow by 0.5–1.0 inch of irrigation.	Make 1st application at time of planting. Make up to 2 additional applications per year at 3-month intervals for maximum control; in most cases a late spring and late summer application should be sufficient.
Bearing		40–80 lb/acre of treated soil surface	Banded application under the canopy. If rain not expected for 3 days, follow by 0.5–1.0 inch of irrigation.	Begin applications during the spring rot flush period. Apply up to 3 times/year on 3-month intervals (late spring, summer, early fall).
UltraFlourish[c,d]	4		Protectant and curative systemic. Do not apply tank mixes of UltraFlourish and residual herbicides to trees less than 3 years old. Apply herbicide first, then wait 3–4 weeks to apply UltraFlourish.	
Nonbearing		2–4 qt/acre of treated soil surface	Surface spray on weed-free area, followed immediately by 0.5 inch irrigation or by microsprinkler in 0.1–0.3 inches of water.	Apply every 3 months for maximum control; in most cases a late spring and late summer application should be sufficient.
		1 pt/grove acre	Through irrigation injection	

(Continued)

TABLE 3.6 (CONTINUED)
Recommended Chemical Controls for Phytophthora Foot Rot and Root Rot—Mefenoxam and Copper Products

Pesticide	FRACMOA[a]	Mature Trees Rate/Acre[b]	Method of Application	Comments
		2–3 oz/100 gal	Soil drench; apply 5 gal of mix in water ring.	Apply every 3 months for maximum control; in most cases a late spring and late summer application should be sufficient.
Bearing		1 qt/acre of treated soil surface <20 propagules/cm³soil; 2 qt/grove acre >20 propagules/cm³ soil	Surface spray on weed-free area, followed immediately by 0.5 inch irrigation or microsprinkler in 0.1–0.3 inch of water.	Apply 3 times/year (late spring, summer, early fall).
		1 pt/grove acre	Through irrigation injection	
Copper-Wettable Powder	M1	0.5 lb (metallic) Cu/1 gal water	Trunk paint[e]	Protectant
Copper-Count-N	M1	1 qt in 3 qt water	Trunk paint[e]	Protectant. Do not apply to green bark; may cause gumming.

Note: This Table is from PP-156, one of a series of the Plant Pathology Department, Florida Cooperative Extension Service, Institute of Food and Agricultural Sciences, University of Florida. Original publication date December 1999. Revised February 2012.

a Mode of action class for citrus pesticides from the Fungicide Resistance Action Committee (FRAC) 2011. Refer to ENY624, Pesticide Resistance and Resistance Management, in the 2012 Florida Citrus Pest Management Guide for more details.

b Lower rates may be used on smaller trees. Do not use less than the minimum label rate.

c Do not exceed the equivalent of 6 lb a.i./acre/year of mefenoxam-containing products.

d Do not apply to bare roots or higher than 1 qt/acre of treated soil surface to citrus resets or trees less than 5 years old to avoid potential phytotoxicity.

e Apply in May prior to summer rains and/or in the fall prior to wrapping trees for freeze protection.

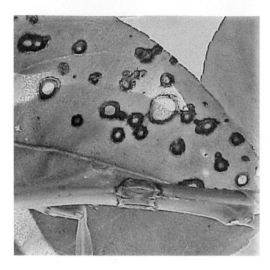

FIGURE 3.19   Anthracnose symptoms on leaf.

the surface of leaves and immature fruit during the growing season. During periods of high moisture, such as after rainfall, heavy dew, or overhead irrigation, the spores germinate to form microscopic "appressoria", which are very hard and resistant to some commonly used fungicides. The appressoria remain dormant for many weeks and months until the fruit is susceptible. The fungus is a very weak pathogen and anthracnose usually only appears on fruit injured by other factors, such as sunburn, chemical treatments or pests, and on fruit that is very mature or held too long in storage.

However, the disorder may develop in certain mandarin varieties without apparent injury. Ethylene de-greening stimulates the appressoria to germinate and increases the susceptibility of the rind to further invasion. Ethylene concentrations, more than those required for optimal de-greening, can significantly increase the incidence of anthracnose.

3.7.4   FAVORABLE CONDITIONS OF DISEASE DEVELOPMENT

The anthracnose fungus usually infects weakened twigs. The disease is most common during springs with prolonged wet periods and when significant rains occur later in the season than normal. During wet or foggy weather, anthracnose spores drip onto fruit, where they infect the rind and leave dull, reddish-to-green streaks on immature fruit and brown-to-black streaks on mature fruit (tear stains). Certain conditions, however, such as applications of insecticidal soaps, which damage the protective wax on the fruit peel, can increase the severity of this disease.

FIGURE 3.20   Anthracnose symptoms on citrus fruit.

FIGURE 3.21   Gray-black lesions on fruit.

3.7.5   DISEASE CYCLE

Anthracnose is a primary colonizer of injured and senescent tissue. The organism grows on dead wood in the canopy, and it spreads short distances by rain splash, heavy dew, and overhead irrigation. Such movement deposits the spores on susceptible tissues of young leaves or immature fruit. Sexual spores, although less numerous, are significant for long-distance dispersal because of their ability to become airborne. Once the spores germinate, they form a resting structure that allows them to remain dormant until an injury occurs or until de-greening. The disease is especially troublesome on fruit that are harvested early and de-greened for over 24 hours because ethylene stimulates the growth of the fungus.

3.7.6   MANAGEMENT

Best control of anthracnose can be achieved by a combination of in-field and postharvest treatments.

- The fungus responsible for anthracnose harbors in deadwood. Good cultural practices to reduce deadwood should be encouraged.
- Field sprays of copper-based fungicides or Mancozeb® may inhibit spore germination. Heavy rain may wash off a copper application and allow infection.
- Ethylene stimulates anthracnose development. Delayed harvesting or selective picking for better color will minimize the amount of time in de-greening.
- Ethylene de-greening should not be above optimal concentrations (5ppm trickle method).
- Harvested fruit should be washed on revolving brushes to remove appressoria or dipped in a benzimidazole fungicide (carbendizim or TBZ) to control anthracnose before de-greening.
- Dipping in guazatine alone will not control anthracnose. Fruit treated with guazatine and a benzimidazole will control molds, sour rot, and anthracnose.
- Immediate cold storage of fruit after packing may assist in reducing the expression of anthracnose.
- Benzimidazole fungicides, such as Tecto®, Spinflo®, or Bavistin®, should be used for dipping or drenching field bins to provide protection from anthracnose. Chlorine alone will not control fungal spores or appressoria on the surface of the fruit. Inappropriately high levels of chlorine and/or by-products may induce weakness in the rind and increase the susceptibility of fruit to anthracnose.

3.8   CITRUS CANKER

Citrus Canker is one of the most feared of citrus diseases, affecting all types of important citrus crops. The disease causes extensive damage to citrus and severity of this infection varies with different species and varieties and the prevailing climatic conditions. The disease is endemic in India, Japan, and other South-East Asian countries, from where it has spread to all other citrus-producing continents except Europe. Generally, canker does not occur in arid citrus growing areas and has been eradicated from some areas. However, widespread occurrence of the disease in many areas is a continuous threat to citriculture, especially in canker-free areas.[5]

Citrus canker is most severe in grapefruit, acid lime, and sweet orange.

3.8.1   CAUSAL ORGANISM

Species	Associated Disease Phase	Economic Importance
Xanthomonas campestris pv. Citri	Raised Necrotic Lesions on Fruits, Leaves, and Twigs.	Severe

3.8.2   SYMPTOMS

3.8.2.1   Leaf

Typical citrus canker lesions on leaves will range from 2 to 10 mm in size and will have raised concentric circles on the underside of the leaf. Frequently, lesions will be surrounded by a water-soaked margin (Figure 3.22) and a yellow halo (Figure 3.23). As a canker lesion ages, it may lose it palpable roughness, but the concentric circles will still be visible with a hand lens (on the underside of the leaf). The yellow halo eventually changes to dark brown or black and the water-soaked margin surrounding the lesion may diminish (Figure 3.24). The middle of the lesion (on the underside of the leaf) will be corky in texture with a volcano or pimple-like point. With the exception of very young lesions, lesions always penetrate through both sides of the leaf. In the presence of damage,

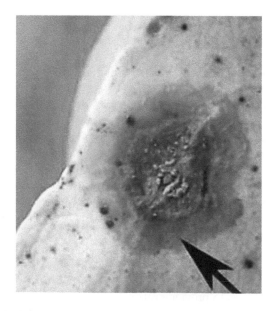

FIGURE 3.22   Water-soaked ring on a citrus leaf.

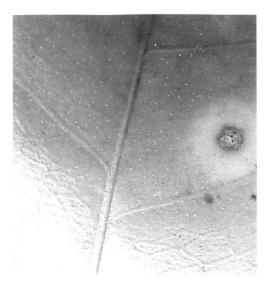

FIGURE 3.23 Yellow halo surrounding leaf lesion.

the lesion may follow the contours of the damage and therefore may not be circular. In older lesions, a saprophytic white fungus may grow over the center of the lesion. The center of a lesion may fall out producing a shot-hole appearance.[5]

3.8.2.2 Fruit

Typical citrus canker lesions on fruit will range from 1 to 10 mm in size. Larger lesions usually penetrate a few mms into the rind. Fruit lesions may vary in size and may coalesce. Fruit lesions consist of concentric circles. On some varieties these circles are raised with a rough texture; on other varieties the concentric circles are relatively flat like the surface of a record. The middle of the lesion will be corky in texture with a volcano or pimple-like point. The center of a lesion may crack and has a crusty material inside that resembles brown sugar (Figure 3.25). Frequently on green fruit, a yellow halo will be visible; however, it will not be visible on ripened fruit. Lesions may have a water-soaked margin and the water-soaked margin is especially evident on smaller lesions. In the presence of damage, the lesion may follow the contours of the damage therefore not being circular. In older lesions a saprophytic white fungus may grow over the center of the lesion. Figures 3.26 and 3.27 refer to citrus canker on acid lime and grapefruit.[5]

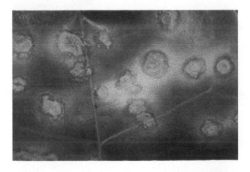

FIGURE 3.24 Dark brown halo surrounding leaf lesion on grapefruit. (Picture courtesy: Dan Robl, USDA.)

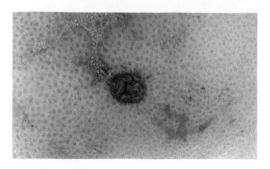

FIGURE 3.25 Fruit lesion with center cracking.

3.8.2.3 Twigs

On twigs and fruit, citrus canker symptoms are similar: raised corky lesions surrounded by an oily or water-soaked margin. No chlorosis surrounds twig lesions but may be present on fruit lesions (Figure 3.28).

Twig lesions on angular young shoots perpetuate *Xac* inoculum in areas where citrus canker is endemic. Twig dieback, fruit blemishes, and early fruit drop are major economic impacts of the disease in advanced stages. If twigs are not killed back by girdling infections, the lesions can persist for many years, causing raised corky patches in the otherwise smooth bark.

3.8.3 Cause and Disease Development

Since *Xanthomonads* have a mucilaginous coat, they easily suspend in water and are dispersed in droplets. Spread of canker bacteria by wind and rain is mostly over short distances, i.e., within trees or neighboring trees. Cankers develop more severely on the side of the tree exposed to wind-driven rain. Long-distance spread more often occurs with the movement of diseased propagating material, such as bud wood, rootstock seedlings, or budded trees.

3.8.4 Favorable Conditions of Disease Development

Optimum temperature for infection falls between 20°C and 30°C. Citrus canker is severe in regions where temperature and rainfall ascend and descend together during the year. Therefore, the disease

FIGURE 3.26 Canker on acid lime.

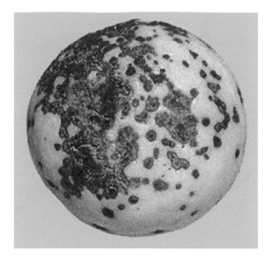

FIGURE 3.27   Citrus canker on grapefruit. (From Based on blog written by Karen Harty, Florida Master Gardener and Citrus Advisor. https://growagardener.blogspot.com/2014/01)

occurs in its most severe form in seasons and/or areas characterized by warm and humid weather conditions.

3.8.5   DISEASE CYCLE

The bacterium propagates in lesions on leaves, stems, and fruit. When there is free moisture on the lesions, bacteria ooze out and can be dispersed (Figure 3.29) to new growth on the plant already infected. Rainwater collected from foliage with lesions contains between 105 and 108 cfu/mL. Wind-driven rain is the main natural dispersal agent, and wind speeds 18 mph (8 m/s) aid in the penetration of bacteria through the stomatal pores or wounds made by thorns, insects such as the Asian leaf-miner, and blowing sand. The serpentine mines under the leaf cuticle caused by the larvae of the Asian citrus leaf-miner, a pest first detected in 1993 in Florida, provide ample wounding on new growth to greatly amplify the citrus canker infection. Water congestion of leaf tissues can be seen following rainstorms with wind. Citrus foliage can hold 7 μL/cm^2 of leaf area. Studies of inoculum associated with water congestion have demonstrated how as few as one or two bacterial cells forced through stomatal openings can lead to infection and lesion formation. Windblown inoculum was detected up to 32 ms from infected trees in Argentina. However, in Florida, evidence for much longer dispersals (up to seven miles) associated with meteorological events such as severe rainstorms and tropical storms has been presented. Pruning causes severe wounding and can be a site for infection.

FIGURE 3.28   Citrus canker lesion on sweet orange stem.

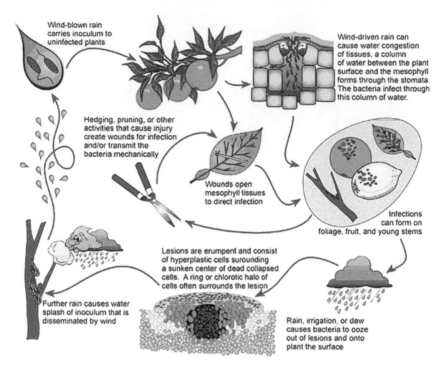

Wind-blown rain carries inoculum to uninfected plants

Wind-driven rain can cause water congestion of tissues, a column of water between the plant surface and the mesophyll forms through the stomata. The bacteria infect through this column of water.

Hedging, pruning, or other activities that cause injury create wounds for infection and/or transmit the bacteria mechanically

Wounds open mesophyll tissues to direct infection

Infections can form on foliage, fruit, and young stems

Lesions are erumpent and consist of hyperplastic cells surounding a sunken center of dead collapsed cells. A ring or chlorotic halo of cells often surrounds the lesion

Further rain causes water splash of inoculum that is disseminated by wind

Rain, irrigation, or dew causes bacteria to ooze out of lesions and onto plant the surface

FIGURE 3.29 Citrus canker disease cycle. Image from *Plant Health Progress* Article—"Citrus Canker: The Pathogen and its Impact".

3.8.6 Management

The following table mentions the citrus cultivars which are highly resistant to highly susceptible to the disease (Table 3.7).

3.8.6.1 Cultural Practices

Cultural practices including windbreaks, and pruning or defoliation of diseased summer and autumn shoots, are recognized throughout the world as important measures for the management of citrus canker. Windbreaks are the most effective measure for the control of the disease on susceptible citrus cultivars. Windbreaks alone or in combination with copper sprays may reduce disease incidence on leaves and fruits to non-detectable levels on field resistant cultivars. Pruning and defoliation of diseased shoots in combination with copper sprays as a complete control have also been effective in light outbreaks. Pruning of the citrus trees is performed during the dry season when the environmental conditions are less favorable for the spread of the bacterium from pruned to adjacent non-infected trees. However, pruning is very labor intensive and therefore expensive.

3.8.6.2 Chemical Control

Worldwide, citrus canker is managed with preventive sprays of copper-based bactericides. Such bactericides are used to reduce inoculum build up on new leaf flushes and to protect expanding fruit surfaces from infection. Effective suppression of the disease by copper sprays depends on several factors, such as the susceptibility of the citrus cultivar, environmental conditions, and adoption of other control measures. As a stand-alone measure, control of citrus canker with copper sprays on resistant or moderately resistant citrus cultivars may be achieved, whereas adequate control on susceptible or highly susceptible cultivars requires the implementation of several control measures.

The timing and number of copper sprays for effective control of citrus canker are not only highly dependent on the susceptibility of the citrus cultivar, but on the age of the citrus trees, environmental

TABLE 3.7

Relative Susceptibility/Resistance to Citrus Canker of Commercial Citrus Cultivars and Species

Rating	Citrus Cultivars
Highly resistant	Calamondin (*C. mitus*); Kumquats (*Fortunella* spp.)
Resistant	Mandarins (*C. reticulata*)—Ponkan, Satsuma, Tankan, Satsuma, Cleopatra, Sunki, Sun Chu Sha
Less susceptible	Tangerines, Tangors, Tangelos (*C. reticulata* hybrids); Cravo, Dancy, Emperor, Fallglo Fairchild, Fremont, Clementina, Kara, King Lee, Murcott, Nova, Minneola, Osceola, Ortanique, Page, Robinson, Sunburst, Temple, Umatilla, Willowleaf (all selections); Sweet oranges (*C. sinenesis*)—Berna, Cadenera, Coco, Folha Murcha, IAPAR 73, Jaffa, Moro, Lima, Midsweet, Sunstar, Gardner, Natal, Navelina, Pera, Ruby Blood, Sanguinello, Salustiana, Shamouti, Temprana and Valencia; Sour oranges (*C. aurantium*)
Susceptible	Sweet oranges—Hamlin, Marrs, Navels (all selections), Parson Brown, Pineapple, Piralima, Ruby, Seleta Vermelha (Earlygold), Tarocco, Westin; Tangerines, Tangelos—Clementine, Orlando, Natsudaidai, Pummelo (*C. grandis*); Limes (*C. latifolia*)—Tahiti lime, Palestine sweet lime; Trifoliate orange (*Poncirus trifoliata*); Citranges/Citrumelos (*P. trifoliata* hybrids)
Highly susceptible	Grapefruit (*C. paradisi*); Mexican/Key lime (*C. aurantiifolia*); Lemons (*C. limon*); and Pointed leaf Hystrix (*C. hystrix*)

conditions, and the adoption of other control measures. In general, three to five copper sprays are necessary for effective control of citrus canker on citrus cultivars with intermediate levels of resistance, whereas, in years with weather that is highly conducive for epidemic development of citrus canker, up to six sprays may be recommended.

3.8.6.3   Integrated Management Programs

In regions where citrus canker is endemic, integrated control measures rely most heavily on the planting of resistant varieties of citrus. In Southeast Asia, where climatic conditions are most favorable for epidemics, the dominant cultivars grown are based on mandarins. Citrus canker has not been a serious problem until more susceptible sweet oranges were introduced into disease prone areas of Japan and China. In Brazil, eradication/control programs have been on-going since the 1950s to control the spread of *Xac* into the largest sweet orange production area in the world: São Paulo State. The strategies of the integrated program for citrus canker control are based on research carried out in the 1960s and early 1970s in Japan, and later in the 1970s in Argentina and 1980s in Brazil.

The most important feature of this program is the shift in planting from susceptible to field resistant citrus cultivars. Regulations in these regions not only address the requirement for more resistant cultivars but also mandate production of *Xac*-free nursery trees and other means for exclusion of canker from orchards. Guidelines also specify management practices for citrus canker and marketing of fresh fruit and nursery stock. Under these regulations, nurseries can only be located in areas free of citrus canker. In orchard production areas designated as citrus canker-free, regulations are designed to prevent or reduce the risk of citrus canker epidemics through the establishment of windbreaks, construction of fences to restrict the access to the orchard, and the use of preventive copper sprays. Fresh fruit for internal and export markets is subject to inspection protocols for freedom of citrus canker symptoms on fruit in orchards and sanitation treatments in the packinghouse.

3.9   CITRUS GREENING

Citrus greening, also known as Huanglongbing (HLB) or yellow dragon disease, is one of the most serious citrus diseases in the world. It is a bacterial disease that greatly reduces production, destroys the economic value of fruit, and can kill trees. It has significantly reduced citrus production in Asia, Africa, the Arabian Peninsula, and Brazil. Once infected, there is no cure for a tree with the citrus greening disease. In areas of the world where citrus greening is endemic, citrus trees decline and die within a few years. The disease specifically attacks citrus plants and presents no threat to humans or animals.

Sweet orange and mandarin orange are highly susceptible to the disease; grapefruit and lemons are moderately susceptible.

3.9.1 CAUSAL ORGANISM

Species	Associated Disease Phase	Economic Importance
Candidatus Liberibacter asiaticus (Asian Strain) *Ca. L. africanus* (African Strain)	Stunting, Sparse Yellow Foliage, Severe Fruit Drop	Most Severe

3.9.2 SYMPTOMS

Symptoms are many and variable: yellow shoots, twig dieback, leaf drop, leaves with blotchy yellow/green coloration similar to the symptoms of zinc nutritional deficiency, enlarged veins that appear corky (Figure 3.30), excessive fruit drop, small and misshapen fruit, fruit that remains green at one end (the stylar end) after maturity, fruit with mottled yellow/ green coloration (Figure 3.31), small dark aborted seed inside fruit, discolored vascular bundles in the pithy center of the fruit, bitter-tasting fruit, and silver spots left on fruits that are firmly pressed.

The time from infection to the appearance of symptoms is variable, depending on the time of year, environmental conditions, tree age, host species/cultivar and horticultural health ranging from less than one year to several years. The three disease agents (*C. Liberibacter* spp.) are not distinguishable from each other based on symptoms produced.

3.9.3 TRANSMISSION OF THE DISEASE

The disease can be transmitted by bud grafting but not at high rates due to necrosis in sieve tubes and uneven distribution of the bacteria. Dodder (*Cuscuta* spp.) also causes the spread of the disease. Citrus psyllid is, however, the primary vector. Also, the disease occurs with high psyllid populations when the host is flushing which is when the psyllid migrations are highest. Two species of citrus psyllid are vectors.

FIGURE 3.30   Citrus greening resulting in a blotchy leaf.

FIGURE 3.31   Fruit with yellow and green coloration. (Picture courtesy: ARS, USDA.)

- The African Citrus Psyllid, *Trioza erytreae*, occurs in Africa, Réunion, and Yemen and vectors the African strain of greening. It survives well in cool upland areas.
- The Asian Citrus Psyllid, *Diaphorina citri*, is in Asia, India, Saudi Arabia, Réunion, and North, South and Central America. It is more resistant to high temperatures and survives in hot lower altitudes.

3.9.4   DISEASE CYCLE

Candidatus Liberibacters are gram-negative bacteria with a double-membrane cell envelope. *Ca. L. asiaticus*, *africanus*, and *americanus* are found only in the phloem cells of plants. The bacteria are transmitted by psyllids, a type of insect, as they feed. *Ca. L. asiaticus* and *Ca. L. americanus* are transmitted by the adults of the citrus psyllid *D. citri Kuwayana*. *Ca. L. africanus* is transmitted by the adult psyllid Trioza erytreae Del Guercio. The bacteria can be acquired by the insects in the nymphal stages and the bacteria may be transmitted throughout the life span of the psyllid.

Eggs are laid on newly emerging leaves and hatch in two to four days. Five nymphal instars complete development in 11–15 days. The entire life cycle takes 15–47 days, depending upon temperature, and adults may live several months with females laying up to 800 eggs in a lifetime.

In an orchard, diseased trees are clustered together, with secondary infections produced 25–50 m away. *Ca. L. africanus* is found at elevations greater than 700 m and is less heat tolerant than *Ca. L. asiaticus*. *Ca. L. americanus* resembles *Ca. L africanus* in being less heat tolerant. Infections of *Ca. L. asiaticus* and *Ca. L. americanus* are more severe than *Ca. L. africanus* and can lead to tree death.

3.10   CITRUS VARIEGATED CHLOROSIS

Citrus variegated chlorosis (CVC) was described as a new disease of citrus in 1987 in Brazil. CVC is also a highly injurious disease of citrus. Caused by a strain of the bacterium *Xylella fastidiosa*, CVC causes severe chlorosis between veins on the leaves of affected plants. Leaves on affected plants frequently have discoloration of the upper leaf coupled with brown lesions underneath. CVC may reduce plant growth and lead to abnormal flowering and fruit production. CVC is currently not known to occur in the United States.

Sweet oranges are the most susceptible. Grapefruit, mandarins, mandarin hybrids, lemons, and limes are moderately susceptible, showing less severe symptoms.

3.10.1   CAUSAL ORGANISM

Species	Associated Disease Phase	Economic Importance
X. fastidiosa	Chlorosis of Leaves and Abnormal Fruits	Severe

FIGURE 3.32   CVC symptoms on upper side of the leaf.

3.10.2 Symptoms

3.10.2.1   Leaf

Foliar symptoms of CVC are very similar to nutrient deficiency and other diseases; therefore, it is difficult to rely on foliar symptoms alone for identification. Early leaf symptoms resemble zinc deficiency with interveinal chlorotic areas on the upper surface (Figure 3.32). Early symptoms may be limited to a single branch. As the leaf matures, gummy lesions become visible on the lower leaf surface (Figure 3.33) corresponding to chlorotic areas on the upper surface of the leaf. The chlorotic areas gradually enlarge toward the leaf margin, and the lesions on the underside of the leaf may become dark brown or necrotic. Leaves may be smaller than normal. Leaf symptoms are most pronounced on mature leaves (behind the new flush).

3.10.2.2   Fruit

Blossom and fruit occur at the normal time, but fruit thinning does not occur. This results in clusters of four to ten early maturing fruit. Fruit size is significantly reduced (Figure 3.34), with increased sugar content and hard rind. Fruits of infected trees may exhibit sunburn damage because of defoliation at branch terminals. In addition, fruit may change color early, have hard rinds, lack juice, and have an acidic flavor. Fruit symptoms of CVC are more easily recognized from a distance (Figure 3.35).

3.10.2.3   Whole Tree

Affected trees may exhibit reduced vigor and growth and show abnormal flowering and fruit set. Newly affected trees may only exhibit symptoms on one limb or branch, and then symptoms may spread to the entire canopy. Older trees may only show symptoms on the extremities of the branches.

FIGURE 3.33   CVC symptoms on lower side of the leaf.

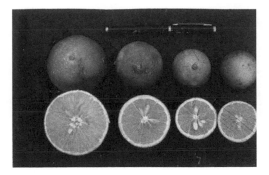

FIGURE 3.34   Reducing fruit size. (Picture by: MAM and Francisco Laranjeira.)

Severely diseased trees frequently possess upper crown branches with defoliation at terminal twigs and small leaves and fruit (Figure 3.36).

3.10.3   FAVORABLE CONDITIONS FOR DISEASE DEVELOPMENT AND DISEASE TRANSMISSION

The fastidious bacteria grow well at 20°C–25°C and pH 6.7–7.0. The bacterium has been found to be transmitted in Brazil by sharpshooter leafhoppers (Cicadellidae). Sharpshooters are present in most citrus growing areas of the United States. The sharpshooter leafhopper, *Oncometopia nigricans Walker*, frequently is found feeding on citrus in Florida.[6]

3.10.4   DISEASE CYCLE

CVC is a systemic disease that only survives in plant xylem or within its vector. *X. fastidiosa* has been shown to move from seed to seedling in sweet orange. CVC has a latency period of nine to twelve months before symptoms occur. Natural spread of *X. fastidiosa* occurs by several species of sharpshooter leafhoppers in the order Hemiptera. At least eleven species of sharpshooter have been shown to vector CVC. Some of these species currently occur in the United States. Sharpshooters are xylem feeders and acquire *X. fastidiosa* within two hours of feeding. Sharpshooters have a high rate of feeding and retain infectivity indefinitely. Sharpshooters do not pass *X. fastidiosa* onto the next generation. Sharpshooters have an extensive host range and may undergo one to several generations per year.

3.10.5   MANAGEMENT

Disease management is costly. It involves insecticide applications, pruning of symptomatic branches and elimination of affected trees. Pruning success is erratic, especially in areas of high CVC

FIGURE 3.35   Fruit of sweet orange affected by CVC. (Picture by: MAM and Francisco Laranjeira.)

FIGURE 3.36   Whole tree affected by CVC.

incidence, and has been recommended only for affected trees over three years of age expressing initial leaf symptoms. Although all commercial sweet orange cultivars are susceptible, the commonly used commercial rootstocks have shown considerable levels of CVC resistance. The strategy consists of removing the entire canopy of the condemned tree by pruning the trunk at soil level, below the graft line. Two to three months later, one healthy bud is grafted on each of the two or three new selected shoots. Around six weeks later, the shoot portion above graft is removed and the new scion is allowed to grow to form a new canopy. The method was tested in four experiments in three distinct locations (north, center, and south São Paulo State) involving 320 four- to five-year-old Pêra Rio/Rangpur lime trees expressing severe and extensive CVC symptoms. The healthy buds used were from the same sweet orange cultivar. In total, CVC symptoms have appeared in 13 (4.1%) canopies that developed on the trunks, 32–48 months after pruning. Probably due to a high xylem sap influx coming from the already established root system, in general, the new canopies grew and started producing fruits faster than the new young trees planted in one same orchard. This study confirms high levels of resistance of Rangpur lime to *X. fastidiosa* and demonstrates the possibility of reusing the rootstock of a tree severely affected by CVC to generate a healthy and productive sweet orange scion.

3.11   GREEN MOLD ON CITRUS

Green mold occurs in all citrus-producing regions of the world. Green mold is the most common and serious postharvest disease of citrus. All types of citrus fruit are susceptible to green and blue mold.

3.11.1   CAUSAL ORGANISM

Species	Associated Disease Phase	Economic Importance
Penicillium digitatum	Rotting of the Fruit	Severe

3.11.2   SYMPTOMS

The first symptom is a tiny soft, watery spot 5–10 mm in diameter. In one day, the spot enlarges until it measures 2–4 cm across. White mycelium appears on the rind surface (Figure 3.37). After it reaches a size of 2.5 cm, the fungus begins to produce green spores (Figure 3.38). These disperse easily if the fruit is handled, or exposed to the wind. The decayed fruit becomes soft and shrinks in size. If the atmosphere is humid, the infected fruit also becomes attacked by other molds and bacteria, and soon collapses into a rotting mass.

3.11.3   CAUSE AND DISEASE DEVELOPMENT

The fungus survives in the orchard from season to season mainly in the form of conidia. Infection is from airborne spores, which enter the peel of the fruit in places where there are small injuries or

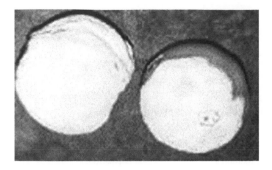

FIGURE 3.37   White mycelium on rind surface.

blemishes. It can also invade fruit which have been damaged on the tree by chilling injury. Infected fruit in storage do not infect the fruit packed around them. However, infected fruit may give off abundant green fungus spores which soil the skin of adjacent fruit. Since it attacks only injured fruit, the best way to prevent green mold is to handle the fruit carefully during and after harvest.

3.11.4   FAVORABLE CONDITIONS OF DISEASE DEVELOPMENT

Infections develop from damaged areas. The growth of mold increases with storage temperatures (up to an optimum of 24°C). Late season fruit is more susceptible. Damaged rind is more susceptible. Green mold develops most rapidly at temperatures near 24°C and much more slowly above 30°C and below 10°C. The rot is almost completely inhibited at 1°C.

3.11.5   DISEASE CYCLE

P. digitatum (green mold) survives in the orchard from season to season primarily as conidia. Infection is initiated by airborne spores, which enter the rind through injuries. Even injuries to the oil glands alone can promote some infection. It can also invade fruit through certain physiologically induced injuries, such as chilling injury, oleocellosis, and stem-end rind breakdown. The fungus does not usually spread from decayed fruit to adjacent, intact healthy fruit in packed containers. The infection and sporulation cycle can be repeated many times during the season in a packinghouse, and inoculum pressure increases as the picking season advances if precautions are not taken.

Green mold develops most rapidly at temperatures near 24°C and much more slowly above 30°C and below 10°C. The rot is almost completely inhibited at 1°C.

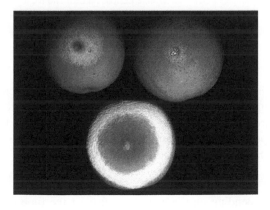

FIGURE 3.38   Production of green spores by the fungus.

TABLE 3.8

Postharvest Fungicides Include Thiophanate-Methyl, Imazalil, Prochloraz, and Guazatine

Active Ingredient	Activity	Rates
Guazatine acetate 2'5%	systemic	5–10%
Imazalil 34%	systemic	0.1–0.125%
Thiophanate-methyl 45%	systemic	0.4%
Prochloraz	systemic	0.2%

3.11.6 MANAGEMENT

Careful picking and handling of fruit minimize injuries to the rind and the risk of molds. Sanitary practices should be applied to prevent sporulation on diseased fruit and the accumulation of spores on equipment surfaces and in the atmosphere of packing and storage facilities. Immediate cooling after packing, significantly delays development of molds, particularly if combined with an effective fungicide.

The following table includes the commonly used fungicides (Table 3.8).

Penicillium spp. can develop resistance to some fungicides. The use of two or more fungicides minimizes resistance problems in addition to the use of stringent sanitary practices.

3.12 CITRUS MELANOSE

Melanose disease can affect young leaves and fruits of certain citrus species or varieties when the tissues grow and expand during extended periods of rainy or humid weather conditions. The symptoms of this widely distributed fungal disease vary from small spots or scab-like lesions to patterns of damage referred to as tear-drop, mudcake, and star melanose.

Grapefruit and lemons tend to be more susceptible than other kinds of citrus.

3.12.1 CAUSAL ORGANISM

Species	Associated Disease Phase	Economic Importance
Diaporthe citri (anamorph: *Phomopsis citri*)	Scab-like Lesions on Leaf and Rind	Severe

3.12.2 SYMPTOMS

3.12.2.1 Leaf

About 1 week after infection foliar symptoms appear as small brown discrete spots (Figure 3.39). These spots become impregnated with a reddish-brown gum and are raised above the leaf surface. Early pustules on leaves are surrounded by a yellow halo. However, this halo quickly disappears leaving only small corky pustules. The numerous small pustules give the leaf a rough sandpaper texture. Distortion and dieback of young shoots are associated with severe infections.

3.12.2.2 Fruit

Fruit symptoms can vary depending on the age of the fruit at the time of infection. Early infections will result in premature petal fall and plants will have relatively large pustules, which in large

FIGURE 3.39   Melanose symptoms on leaf.

numbers may coalesce to form extensive areas that often crack to produce a pattern described as "mudcake melanose" (Figure 3.40). Infections during later stages of fruit development produce small discrete pustules distributed by spore-laden rain or dew which flows over the fruit surface creating the "tear-stain melanose" pattern. These injuries to the fruit rind are superficial and are not important if the crop is processed. If copper fungicides are applied for control, stippling, copper fungicide damage may occur, which can resemble the disease and is often called star melanose.

3.12.3   Cause and Disease Development

The fungus reproduces in dead twigs, particularly those which have died recently. For this reason, pruning away deadwood is an essential part of controlling melanose.

The fungus produces two kinds of spores. One kind are numerous but do not travel far. They are mainly carried in splashes of water such as raindrops. Infected dead twigs in the upper part of the tree may cause heavy infestations in lower branches.

The other kind of spore is fairly scarce. These spores are airborne and can be carried on the wind. They are likely to cause outbreaks if heaps of dead citrus wood are left lying on the ground in or around the citrus orchard.

3.12.4   Favorable Conditions of Disease Development

The causal pathogen is the fungus *D. citri*. Trees are susceptible only for a few months after petal fall. The severity of the disease is determined mainly by the amount of dead wood in the canopy, and by the length of time the fruit remains wet after rainfall or sprinkler irrigation.

FIGURE 3.40   A combination of mudcake and tear-staining of melanose on grapefruit rind

Temperature also plays a role. After a spore lands on wet leaves or fruit tissue, it must remain moist for some hours to transmit the disease. At 15°C, 18–24 hours of wetness are needed for infection, but at 25°C, only 10–12 hours are needed. Young trees suffer less from melanose than older ones because they have less dead wood.

3.12.5   DISEASE CYCLE

3.12.5.1   Dispersal

Rain or overhead irrigation water spreads the anamorph spores over short distances to susceptible tissues in the citrus canopy. The ascospores are dispersed by wind over longer distances. The more dead wood that exists in a canopy, the more ascospores will be produced. Most fruit infections probably start other infections caused by conidia.

3.12.5.2   Infection

Fruits are susceptible to infection from about three to five months after petal drop, depending on the area. Approximately 8–24 hours of continuous moisture on leaf or fruit surfaces is required for infection to occur, depending on air temperature (shorter periods at higher air temperature). Therefore, periods of extended rainfall at warm locations are most likely to initiate rapid and severe melanose disease development.

3.12.6   MANAGEMENT

The disease may not severely impact fruit yield, and if fruits are grown for juicing or other processing, melanose disease management may not be warranted.

- Pruning—Periodically prune away dead branches. This will reduce pathogen survival, increase air circulation to dry out the canopy, and allow for more effective fungicide penetration and coverage of the foliage.
- Fungicides—Sprays of fungicides to young fruits and leaves may be necessary for disease management. Where the disease tends to be severe, frequent fungicide applications may be required (refer to Table 3.9). Worldwide, copper fungicides are the most commonly applied. After application of copper sprays to citrus fruits, star melanose symptoms may appear which differ from the symptoms described above on unsprayed fruits. Postharvest treatments and storage conditions of fruits are not effective in reducing melanose disease damage to citrus rinds.
- Citrus variety—Avoid planting very susceptible citrus varieties or species (sweet orange, grapefruit) in high-rainfall areas.
- Choice of planting location—Plant citrus in sunny, low-rainfall regions.
- Cropping system—Interplant citrus with non-susceptible hosts (avoid monocrops).
- Sanitation—Pick up and destroy plant materials that have fallen from the citrus canopy.

Rates for pesticides are given as the maximum amount required to treat mature citrus trees unless otherwise noted. To treat smaller trees with commercial application equipment including handguns, mix the per acre rate for mature trees in 125 gallons of water. Calibrate and arrange nozzles to deliver thorough distribution and treat as many acres as this volume of spray allows (Table 3.9).

3.13   POSTBLOOM FRUIT DROP

Postbloom fruit drop (PFD) caused by *Colletotrichum acutatum*, affects citrus flowers and produces abscission of young fruitlets. PFD is a serious problem in most humid citrus production areas of the

TABLE 3.9
Recommended Chemical Controls for Melanose

Pesticide	FRAC MOA[a]	Mature Trees Rate/Acre[b]
Copper fungicide	M1	Use label rate.
Abound 2.08F[c]	11	12.0-15.5 fl oz. Do not apply more than 92.3 fl oz/acre/season for all uses.
Gem 25WG[c]	11	4.0–8.0 oz. Do not apply more than 32 oz/acre/season for all uses.
Gem 500 SC[c]	11	1.9–3.8 fl oz. Do not apply more than 15.2 fl oz/acre/season for all uses.
Headline[c]	11	12–15 fl oz. Do not apply more than 54 fl oz/acre/season for all uses.
Pristine[c]	11/7	16–18.5 oz. Do not apply more than 74 oz/acre/season for all uses.
Quadris Top[c]	11/3	15.4 fl oz. Do not apply more than 61.5 fl oz/acre/season for all uses. Do not apply more than 0.5 lb ai/acre/season difenconazole. Do not apply more than 1.5 lb ai/acre/season azoxystrobin

Note: This table is from PP-145, one of a series of the Plant Pathology Department, Florida Cooperative Extension Service, Institute of Food and Agricultural Sciences, University of Florida. Original publication date December 1999. Revised February 2012.

[a] Mode of action class for citrus pesticides from the Fungicide Resistance Action Committee (FRAC) 2011. Refer to ENY624, Pesticide Resistance and Resistance Management, in the *2012 Florida Citrus Pest Management Guide* for more details.

[b] Lower rates can be used on smaller trees. Do not use less than the minimum label rate.

[c] Do not use more than 4 applications of strobilurin-containing fungicides/season. Do not make more than 2 sequential applications of strobilurin fungicides.

world. PFD must be controlled on processing and fresh market fruit. PFD affects all species and cultivars of citrus, but severity on a given cultivar varies according to the time of bloom in relation to rainfall.

3.13.1 CAUSAL ORGANISM

Species	Associated Disease Phase	Economic Importance
C. acutatum	Infection on Citrus Flowers	Severe

3.13.2 SYMPTOMS

The pathogen infects flower petals causing peach to orange-colored lesions on open flowers (Figure 3.41) and flower buds (Figure 3.42). Subsequently, the fruitlet abscises leaving the calyx and floral disk

FIGURE 3.41   Peach to orange-colored PFD lesions on flower.

FIGURE 3.42   PFD lesions on flower buds.

attached to the twig. These persistent structures remain attached for the life of the twig, and the leaves around inflorescences are usually twisted and distorted. Necrotic spots on petals often coalesce, producing a blight of the entire inflorescence. Senescent petals on healthy flowers usually are light tan in color or dry from the tip downward, but diseased petals are dark brown to orange and dry first in the areas affected (Figure 3.43). Affected petals become hard and dry, persisting for several days after the healthy flowers have fallen. After petal fall, the young fruit show a slight yellowish discoloration and usually abscise, leaving the calyx and floral disc intact. These structures are commonly called buttons (Figure 3.44) and stay green for a year or more and callous tissue begins to form around the abscission zone. Occasionally, young fruit remain attached to the button but never develop.[7]

3.13.3   CAUSE AND DISEASE DEVELOPMENT

PFD severity on a given cultivar may vary according to the time of bloom in relation to rainfall. Most spores of this fungus are produced directly on the surface of infected petals. Spores are splash-dispersed by rains to healthy flowers where they infect within 24 hours. Long-distance spread may

FIGURE 3.43   PFD lesions on senescent petals.

FIGURE 3.44   PFD leaves a persistent calyx or button after the petals fall.

occur by windblown rain, by bees or other insects that visit flowers, or by plant debris carried on equipment or in picking sacks or boxes.[7]

3.13.4   FAVORABLE CONDITIONS OF DISEASE DEVELOPMENT

3.13.4.1   Rainfall

Rain is needed for epidemic development to supply moisture for infection and, as importantly, to disperse conidia by the force of droplets impacting on spore-laden petals. The amount of rain is considered in the system, but the force of the rain is as important. Fortunately, these two factors are highly related, and the amount of rainfall is a good indicator of conidial dispersion.

3.13.4.2   Leaf Wetness

At least eight hours of moisture is needed for infection and the amount of infection increases as the duration of wetness increases. Only the number of hours of wetness that occurred during and after a rain in the last five days is considered. Without the force of rainfall, conidia are not dispersed. Dews and fogs are not considered since only localized infection results from even extended wetting periods. Few aids in the growth of inoculum by spreading from infected leaves and flowers to uninfected leaves and flowers. This is considered under inoculum level.

3.13.5   DISEASE CYCLE

Asexual spores of *C. acutatum* are produced in abundance on infected petals. These spores are dispersed by rain splashing onto adjacent healthy flowers; they germinate in the presence of moisture in 12–24 hours, infect petals in 24–48 hours and produce symptoms and new spores in four to five days. The fungus survives between blooms on the surface of leaves, twigs, and buttons as appressoria (single-celled, thick-walled structures). The following spring, these structures germinate in the presence of moisture and substances present in flower petals to produce new asexual spores. These spores may then be dispersed to the fresh flowers by splashing rain (Figure 3.45).

3.13.6   MANAGEMENT

The increase in the incidence of PFD is very high under optimum conditions making the disease difficult to control. Overhead irrigation should be avoided during bloom, if possible, or trees should be irrigated at night and allowed to dry during the day. Trees declining because of blight, tristeza, or other factors often flower out of season, thus maintaining high levels of inoculum. Such trees should be removed prior to bloom in PFD-affected blocks.

Refer to the following table for fungicides registered for citrus which control PFD (Table 3.10).

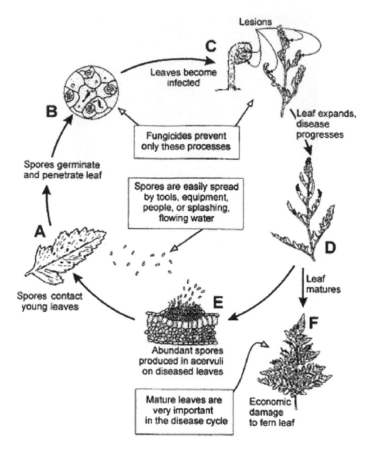

FIGURE 3.45   Disease cycle of *C. acutatum* (hypothetical representation for citrus PFD).

3.14   SOOTY BLOTCH OF CITRUS

Sooty blotch is a fungal disease in which dark, round smudges, that may expand to form irregular shaped blotches that cover most of the fruit, occur. The smudges are initially circular but may form irregular shapes as they overlap.

3.14.1   CAUSAL ORGANISM

Species	Associated Disease Phase	Economic Importance
Gloeodes pomigena	Rind Blemish	Minor

3.14.2   SYMPTOMS

Sooty blotch is a rind blemish caused by the development of the fungus *G. pomigena* over the fruit surface. The gray to black fungal strands give the appearance of a light dusting of soot (hence the name), which can be rubbed vigorously and removed (Figure 3.46). This is convenient for smooth-skinned varieties. Rubbing off the soot of rough-skinned varieties is particularly difficult because of the rough texture of the rind. The growth of the fungus is generally superficial, and it can only rupture the cuticle, not the epidermis. Fruits can be eaten, but it looks unsightly. It usually makes the fruit unmarketable for the fresh trade and its storage life is reduced.

TABLE 3.10

Recommended Chemical Controls for Post-Bloom Fruit Drop

Pesticide	FRAC MOA[a]	Mature Trees Rate/Acre[b]
Abound 2.08 F	11	12.0–15.5 fl oz. Do not apply more than 92.3 fl oz/acre/season for all uses.
Abound 2.08 F + Ferbam	11, M3	12.0 fl oz + 5 lb
Gem 500 SC	11	1.9–3.8 fl oz. Do not apply more than 15.2 fl oz/acre/season for all uses.
Gem + Ferbam	11, M3	1.9 fl. oz. + 5 lb
Headline	11	12 fl oz. Do not apply more than 54 fl oz/acre/season for all uses.
Headline + Ferbam	11, M3	12-15 fl oz + 5 lb

Note: This table is from PP-45, one of a series of the Plant Pathology Department, Florida Cooperative Extension Service, Institute of Food and Agricultural Sciences, University of Florida. Original publication date December 1995. Revised January 2012.

[a] Mode of action class for citrus pesticides according to the Fungicide Resistance Action Committee (FRAC) 2011. Refer to ENY-624, Pesticide Resistance and Resistance Management, in the *2012 Florida Citrus Pest Management Guide* for more details.

[b] Lower rates can be used on smaller trees. Do not use less than the minimum label rate.

3.14.3 FAVORABLE CONDITIONS OF DISEASE DEVELOPMENT

The disease is a frequent problem under wet and shaded conditions and is commonly observed during autumn and winter in coastal regions of the world. Mature fruit on the inside of the tree canopy are most affected.

3.14.4 MANAGEMENT

Application of copper fungicides can help control Sooty Blotch to some extent. Since wet conditions are favorable for disease development, an additional application of copper fungicides during autumn will help reduce the disease.

3.15 SWEET ORANGE SCAB

Sweet orange scab (SOS) is a fungal disease that causes unsightly corky lesions on fruit. This disease is so named because it results in scab-like lesions that develop primarily on the fruit rind (thick outer skin) and infrequently, on leaves and twigs. These lesions do not generally affect yield as there

FIGURE 3.46 Soot dust on the rind giving a gray to black appearance.

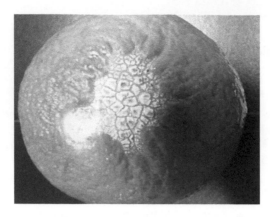

FIGURE 3.47   Early lesions on the fruit.

is little effect on the internal fruit quality but can reduce the marketability of the fruit. However, the disease can cause stunting of the young nursery trees and new field planting and premature fruit drop. This disease is known to occur principally in South America.

3.15.1 CAUSAL ORGANISM

Species	Associated Disease Phase	Economic Importance
Elsinoe australis	Lesions on Fruit	Low

3.15.2 SYMPTOMS

The initial scab is observed during the early stages of fruit development. The lesion is slightly raised and pink to light brown in color. These are the early lesions (Figure 3.47). As the lesion expands, it takes on a cracked or warty appearance and may change color to a yellowish-brown and eventually to a dark gray, forming advanced lesions (Figure 3.48). The scabs typically form a pattern on the fruit similar to water splashes.

On young stems, the lesions resemble an area of dieback that has been scabbed over. Lesions begin on the underside of leaves as water-soaked spots. They typically form along the edge of the leaf or the mid-vein.

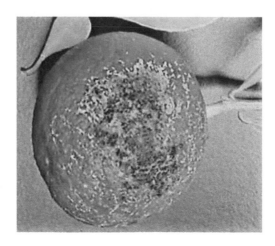

FIGURE 3.48   Advanced lesions on the fruit.

3.15.3   Cause and Disease Development

The SOS disease is spread by spores which are dispersed by splashing water and wind-driven rain. The spores are produced when scab lesions are wet for at least one or two hours, the humidity is very high, and the temperatures are from 20°C–28°C. Spores can spread the disease to susceptible plants if there is a sufficient level of moisture in the environment.

3.15.4   Favorable Conditions of Disease Development

Diseases develop over a range of temperatures as long as there is sufficient moisture, and disease can develop quite rapidly (in less than four hours) under favorable temperature range of 20°C–28°C. The fungus can live through the winter in the tree canopy on limbs and on fruit that were infected during the previous season.

3.15.5   Disease Cycle

SOS forms spores on the surface of the scab pustules. This species of scab attacks mainly fruits. The conidia (asexual spores) are similar to those of *E. fawcettii*, require moisture for spore production and are primarily spread by splashing rain. Fruits are susceptible for six to eight weeks after petal fall. The role of ascospores (sexual spores) is uncertain.

3.15.6   Management

- It is important that the trees are sprayed with a copper fungicide from petal fall until the fruit is two months old. This, preferably, should be done in three sprays—the first spray should be applied prior to bloom to protect the foliage, the second spray three to four weeks later, and the third spray three to four weeks after the second spray. The copper-based fungicides should not be applied during bloom.
- The removal of infected fruit, leaves, and plant parts is also recommended to help control the disease. With proper spraying of a copper fungicide and with proper removal of plant parts the SOS disease can be controlled and possibly eliminated.
- Maintaining cleanliness, good sanitation, and hygiene practices also help in keeping the disease away.
- The symptoms of the disease should be well known so as to detect it at an early stage.

REFERENCES

1. Dewdney, M. M., and Timmer, L. W. 2009. *Alternaria Brown Spot.* University of Florida, 2010. http://gardener.wikia.com/wiki/Alternaria_brown_spot
2. Dewdney, M., *UF/IAF Citrus Extension*, University of Florida. http://www.crec.ifas.ufl.edu/extension/black_spot/citrus_black_spot.shtml
3. Menge, J. A., and Ohr H. D., Plant Pathology, UC Riverside, *UC IPM Pest Management Guidelines: Citrus*, UC ANR Publication 3441. http://ipm.ucanr.edu
4. Adaskaveg, J. E., *Management of Citrus Brown Rot*, Department of Plant Pathology, University of California Riverside, C. A. http://www.calcitrusquality.org/wp-content/uploads/../Citrus-Brown-Rot-JA-9-29-11.pdf
5. Based on blog written by Karen Harty, Florida Master Gardener and Citrus Advisor. https://growagardener.blogspot.com/2014/01
6. Brlansky, R. H., Damsteegt, V. D., and Hartung, J. S. 2002. Transmission of the citrus variegated chlorosis bacterium Xylella fastidiosa with the sharpshooter Oncometopia nigricans. *Plant Dis.* 86:1237–1239. https://pubag.nal.usda.gov/pubag/downloadPDF.xhtml?id=741&content=PDF
7. Mossler, M. A., *Florida Crop/Pest Management Profiles: Citrus (Oranges/Grapefruit)*, Cir 1241, University of Florida IFAS Extension. http://manatee.ifas.ufl.edu/lawn_and_garden/master../c/citrus-pest-management.pdf

4 Apple

The apple tree is a deciduous tree in the rose family best known for its sweet, pomaceous fruit, the apple. Apples are cultivated worldwide as a fruit tree and is the most widely grown species in the genus *Malus*. The apple tree grows approximately 1.8 to 4.6 m (6 to 15 ft) tall in cultivation and up to 12 m (39 ft) in the wild. The leaves are characterized by alternate arrangements of dark green-colored, simple ovals with serrated margins and slightly downy undersides. The apple fruit matures during late summer or autumn, and cultivars exist in a wide range of sizes. World production of apples in 2014 was 84.6 million tons from which China contributing to 48% of the world total. Other major producers with 6% or less of the world total were the United States, Poland, Turkey, and Italy (Figure 4.1).

Apple trees are susceptible to a number of fungal and bacterial diseases and insect pests. Many commercial orchards pursue a program of chemical sprays to maintain high fruit quality, tree health, and high yields. Many organic methods are also used in orchard management. This chapter provides valuable information about various apple diseases and their management.

4.1 ALTERNARIA BLOTCH OF APPLE

Alternaria blotch has been a severe problem in apple growing areas. Disease severity is aggravated by severe mite infestation. Maintaining good mite management is a key factor in preventing severe disease development.

4.1.1 CAUSAL ORGANISM

Species	Associated Disease Phase	Economic Importance
Alternaria mali	Lesions on Leaves and Fruit	Increasing Toward Major

4.1.2 SYMPTOMS

Lesions first appear on leaves in late spring or early summer as small, round, purplish or blackish spots, gradually enlarging to 1/16 to 3/16 inch (1.5-5 mm) in diameter, with a brownish purple border (Figure 4.2). Lesions may coalesce or undergo secondary enlargement and become irregular and much darker, acquiring a "frog-eye" appearance. When lesions occur on petioles, the leaves turn yellow and 50% or more defoliation may occur. Severe defoliation leads to premature fruit drop. Alternaria leaf blotch is most likely to occur on "Delicious" strains and should not be confused with frogeye leaf spot, captan spot, or with "Golden Delicious" necrotic leaf blotch. Frogeye leaf spot usually appears earlier in the season and is associated with nearby dead wood or fruit mummies. Captan spot spray injury occurs when captan fungicide is applied under wet conditions; it is usually worse near the sprayer, and regularly appears on leaves of the same age on the terminal shoots. "Golden Delicious" necrotic leaf blotch commonly occurs in July and August because of physiological stress caused by fluctuating soil moisture.[1]

A. mali also causes a rather inconspicuous fruit symptom (Figure 4.3) like cork spot (calcium deficiency). Typically, the incidence of fruit infection is relatively low, but in heavily defoliated orchards, fruit infection as high as 60% has been reported.

FIGURE 4.1   Apple.

FIGURE 4.2   Symptoms of alternaria blotch on apple leaf.

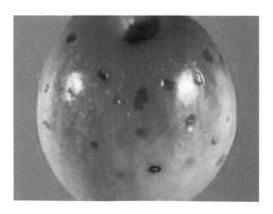

FIGURE 4.3   Symptoms of alternaria blotch on apple fruit.

4.1.3 CAUSE AND DISEASE DEVELOPMENT

Primary infection is initiated in the spring by spores produced on overwintered shoot infections or infected leaf debris from the previous year. On apple, leaves, shoots, and fruits are infected only when young. On all hosts, tissues of susceptible cultivars are killed by toxins produced by the pathogens. Secondary spread occurs as further spores are produced on infected tissue and dispersed by wind or rain.[2]

4.1.4 FAVORABLE CONDITIONS OF DISEASE DEVELOPMENT

Disease development is favored by wet weather and elevated temperatures (25°C–30°C).

4.1.5 DISEASE CYCLE

The fungus can overwinter as mycelium on dead leaves on the orchard floor, in mechanical injuries in twigs, or in dormant buds. Primary infection takes place about one month after petal fall. The disease advances rapidly in the optimum temperature range of 25°C–30°C (77°F to 86°F) and wet weather. At optimum temperatures, infection occurs within 5.5 h of wetting, and lesions can appear in the orchard two days after infection, causing a serious outbreak. The fungus produces a chemical toxin which increases the severity of the disease on susceptible cultivars.

4.1.6 MANAGEMENT

It is very important to maintain mites below 10 per leaf in orchards with a history of the disease to minimize defoliation. Shredding leaves or applications of urea to leaves just before or after leaf fall may reduce the inoculum. The strobilurin fungicides Flint and Sovran have shown good activity on the disease when applied in a sequence of three sprays. A threshold for beginning the applications has not been established, but the disease incidence should not exceed 40% before the first application is made.

4.2 APPLE SCAB

Apple scab occurs everywhere in the world where apples are grown and causes more losses than any other apple disease. It is most serious in areas that have cool, wet weather during the spring and may not be economically important in warm or dry climates. If left untreated, it will cause green, brown, or black lesions on the leaves and corky areas on the fruit. It can defoliate susceptible varieties mid-season, which will weaken trees, reduce yields, and blemish fruit so that they crack and split, making them only fit for cider. It is very widespread and can occur wherever apples are grown.

4.2.1 CAUSAL ORGANISM

Species	Associated Disease Phase	Economic Importance
Venturia inaequalis	Lesions on Leaves and Fruits	Severe

4.2.2 SYMPTOMS

Apple scab can be observed on leaves, petioles, blossoms, sepals, fruit, pedicels, and less frequently, on young shoots and bud scales. The first lesions are often found on the lower surfaces of leaves as they emerge and are exposed to infection in the spring. Later, as the leaves unfold, both surfaces are exposed and can become infected. Young lesions are velvety brown to olive green and have feathery, indistinct margins. With time, the margins become distinct, but they may be obscured if several lesions coalesce. As an infected leaf ages, the tissues adjacent to the lesion thicken, and the leaf surface becomes deformed. Young leaves may become curled, dwarfed, and distorted when infections are numerous. The lesions may remain on the upper and lower leaf surface for the entire growing season; occasionally, the underlying cells turn brown and die, so that brown lesions are visible on both surfaces. The number of lesions per leaf may range from one or two to more than a 100. The term "sheet scab" is often used to refer to leaves with their entire surfaces covered with scab. Young leaves with sheet scab often shrivel and fall from the tree. Infections of petioles and pedicels result in premature abscission of leaves and fruit, respectively. In late summer or early fall, lesions may appear whitish due to the growth of a secondary fungus on the lesion surface.[3]

Lesions on young fruit appear like those on leaves, but as the infected fruit enlarge, the lesions become brown and corky. Infections early in the season can cause fruit to develop unevenly as uninfected portions continue to grow. Cracks then appear in the skin and flesh, or the fruit may become deformed. The entire fruit surface is susceptible to infection, but infections early in the season are generally clustered around the calyx end. Fruit infections that occur in late summer or early fall may not be visible until the fruit is in storage. This syndrome is called "pinpoint" scab, with rough circular black lesions ranging from .004 to 0.16 inches (0.1–4 mm) in diameter.[3] (Figures 4.4 and 4.5)

FIGURE 4.4   Symptoms of apple scab on apple leaves.

FIGURE 4.5   Symptoms of apple scab on apple fruit.

4.2.3   CAUSE AND DISEASE DEVELOPMENT

Apple scab is caused by the fungus, *V. inaequalis*. It survives the winter in the previous year's diseased leaves that have fallen under the tree. In the spring, the fungus in the old diseased leaves produces millions of spores. These spores are released into the air during rain periods in April, May, and June. They are then carried by the wind to young leaves, flower parts, and fruits. Once in contact with susceptible tissue, the spore germinates in a film of water and the fungus penetrates the plant. Depending upon weather conditions, symptoms (lesions) will show up in nine to seventeen days. The fungus produces a different kind of spore in these newly developed lesions. These spores are carried and spread by splashing rain to other leaves and fruits where new infections occur. The disease may continue to develop and spread throughout the summer. Because a film of water on leaves and fruit is required for infection to occur, apple scab is most severe during years with frequent spring rains.

4.2.4   FAVORABLE CONDITIONS OF DISEASE DEVELOPMENT

Apple scab is most severe during spring and early summer when the humidity is high, and the temperature is moderate. Overwintering fungal spores (ascospores) are produced in the diseased leaves on the ground. In most years, the first fungal spores (primary inoculum) mature and are capable of causing infections in the spring at about the time of bud break (leaf expansion). Fungal spores are expelled into the air following rainfall and continue to be discharged over a period of one to three months. The peak period of spore dispersal often occurs near the end of bloom (pink to full-bloom stages). Whether infection occurs or not depends on the period of wetness and the temperature. Fewer hours of wetness are required for infection at elevated temperatures than at low temperatures.

4.2.5   DISEASE CYCLE

The infection cycle begins in the springtime when suitable temperatures and moisture promote the release of *V. inaequalis* ascospores from leaf litter around the base of previously infected trees. These spores rise into the air and land on the surface of a susceptible tree, where they germinate and form a germ tube that can directly penetrate the plant's waxy cuticle. A fungal mycelium forms between the cuticle and the underlying epidermal tissue, starting as a yellow spot that grows and ruptures to reveal a black lesion bearing asexually as the conidia are released and germinate on fresh areas of the host tree, which in turn produce another generation of conidial spores. This cycle of secondary infections continues throughout the summer until the leaves and fruit fall from the tree at the onset of winter. Over the winter, *V. inaequalis* undergoes sexual reproduction in the leaf litter around the base of the tree, producing a new generation of ascospores that are released the following spring. Scab lesions located on the woody tissues may also overwinter in place but will not undergo a sexual reproduction cycle; these lesions can still produce infective conidial spores in the spring (Figure 4.6).

4.2.6   MANAGEMENT

In affected orchards, new infections can be reduced by removing leaf litter and trimmings containing infected tissue from the orchard and incinerating them. This will reduce the amount of new ascospores released in the spring. Additionally, scab lesions on woody tissue can be excised from the tree if possible and similarly destroyed.

Chemical controls can include a variety of compounds. Benzimidazole fungicides (e.g. Benlate [now banned in many countries due to its containing the harmful chemical Benzene]) work well but resistance can arise quickly. Similarly, many chemical classes including sterol inhibitors such as Nova 40, and strobilurins such as Flint among others, were used extensively but are slowly being phased out because of resistance problems.

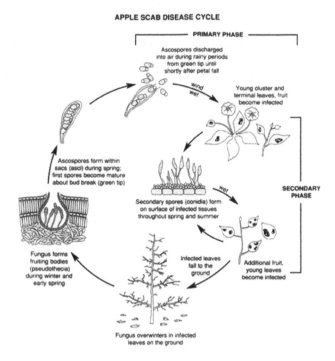

APPLE SCAB DISEASE CYCLE

FIGURE 4.6   Disease cycle of V. inaequalis.

Contact fungicides not prone to resistance such as Captan are viable choices. Copper or Bordeaux mixture are traditional controls but are less effective than chemical fungicides and can cause russeting of the fruit. Wettable sulfur also provides some control. The timing of application and concentration varies between compounds.

Fifteen genes have been found in apple cultivars that confer resistance against apple scab. Researchers hope to use cisgenic techniques to introduce these genes into commercial cultivars and therefore create new resistant cultivars. This can be done through conventional breeding but would take over 50 years to achieve.

Tables 4.1 and 4.2 provide some information on temperature and wetting periods required for infection to occur and resistant, susceptible and immune varieties of apples.

4.3   FIRE BLIGHT OF APPLE

Fire blight is a common and very destructive bacterial disease of apples and pears. The disease is so named because infected leaves on very susceptible trees will suddenly turn brown, appearing as though they had been scorched by fire. The disease is also referred to as blossom blight, spur blight, fruit blight, twig blight, or rootstock blight—depending on the plant part that is attacked. Economic losses to fire blight occur due to a loss of fruit-bearing surface and tree mortality. Trees may need to be removed and replanted or, in severe cases, whole blocks of trees may need to be replaced.

4.3.1   CAUSAL ORGANISM

Species	Associated Disease Phase	Economic Importance
Erwinia amylovora	Blossom Blight, Shoot Blight, and Branch and Trunk Canker.	Severe

TABLE 4.1

Approximate Number of Hours of Wetting Required for Primary Apple Scab Infection at Different Air Temperatures and the Length of Time Required for the Development of Conidia

Average Temperature (deg. F)	Light (hrs)[b]	Moderate (hrs)	Heavy (hrs)	Incubation (days)[a]
78	13	17	26	
77	11	14	21	
76	9.5	12	19	
64 to 75	9	12	18	9
62	9	12	19	10
61	9	13	20	10
60	9.5	13	20	11
59	10	13	21	12
58	10	14	21	12
57	10	14	22	13
56	11	15	22	13
55	11	16	24	14
54	11.5	16	24	14
53	12	17	25	15
52	12	18	26	15
51	13	18	27	16
50	14	19	29	16
49	14.5	20	30	17
48	15	20	30	17
47	17	22	35	
46	19	25	38	
45	20	27	41	
44	22	30	45	
43	25	34	51	
42	30	40	60	
33 to 41[c]				

Source: From Mills, W.D., *N.Y. Agr. Exp. Sta. Ithaca Ext. Bull.,* 630, 4 pp, 1944.

[a] Approx. no. days required for conidial development after primary scab infection.

[b] The infection period is considered to start at the beginning of the rain

[c] Data are incomplete at low temperatures.

4.3.2 SYMPTOMS

Fire blight is a systemic disease. The term "fire blight" describes the appearance of the disease, which can make affected areas appear blackened, shrunken and cracked, as though scorched by fire. Primary infections are established in open blossoms and tender new shoots and leaves in the spring when blossoms are open.[4]

The fire blight bacterium can infect any portion of a susceptible plant. The common types of infection are blossom blight, shoot blight, and branch and trunk canker. Blossom blight is most common in apple.

Overwintering cankers harboring the fire blight pathogen are often clearly visible on trunks and large limbs as slightly to deeply depressed areas of discolored bark, which are sometimes cracked about the margins. The largest number of cankers, however, is much smaller and not so easily distinguished. These occur on small limbs where blossom or shoot infections occurred the previous year

TABLE 4.2
Apple Cultivar Resistance or Susceptibility to Apple Scab

Cultivar	Resistance
Akane (W)[a]	Resistant
Chehalis (W)[b]	Resistant
Criterion (E)	Susceptible
Fuji (E)	Susceptible
Gala (E,W)	Susceptible
Golden Delicious (E,W)	Susceptible
Granny Smith (E)	Susceptible
Gravenstein (W)	Susceptible
Idared (W)	Susceptible
Jonagold (E,W)	Susceptible
Liberty (W)[a]	Immune
Mcintosh (E)	Susceptible
Melrose (W)	Susceptible
Mutsu (E,W)	Susceptible
Paulared (W)[a]	Resistant
Prima (W)[a]	Immune
Red Delicious (E)	Susceptible
Rome (E)	Susceptible
Spartan (W)	Susceptible
Summer Red (W)	Susceptible
Tydeman's Red (W)[a]	Resistant
Yellow transparent (E,W)	Susceptible

Source: E, Cultivars commonly grown east of the Cascade Mountains; W, Cultivars commonly grown west of the Cascade Mountains.

[a] Resistant cultivars not usually requiring fungicide applications even in western Washington's wet climate.

and often around cuts made to remove blighted limbs. Since many of these cankers are established later in the season, they are not often strongly depressed and seldom show bark cracks at their margins[4] (Figure 4.7).

Blossom blight symptoms most often appear within one to two weeks after bloom and usually involve the entire blossom cluster, which wilts and dies, turning brown on apple. When weather is favorable for pathogen development, globules of bacterial ooze can be seen on the blossoms (Figure 4.8). The spur bearing the blossom cluster also dies and the infection may spread into and kill portions of the supporting limb. The tips of young infected shoots wilt, forming a very typical "shepherd's crook" symptom. Older shoots that become infected after they develop about 20 leaves may not show this curling symptom at the tip. As the infection spreads down the shoot axis, the leaves first show dark streaks in the mid-veins, then wilt and turn brown, remaining tightly attached to the shoot throughout the season. As with blossom infections, the pathogen often invades and kills a portion of the limb supporting the infected shoot. The first symptom on water sprouts and shoots that are invaded systemically from nearby active cankers is the development of a yellow to orange discoloration of the shoot tip before wilting occurs. In addition, the petioles and mid-veins of the basal leaves on such sprouts usually become necrotic before those at the shoot tip.[4]

Depending on the cultivar and its stage of development at the time infection occurs, a single blossom or shoot infection can result in the death of an entire limb, and where the central leader or trunk of the tree is invaded, a major portion of the tree can be killed in just one season. In general,

FIGURE 4.7   Fire blight canker.

infections of any type that occur between petal fall and terminal bud set usually lead to the greatest limb and tree loss. In addition, heavily structured trees tend to suffer less severe limb loss than those trained to weaker systems for high productivity. Where highly susceptible apple rootstocks become infected, much of the scion trunk and major limbs above the graft union very typically remain symptomless, while a distinct dark brown canker develops around the rootstock. As this rootstock canker girdles the tree, the upper portion shows symptoms of general decline (poor foliage color, weak growth) by mid to late season. In some instances, the foliage of trees affected by rootstock blight develop an early fall red color in late August to early September, not unlike the color change that often occurs with collar rot disease caused by a soilborne fungus. Some trees with rootstock

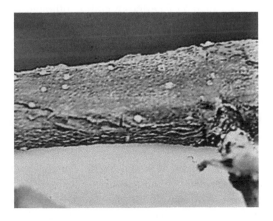

FIGURE 4.8   Fire blight–infected apple with bacterial ooze.

infections may not show decline symptoms until the following spring, at which time cankers can be seen extending upward into the lower trunk.

4.3.3 FAVORABLE CONDITIONS OF DISEASE DEVELOPMENT

Fire blight blossom infection is favored by moderately high temperatures and rainy or humid weather. Temperatures of 18°C and above favor rapid infection whenever moisture or dew is present. Temperatures below 15.5°C retard blight development.

4.3.4 DISEASE CYCLE

The bacterial pathogen causing fire blight overwinters almost exclusively in cankers on limbs infected the previous season. The largest number of cankers, and hence, those most important in contributing inoculum, occur on limbs smaller than 1.5 inches (38 mm) in diameter, especially around cuts made the previous year to remove blighted limbs. During the early spring, in response to warmer temperatures and rapid bud development, the bacteria at canker margins begin multiplying rapidly and produce a thick yellowish to white ooze that is elaborated onto the bark surface up to several weeks before the bloom period. Many insect species (predominantly flies) are attracted to the ooze, and subsequently, disperse the bacteria throughout the orchard. Once the first few open blossoms are colonized by the bacteria, pollinating insects rapidly move the pathogen to other flowers, initiating more blossom blight. These colonized flowers are subject to infection within minutes after any wetting event caused by rain or heavy dew when the average daily temperatures are equal to or greater than 16°C (60°F) while the flower petals are intact (flower receptacles and young fruits are resistant after petal fall). Once blossom infections occur, early symptoms can be expected with the accumulation of at least 57 degree days (DD) greater than 13°C (103 DD greater than 57°F) which, depending upon daily temperatures may require five to thirty calendar days.

With the appearance of blossom blight symptoms, the number and distribution of inoculum sources in the orchard increase greatly. Inoculum from these sources is further spread by wind, rain, and many casual insect visitors to young shoot tips, increasing the likelihood of an outbreak of shoot blight. Most shoot tip infections occur between the time that the shoots have about nine to ten leaves and terminal bud set when sources of inoculum and insect vectors are available and daily temperatures average 16°C (60°F) or more.

In years when blossom infections do not occur, the primary sources of inoculum for the shoot blight phase are the overwintering cankers and, in particular, young water sprouts near these cankers, which become infected as the bacteria move into them systemically from the canker margins. Such systemic shoot infections, called canker blight, are apparently initiated after about 111 DD greater than 13°C (200 DD greater than 55°F) after green tip, although visible symptoms may not be apparent until the accumulation of at least 167 DD greater than 13°C (300 DD greater than 55°F) after green tip. In the absence of blossom infections, the development of shoot blight infections is often localized around areas with overwintering cankers.

Although mature shoot and limb tissues are generally resistant to infection by *E. amylovora*, injuries caused by hail, late frosts of -2°C (28°F) or lower, and high winds that damage the foliage can create a trauma blight situation in which the normal defense mechanisms in mature tissues are breached and infections occur. Instances of trauma blight are known to occur even on normally resistant cultivars like "Delicious".

Rootstock blight, yet another phase of fire blight, has been recognized recently and is associated primarily with the highly susceptible M.26, M.9, and Mark rootstocks. On these trees, just a few blossoms or shoot infections on the scion cultivar can supply bacteria that then move systemically into the rootstock where a canker often, but not always, develops and eventually girdles the tree. Trees affected by rootstock blight generally show symptoms of decline and early death by mid to late season but may not be apparent until the following spring (Figure 4.9).

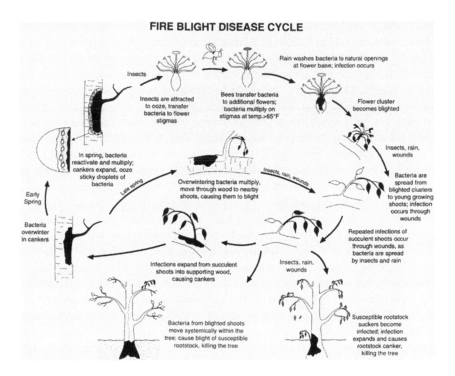

FIGURE 4.9 Disease cycle of fire blight.

4.3.5 MANAGEMENT

No single method is adequate to effectively control fire blight. A combination of practices is needed to reduce the severity of the disease.

1. Choose the proper cultivars. Apple cultivars differ widely in their susceptibility to fire blight. During warm and rainy weather, cultivars rated moderately susceptible or moderately resistant will develop shoot infections; however, the extent to which shoot infections progress will be less in resistant cultivars than in susceptible cultivars. Commercial growers should select rootstocks that are less susceptible to fire blight.
2. Select planting sites with good soil drainage. Trees are more susceptible to fire blight in poorly drained sites than in well-drained ones. Tree productivity will also be lower on such sites. Drainage can often be improved by tiling.
3. Follow proper pruning and fertilization practices. Using nitrogen-containing fertilizer and/or doing heavy pruning promotes vigorous growth and increases susceptibility. Fertilization and pruning practices on susceptible cultivars should be adjusted to limit excessive growth. For bearing trees, moderate shoot growth is six to twelve inches (15 to 30 cms) per year. If the growth is more than twelve inches, do not apply fertilizer until shoot growth is reduced to less than 6 inches. Apply fertilizer in the early spring (six weeks before bloom) or apply in late fall after growth has ceased. Applications in midseason prolong the time during which shoots are susceptible to infection and increase the likelihood of winter injury to tender wood.
4. Prune out fire blight cankers during the dormant season. Delay the removal of infected shoots until the dormant season to avoid spreading infection to healthy shoots. Make pruning cuts at least six inches (15 cms) below the last point of visible infection. After each pruning cut, sterilize the pruning shears by dipping them in a freshly made solution of one part liquid bleach (Clorox, Purex, Saniclor, Sunny Sol) added to four parts water. Examine the larger branches and trunks carefully for cankers, since these are likely to overwinter and produce new

infections in the spring. Root suckers and water sprouts should also be removed because infection of these parts can lead to infection and death of entire trees. Certain dwarfing rootstocks used for apples are prone to suckering. Commercial growers should select rootstocks that are resistant to fire blight or that show little tendency to produce root suckers. Examine wild, neglected, ornamental hosts of the fire blight bacterium growing near home or commercial orchards for cankers. In addition to seedling apples, crabapples, pears, and quince, check hawthorns (*Crataegus* spp.), firethorns (*Pyracantha* spp.), cotoneasters, mountain-ashes (*Sorbus* spp.), and spiraea. Remove the cankers when found or destroy the entire plant where feasible.

5. Follow a bactericide spray program. Like most bacterial diseases of plants, fire blight is very difficult to control; however, it can be reduced by spraying. The antibiotic streptomycin is the most effective herbicide for controlling fire blight; timely sprays will reduce the incidence of fire blight but must be applied before the appearance of symptoms. Temperatures at the pre-bloom and bloom stages are important in determining whether fire blight will occur in any given year. The bacteria reproduce only when the temperature is warmer than 18°C (65°F). The following concept was developed for predicting outbreaks of blossom blight in Illinois. The first idea to understand is that of a "heating degree day". A "degree day" occurs when the maximum daily temperature reaches 19°C (66°F). Start counting DDs after each spring frost. A freeze greatly reduces the number of fire blight bacteria in holdover cankers and on tree surfaces. Bacteria reach dangerous population levels ONLY after 30 degree days have elapsed since the last frost. Degree days may be accumulated in a variety of ways; for example, two days with a maximum daily temperature of 27°C (80°F), three days of 24°C (75°F), or six days of 21°C (70°F) following a freeze will provide enough accumulated warmth to allow bacterial populations to increase greatly in number and present a serious fire blight threat to blossoms. When 30 degree days have occurred and when blossoms are still present (including secondary bloom), apply the first streptomycin spray. Repeat the spray at four-day intervals through the bloom period. At temperatures above 30°C (86°F), bacteria will not multiply. Therefore, it is not necessary to apply streptomycin when the temperatures average below 18°C or above 30°C. Streptomycin can effectively protect the susceptible apple and pear flowers, but for maximum effect, it must be applied the day of or the day before infection event occurs. Missing the critical window of effectiveness by even 24 hours can result in plant infection and buildup of a significant number of bacteria for later infections. If the blossom blight is well controlled, the subsequent increase of fire blight in summer is often prevented. To prevent the development of streptomycin-resistant strains of the pathogen, no more than four applications of streptomycin per season is recommended. Streptomycin is more effective in preventing blossom infection, and the management of the shoot blight phase of fire blight should not be attempted with streptomycin. However, application of streptomycin immediately following hail storms is highly recommended. Streptomycin is most effective when applied alone, as a dilute spray, under slow drying conditions (generally between 10 p.m. and 3 a.m.), and when daytime temperatures reach 18°C or above. Apply 100 parts per million (ppm) of streptomycin if the temperature is below 18°C, and 50 ppm if the temperature is above 18°C. Bordeaux mixture (8–8–100), made by mixing 8 pounds of crystalline copper sulfate (bluestone or blue vitriol) and 8 pounds of fresh hydrated spray lime in 100 gallons of water, will help control fire blight but may cause russeting of the fruit. Bordeaux mixture is recommended for use by growers who had a severe epidemic the previous year. Bordeaux mixture should be applied at the silver tip stage of flower bud growth. Do not mix Bordeaux with other chemicals and use it as soon as it is prepared. Do not follow Bordeaux mixture with streptomycin, and do not concentrate Bordeaux mixture greater than two times. Copper sulfate (4 lb/100 gal or 2kg/400 liters), applied when trees are dormant in early spring also helps reduce the number of bacteria present in ooze on cankers and thus slows the buildup of bacteria in the orchard prior to bloom.

6. Control sucking insects. Good control of aphids, leafhoppers, plant bugs, and psylla on pears helps prevent shoot infection (Table 4.3).

4.4 APPLE POWDERY MILDEW

Powdery mildew of apples, caused by the fungus *Podosphaera leucotricha*, affects leaves, buds, shoots, and fruits, and forms a dense white fungal growth (mycelium) on the host tissue. The disease stunts the growth of trees and is found wherever apples are grown. It interferes with the proper functioning of leaves, reduces shoot growth and fruit set, and produces a netlike russet on the fruit of some varieties. It often is a serious problem in apple nurseries. It is the only fungal apple disease that is capable of infecting without wetting from rain or dew. Mildew severity and the need for control measures are related to cultivar susceptibility and intended fruit market.

4.4.1 CAUSAL ORGANISM

Species	Associated Disease Phase	Economic Importance
P. leucotricha	Reduced Vigor, Leaf Malformation, and Reduced Viability of Buds	Major

TABLE 4.3

Relative Susceptibility of Common Apple Cultivars and Rootstocks to Fire Blight

Highly Susceptible	Moderately Susceptible	Moderately Resistant
• Beacon	• Baldwin	• Arkansas Black
• Braeburn	• Ben Davis	• Britemac
• Burgundy	• Empire	• Carroll
• Cortland	• Golden Delicious	• Delicious
• Fuji	• Granny Smith	• Liberty
• Gala	• Gravenstein	• Northwestern
• Ginger Gold	• Grimes Golden	• Greening
• Idared	• Jerseymac	• Liberty
• Jonathan	• Jonafree	• Melba
• Liberty	• Jonagold	• Priam
• Lodi	• Jonamac	• Prima
• Molly's Delicious	• Julyred	• Priscilla
• Niagara	• Macoun	• Quinte
• Nittany	• Maiden Blush	• Redfree
• Paulared	• McIntosh	• Sir prize
• Red Yorking	• Milton	• Stark Bunty
• R.I. Greening	• Monroe	• Stark Splendor
• Rome Beauty	• Mutsu	• Turley
• Spigold		
• Starr	• Northern Spy	• Viking
• Twenty Ounce	• Scotia	• Wellington
• Tydeman Early	• Spartan	• Winesap
• Wayne	• Spijon	
• Wealthy	• Starkspur Earliblaze	
• Winter Banana	• Stayman	
• Yellow Newton	• Summer Rambo	
• Yellow Transparent	• Summer Red	
• York Imperial		

4.4.2 SYMPTOMS

Primary mildew infection results from the growth of infected, overwintered leaf or fruit buds. These buds may be killed, or they may grow abnormally; leaves become narrow, brittle, curled, or longitudinally folded and covered with a white powdery layer, stunted whitish-gray twig growth is evident on dormant shoots, while flowers may be stunted and fail to develop (Figure 4.10). Secondary mildew infections may appear as a powdery mottling on either side of the leaves. Early fruit infection causes a web-like russet on the skin that may be difficult to distinguish from early spray damage. Less commonly, fruit may be distorted and partly covered with a white powdery coating of spores (Figure 4.11). Economic damage occurs in the form of aborted blossoms, reduced fruit finish quality, reduced vigor, poor return bloom and yield of bearing trees, and stunted growth and poor form of nonbearing trees.

4.4.3 CAUSE AND DISEASE DEVELOPMENT

Apple powdery mildew overwinters in buds which were infected the previous summer. When conditions warm up, the resulting shoots are stunted and whitened. The white powder consists of spores or "conidia" which are spread on the breeze infecting shoots, leaves, and occasionally, fruit during the summer. The flowers, too, can be infected by overwintering fungus in the buds. Infected flowers are pale with narrow petals and they don't set any fruit.

FIGURE 4.10   Powdery mildew symptoms on apple leaf.

FIGURE 4.11   Powdery mildew symptoms on apple fruit.

4.4.4 FAVORABLE CONDITIONS OF DISEASE DEVELOPMENT

Powdery mildew infections occur when the relative humidity is greater than 90% and the temperature is between 10°C–25°C. The optimum temperature range for the fungus is 19°C–22°C. Although high relative humidity is required for infection, the spores will not germinate if immersed in water. Leaf wetting is, therefore, not conducive to powdery mildew development. Under optimum conditions, powdery mildew can be obvious to the naked eye 48 hours after infection. About five days after infection, a new crop of spores is produced.

4.4.5 DISEASE CYCLE

The mildew fungus overwinters mainly as mycelium in dormant blossoms and shoots or buds produced and infected during the previous growing season. Conidia are produced and released from the unfolding leaves as they emerge from infected buds at about tight cluster stage. Conidia germinate in the high relative humidity usually available on the leaf surface at 10°C–25°C (50°F–77°F) with an optimum of 19°C–22°C (66°F–72°F). Germination does not occur in free moisture. Early season mildew development is affected more by temperature than by relative humidity. Abundant sporulation from overwintering shoots and secondary lesions on young foliage leads to a rapid buildup of inoculum. Secondary infection cycles may continue until susceptible tissue is no longer available. Since leaves are most susceptible soon after emergence, infection of new leaves may occur as long as shoot growth continues. Fruit infection occurs from pink to bloom. Overwintering buds are infected soon after bud initiation. Heavily infected shoots and buds are low in vigor and lack winter hardiness, resulting in a reduction of primary inoculum at temperatures below -24°C (-11°F). This phenomenon has been more commonly observed in other areas with lower winter temperatures than those commonly experienced in the mid-Atlantic region.

4.4.6 MANAGEMENT

Carefully removing affected leaves when they first appear in spring, so that the powdery spores are not shaken on to other leaves, can be effective with a small tree. Later, cut off and dispose of the worst-affected shoots, and try to keep the tree well-watered in dry spells.

Healthy trees are less susceptible to this disease. Make sure they are not going short of water—this is a frequent problem if they're planted near walls or fences. Irrigation is the only fail-safe way of making sure that trees have enough water. However, one may find that a mulch of bark, compost, or straw, applied in April, will help retain the winter rain and reduce mildew attacks. In winter, make sure to remove any distorted or "silvered" shoots or buds. These reveal where the fungus is overwintering. Cut back the branch to a point several buds below where the visible distortion or silvering stops. There are chemicals approved for spraying apple powdery mildew such as Systhane Fungus Fighter. Vitax Green/Yellow sulfur is acceptable to organic growers and is a traditional mildew control. However, there are some apple varieties that are sensitive to sulfur or "sulfur shy". Examples of these are "Lane's Prince Albert" and "Stirling Castle". Remember that sulfur is a preventative fungicide and will need repeated applications for the tree to be protected.

Any apple variety can get apple powdery mildew, but some varieties show above average resistance. Dessert varieties include: "Falstaff", "Ashmead's Kernel", "Egremont Russet", "Ellison's Orange", "Pixie", "Golden Delicious", "Jupiter", "Red Devil", "Scrumptious", "Winston", "Winter Gem", "Saturn", "Laxton's Fortune", and "St Edmund's Pippin". Culinary varieties: "Bountiful", "Bramley's Seedling", "Grenadier", "Edward VII", "Golden Noble", and "Reverend W Wilks". The popular "Cox's Orange Pippin" though, is particularly susceptible.

4.5 SOOTY BLOTCH AND FLYSPECK OF APPLE

Sooty blotch and flyspeck are the two most common "summer diseases" of apples. Although caused by two different organisms, the diseases often occur together since both are confined to the fruit surface and are favored by similar environmental and horticultural conditions. Disease incidence and severity can be highly variable among production regions, growing seasons, and even individual orchards. Economic losses result primarily from the diminished appearance and commercial quality of infected fruit.

4.5.1 Causal Organism

Species	Associated Disease Phase	Economic Importance
Sooty Blotch is Caused by a Complex of Fungi:	Sooty Blotches on Fruit Surface	Common but Minor
Peltaster fructicola,	Black and Shiny Round Dots on Fruit	
Geastrumia polystigmatis,	Surface	
Leptodontium elatius		
Flyspeck is Caused by:		
Zygophiala jamaicensis		

4.5.2 Symptoms

4.5.2.1 Sooty Blotch

Brown to dull black, sooty blotches with an indefinite outline, form on the fruit surface. Blotches may be 1/4 inch in diameter or larger. Patches of dark-green to black, sooty fungal growth develop on the surface of the fruit. Individual patches are 5 mm–1 cm in diameter, circular to more irregular in shape, with a diffuse margin. The blotches may coalesce to cover practically the entire fruit. The sooty blotch fungus is restricted to the outer surface of the fruit, and in many cases, the blotches can be easily rubbed off. However, if infection occurs early in the season, you may need to rub or bleach the fruit vigorously to remove it.

4.5.2.2 Flyspeck

Flyspeck generally appears along with sooty blotch, but it can appear individually as well. Groups of 6 to 50 or more black and shiny round dots that resemble fly excreta appear on the surface of the fruit. The individual "fly specks" are clearly separated and can be easily distinguished from sooty blotch. The clusters are often irregular in shape, and usually about 1–3 cm (3/4–1 1/4 inches) diameter. The clusters may merge to affect a large percentage of the surface area. Like sooty blotch, flyspeck infections are superficial; however, they are usually harder to rub off than sooty blotch.

4.5.3 Cause and Disease Development

4.5.3.1 Sooty Blotch

The "sooty blotch" or "smudge" appearance on affected fruit results from the presence of 100s of minute, dark fungal fruiting bodies (pycnidia) that are interconnected by a mass of loose, interwoven dark hyphae (fungal filaments). In spring, pycnidia on wild plants produce large numbers of spores (conidia) that ooze out and collect in a gelatinous mass. The conidia are then spread by water splash or windblown mists into orchards from late May or early June until fall. The fungus first affects apple twigs, then secondary colonies are initiated on the fruit.[5]

4.5.3.2 Flyspeck

The individual "fly specks" are sexual fruiting bodies (ascocarps) of the fungus. Starting in late spring, the fungus produces spores on wild hosts. These spores are carried by wind into the orchard. When spores meet the fruit under the proper environmental conditions, they germinate and infect.[5]

FIGURE 4.12   Sooty blotch on apple.

4.5.4   Favorable Conditions of Disease Development

4.5.4.1   *Sooty Blotch*
Cool, humid weather (optimum 65°F or 18°C) is essential for disease development. The disease does not develop when temperatures reach 30°C (85°F).

4.5.4.2   *Flyspeck*
Symptoms can develop within 15 days under favorable environmental conditions (65°F or 18°C).

4.5.5   Disease Cycle

These fungi are commonly found on the stem surfaces of many woody plants, including apple shoots. Infections may occur on fruit as early as two to three weeks after petal fall and are highly favored by frequent rain periods and poor drying conditions. Mycelial growth that forms the sooty blotches can occur in the absence of free water at relative humidity greater than 90%. Symptom development of both diseases is relatively slow, typically requiring 20 to 25 days in the orchard, but may occur in

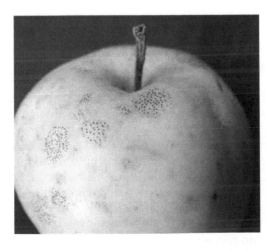

FIGURE 4.13   Flyspeck on apple.

FIGURE 4.14   ApMV symptoms on apple leaves.

FIGURE 4.15   Abnormal graft union in Virginia crab.

eight to twelve days under optimum conditions. Optimum conditions for conidial production for the flyspeck pathogen are 18°C–21°C (65°F–70°F) and relative humidity greater than 96%.

4.5.6   MANAGEMENT

Control of sooty blotch and flyspeck should integrate horticultural practices designed to reduce the chances of disease development together with fungicide sprays to protect against infection when necessary.

Because both diseases are so dependent on long periods of extreme humidity around the fruit, annual pruning to open tree canopies and promote air circulation will minimize the periods favorable for their development. Supplemental summer pruning in dense-canopied trees can provide significant additional benefits in some years. Proper fruit thinning is also important for reducing the development of high-humidity microclimates around clustered fruit; like good pruning, thinning will furthermore improve the spray coverage for any fungicides that may be applied. Mowing of grass middles and good within-row weed control will provide additional help in reducing overall humidity levels within orchards during the summer.

The removal of hedgerows or surrounding woodlots is not always practical but can substantially improve airflow and reduce humidity within the orchard. Destruction of the many woody reservoir hosts in these sites will also reduce some of the inoculum that initiates fruit infections. Because of their importance as an inoculum source, it is particularly important to eliminate brambles in hedgerows and within the orchard itself should they occur there.

The need for and timing of fungicide sprays to control these diseases is variable among orchards and years. In regions where they occur regularly, sprays should start around the first cover and be repeated as necessary according to the prevailing weather conditions and material being used. Where the diseases occur more sporadically, fungicide programs should be initiated and continued based on weather conditions, specific orchard factors, and previous experience. In general, fungicide programs will need to be most intense for orchards in low fog pockets, surrounded by woods, or with dense tree canopies. Minimizing these factors at the time of planting will help reduce the intensity of sooty blotch and flyspeck control programs required in subsequent years (Table 4.4).

4.6 APPLE MOSAIC VIRUS

Apple mosaic virus (ApMV) is one of the oldest known and most widespread apple viruses. The same virus can cause line pattern symptoms in plum and rose mosaic disease. ApMV is related to Prunus necrotic ringspot virus. The ApMV is the most widespread apple virus. An apple tree infected with the ApMV will display symptoms of pale to bright cream spots on the leaves. The infected leaves may be depicted throughout the whole tree or only on a single tree limb. The apple is the most widely grown fruit crop; therefore, an understanding of the pathogen is economically important to commercial apple cultivars. An infected apple tree may have a crop yield reduction of up to 60%.[6]

4.6.1 CAUSAL ORGANISM

The virus is classified as:

Family: Bromoviridae
Genus: *Ilarvirus*
Species: Apple mosaic virus

4.6.2 SYMPTOMS

Apple trees infected with ApMV develop pale to bright cream spots on spring leaves as they expand. These spots may become necrotic after exposure to summer sun and heat. Most commercial cultivars are affected but vary in severity of symptoms. "Golden Delicious" and "Jonathan" are severely affected, whereas "Winesap" and "McIntosh" are only mildly affected. Except in severe cases, a crop can still be produced by infected trees; yield reductions vary from 0 to 50%. In some cultivars, bud set is severely affected.

4.6.3 MEANS OF MOVEMENT AND TRANSMISSION

The natural vector of ApMV is unknown. The virus is transmitted by mechanical inoculation and root grafting. It is also transmitted by the seed and pollen of hazelnut. ApMV is usually transmitted with difficulty by mechanical inoculation of sap from woody plants to several herbaceous species in vitro. As such, it is not very common and can easily be prevented. Infected trees are slow growing and produce low fruit yields. Leaves develop a distinctive, random pattern of chlorotic (yellow) tissue.[6]

TABLE 4.4

Fungicide Effectiveness for Sooty Blotch and Flyspeck

Fungicide	Rate Per 100 Gallons Dilute	Sooty Blotch/Flyspeck
Bayleton 50DF	1 oz.	—
Benlate 50W + Captan 50W	2–3 oz. + 1 lb.	E
Benlate 50W + Mancozeb 75DF	2–3 oz + 1 lb.	E
Benlate 50W + Mancozeb 75DF + Superior oil	2–3 oz. + 1 lb. + 1 qt.	E
Benlate 50W + Polyram 80DF	2–3 oz. + 1 lb.	G
Benlate 50W + Ziram 76DF or WDG	2–3 oz. + 1 lb.	E
Captan 50W	2 lb.	G
Captan 50W	1.5 lb.	F
Captan 50W + Ziram 76DF or WDG + Benlate 50W or Topsin M 70W	1 lb. + 1 lb. + 2 oz. or 2 oz.	E
Dodine 65W	8 oz.	F
EBDC + Captan or Ziram 76DF or WDG	1 lb. + 1–2 lb. or 1–2 lb.	G–E
Ferbam 76W	2 lb.	F
Flint 50WG	2.0 oz./A	E
Mancozeb 75DF or equiv.	1 lb.	G
Nova 40W + Captan 50W	1.25 + 1 lb.	—
Nova 40W + Dodine 65W	1.25 oz. + 4 oz	—
Nova 40W + Mancozeb 75DF	1.25 oz + 1 lb.	—
Nova 40W + Polyram 80DF	1.25 oz + 1 lb.	—
Nova 40W + Ziram 76DF or WDG	1.25 oz. + 1 lb.	—
Polyram 80DF	1 lb.	F
Procure 50WS + Captan 50W	3 oz. + 1 lb.	—
Procure 50WS + Mancozeb 75DF	3 oz. + 1 lb.	—
Procure 50WS + Polyram 80DF	3 oz. + 1 lb.	—
Procure 50WS + Ziram 76DF or WDG	3 oz. + 1 lb.	—
Rubigan 1E + Polyram 80W	9–12 fl. oz. + 3.25 lb. per acre	—
Rubigan 1E + Captan 50W	9–12 fl. oz. + 3.25 lb. per acre	—
Rubigan 1E + Dodine 65W	9–12 fl. oz. + 4 oz. per acre	—
Rubigan 1E + Mancozeb 75DF	9–12 fl. oz. + 3.25 lb. per acre	—
Rubigan 1E + Polyram 80DF	9–12 fl. oz. + 3.25 lb. per acre	—
Rubigan 1E + Ziram 76DF or WDG	9–12 fl. oz. + 3.25 1 lb. per acre	—
Sovran 50WG	4.0 oz./A	E
Sulfur 95W	2–3 lb.	S
Sulfur 95W	5 lb.	S
Thiram 65W	2 lb.	F
Topsin M 70W + Captan 50W	2–3 oz. + 1 lb.	E
Topsin M 70W + Mancozeb 75DF	2–3 oz. + 1 lb.	G
Topsin M 70W + Ployram 80DF	2–3 oz. + 1 lb.	G
Topsin M 70W + Ziram 76DF or WDG	2–3 oz. + 1 lb.	E
Vangard 75WG + Mancozeb 75DF	3 oz./A + 3 lb./A	—
Ziram 76DF or WDG	2 lb.	G

Caution:

Combinations involving Benlate or Topsin M have become ineffective where resistant strains of the apple scab fungus have developed. Resistance to Benlate or Topsin M has been confirmed in the Virginia county of Botetourt, Clarke, Floyd, Frederick, Madison, Roanoke, Rockingham, Shenandoah, Warren, Wise, and Smith, and West Virginia. If resistance is suspected, use of Benlate or Topsin M should be discontinued and replaced by full rates of other scab fungicides. Dodine

(Continued)

TABLE 4.4 (CONTINUED)
Fungicide Effectiveness for Sooty Blotch and Flyspeck

will become less effective where resistance occurs. Scab resistance to Dodine has been confirmed in Clarke and Warren counties, Virginia.

Effectiveness ratings of fruit fungicides for disease control are based on research at Winchester, Virginia, and Kernersville, West Virginia, and also on research from surrounding states. The ratings also are based on severe disease pressures at the test sites. Although the ratings are compiled from the results of 5 to 10 years of research, they may not hold true for all orchard conditions within Virginia, West Virginia, and Maryland. Results can vary by location depending upon the weather conditions, how well trees were sprayed the previous year, concentration of disease inoculum in orchards, tree size and age, formulation of a given fungicide, and how the fungicide was applied (high or low volume). Under certain environmental and cultural practice, the effectiveness ratings could change from good to excellent or vice versa. The ratings, however, are intended as a general guide to assist the grower in fungicide selection for disease control.

Ziram may be less effective than captan on apple cultivars that are more susceptible to scab.

3 Procure at 8–12 oz/acre may be combined with standard protectant fungicides for improved broad-spectrum disease control, improved fruit scab control, and to offset the development of resistance to sterol-inhibiting fungicides.

Credits: Table of fungicide effectiveness from the 2001 Va./W.Va./Md. Spray Bulletin for Commercial Fruit Growers, table compiled by K. S. Yoder and A. R. Biggs.

Rating scale: E, excellent, generally good disease control under heavy disease pressure; G, good, good control under moderate disease pressure; F, fair, fair control under moderate disease pressure; S, slight, some control under light disease pressure; N, little or no effect on indicated disease; ?, information lacking or—not applicable.

4.6.4 PREVENTION AND CONTROL

- Plant virus-free varieties.
- Since this virus is transmitted by root grafting, use a virus-free grafting scion.
- Diseased trees do not need to be removed but should not be used as a source for scion material.

4.7 APPLE STEM GROOVING VIRUS

Apple stem grooving virus (ASGV), the type member of the genus *Capillovirus* is the causal agent of decline and graft union necrosis diseases in apple. Virginia crab is the only apple cultivar known to show symptoms. ASGV infection is frequently symptomless in apple cultivars.

4.7.1 CAUSAL ORGANISM

The virus belongs to:

Family: Flexiviridae
Genus: *Capillovirus*
Species: Apple stem grooving virus

4.7.2 SYMPTOMS

- Stem grooves
- Abnormal graft union

4.7.3 MEANS OF MOVEMENT AND TRANSMISSION

Transmission occurs by means not involving a vector. The virus is transmitted by mechanical inoculation (from apple, especially in spring, by inoculating extracts from buds, young leaves, or petals ground in 0.05 M phosphate buffer, pH 7–8. Additives such as 2% (w/v) nicotine base or 2% (w/v) PVP or hide powder help but are not essential) and is transmitted by grafting.

4.7.4 PREVENTION AND CONTROL

Studies reported by Mink and Shay (1962) suggest that the disease has been eliminated from some apple scions by heat therapy (30 days at 36°C; de Sequeira and Posnette, 1969).

4.8 APPLE STEM PITTING VIRUS

Apple stem pitting virus (ASPV) is a latent virus of apple, common in commercial cultivars. Symptoms occur only on woody indicators and some ornamental *Malus* species and comprise of xylem pits in the stem of Virginia crab.

4.8.1 CAUSAL ORGANISM

The virus belongs to:

   Family: Flexiviridae
   Genus: *Foveavirus*
   Species: Apple stem pitting virus (ASPV)

4.8.2 SYMPTOMS

- Dieback
- Inner bark necrosis
- Decline
- Epinasty
- Vein yellowing
- Latent infection

4.8.3 MEANS OF MOVEMENT AND TRANSMISSION

The virus is transmitted by means not involving a vector. The virus is transmitted by mechanical inoculation; transmitted by grafting; not transmitted by contact between plants; not transmitted by seed; not transmitted by pollen.

4.9 APPLE UNION NECROSIS AND DECLINE

Apple union necrosis and decline (AUND) is caused by tomato ringspot virus (TmRSV). AUND is of economic importance in commercial apple orchards where the virus is most often isolated from clonally propagated, size-controlling rootstocks. This disease is only a problem on grafted trees where the fruiting variety is resistant to TMRSV and the rootstock is tolerant. Apple cultivars vary in resistance or tolerance to TMRSV. Rootstocks tolerant to TMRSV include MM. 106, EM7A, EM26, EM9, MAC39, MAC9, P2, and Budogovsky 9, while resistant rootstocks include M.4, M.7, Ottawa 3, and Novole. Fruiting varieties resistant to TMRSV include "Delicious", "Quinte", "Tydeman's Red", "Jerseymac", and "Jonathan", while "Golden Delicious", "Empire", and "York Imperial" are tolerant. Ornamental crab apples and other *Malus* species appear unaffected, as are most apple cultivars on seedling rootstocks.

4.9.1 Causal Organism

TmRSV

4.9.2 Symptoms

Symptoms of AUND appear as infected trees reach bearing age. Bud break is often delayed in the spring, and leaves are small and sparse, their color a dull, pale green. Terminal shoot growth is reduced, with shortened internodes. Infected trees flower heavily and set large numbers of small, highly colored fruit. Leaf discoloration and leaf drop occur prematurely in infected trees. Affected trees often produce large numbers of sprouts from the rootstock. Swelling may occur above the graft union. Partial to complete separation of the graft union is common on severely affected trees; sometimes the top breaks off at the union in strong winds. Decline and death are possible, although infection is not always lethal. Removal of the bark above and below the graft union reveals it to be abnormally thick, spongy, and orange-colored, and there is a distinct necrotic line at the scion/rootstock union. Symptom severity is influenced by scion/rootstock combination; "Delicious" on MM. 106 rootstock produces extreme symptoms when infected, while "Golden Delicious", "Empire", and "York" seldom show symptoms even when the rootstock is infected with TmRSV.

4.9.3 Means of Movement and Transmission

The virus is present in common broadleaf orchard weeds, such as dandelion, and dagger nematodes may spread the virus from these weeds to apple trees. Also, the virus may be spread through grafting and from orchard to orchard by seeds from infected dandelion.

4.9.4 Prevention and Control

- To avoid introducing the virus into new plantings, purchase certified virus-free trees grown in soil fumigated to control nematodes.
- Reduce populations of nematode vectors and weed hosts by cultivating the future orchard sites for two years before planting.
- Soil fumigation can be used to reduce nematode and weed levels but is neither economically efficient nor environmentally desirable.

4.10 BLISTER SPOT OF APPLE

This bacterial disease is of economic importance mainly on the cultivar Mutsu (Crispin) but can be seen on Golden Delicious when grown adjacent to Mutsu. Even though fruit grow to maturity and no detectable yield loss occurs, severe infection results in ugly fruit and greatly reduces fresh market quality.

4.10.1 Causal Organism

Species	Associated Disease Phase	Economic Importance
Pseudomonas syringae pv. *papulans*	Blister like Spots on Fruit and Lesions on Leaves (Sometimes)	Severe

4.10.2 Symptoms

Blister spot is first observed as small raised green blisters, which develop on fruit from early- to mid-July. Infections of blister spot are first noticeable two to three months after petal fall as small,

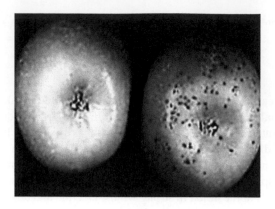

FIGURE 4.16   Blister spots on apple.

green, water-soaked, raised blisters that develop at fruit stomata. These spots result in purplish black lesions associated with fruit lenticels. As the fruit increase in size, the lesions expand to about 3/16 inches (5 mm) and become darkened (Figure 4.16). These blisters are associated with stomata and continue to expand during the growing season. A few to 100s of lesions may develop per fruit. The decay will rarely extend more than 1–2 mm into the fruit. No leaf spot or wood canker symptoms have been associated with this disease; however, a mid-vein necrosis of Mutsu apple leaves has been observed prior to fruit lesion development. The disease causes superficial blemishes on the fruit, making them unsuitable for fresh market. Occasionally, other fungi enter the blister spot lesion resulting in decayed areas on the fruit.

4.10.3   Cause and Disease Development

P. syringae pv. *papulans* is very widespread in Mutsu orchards. The bacterium survives the winter in dormant buds. Up to 40% of the dormant buds in an orchard may harbor the pathogen. The bacterium may also overwinter in infected fruit on the orchard floor.

4.10.4   Favorable Conditions of Disease Development

Warm, humid, or wet conditions during the spring and early summer favor a buildup of bacterial populations and subsequent infections of leaf veins and fruit. During late spring or early summer, a brief shower is all that is required to distribute the bacteria onto the developing fruit where they infect through the fruit pores or lenticels.

4.10.5   Disease Cycle

The bacterium overwinters in a high percentage of apple buds, leaf scars, and diseased fruit on the orchard floor. Throughout the growing season, the bacterium can survive as an epiphyte on foliage and fruit in the orchard. Even though the highest populations of the pathogen have been found on Mutsu, the bacterium have also been detected on foliage and fruit of other apple cultivars. Young Mutsu fruit show an increased susceptibility to infection for about six weeks, beginning about two weeks after petal fall[7] (Figure 4.17).

4.10.6   Management

The disease is mainly a problem on the apple cultivar Mutsu. When Mutsu is inter-planted with other (normally) resistant apple cultivars (i.e. Red Delicious, Cortland, and others), the pathogen may

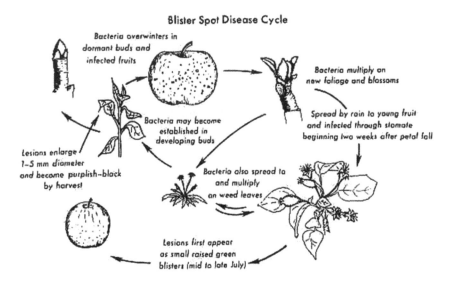

FIGURE 4.17   Disease cycle of blister spot.

spread into these, also. Prior to the development of streptomycin-resistant strains of the pathogen, the disease could be controlled with three well-timed antibiotic sprays, the first applied no later than two weeks after petal fall, and the others applied weekly thereafter. This strategy is still employed in orchards without resistant strains; however, resistant strains may develop after only a few years of antibiotic use. Once resistance to the antibiotic develops, further use of antibiotic is ineffective.

4.11   BLUE MOLD OF APPLE

Blue mold, a common rot of stored apples and pears, is caused by the fungus *Penicillium* spp. like *P. expansum* and *P. italicum*. Other names for the disease are "soft rot" and "penicillium rot". Soft rot is a disease of ripe fruit and develops mostly on apples that are picked before they are mature. Firm fruit in the same container as decaying fruit might absorb a moldy odor and flavor.

4.11.1   CAUSAL ORGANISM

Species	Associated Disease Phase	Economic Importance
Penicillium spp. like:	Soft Rot of Fruit	Major
P. expansum		
P. italicum		

4.11.2   SYMPTOMS

The rot appears as soft, watery spots. The decayed portions are sharply separated from the healthy tissues. If the margin of the rot is rubbed lightly, a sharp margin of healthy flesh can readily be detected. Blue mold originates primarily from infection of wounds such as punctures, bruises, and limb rub on the fruit. Blue mold can also originate from infection at the stem of fruit. Stem-end blue mold is commonly seen in apples such as Red Delicious. Calyx-end blue mold occurs on Red Delicious apples but is usually associated with fruit that are drenched prior to storage.

The decayed area appears light tan to deep brown. The decayed tissue is soft and watery, and the lesion has a very sharp margin between diseased and healthy tissues. Decayed tissue can be readily separated from the healthy tissue, leaving it resembling a "bowl". Blue or blue-green spore masses may appear on the decayed area, starting at the infection site. Decayed fruit has an earthy, musty odor. The presence of blue-green spore masses at the decayed area and associated musty odor are the positive diagnostic indication of blue mold. Without the presence of spore masses of blue mold, blue mold can be misdiagnosed as Mucor rot, but a sweet odor is commonly associated with Mucor rot. The spots range in color from brown to pale straw and show all possible variations in size. They may occur on any part of the fruit. The spots are shallow at first, but they extend deeper very rapidly—in fact, just about as rapidly as they increase in diameter on the surface—so that by the time the rot reaches the core it has involved a third or more of the fruit. Internally the decayed tissue is watery and has a glassy appearance. It can readily be scooped out from the healthy tissue. Development of a surface growth of blue mold is determined more by temperature and moisture conditions than by the size of the spot. In cool, dry air, surface mold rarely appears, even when the fruits are totally decayed. In air that is moist and warm, surface mold is almost sure to appear on spots of any size. Usually small spore-bearing fungus tufts appear on the surface. These are white at first and bluish-green later. The blue-green color is due to the spores (Figure 4.18).

4.11.3 CAUSE AND DISEASE DEVELOPMENT

In the orchard, *Penicillium* spp. survives in organic debris on the orchard floor, in the soil, and perhaps on the dead bark on the trees. Conidia are also present in the air and on the surface of the fruit. In the packinghouse facility, DPA—or fungicide-drench solutions—flume water and dump-tank water are common sources of *Penicillium* spores for fruit infection during the handling and packing processes. Spores of *P. expansum* are also commonly present in the air and on the walls of storage rooms (Figure 4.19).

Blue mold sometimes produces a superficial growth even where it is not causing rot. Where rot exists, the fungus can be found in the rotted tissues, even in the smallest spots; later the fungus may become evident externally in the form and color already described.

Blue mold rot occurs on mashed or overripe fruits around the packinghouse, where they become an important source of infection. Despite careful handling methods and packinghouse sanitation,

FIGURE 4.18 Decayed area—brown, soft, and watery, with sharp margin; mostly originating from infection of wounds; blue-green spore masses.

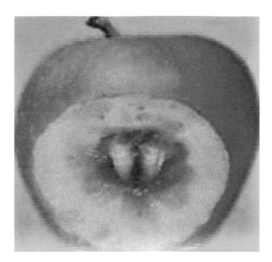

FIGURE 4.19   Decayed tissue completely separable from the healthy tissue, leaving it like a "bowl."

most if not all the fruits carry blue mold spores on their surface when they are packed. If conditions in storage or transit are favorable, this spore load can eventually give rise to fruit decay.

4.11.4   FAVORABLE CONDITIONS OF DISEASE DEVELOPMENT

Environmental conditions such as moisture, ventilation, and temperature directly influence the development of decay. The atmospheric moisture necessary to prevent apples from shriveling is sufficient for blue mold development. Lack of ventilation due to tight packing and to lack of air space in stowing increases the moisture around the fruits and slows the rate of cooling, making conditions usually favorable for rot development. Blue mold develops more rapidly at temperatures higher than the usual storage temperature for apples. Apples wounded by small pin punctures and inoculated by a spore-suspension dip, developed small but visible lesions in 30 days at -0.5°C to 0°C. By the end of 60 days such lesions had enlarged to 3/4 to 1 inch in diameter.

4.11.5   DISEASE CYCLE

Spores of the soft rot fungus are present almost everywhere and can survive prolonged periods of unfavorable conditions. Injuries to fruit, especially during picking and handling operations, are the primary points of entry. At ordinary temperatures, infected fruit can rot in two weeks or less.

4.11.6   MANAGEMENT

Orchard sanitation to remove decayed fruit and organic debris on the orchard floor helps reduce inoculum levels of *Penicillium* spp. in the orchard. Good harvest and handling management to minimize punctures and bruises on the fruit helps prevent the fruit from infection in wounds by *P. expansum* and other *Penicillium* species.

Thiabendazole is commonly used as either a pre-storage drench treatment or a line spray to control gray mold and blue mold. TBZ is effective to control gray mold but is not effective to control TBZ-resistant *Penicillium*. Two new postharvest fungicides, fludioxonil (Scholar) and pyrimethanil (Penbotec), can be used as drenches, dips or line sprays and have been reported to be effective to control blue mold originating from wound infections.

Biocontrol agent BioSave 110 (*P. syringae*) applied on the packing line helps control blue mold from infection of wounds.

Sanitizing dump-tank and flume water is an essential practice to reduce infection of fruit by *Penicillium* spp. during the packing process. Fruit bins and storage rooms can harbor TBZ-resistant isolates. Bin and storage room sanitation may be beneficial in reducing TBZ-resistant populations in the packing facility.

- Benomyl, Thiophanate-methyl
- 2% bleaching powder—5 minutes and 0.2% flit 406—10 minutes
- TBZ—0.1% incorporated in waxol-0-12—under refrigeration
- Pre-storage dip in TBZ (500ppm) for 2–3 minutes

4.12   BOTRYTIS ROT OF APPLE

Botrytis rot caused by the fungus *Botrytis cinerea* is one of the most common causes of rotting in stored apples. The disease is also known as gray mold, dry eye rot, or blossom-end rot. Gray mold (*B. cinerea*) is a common post-harvest disease on apples worldwide. This fungus can spread from decayed fruit to surrounding healthy fruit through fruit-to-fruit contact during storage. The disease develops more rapidly at cold storage temperatures than any other rot. Losses can be significant with up to 12% in untreated fruit. The rot occurs mainly in storage both as a wound rot and as a primary rot. It is rarely seen in the orchard as an extensive rot pre-harvest but occurs as a blemish or slight rot at the calyx end of the fruit which dries to form dry eye rot.

4.12.1   CAUSAL ORGANISM

Species	Associated Disease Phase	Economic Importance
B. cinerea	Dry Eye Rot at the Calyx End of the Fruit	Major

4.12.2   SYMPTOMS

4.12.2.1   In the orchard

Botrytis fruit rot is rarely seen in orchards as a rot. On apple it may be visible as dry eye rot at the calyx end of the fruit (Figure 4.20). The symptoms range from a slight skin red blemish on one side of the calyx to a distinct one-sided rot which has dried and shrunk to form the typical dry eye rot. The presence of such symptoms in the orchard usually bears no relationship to the subsequent incidence of botrytis rot in store.

4.12.2.2   In Store

Botrytis rot associated with wounds tends to be regular in shape, firmish, pale to mid-brown in color, often with a darker area around the calyx and lenticels (sometimes reddish spots), giving the fruit a freckled appearance. botrytis rot associated with calyx (eye) infection varies in color from pale to dark brown and is irregular in shape, often appearing as fingers of rot extending down from the calyx (Figure 4.21). Similar rotting may also originate from the stalk end or on the cheek, which may suggest a core rot origin. This fungus can spread from decayed fruit to surrounding healthy fruit through fruit-to-fruit contact during storage (Figure 4.22).

4.12.2.3   On Other Apple Varieties Botrytis Rot is Mainly Mid-Brown

Infected fruit initially remain moderately firm becoming softer with time. Mycelium with gray spore masses may be visible particularly on the calyx or around the wound. Once out of the store, these spore masses become more abundant and are a useful aid to identification. Very occasionally on apple, large black resting bodies (sclerotia) may be seen, particularly at the wound where the rot originated.

FIGURE 4.20   Early stage of dry eye rot.

Botrytis rot spreads in store by contact and nests of rot may therefore be visible in later stored fruit.

4.12.3   CAUSE AND DISEASE DEVELOPMENT

The dry eye rot pathogen (*B. cinerea*) overwinters as sclerotia, usually in infected apples left on the orchard floor from the previous season. In the spring, sclerotia produce spores that infect the sepals or petals during bloom. The fungus remains quiescent until fruit begins to mature later in the summer.

4.12.4   FAVORABLE CONDITIONS OF DISEASE DEVELOPMENT

Infections usually occur during bloom or shortly after, although frequently symptoms are not visible until several weeks later. Rot symptoms usually become visible about 1 month after petal fall. Dry eye rot often develops later in the season (late July to early August).

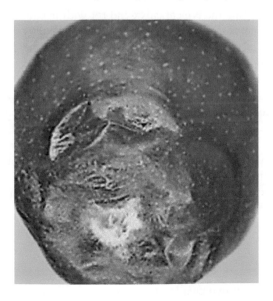

FIGURE 4.21   Matured symptoms.

FIGURE 4.22   Spread of the disease through fruit-to-fruit contact.

4.12.5   DISEASE CYCLE

B. cinerea is ubiquitous in the orchard being present as sclerotia in soil on plant debris, weeds, windbreak trees, mummified fruits, and bark. In wet, windy weather, at most times of the year, the sclerotia sporulate and the spores (conidia) are spread by wind and rain. At blossom time, spores will infect dying blossom and remain as latent infections in the remains of the flower parts still attached to the developing fruits, or become established as latent infections in the calyx. Occasionally the fungus continues to develop and form a small rot or blemish around the calyx. This does not usually progress far and then dries forming the dry-eye rot lesion.

Usually though, the fungus does not start to rot fruit from these blossom infections until the fruit has been in store for several months, generally from December onward. Then the fungus will invade the fruit at the calyx end forming the typical calyx-end rot with irregular fingers of rot spreading down from the calyx. Once developed, the fungus can spread to healthy fruit in the bin by contact spread, forming large nests of rotted fruit. Research has shown that most apples are symptomless when infected with botrytis during flowering.

However, not all infected fruit subsequently rot in store. The factors that affect the development of botrytis eye rot in storage are not fully understood. Controlled atmosphere storage, especially low oxygen, appears to encourage rot development but further research is needed to determine other factors that may be involved. Botrytis may also act as a wound pathogen, where it behaves more like penicillium rot.

Fruit becomes infected via wounds sustained during harvesting and handling, particularly from botrytis spores contaminating drench tanks and water flumes on grading machines. Botrytis that invades via wounds starts rotting immediately in store and the rot readily spreads to healthy fruit in the bin causing extensive nesting of rots.

4.12.6   MANAGEMENT

4.12.6.1   Cultural Control
- Since botrytis is ubiquitous in orchards, elimination of inoculum sources is impossible and cultural methods of control are not appropriate for control of botrytis eye rot.

- However, successful prevention and control of botrytis as a wound rot, similar to the prevention and control of penicillium rot, is dependent on good crop handling and hygiene.
- In the orchard, throw discarded fruit into the alleyway where they can be macerated and more rapidly broken down.
- Remove old rotted fruit from bulk bins and scrub and clean as they come off the grader.
- Keep pack-house areas clean to minimize contamination of water flumes in pack-houses.
- Supervise pickers at harvest to minimize fruit damage and ensure damaged fruit is not stored.
- Ensure that only fruit of the correct mineral status is stored long term.

4.12.6.2 Biological Control

- Research in other countries, particularly the United States has identified various microbial antagonists of botrytis which have been developed as biocontrol agents for use as post-harvest treatments.
- These are generally also active against penicillium rot. Examples include Yield Plus (yeasts ex South Africa).
- These appear to be effective against wound fungi such as botrytis and penicillium, but not against orchard fungi, or botrytis rot arising from latent infections of the calyx.
- Currently, there are no commercially available biocontrol agents approved for use on apples.

4.12.6.3 Chemical Control

Pre-harvest orchard sprays:

- In Europe and the United States, fungicide sprays applied at blossom time have given some control over botrytis eye rot in store. In the United Kingdom, in trials, similarly timed sprays were ineffective.
- A pre-harvest spray of Bellis (pyraclostrobin + boscalid), Switch (cyprodonil + fludioxonil) or ThianosanDGor Triptam (thiram) will give some control over the botrytis wound rot.

Post-harvest treatment:

- On Bramley, the use of diphenylamine for control of scald will also give some protection against botrytis rot.
- Treatment of water with chlorine will reduce inoculum levels of botrytis present in the drench tank water and reduce wound infections due to *B. cinerea*.
- It will not control botrytis present as a latent infection.

Avoiding fungicide resistance:

- Tests have shown that 60% of *B. cinerea* isolates from apple are resistant to benzimidazole fungicides.
- It is not known whether there are isolates resistant to Bellis (pyraclostrobin + boscalid) or Switch (cyprodinil + fludioxonil).

4.13 DISORDERS OF APPLE

4.13.1 BITTER PIT

This is a physiological disorder, which reduces the fresh market quality of fruit. Bitter pit is a physiological disorder believed to be induced by a calcium deficiency in apple fruits. Young trees that are

FIGURE 4.23   Symptoms of bitter pit.

just coming into bearing are the most susceptible. Immature fruit are more susceptible to bitter pit than fruits harvested at the proper maturity stage.

The incidence of bitter pit usually occurs during storage, but in some cases, it can also develop at harvest. With the right conditions, bitter pit can occur on Delicious, Idared, Crispin, Cortland, Empire, Honeycrisp, and other varieties.

Bitter pit starts internally and eventually causes external blemishes. Internal lesions occur anywhere in the tissue from the core line to the skin, but more commonly just below the skin. The lesions are small, brown, dry, slightly bitter-tasting and about 3–5 mm in diameter (Figure 4.23). As the affected cells die, they lose moisture, become "corky", and the skin over the area sinks in a round or slightly angular pattern. The skin over the pits gradually becomes slightly brown or sometimes black. When cut, the tissue starts turning brown beneath the skin.[8]

Even if calcium is present in the soil, it does not readily move into the plant, it requires moisture and many trees are not irrigated. Once the calcium enters the plant by root tips, it does not move well in the plant. When calcium does move, it is directed to the growing tips of the tree and fruit. All the growing tips compete with the fruit for calcium, therefore, fewer tips more calcium for the fruit and less bitter pit.

4.13.1.1   Management

Avoid and remove excessive tree growth, which in many cases are caused by excessive nitrogen fertilizer application and excessive pruning. In winter this can minimize the frequency and severity of bitter pit. Avoid irregular watering, large fruit, light crop load, and elevated levels of potassium and magnesium.

Add calcium in the form of agricultural lime, it does not contain as much magnesium as dolomite lime. The incidence of bitter pit may also be reduced and sometimes controlled by several applications of calcium in post-harvest calcium drenches. One product, Nutra-Plus Calcium 8% can be applied every ten to fourteen days during the growing season. This is how commercial growers control bitter pit.

Also, there are a few ways of preventing bitter pit from affecting apple and plant trees. The most effective way of prevention is spraying the trees with calcium chloride or calcium nitrate. This is so effective because bitter pit becomes a problem when trees undergo periods of stress, usually caused by severe weather and they sacrifice their current crop to survive. To do this, the trees extract calcium from the fruit. The result is calcium-depleted fruit because the trees do not replenish the calcium lost in the fruit.

4.13.2   Water Core

Water core is a physiological disorder of apple fruit characterized by water-soaked tissue around the vascular bundles or core area due to the spaces between cells becoming filled with fluid instead of air. The fruit tissue's abnormal condition is a response of some apple cultivars to adverse environmental

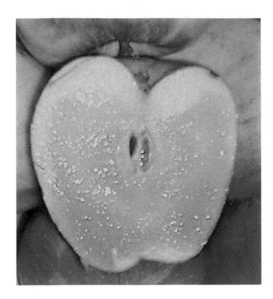

FIGURE 4.24   Water core in apple.

conditions—not to a disease-causing organism. Affected tissue is water-soaked and glassy looking. Generally, the damage is not visible on the skin, and it is only when fruit are cut that the damage is seen. In severe cases, flesh can be affected right up to the skin, which then darkens over affected areas. It is especially a problem where the weather is very hot followed by rain and cold nights around the time of fruit maturation. The lack of calcium, too much nitrogen, girdling of trunks and limbs, soil conditions, and heavy pruning around the time the fruit is maturing also plays a part in causing this effect. Certain varieties, like Delicious, Braeburn, Sundowner, Lady Williams, Granny Smith, and Fuji are more prone to apple water core.

The symptoms are water-soaked regions in the flesh of the apple, which is only seen when the apple is cut (Figure 4.24). In very severe cases it is visible on the skin of the apple. The water-soaked appearance of affected fruit results from the accumulation of sorbitol-rich solutions and water in the intercellular spaces of the apple flesh. It is believed that the sorbitol and water is forced into the fruit by sap pressure when too many leaves are pruned off the tree at fruit maturity or the weather is very hot with cold nights. Sorbitol is the carbohydrate source translocated into the fruit and is converted to fructose by the apple fruit.

Affected apples usually do not store well and may rot by secondary fungi. They may also give off a sweetish, fermented odor. The fruits are edible, but the texture is unpleasant. The extent of apple water core is varied and in mild cases, it is concentrated around the growing points of the fruit such as the core, and in severe cases, it radiates out to the skin. Well-drained soil is usually not affected by this condition.

4.13.2.1   Management

The most effective way to reduce the incidence is to avoid delayed harvests. As fruits approach maturity stage, samples of fruit should be examined for water core development. Fruit should be harvested before water core develops extensively. Fruit lots with moderate to severe water core symptoms should not be placed in controlled atmosphere (CA) storage but should be marketed quickly.

4.13.3   BROWN HEART

This physiological disorder is associated with large and overmature fruits. It can also occur when the CO_2 concentration in storage increases above 1%. The symptoms appear as brown discolorations in the flesh, usually originating in or near the core. Brown areas have well-defined margins and may include dry cavities developed due to desiccation. Symptoms range from a small spot of brown flesh to entire browning of flesh with a margin of healthy white flesh remaining just below the skin. Symptoms develop early in storage and may increase in severity with extended storage time.

4.13.3.1 Management

Harvesting of overmature fruits should be avoided. In case of storage in CA the fruits should be harvested at optimum maturity. The CO_2 concentrations in CA should be below 1% to reduce the development of brown heart incidence.

4.13.4 APPLE SUNBURN (SUNSCALD)

Sunburn is the name applied to a golden or bronze skin discoloration of apples caused by exposure of 1 side of the fruits to intense sunlight on the tree. This physiological disorder occurs due to intense heat of the sun. Fruit on the southwest side of the tree are generally affected. Water stress can also increase the incidence of sunburn. Inadequate ventilation in storage rooms or in packaging boxes also promotes this disorder. It detracts from the appearance of the fruits, but normally the skin is not killed, and the tissues show no sign of breakdown. In fact, the flesh in the sunburned area may be firmer than that of the rest of the fruit at harvest, but it tends to soften rapidly in storage.

Initial symptoms are white, tan or yellow patches on the fruits exposed to the sun (Figure 4.25). With severe skin damage, injured areas of the fruit can turn dark brown before harvest. These areas may become spongy and sunken. Fruit exposed to the sun after harvest can develop severe sunburn.

Sunburn results from heat stress to the fruit leading to injury of the affected cells. The temperature of the surface of the fruit can be as much as 18°C (32°F) above air temperatures when the fruit is exposed to solar radiation and is 8°C to 9°C (14°F to 16°F) warmer than the shaded side of the fruit. The transpiration or evaporation of water from the apple fruit helps to cool the fruit while it is attached to the tree.

4.13.4.1 Management

The best method of control is to avoid sudden exposure of fruit to intense heat and solar radiation. Proper tree training and pruning are critical. Summer pruning must be carefully done to avoid

FIGURE 4.25 Sunburn on apple.

FIGURE 4.26   Apple scald.

excessive sunburn. Pruned orchards should be regularly irrigated to reduce heat stress. Careful sorting to remove affected fruit upon packing is the only solution once the injury has occurred.

4.13.5   SCALD

This physiological disorder is a critical concern for apple growers. Susceptibility to this storage disorder varies with the variety of apple, environment, and cultural practices. Incidence and severity of scald are favored by hot, dry weather before harvest, immature fruit at harvest, high nitrogen, and low calcium concentrations in the fruit. Inadequate ventilation in storage rooms or in packaging boxes also promotes this disorder. Irregular brown patches of dead skin develop within three to seven days due to warming of the fruit after removal from the cold storage.

The warm temperatures do not cause the scald but allow symptoms to develop from previous injury, which occurred during cold storage. Symptoms may be visible in cold storage when injury is severe (Figure 4.26).

4.13.5.1   Management

Harvesting at proper maturity and ventilation in cold storage help reduce the scald incidence. The most common method used to control scald is the application of an antioxidant immediately after harvest. Diphenylamine (DPA) is commonly used. Ethoxyquin is also effective for some varieties but can cause damage to other apple varieties. Antioxidants should be applied within one week of harvest for maximum control.

REFERENCES

1. Peter, K. A., PH.D., *"Apple Disease - Alternaria Leaf Blotch"*, Pennsylvania State University Extension, October 17, 2017. https://extension.psu.edu/apple-disease-alternaria-leaf-blotch
2. Central Science Laboratory, Sand Hutton, York, *"Alternaria Blotches on Apple and Pear EC Listed Diseases"*. https://www.adlib.ac.uk/resources/000/193/618/alternaria-defra.pdf
3. Biggs, A. R., *"Apple Scab"*, West Virginia University, August 14, 2013. http://articles.extension.org/pages/66202/apple-scab
4. Steiner, P. W., University of Maryland and Biggs A. R., West Virginia University, *"Fire Blight"*, 1998. http://www.caf.wvu.edu/Kearneysville/disease_month/fireblight.html
5. Ellis, M. A., *"Sooty Blotch and Fly Speck of Apple"*, Department of Plant Pathology, Ohio State University Extension, April 13, 2016. https://ohioline.osu.edu/factsheet/plpath-fru-41

6. Grimova, Lenka, Winkowska, Lucie, Konrady, Michal and Rysanek, Pavel, *"Apple Mosaic Virus"*, Phytopathologia Mediterranea (2016) 55, 1, pp. 1–19.

7. Zwet, T. van der, Yoder, K. S., and Biggs, A. R., *"Blister Spot of Apple"*, Mid-Atlantic Orchard Monitoring Guide (NRAES-75), August 30, 2011. http://articles.extension.org/pages/60624/blister-spot-of-apple

8. Home Orchard Society, *"Bitter Pit: Cause And Control"*. http://www.homeorchardsociety.org/growfruit/trees/bitter-pit-cause-and-control.

9. de Sequeira, O.A. and Posnette, A.F. (1969). *Commonw. Bur. Hort. Pl. Crops. Tech. Commun.* No. 30, Suppl. 2/3/4, 76a.

10. Mink, G. I., and Shay, J. R., 1962. Latent viruses of apple. *Purdue Agric. Exp. Stn. Res. Bull.* 756.

5 Banana

The word banana is believed to be originated from West Africa, possibly from the Wolof word banana, and passed into English via Spanish or Portuguese. Banana is a fruit which is botanically a berry and produced by largest herbaceous flowering plants in the genus Musa. The fruits grow in clusters hanging from the top of the plant.

Banana is a staple food in East and Central Africa (ECA), where it provides approximately 20% of the total calorie consumed per capita. Production of bananas in ECA, however, has declined since the 1970s, and now yields a fraction of its potential (Van Asten et al., 2005). While low yields are partly due to poor soil fertility in the region, pests and diseases have played a significant role in reducing banana production. Breeding for resistance in banana is complicated by a number of constraints including pathogen and pest diversity. Hence, screening against the different key races and species is essential. This chapter provides information about diseases which are affecting banana produce and possible effective remedies for those diseases.

5.1 BANANA BACTERIAL WILT

Banana bacterial wilt (BBW) is a bacterial disease and all banana cultivars in affected areas are susceptible to BBW. It has been found to be very destructive with incidence of 70–80% in many plantations and yield losses of 90% have been reported on some farms.[1]

5.1.1 CAUSAL ORGANISM

Banana xanthomonas wilt (BXW) is caused by the bacterium Xanthomonas campestris pv. Musacearum (Xcm). BXW was first observed on a close relative of banana Ensete ventricosum about 90 years ago in Ethiopia. In 1974, it was reported for the first time on bananas also in Ethiopia.

5.1.2 SYMPTOMS

First symptoms are a dull green color of the lamina which gradually it assumes a scalded appearance and wilting back on its midrib (Figure 5.1). Cross-sections of diseased petioles or pseudo stems reveal a yellowish coloration (Figure 5.2). There is uneven and premature ripening of the fruit, and when fruit is cut, the sections show unique yellowish blotches in the flesh fingers, dark-brown placental scars, and uneven and premature ripening which leads to rotting (Figure 5.3) and discoloration of the fresh fruits (Figure 5.4).

5.1.3 CAUSE AND DISEASE DEVELOPMENT

The Xanthomonas pathogen can infect banana plants in numerous ways. The most efficient natural pathway of disease spread is through insect vectors that pick up the pathogen when visiting plants to collect nectar or pollen. Insect vectors are of highest importance in mid-altitude agroecological conditions (1100–1600 m above sea level), where vectors are thought to be more active and to occur more abundantly.

The other main pathway of disease spread is through human activity: mainly using tools that are contaminated by contact with infected plants and by utilizing infected suckers to establish new banana plantations in distant locations. Contamination through tools is a crucial factor where banana farms are more intensely managed, e.g. where farmers regularly remove excess suckers, dry fibers, or green leaves.[2]

FIGURE 5.1   Early wilt symptoms in a young banana.

FIGURE 5.2   Pseudo stems reveal yellowish coloration.

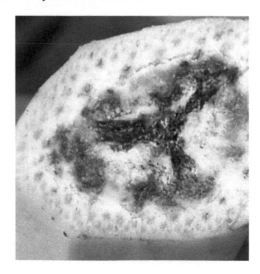

FIGURE 5.3   Internal rotting of banana fruit.

FIGURE 5.4   Uneven and premature ripening of the fruit and discoloration of the fresh fruit.

5.1.4   MANAGEMENT

Decapitation of male flower buds after flowering, disinfecting the farm tools after use and disinfecting hands, and destruction of affected banana stools and burying them reduced the inoculum. This has been found to reduce infection. Restriction of movement of banana plant material from infected areas to other areas also reduces the spread of BBW (through quarantine). There are, however, problems with the practicability of implementation.

To reduce insect-transmitted infections, farmers are advised to remove the male flower immediately after the last cluster of fruits has formed. Removing the male flower, a process called "debudding", ensures there is no nectar or pollen to attract insects and no entry points for the bacteria into the plant (Figure 5.5).

5.1.4.1   Disease Cycle

When farmers consistently remove the male flowers the impact of insect vectors is reduced drastically, even when ecological conditions are favorable.[2]

The risk of tools spreading disease can be minimized by disinfecting the knife (e.g. by dipping in bleach) and using pesticides or fire after working on each different mat, while the use of suckers from sources certified to be healthy reduces further spread.

The problem associated with BXW is that once plants have been infected there is no remedy other than to uproot them since they will eventually die off. A new banana crop cannot be planted immediately after uprooting and removing infected mats because pathogen cells released into the soil can survive for up to six months. If there are any remaining plant parts in the soil the pathogen can survive for up to a year. After uprooting infected mats, it is recommended to leave the field fallow or grow a crop that is not closely related to banana, and thus, cannot be infected by the Xanthomonas pathogen.

5.2   BANANA BRACT MOSAIC VIRUS

Banana bract mosaic virus (BBrMV) causes significant yield losses in banana and abacá in the Philippines, India, and Sri Lanka. BBrMV could potentially be confused with sugarcane mosaic

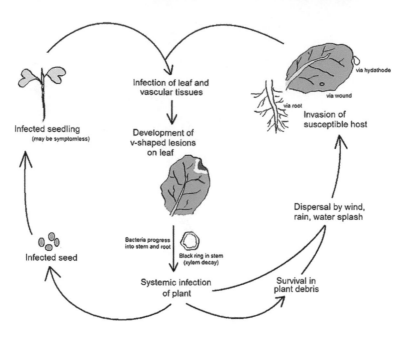

FIGURE 5.5   Disease cycle of X. campestris.

virus (syn. abaca mosaic virus), another potyvirus infecting both banana and abacá in the Philippines and inciting similar foliar symptoms. (Courtesy: Dr. Andrew Geering.)

5.2.1   CAUSAL ORGANISM

BBrMV is a plant pathogenic virus of the

   Family: Potyviridae
   Genus: Potryvirus
   Species: Banana bract mosaic virus.

The virus is known to infect only banana and can be transmitted in a non-persistent manner (i.e. retained by the vector for a brief period) by at least three species of aphid, Pentalonia nigronervous, Aphis gossypii, and Rhopalosiphum maidis.

5.2.2   SYMPTOMS

BBrMV causes yellow streaks on leaves, splitting of the pseudo stem, abnormal emergence of the bunch from the middle of the pseudo stem, production of fewer fruits of smaller size, dark reddish-brown mosaic symptoms, which are especially obvious in bracts (Figure 5.6) and spindle-shaped chlorotic streaks running parallel to the veins (Figure 5.7). BBrMV also induces red diamond-shaped patches on the pseudo stem (Figure 5.8) with splitting at the base and midrib of leaves and dark patches on the flower of the banana flower.[3]
   Distinctive dark colored mosaic patterns, stripes, or spindle-shaped streaks are also visible.
   The symptoms can be summarized as,

   • Fruit—reduced size, abnormal patterns
   • Inflorescence—abnormal leaves (phyllody), mosaic, lesions; flecking; streaks, discoloration

FIGURE 5.6   Characteristic mosaic on bract.

FIGURE 5.7   Chlorotic streaks running parallel to the veins.

FIGURE 5.8   Red to dark-brown patches on pseudo stem.

- Leaves—abnormal colors, abnormal patterns, abnormal forms
- Stems—discoloration of bark
- Whole plant—dwarfing

5.2.3   MEANS OF MOVEMENT AND TRANSMISSION

The virus can be transmitted in a non-persistent manner (i.e. retained by the vector for a short period) by at least three species of aphid, P. nigronervous, A. gossypii, and R. maidis. The virus can also be transmitted by infected planting material.

5.2.4   PREVENTION AND CONTROL

Eradication of the mealy bugs that transmit the disease is essential. But research has shown that aphid-transmitted viruses of Musa cannot be completely controlled using aphicides or biocontrol agents that are effective against banana aphids. However, these methods may reduce inoculum levels and slow the rate of the spread.

The symptoms of the disease should be studied carefully for early detection and immediate eradication of infected plants. Virus-free propagating materials should be planted.

5.3   BANANA MILD MOSAIC VIRUS

Banana mild mosaic virus (BanMMV) is a recently characterized virus of the family Flexiviridae, which infects banana and plantains. BanMMV has a very wide geographical distribution.

5.3.1   CAUSAL ORGANISM

BanMMV is a recently described filamentous virus found in symptomatic plants of banana. The symptoms associated with BanMMV in banana have been reported to be generally mild, but increased symptom severity has been suggested in cases of mixed infection with other banana infecting viruses, such as cucumber mosaic virus (CMV) (Caruana and Galzi, 1998) and banana streak virus (BSV) (Thomas et al., 1999).

5.3.2   SYMPTOMS

BanMMV infection causes transitory mild chlorotic mosaic and streak symptoms on leaves of highly susceptible cultivars and is a symptomless infection otherwise. Mixed infections of BanMMV and BSV are common, and in such cases, symptoms are reminiscent of those caused by BSV alone (Thomas et al., 1999).

5.3.3   MEANS OF MOVEMENT AND TRANSMISSION

BanMMV is spread by the movement of infected propagation material. In the case of banana, two vegetative propagation techniques have the potential to give rise to infected plants and contribute to the spread of the virus: the use of suckers or rhizome pieces and shoot-tip culturing to mass produce in-vitro plants.

Although BanMMV is a pathogenic agent for banana, it can be concluded that the economic consequences of its further entry in the banana plantations can be considered low since BanMMV, which is present in most if not all banana growing areas of the world, does not appear to be associated with significant yield loss in banana.

5.3.4 Prevention and Control

Removal of infected plants and the use of pathogen-free propagation material seem to be the only effective control method known to science now. It also indicates that no resistant plant has been described to date.

5.4 BANANA STREAK VIRUS

BSV is an important disease of Musa and affects the productivity of both bananas and plantains. The disease is reported in nearly all countries where this crop is grown including Mauritius, India, and many countries of the African continent.

BSV infection includes yield losses and restricts the international exchange of banana germplasm. In recent years, the virus has caused increasing concern worldwide as infection of new banana hybrids with many desirable traits frequently occurs, curtailing their exploitation. Recent reports indicate that BSV infection increase from the activation of viral sequences that are integrated into the Musa genome.

5.4.1 Causal Organism

BSV is the causal agent of viral streak of banana and plantains and is the most widely distributed virus of these crops. The virus is a para-retrovirus of

Family: Caulimoviridae
Genus: Badnavirus
Species: Banana streak virus

with non-enveloped bacilliform particles (episomal virus) containing a 7.4 kb circular dsDNA genome that has been replicated by reverse transcription. The different virus strains show a high degree of serological and genomic heterogenicity.

5.4.2 Symptoms

Foliar symptoms caused by BSV initially resemble those caused by CMV. However, the leaves develop chlorotic (light green/yellow) stripes or flecks (Figure 5.9), which become progressively darker with age and turn necrotic (black) (Figure 5.10). The symptom's expression is periodic. Streaks or flecks are not evident on all leaves and for months at a time, emerging leaves may not show any symptoms. A characteristic of BSV infection is the periodicity of symptom expression. Plants may not show streaks on all leaves and, for several months at a time, emerging leaves may be

FIGURE 5.9   Yellow/light green striations (initial symptom).

FIGURE 5.10   Necrotic streaks (mature symptom).

symptomless or show only slight symptoms. For this reason, the suspected banana plants must be observed for at least nine months. Infected plants have

- Narrow thicker leaves
- General leaf distortion
- Leaf twisting
- Broad yellow lines on the leaf lamina parallel to the midrib
- Small leaves with lamina absent
- Purple margin on leaf lamina
- Abnormal arrangement of leaves on pseudo stem
- Grooves in bases of petioles
- Choking
- Death of the growing point inside pseudo stem
- Splitting/cracking on leaf bases on the pseudo stem
- Small bunches of very short fruits

5.4.3   MEANS OF MOVEMENT AND TRANSMISSION

BSV is transmitted in a semi-persistent manner by the citrus mealy bug Planococcus citri from banana to banana. Other vectors may also be involved. In experiments, the pink sugarcane mealy bug Saccharicoccus sacchari transmits Sugarcane Bacilliform Virus (ScBV) from sugarcane to banana (Figures 5.11 and 5.12).

5.4.4   PREVENTION AND CONTROL

Banana streak disease can be controlled by the eradication of affected plants and the use of BSV-free planting material. BSV can be carried in in-vitro plantlets as it is not eliminated by shoot-tip culture. Virus particles can only be detected in areas of leaf tissue with symptoms. Parts of leaves with pronounced symptoms should be used for serological indexing.

5.5   CMV

Banana mosaic or infectious chlorosis is one of the important and widely distributed viral diseases of banana. Banana mosaic is cosmopolitan and is found wherever bananas are grown. The virus is called CMV because this virus was first found on cucumber showing mosaic symptoms.

FIGURE 5.11   Citrus mealy bug—P. citri.

FIGURE 5.12   Pink sugarcane mealy bug—S. sacchari.

5.5.1 CAUSAL ORGANISM

This virus is a member of the

Family: Bromoviridae
Genus: Cucumovirus
Species: Cucmber mosaic virus (banana mosaic).

This virus has a worldwide distribution and a very wide host range, banana being one of them.

5.5.2 SYMPTOMS

CMV causes chlorosis, mosaic, and heart rot in banana and has been found in most banana growing areas of the world. The most characteristic symptom is the loss of leaf color, rendering leaves variegated in appearance. The variegations may be roughly parallel to the lateral veins, but not always, giving leaves a striped appearance (Figure 5.13). As the disease progresses, leaves emerge, having perhaps one or both sides of the lamina not fully developed so that the leaf margin instead of being smoothly curved is irregularly wavy, often with blotches of necrotic tissue, and the lamina is reduced in width. The flower deforms, and streaking develops (Figure 5.14). Sometimes rotten areas are found throughout the leaf sheaths and the pseudo stem.

This disease may be confused with banana bunchy top virus (BBTV). CMV has a very wide host range and is considered an economically significant disease of bananas in various countries of the world.

FIGURE 5.13   Interveinal chlorotic variegations on banana leaf.

5.5.3   MEANS OF MOVEMENT AND TRANSMISSION

CMV is easily transmitted by sap inoculation and by aphids in a non-persistent manner. Banana mosaic disease is transmitted successfully by A. gossypii and poorly by Aphis craccivora but not at all by Pentalonia nigronervosa. A single aphid can produce infection, although in a low percentage, and the percentage infection increases with an increase in the number of aphids.

5.5.4   PREVENTION AND CONTROL

The infected plants must be destroyed immediately by digging them up. Eradication of all the suckers in the mat where infection has been identified, even if they appear healthy, is necessary. Destruction of aphids using sprays like Malathion, or other available aphicides, must be used. The best feasible way is to use infection-free plantlets.

5.6   BURROWING NEMATODE

Scientific name: Radopholus similis
Affected plant stages: All plant stages
Affected plant parts: Corms and primary roots

5.6.1   SYMPTOMS

The nematode burrows the root tips and other underground parts of their hosts forming extensive cavities, hence the name.

FIGURE 5.14   Flower deformity and streaking.

On banana, the damage starts when the nematode enters the primary roots and attacks the corm that cause reddened spots around the feeding sites. This infection is called as "blackhead toppling disease", wherein the entire feeding site is exposed, showing the blackened and broken primary roots. Infected plants are uprooted at any growth stages, but the damage is more common during the fruiting stage. The uprooted plant falls together with the adjoining sucker, which could be the succeeding fruit-bearing stem. During a severe infestation, the plants are weakening, have poor growth, low yields, and are more susceptible to fusarium wilt disease.

5.6.2 Life Cycle

The life cycle of Radopholus similis in banana roots from egg to egg occurs in 20 to 25 days at 24°C to 32°C (75°F to 90°F). Eggs are reported to hatch in eight to ten days with juvenile stages completed between ten and thirteen days. As with all nematode species, there are four juvenile stages: the first-stage (J1) develops within the egg, then molts and emerges as a J2 nematode (Figure 5.15).

5.6.3 Conditions that Favor Development

- Infected planting materials.
- Presence of weeds that serves as the alternate hosts.

5.6.4 Prevention and Control

- Remove and peel discolored spots on corms and disinfect these with warm water before planting (55°C for 15–25 minutes).
- Use disease-free suckers.
- Mulch the plants with organic matter.
- Support fruiting pseudo stem with poles or rope to prevent uprooting.
- Plant sugarcane as rotational crop to banana.

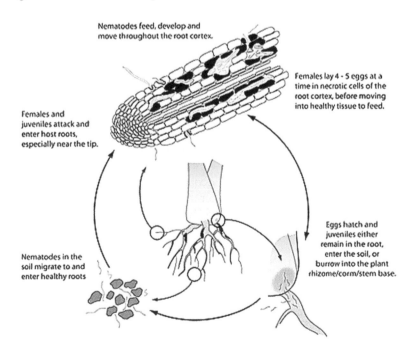

FIGURE 5.15   Life cycle of R. similis.

5.7 ROOT KNOT NEMATODE

The root knot nematodes Meloidogyne incognita and Meloidogyne javanica occur on banana and plantain roots wherever this crop is grown. Meloidogyne arenaria and occasionally some other Meloidogyne species may also be found associated with banana and plantain. Despite their widespread occurrence and high abundance, root knot nematodes are not considered important pathogens of banana and plantain. Root knot nematodes often occur on banana roots together with the other pathogenic nematode species Radopholus similis and Pratylenchus spp. The damage caused by these other nematode species is more visible (root necrosis) and more destructive (toppling of plants) than the symptoms (galling) and other adverse effects caused by Meloidogyne spp. Moreover, R. similis, and to a lesser extent, Pratylenchus spp., tend to outnumber and eventually replace root knot nematode populations.

M. incognita and M. javanica occur worldwide. On banana and plantain, they can become abundant in areas where either the climate is too cold for R. similis, R. similis has not been introduced (as it commonly has been with dessert bananas of the Cavendish (AAA) subgroup), or Pratylenchus goodeyi is not present. Such areas can be found in the Mediterranean (e.g. Crete and Lebanon), in the subtropics (e.g. South Africa), in the tropics at higher altitude in South America (e.g. the Andes in Colombia) and in the lowlands in Africa (e.g. Ghana, Tanzania) where dessert bananas of the Cavendish subgroup were not introduced on a large scale and R. similis was therefore not spread with infected planting material.

5.7.1 Life Cycle, Symptoms, and Damage

Meloidogyne spp. are sedentary endoparasites. Mobile second-stage juveniles (J2) emerge from the eggs, move toward the roots, and penetrate the roots either at the root tip, in the regions of previous penetration, or where minor wounds are present. In the root, the J2 invade the endodermis, and on entering the stele, induce multinucleate giant cells derived from vascular parenchyma or differentiating vascular cells in the central part of the stele. The formation of these giant cells disturbs or blocks the surrounding xylem vessels. Multiplication of cortical cells is also induced, resulting in the formation of the characteristic galls. The J2 feed on these giant cells and molt three times to form the adult females which enlarge rapidly. Dissection of galls reveals the typical swollen females in various stages of maturation. At maturity the females are saccate. Reproduction is parthenogenetic. A high percentage of males are produced only in adverse conditions. The eggs are laid within a gelatinous matrix to form an external egg sac or egg mass. A single egg sac can contain several hundred eggs. In thick, fleshy primary roots the egg masses may not protrude outside the root surface. On banana and plantain, the complete life cycle takes between four and six weeks. Different Meloidogyne species can be observed in the same gall. Root knot nematodes may also colonize the outer layers of the corm up to 7 cm deep.

On banana and plantain, the most obvious symptoms of Meloidogyne spp. infection are swollen, galled primary and secondary roots. Sometimes the root tips are invaded and there is little or no gall formation, but root tip growth ceases and new roots proliferate just above the infected tissues. Infected plants may have a much lower number of secondary and tertiary roots. Aboveground symptoms caused on banana by M. javanica in Pakistan included yellowing and narrowing of leaves, stunted plant growth, and reduced fruit production (Figure 5.16).

5.7.2 Management

- Meloidogyne spp. may be spread with infected planting material. Infected corms can be disinfected of root knot nematodes by peeling, followed by a hot-water treatment (at 53°C to 55°C for 20 minutes), or treatment with a nematicide before planting.
- Crop rotation trials with Pangola grass, maize, and sugarcane in Cuba and with Tagetes patula in South Africa were found to be successful. Because of the period of flooding, rotation with paddy rice may also drastically reduce root knot nematode populations.

FIGURE 5.16   Banana roots infected with root knot nematode.

- Dipping the corms for ten minutes in a nematicide solution as a preplanting treatment may protect the plants for a few months against nematode infection. Nematicides found effective included dibromochloropropane (DBCP or Nemagon, today forbidden in numerous countries), the organophosphates: ethoprophos and fenamiphos, and the carbamates: aldicarb and carbofuran. Immersion of peeled corms in 1% sodium hypochlorite (NaOCl) for five or ten minutes also controlled Meloidogyne spp. and is considered as an effective, low cost, and non-toxic preplanting treatment. Likewise, it was shown that preplanting fumigation with ethylene dibromide (EDB, today forbidden in numerous countries), dichloropropane-dichloropropene (D-D), or methyl bromide and post-planting soil treatment with most organophosphates (ethoprophos, cadusaphos, fenamiphos, isazofos, terbufos) and carbamates (aldicarb, carbofuran, oxamyl) applied several times a year, may significantly control root knot nematodes in established banana plantations and improve plant growth and yield. By studying the seasonal fluctuation pattern of the nematodes, an effective nematicide control programme can be developed with nematicides being applied when the population approaches a critical level, usually at the onset of the rainy season. In Puerto Rico, oxamyl applied four times with 30-day intervals during the growing season to the leaf axils of Giant Cavendish bananas effectively controlled M. incognita.

5.8   CIGAR-END ROT OF BANANA

Cigar-end rot of banana and plantain is an important post-harvest disease, which is associated with one of the two fungi mentioned below. The disease causes a black necrosis spreading from the perianth into the tip of immature fingers, and the pulp undergoes a dry rot. The disease causes severe losses in the marketability of the fruit.

5.8.1 Causal Organism

Species	Associated Disease Phase	Economic Importance
Verticillium theobromae (V. theobromae) or Trachyspaera fructigena (T. fructigena)	Tip Rot	Severe

5.8.2 Symptoms

Cigar-end rot is essentially a plantain disease (but is also found in banana). The fruits are apparently most subject to attack in their more immature stages. The disease causes a black necrosis spreading from the perianth into the tip of immature fingers, and the pulp undergoes a dry rot. The infected tissues are covered with fungal mycelia that resemble the grayish ash of a cigar end (Figure 5.17). The rot spreads slowly and rarely affect more than the first 2 cm of the fingertip. Incidence is highest during the rainy season. The pathogen colonizes banana leaf trash and flowers, from where spores are disseminated in air currents to other drying flower parts. T. fructigena causes premature ripening of fruits and can attack fruits after harvest, invading freshly cut crown surfaces and wounds in the peel caused by improper handling.

5.8.3 Favorable Conditions of Disease Development

Moist rainy weather favors the development of the disease.

5.8.4 Management

The principal method of control is frequent manual removal and burning of dead flower parts and infected fruits. Use of fungicides to control the disease is also recommended. The following methods should be followed:

* Early removal of dead leaves and flowers to eliminate inoculum before it reaches developing fruits. Infected fingers should be cut off before harvesting and packaging.
* With improved packaging, fruits can be washed in disinfectant, such as bleach, to remove T. fructigena or V. theobromae, and at the same time kill any Xanthomonas cells that may be smeared on the surface of fruits.

FIGURE 5.17   Cigar-end rot on bananas.

- Early removal of the male bud (after formation of last fruit cluster) should be promoted as it prevents the spread of cigar-end rot.
- Spray the bunches with 0.25% Indofil M-45 or 0.25% Kavach or 0.05% Bavisitin or 0.1% Thiophanate-methyl once in 15 days and later cover the bunches with polythene sleeve.

5.9 DISORDERS

5.9.1 CHOKE THROAT

It is due to low temperature affecting active growth of the plant. Leaves become yellow and in severe cases, the tissue gets killed. In the case of normal flowering plants, the stalk carrying bunches elongates freely so that the entire inflorescence comes out of the pseudo stem and hangs down. Bunch development is normal, but when the time of flowering synchronizes with low temperature, the bunch is unable to emerge from the pseudo stem properly. The distal part of the inflorescence comes out and the basal part gets stuck up at the throat. Hence, it is called choke throat. Maturity of the bunch is delayed by taking five to six months instead of three and a half to four months for harvest.[5]

Choke throat is seasonal in nature. It is usually worst in the winter and early spring following cold weather. However, it can also occur following periods of waterlogging or severe water stress and following wind storms. Two factors contribute to the actual difficulty in bunch emergence. Firstly, there is a reduction in the elongation of internodes of the true stem bearing the bunch inside the pseudo stem (pseudo = false). Secondly, the stiffness of the leaf bases at the top of the pseudo stem can prevent proper bunch emergence. Some varieties are more susceptible to this disorder than others. Notably, Williams and Mons Mari are far less susceptible than Dwarf Cavendish. Choke throat is often associated with dwarf varieties. Dwarf off types from tissue culture are particularly susceptible (Figure 5.18).

5.9.2 MANAGEMENT

- Select taller varieties, which are less susceptible to choke throat.
- Choose a warm environment, one which is well protected from frosts and strong winds. Slopes facing the north and north west are usually warmer.
- Provide shelter belts.
- Plant low temperature tolerant varieties.
- Control time of bunching to avoid cold weather prior to bunching. Plants bunching in the late spring to mid-autumn are less affected.

FIGURE 5.18   Severe form of choke throat where banana bunch bursts out through the side of a pseudo stem.

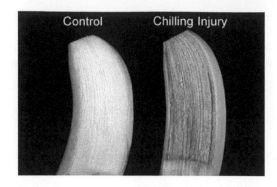

FIGURE 5.19   Banana fruit (left) normal and (right) affected by chilling injury.

- Good on-farm drainage measures including mounding of rows.
- Regular irrigation to avoid water stress particularly during hot-dry weather.
- Higher nitrogen rates are thought to be beneficial.
- Adopt corrective measures for calcium and Boron deficiency, if necessary.

5.9.3 CHILLING INJURY

Chilling occurs when pre-harvest or post-harvest temperatures fall below 14°C for various time periods. Chilling injury is the permanent or irreversible physiological damage to bananas, which results from exposure to low temperature. It is a physiological disorder that occurs in most fruits of tropical origin. Banana chilling injury may occur below 12°C. Even a few hours of chilling temperature can be sufficient to induce permanent damage. It can occur to ripe and unripe fruits. It can cause lower market quality and a total loss.

The peel of the banana becomes dark and the fruit exhibit uneven ripening. Ripening fingers show a dull yellow to smoky yellow color and watery dark patches are observed on the skin. Brittleness of the fruit and fungal invasion is also observed. The vascular bundles of the sub-epidermal layer show brown streaks (Figure 5.19). The discoloration is ascribed to the enzymatic oxidation of dihydroxy phenylalanine.

5.9.4 POTASSIUM DEFICIENCY

Chlorosis—The most characteristic of the K deficiency symptoms is the yellowing of older leaf tips followed by inward leaf curling and death.

Stunted growth—Usually, a K deficient banana plant will grow slowly and have a sturdy appearance due to the shortening of internodes.

Bunch deformation—The banana bunches in K deficient plants are short, slim, and deformed because of poor fruit filling caused by reduced photosynthesis and sugar transportation.

REFERENCES

1. Proceedings of the workshop on review of the strategy for the *"Management of banana Xanthomonas wilt"*, Bioversity International pp. 84–87, 2009
2. Mwangi, M., *"Responding to Banana Xanthomonas Wilt Amidst Multiple Pathogens and Pests"*, Crop Crisis Control Project, 2007
3. Gaur, R. K., Hohn, T., and Sharma, P., *"Plant Virus–Host Interaction Molecular Approaches and Viral Evolution"*, Cambridge: Academic Press, 2014.
4. Lockhart, B. E., *"Banana Streak Badnavirus Infection in Musa: Epidemiology, Diagnosis and Control"*, Department of Plant Pathology, St. Paul, USA: University of Minnesota.

5. Jeyakumar, P., Paper on *"Physiological Disorders in Fruit Crops"*, Assistant Professor (Crop Physiology), Department of Pomology Horticultural College and Research Institute Tamil Nadu Agricultural University, Coimbatore, India.
6. Viljoen, A., Mahuku, G., Massawe, C., et al., *"Banana Pests and Diseases Field Guide for Disease Diagnostics and Data Collection"*, International Institute of Tropical Agriculture, 2016.
7. Richard, B., and David, C., "Scientific Opinion of the Panel on Plant Health", *The EFSA Journal* 652, 5–21, 2008.

6 Pepper

The bell pepper (also known as sweet pepper or pepper in the United Kingdom, Canada, and Ireland, and capsicum in Australia, India, Pakistan, Bangladesh, Singapore, and New Zealand) is a cultivar group of the species Capsicum annuum. Peppers are native to Mexico, Central America, and northern South America. Pepper seeds were imported to Spain in 1493, and from there were spread to other European, African, and Asian countries. Today, China is the world's largest pepper producer, followed by Mexico and Indonesia. The terms "bell pepper" (the United Srares), "pepper" (United Kingdom), and "capsicum" (Pakistan, India, Australia, and New Zealand) are often used for any of the large bell-shaped vegetables, regardless of their color. In 2013, global production of both green and dried chili pepper was 34.6 million tonnes, with 47% of output coming from China alone. The fruit of most species of Capsicum contains capsaicin (methyl-n-vanillyl nonenamide), a lipophilic chemical that can produce a strong burning sensation (pungency or spiciness) in the mouth of the unaccustomed eater. There are many bacteria and viruses which are affecting the production of pepper. This chapter gives a better understanding of identification of pepper diseases and its control measures.

6.1 ANTHRACNOSE OF PEPPER

Several species of plant pathogenic fungi in the genus Colletotrichum cause anthracnose in peppers and many other vegetables and fruits. Until the late 1990s, anthracnose of peppers and tomatoes was only associated with ripe or ripening fruit. This form attacks peppers at any stage of fruit development and may threaten the profitability of pepper crops in areas where it becomes established. This disease can also affect tomatoes, strawberries, and possibly other fruit and vegetable crops.

6.1.1 CAUSAL ORGANISM

Species	Associated Disease Phase	Economic Importance
Colletotrichum spp.	Fruit Rot	Severe
• C. gloeosporiodes		
• C. capsici		
• C. coccodes		

6.1.2 SYMPTOMS

Circular or angular sunken lesions develop on immature fruit of any size. Often multiple lesions form on individual fruit. When the disease is severe, lesions may coalesce. Often pink to orange masses of fungal spores form in concentric rings on the surface of the lesions (Figure 6.1). In older lesions, black structures called acervuli may be observed. With a hand lens, these look like small black dots; under a microscope, they look like tufts of tiny black hairs. The pathogen forms spores quickly and profusely and can spread rapidly throughout a pepper crop, resulting in up to 100% yield loss. Lesions may also appear on stems and leaves as irregularly shaped brown spots with deep brown edges (Figure 6.2).

6.1.3 CAUSE AND DISEASE DEVELOPMENT

Anthracnose is caused by fungi in the genus Colletotrichum, which is a very common group of plant pathogens, and they are responsible for diseases on numerous plant species worldwide. Identification of

FIGURE 6.1   Pepper anthracnose.

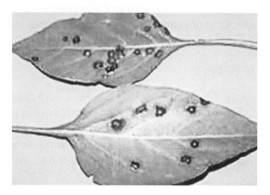

FIGURE 6.2   Anthracnose lesions on pepper leaves.

Colletotrichum to species is usually based on more than one characteristic, such as physical appearance and pathogenicity on the host(s). Many species of Colletotrichum infect more than one host and, to confound identification, more than one Colletotrichum sp. may be present on one host. At least three species of Colletotrichum (C. gloeosporiodes, C. capsici, and C. coccodes) are reported to cause this disease.

6.1.4   FAVORABLE CONDITIONS OF DISEASE DEVELOPMENT

The anthracnose organism can be present in crop debris, weeds, and pepper foliage. It prefers wet conditions and temperatures of 10°C–30°C. At 20°C–25°C, it only takes 12 hours of leaf wetness (on unprotected leaves or fruit) for infection to occur. Within five to six days of a spore landing on a ripe fruit, a lesion can be seen.

6.1.5   DISEASE CYCLE

This form of pepper anthracnose is caused by the fungus Colletotrichum acutatum. The pathogen survives on plant debris from infected crops and on other susceptible plant species. The fungus is not soilborne for prolonged periods in the absence of infested plant debris. The fungus may also be introduced into a crop on infested seed. During warm and wet periods, spores are splashed by rain or irrigation water from diseased to healthy fruit. Diseased fruit act as a source of inoculum, allowing the disease to spread from plant to plant within the field.[1]

6.1.6 Management

- Control of the disease is through integrated management techniques. The disease should not be introduced on infected plants.
- Only seeds that are pathogen-free should be planted. Transplants should be kept clean by controlling weeds and solanaceous volunteers around the transplant houses.
- The field should have good drainage and be free from infected plant debris. If disease was previously present, crops should be rotated away from solanaceous plants for at least two years. Sanitation practices in the field include control of weeds and volunteer peppers plants.
- Resistance is available in some varieties of chili peppers but not in bell peppers. For bell pepper production, choose cultivars that bear fruit with a shorter ripening period which may allow the fruit to escape infection by the fungus.
- Wounds in fruit from insects or other means should be reduced to the extent possible because wounds provide entry points for Colletotrichum spp. and other pathogens like bacteria that cause soft rot.
- For late-maturing peppers, when disease is present, apply a labeled fungicide several weeks before harvest.
- The disease can be controlled under normal weather conditions with a reasonable spray program. At the end of the season, remove infected plant debris from the field or deep plow to completely cover crop diseases.

6.2 BACTERIAL SPOT OF PEPPER

Bacterial spot is one of the most devastating diseases of pepper and tomato grown in warm, moist environments. Once present in the crop, it is almost impossible to control the disease and prevent major fruit loss when environmental conditions remain favorable. Bacterial leaf spot of pepper is caused by Xanthomonas campestris pv. vesicatoria. This bacterium survives in crop debris in the soil and on seed. Disease develops only when there is plenty of moisture and warm temperatures.[2]

6.2.1 Causal Organism

Species	Associated Disease Phase	Economic Importance
X. campestris pv. vesicatoria	Lesions on Leaves and Water-Soaked Spots on Fruits	Severe

6.2.2 Symptoms

Necrotic spots may appear on leaves, stems, and fruits. Leaf symptoms appear first on the undersides of leaves as small water-soaked areas. These spots enlarge up to 1/4 inch in diameter, turn dark brown, and are slightly raised. On the upper leaf surface, the spots are depressed with a brown border around a beige center. Several lesions may coalesce, resulting in large necrotic areas, and large numbers of lesions can occur on leaf margins and tips where moisture accumulates (Figure 6.3). Eventually the leaves yellow and drop off, increasing the chance for sunscald. Spots on fruits become raised, scab-like areas that make the product unmarketable (Figure 6.4).

6.2.3 Cause and Disease Development

The bacterial spot pathogen may be carried as a contaminant on pepper seed. It is also capable of overwintering on plant debris in the soil and on volunteer host plants in abandoned fields. Because

FIGURE 6.3   Bacterial spot lesions on pepper leaves.

FIGURE 6.4   Bacterial spot on pepper fruit.

the bacteria have a limited survival period of days to weeks in the soil, contaminated seed is a common source of primary infection in nursery and home gardens. In commercial fields, volunteer host plants are the main source of initial inoculum because the bacterial pathogen survives in lesions on those plants.[3]

6.2.4   FAVORABLE CONDITIONS OF DISEASE DEVELOPMENT

The optimum temperature for infection is between 24°C and 30°C (75°F and 86°F). Disease development is favored by temperatures that fluctuate between 20°C and 35°C (68°F and 95°F). Night temperatures of 23°C to 27°C (75°F to 82°F) favor disease development, while low night temperatures (61°F or 16°C) suppress disease development independent of daytime temperatures. High nitrogen levels increase the severity of bacterial spot. Symptoms can appear on pepper leaves five to fourteen days after infection takes place.

6.2.5   DISEASE CYCLE

Bacteria enter through stomata on the leaf surfaces and through wounds on the leaves and fruit caused by abrasion from sand particles and/or wind. Prolonged periods of high relative humidity favor infection and disease development. In a 24-hour period, the bacteria can multiply rapidly and produce millions of cells. Symptom development is delayed when relative humidity remains low for several days after infection. Although bacterial spot is a disease of warm, humid regions, it can develop in arid but well-irrigated regions.

6.2.6 MANAGEMENT

Control is based on preventive steps taken during the entire season. Once the disease has started in a field, control is very difficult, especially during wet weather.

- Obtain seed that has been grown in regions without overhead irrigation and is certified free from the disease-causing bacteria. This is by far the most crucial step. Seed may be treated by washing for 40 minutes with continuous agitation in 2 parts Clorox liquid bleach (5.25% sodium hypochlorite) plus 8 parts water (e.g. 2 pints Clorox plus 8 pints water). Use one gallon of this solution for each pound of seed. Prepare fresh solution for each batch of seed treated. Rinse seed in clean water immediately after removal from the Clorox solution and promptly allow to dry prior to storing or treating with other chemicals. This treatment will likely reduce seed germination. Thus, before attempting to treat an entire seed lot perform a test using 50–100 seed and check for the effect on germination.
- Produce plants in sterilized soil or commercially prepared mixes.
- Avoid fields that have been planted to peppers or tomatoes within one year, especially if they had bacterial spot.
- Do not plant diseased plants. Inspect plants very carefully and reject infected lots—including your own! Use certified plants.
- Prevent bacterial leaf spot in the plant beds: Keep the greenhouse as dry as possible and avoid splashing water; spray with a fixed copper (Tribasic Copper Sulfate 4 lb, Copper-Count N, or Citcop 4E 2 to 3 qt, or Kocide 101 1.0 to 1.5 lb per 100-gal water). The addition of 200 ppm streptomycin (Agri-mycin 17-1.0 lb in 100 gal of the copper spray with a spreader-sticker) will improve the effectiveness of the spray program. Make applications on a seven- to ten-day schedule if spots appear, and one day before pulling plants.
- In the field, start the spray schedule when the disease first appears. However, do not use streptomycin in the field.
  - Mix fixed copper and 1.5 lbs of mancozeb (Manzate 200DF or Dithane M-45) or maneb (Maneb 80WP, Maneb 75DF, Manex) in 100 gal of water. Do not use mancozeb on tomatoes within five days of harvest. Mancozeb is not registered for use on peppers; Use maneb.
  - Adjust sprayer and speed of tractor to obtain complete coverage of all plant surfaces. Spray pressure of 200–400 psi is recommended and the use of at least three nozzles per row for pepper and five drop nozzles per row for tomato. Depending on plant size, use 50–150 gal/A of finished spray.
  - Adjust spray schedules according to the weather and presence of disease: (a) Spray one week after plants are set; (b) spray every five to seven days during rainy periods; spray on ten-day intervals during drier weather; (c) spray before rain is forecast but allow time for spray to dry.

6.3 CERCOSPORA LEAF SPOT

Cercospora leaf spot, also known as, frogeye leaf spot, is encountered on a regular basis on pepper. This fungus survives the winter and is most troublesome in warm, wet weather.

6.3.1 Causal Organism

Species	Associated Disease Phase	Economic Importance
Cercospora capsici	Spots on Leaves, Defoliation, and Exposure of Fruits to Sunscald	Minor

6.3.2 Symptoms

Circular spots appear with a light gray center and a reddish-brown margin, growing up to 1 cm in diameter. Spots later become tan with a dark ring and a yellowish halo around the ring, resulting in a "frog-eye" appearance (Figure 6.5). Under conditions of high humidity, and using a good high magnification hand lens, thin, needle-like spores may be seen in the center of the spots arising from small black fungal tissue. The affected centers of lesions dry and often drop out as they age. When numerous spots occur on the foliage, the leaves turn yellow and may drop or wilt (Figure 6.6). Defoliation is often serious, exposing fruits to sunscald. Spots also develop on stems and petioles, but they are oblong rather than circular. Fruits are not infected.

6.3.3 Cause and Disease Development

The fungus survives in or on seed and as tiny black fungal tissue known as stromata in old affected leaves in the soil. Spores will survive in infected debris for at least one season. Foliar infection occurs by direct penetration of the leaf. The fungus spores require water for germination and penetration of the host; however, heavy dew appears to be sufficient for infection.

6.3.4 Favorable Conditions of Disease Development

The disease is most severe during periods of warm temperatures; for example, 20–25°C during the day and excessive moisture (either from rain or overhead irrigation). Fungal growth is limited if the temperature is <5°C or >35°C. The fungus is spread by splashing water, wind-driven rain, wind, on implements, tools, workers, and by leaf-to-leaf contact. It is not known whether the fungus will infect solanaceous weeds.

6.3.5 Disease Cycle

Development of Cercospora leaf spot disease begins when the fungus' spores are dispersed by rain, irrigation water, and wind. Germination occurs in humid conditions, usually during late spring and summer, and fungus growth is encouraged by frequently damp leaves. Plants that mature in the

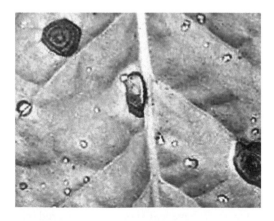

FIGURE 6.5   Circular tan spots with dark rings and yellow halos. Centers of spots are light gray with black spores.

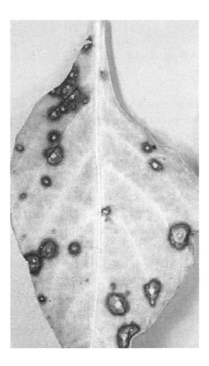

FIGURE 6.6 Numerous spots create yellowing, defoliation.

fall may escape acute infection. In general, the more rain, the worse the disease spreads. C. capsici infests seeds and crop residue where it harbors over the winter.

6.3.6 MANAGEMENT

Use seed from disease-free areas. Treat seed with hot water at 52°C for 30 minutes. Alternatively, use a seed disinfectant if seed comes from infected plants. Check seedbeds and young plants or transplants for any symptoms of the disease. Remove the affected plants and one or two neighboring plants that may be infected already but do not yet show symptoms.

Space plants properly in the field to allow for good air circulation and to avoid extended periods of leaf wetness. After harvest, promptly destroy infected pepper tissues by burning or deep-plowing. Rotate crops using a two-year rotation period. Control solanaceous weeds during the rotation period. Check older plants carefully for the first incidence of the disease particularly after extended periods of leaf wetness and warm temperatures. If symptoms appear, apply a protectant fungicide as soon as possible. Resistant varieties are available.[4]

6.4 CMV, PVY, TEV

6.4.1 CUCUMBER MOSAIC VIRUS ON PEPPER

Cucumber mosaic virus (CMV) can occur wherever peppers are grown. The host range of the virus consists of more than 750 plant species including many vegetables (such as pepper, tomato, cucurbits, and legumes), weeds, and ornamentals. Strains of CMV have been reported that are specific for pepper.

6.4.1.1 Causal Organism
This virus is a member of the

Family: Bromoviridae
Genus: Cucumovirus

Species: Cucumber mosaic virus

This virus has a worldwide distribution and a very wide host range, pepper being one of them.

6.4.1.2 Symptoms
Pepper infected with CMV often are stunted and bushy (shortened internodes) and may have distorted and malformed leaves. Leaves may appear mottled (intermingling of dark green, light green, and yellow tissue), a similar symptom to those caused by other viruses. The most characteristic symptom of CMV is extreme filiformity, or shoe-stringing, of leaf blades. CMV symptoms can be transitory, that is, the lower or upper leaves can show symptoms while those in the midsection of the plant appear normal. The effect of CMV on yield depends on several factors, including plant age when infected and environmental conditions. Severely affected plants produce few fruit, which are usually small.

6.4.1.3 Means of Movement and Transmission
CMV is the second most important virus disease of pepper. CMV has an extensive host range and is transmitted by aphids in a non-persistent manner. Unlike TMV, CMV is not seedborne in pepper and does not persist in plant debris in the soil or on workers' hands. CMV has been found in greenhouse planting. Seedlings grown outdoors and left unprotected by isolation before moving indoors are 1 likely source of infection. Other sources of inoculum are the spread of CMV by aphids from infected plants in adjoining greenhouses (weeds under benches, ornamentals, or other vegetables) and by viruliferous aphids entering through non-insect-proof vent windows.

6.4.2 Potato Virus Y

Potato virus Y (PVY) occurs worldwide but has a narrow host range, affecting plants in the Solanaceae family (that is, potatoes, tomatoes, and peppers). It is transmitted by aphids. Near total crop failures have been reported when PVY was detected early in the season and high aphid populations were present.

6.4.2.1 Symptoms
Symptoms on pepper vary according to PVY strain, plant age, varieties infected, and environmental conditions. General symptoms on pepper are faint mottling and slight distortion of the leaves. Severe symptoms include deep brown, dead areas in the blade of nearly mature leaflets. Leaflets at the terminal end of a leaf usually are most adversely affected, often showing severe necrosis. In many cases, all leaflets are affected. Leaves formed after the onset of PVY exhibit mild wrinkling, slight distortion, and mild mottling. Leaflets of plants infected for some time are rolled downward with curved petioles, giving the plant a drooping appearance. Stems often show a purplish streaking but no symptoms are produced on the fruit. Mature plants are stunted and unthrifty and yield is reduced.

6.4.2.2 Means of Movement and Transmission
PVY is transmitted in a non-persistent manner by many aphid species. Aphids can acquire the virus in less than 60 seconds from an infected plant and transmit it to a healthy plant in less than 60 seconds. The virus may be retained by the aphid for longer than 24 hours if feeding does not occur. PVY can also be transmitted mechanically. Potato is an important source of the virus for pepper and other solanaceous crops. The virus does not appear to be seed-transmitted.

6.4.3 Tobacco Etch Virus

Tobacco etch virus (TEV) infects pepper and tomatoes along with other plants in the Solanaceae family. The occurrence of TEV in pepper fields is closely associated with other infected solanaceous crops, especially pepper, and natural weed hosts, which serve as virus reservoirs.

6.4.3.1 Symptoms

Leaves of infected plants are severely mottled, puckered, and wrinkled. Plants infected at an early age are severely stunted. Fruit from infected plants are mottled and never achieve marketable size. The younger the plants are when infected, the greater the reduction in yield.

6.4.3.2 Means of Movement and Transmission

The virus can be transmitted by at least ten species of aphid in a non-persistent manner. There are no reports of seed transmission in any host plant.

6.4.4 Prevention and Control of CMV, PVY, and TEV

There are no good sources of resistance in pepper for CMV, PVY, or TEV, so other control strategies must be used. These include the following:

- Eradicate all biennial and perennial weeds and wild reservoir hosts in and around fields. Maintain a distance of at least 30 ft between susceptible crops, weeds, or other susceptible plants, including those in ditch banks, hedge or fence rows, and other locations.
- Plant earlier to avoid high aphid populations that occur later in the season.
- Plant late settings as far as possible from fields used to produce early peppers. These areas can act as sources of viruses and aphids for subsequent crops.
- Scout fields for the first occurrence of virus disease. Where feasible, pull up and destroy infected plants but only after spraying them thoroughly with an insecticide to kill any insects they may be harboring.
- Use reflective mulches to repel aphids, thereby reducing the rate of spread of aphid-borne viruses.
- Monitor aphid populations early in the season and apply insecticide treatments when needed.
- Minimize plant handling to reduce the amount of virus spread mechanically.
- Avoid planting peppers near potato fields to control PVY.

6.5 PEPPER MILD MOTTLE VIRUS

Pepper mild mottle virus (PMMoV) occurs worldwide in field-grown bell, hot, and ornamental pepper species. The presence of the virus is difficult to detect in the greenhouse until the plants begin to bear fruit.

6.5.1 Causal Organism

PMMoV is in the tobacco mosaic virus (Tobamovirus) family. It is spread by mechanical transmission and by infected seed but cannot be transmitted by insects. PMMoV infects mainly Capsicum species (peppers). It is one of at least four distinct species of Tobamovirus that infect peppers. The others include Tobacco mosaic virus (TMV), Tomato mosaic virus (ToMV) and Tobacco mild green mosaic virus (TMGMV).

6.5.2 Symptoms

Symptoms caused by PMMoV on pepper plants may vary between cultivars. Infected leaves are frequently puckered and mottled yellow or light green (Figure 6.7). Leaf symptoms are more evident on younger leaves. Plants can be stunted, especially when the infection occurs early in the plant's development. Although infected fruit can be somewhat reduced in size and show variations

FIGURE 6.7   Infected leaves are frequently puckered and mottled yellow or light green.

FIGURE 6.8   Distorted or lumpy appearance of the fruit.

in color (mottling and color changes at maturity), the most obvious symptom is the distorted or lumpy appearance of the fruit (Figure 6.8). Older fruit may develop brown streaks or splotches.

6.5.3   Means of Movement and Transmission

It is highly infectious, however, it does not spread via insects. It is spread by infected seed or through mechanical means. Mechanical transmission refers to daily crop maintenance: tools that might be used while tending, shoes carrying infected soil, or any debris from infected plants left behind (dead leaves or stems). The virus can be passed on through this sort of debris for up to several months after the plant that left it behind has passed. A major source of seed infection is itself by mechanical means; the seed coats can become infected easily by being exposed during transplanting. Even if only one piece of equipment or a small amount of debris carrying the virus is in a facility where commercial seeds are harvested, and even if only a few seeds meet the virus, they presumably would mix in with the rest of the seeds that weren't exposed, infecting them through the outer shells.

6.5.4   Prevention and Control

- Avoidance is the best means of control. Only seed that has been tested and determined to be free of the virus should be planted. Infected seed can be treated with heat, acid, or trisodium phosphate, but virus both on the seed surface and inside the seed must be removed to ensure freedom from disease. Seed treatments can reduce the seed germination percentage even if done accurately.

- The virus enters the plant through microscopic abrasions or wounds. There are no chemical or biological control methods that can be used to control the disease once the plant is infected. Like other members of this family of viruses, the virus is very stable and can be present on skin, clothing, tools, and equipment so infected plants should be handled as little as possible and infected fields should be worked (staked, tied, harvested, sprayed, etc.) last to avoid spreading the virus to uninfected areas. Infected plants should only be removed if elimination can be done without contact with healthy plants. A symptomless plant on either side of those removed should also be rogued, as it is likely that they are also infected. Some viruses can be spread through smoke so diseased plants should be disposed of by composting or burying them away from fields where peppers will be grown rather than by burning them.
- Viruses in the TMV family are notoriously easy to spread and difficult to eliminate. To reduce spread of the disease, anyone working with the plants should wash their hands with 70% alcohol or strong soap, also cleaning under the nails. Clothing should be washed as frequently as possible. Equipment should be washed and then cleaned with 3% trisodium phosphate and not rinsed. Stakes from infected areas of the field should be discarded or soaked in 3% trisodium phosphate before being reused. Household bleach can also be used to clean equipment or stakes.
- Diseased plant material will remain infectious until completely broken down. Tillage, increased irrigation, and high temperatures encourage the breakdown of plant material in the soil. Any infected plant material in the soil can serve as a source of inoculum for subsequent crops so crop rotation should be practiced, if possible. Volunteer peppers and weeds, particularly those in the Solanaceae family (such as nightshades) should be removed to reduce possible sources of infections.
- Accurate identification is important to avoid yield loss. Other pepper viruses can have similar symptoms but may be spread and controlled through different means. For example, pepper mottle virus, which has somewhat similar symptoms and is also found in Florida, is transmitted by aphids and not by mechanical means.

6.6 TMV ON PEPPER

TMV is distributed worldwide and may cause significant losses in the field and greenhouse. TMV is one of the most stable viruses known, able to survive in dried plant debris for 100 years. Many strains of TMV have been reported and characterized. TMV can be seedborne in tomato, is readily transmitted mechanically by human activities, and may be present in tobacco products. The virus is not spread by insects commonly occurring in the greenhouse or field.

6.6.1 CAUSAL ORGANISM

The virus belongs to

Genus: Tobamovirus
Species: Tobacco Mosaic Virus (TMV)
Several strains are known.

6.6.2 SYMPTOMS

The symptoms in pepper vary greatly in intensity depending upon the variety, virus strain, time of infection, light intensity, and temperature. Hot temperatures, for example, may mask foliar symptoms. The most characteristic symptom of the disease on leaves is a light- and dark-green mosaic pattern. Some strains may cause a striking yellow mosaic, whereas other strains may cause leaf

FIGURE 6.9   Symptoms of ToMV on pepper plant.

malformation and "fern-leafing", suggestive of CMV infection. With the use of ToMV resistant or -tolerant varieties, plants may be infected by some strains whose symptoms are latent. Ordinarily, the fruit from infected plants do not show mosaic symptoms but may be reduced in size and number. Under certain environmental conditions, some varieties with resistance (heterozygous) to ToMV will show necrotic streaks or spots on the stem, petiole, and foliage as well as on the fruit. This disorder is particularly evident in well-developed but unripe fruit of the first cluster, involving a collapse of cells in the fleshy parenchyma. The cause is attributed to a "shock reaction" following ToMV infection, with a number of other factors contributing to the severity of symptoms (high soil moisture, low nitrogen and boron, and sensitivity of the variety) (Figure 6.9).

6.6.3   Means of Movement and Transmission

The virus may be introduced on infected seed. Only a small number of seedlings need to be infected for the virus to spread rapidly. It can also be spread on contaminated tools and the clothing and hands of workers during routine activities. It is readily transmitted by machinery or workers from infected to healthy plants during handling. Infested debris from a previous crop can lead to infection when the roots of the new tomato plants meet the debris. Chewing insects can transmit the virus but are not considered a major source of infection. Pepper seed can carry the virus, but actual infection is thought to occur when plants are thinned or transplanted.

6.6.4   Prevention and Control

ToMV is spread readily by touch. The virus can survive on clothing in bits of plant debris for about two years and can easily enter a new plant from a brief contact with a worker's contaminated hands or clothing. Tobacco products can carry the virus, and it can survive on the hands for hours after touching the tobacco product. Ensure that workers do not carry or use tobacco products near the plants and wash well (with soap to kill the virus) after using tobacco products. Ensure that workers wear clothing not contaminated with tomato, tobacco or another host-plant material. Exclude non-essential people from greenhouses and growing areas.

   Choose resistant varieties. Use disease-free seed and transplants, preferably certified ones. Avoid the use of freshly harvested seed (two years old is best if non-certified seed is used). Seed treatment with heat (two to four days at 70°C using dry seed) or trisodium phosphate (10% solution for 15 minutes) has been shown to kill the virus on the outside of the seed, and often, most of the virus inside the seed as well. Care must be taken to not kill the seeds, though. Use a two-year rotation away from susceptible species. In greenhouses, it is best to use fresh soil, as steaming soil is not 100% effective in killing the virus. If soil is to be steamed, remove all parts of the plant from the soil, including roots. Carefully clean all plant growing equipment and all greenhouse structures that come into contact with plants.

When working with plants, especially when picking out seedlings or transplanting, spray larger plants with a skim milk solution or a solution made of reconstituted powdered or condensed milk. Frequently dip hands, but not seedlings, into the milk. Wash hands frequently with soap while working with plants, using special care to clean out under nails. Rinse well after washing. Tools should be washed thoroughly, soaked for 30 minutes in 3% trisodium phosphate and not rinsed.

Steam sterilizing the potting soil and containers as well as all equipment after each crop can reduce disease incidence. Sterilizing pruning utensils or snapping off suckers without touching the plant instead of knife pruning helps reduce disease incidence. Direct seeding in the field will reduce the spread of ToMV. Another method for control of this disease is to artificially inoculate plants with a weak strain of the virus. This will not cause symptoms on the plants but protect them against disease-causing strains of the virus.

6.7 TOMATO SPOTTED WILT VIRUS

Tomato spotted wilt virus (TSWV) causes serious diseases of many economically important plants representing 35 plant families, including dicots and monocots. A unique feature is that TSWV is the only virus transmitted in a persistent manner by certain thrips species. At least six strains of TSWV have been reported; the symptoms produced and the range of plants infected vary among strains. Although previously it was only a threat to crops produced in tropical and subtropical regions, today the disease occurs worldwide, largely because of wider distribution of the western flower thrips and movement of virus-infected plant material. Early and accurate detection of infected plants and measures to reduce the vector population are discussed as critical steps for disease control.

6.7.1 Causal Organism

Pathogen: TSWV in the tospovirus group.

6.7.2 Symptoms

The symptoms of tomato spotted wilt in pepper vary depending on the stage of growth in which the plant is infected, the cultivar, co-infections with other viruses, and other factors such as environmental conditions. Certain symptoms of TSWV infection—the spotting, bronzing, and necrosis of leaves and the ringspots on fruit—are typical (Figure 6.10).

6.7.3 Means of Movement and Transmission

The virus is transmitted by insects called thrips. It is not easily mechanically transmitted by rubbing. Juvenile thrips acquire the virus from infected weed hosts. They disperse it after becoming

FIGURE 6.10   Symptoms of TSWV on pepper fruit.

winged adults that feed on host tomato and other target plants. Consequently, this disease is more of a problem in spring crops and in certain favorable years.

Weed hosts identified as potential virus carriers include spiny amaranthus, wild lettuce (Lactuca sp.), pasture buttercup (Ranunculus sp.), Solanum sp., and sowthistle (Sonchus sp.).

6.7.4 PREVENTION AND CONTROL

Currently, there is no effective way to control TSWV. To reduce the source of infection

- Control weeds adjacent to the field. (TSWV can overwinter in weeds.)
- Apply systemic insecticides to the soil at planting to slow initial spread of the virus into the field.
- Spray bordering weeds and the tomato crop with insecticides to suppress thrip populations and spread of TSWV.
- Remove and destroy infected plants as soon as symptoms appear to further reduce virus spread.

6.8 BLOSSOM-END ROT OF PEPPER

Blossom-end rot (BER) is a physiological disorder, not a disease. The disorder is often prevalent in pepper, and severe losses may occur if preventive control measures are not undertaken. BER is caused by calcium deficiency, usually induced by fluctuations in the plant's water supply. Because calcium is not a highly "mobile" element in the plant, even brief changes in the water supply can cause BER. Droughty soil or damage to the roots from excessive or improper cultivation (severe root pruning) can restrict water intake preventing the plants from getting the calcium they need. Also, if plants are growing in highly acidic soil or are getting too much water from heavy rain, over-irrigation, or high relative humidity, they can develop calcium deficiency and BER.

6.8.1 SYMPTOMS

On peppers, the affected area appears tan (Figure 6.11) and is sometimes mistaken for sunscald, which is white. Secondary molds often colonize the affected area, resulting in a deep brown or black appearance. BER also occurs on the sides of the pepper fruit near the blossom end.

FIGURE 6.11   BER on pepper.

6.8.2   MANAGEMENT

Control of BER is dependent upon maintaining adequate supplies of moisture and calcium to the developing fruits. Peppers should not be excessively hardened nor too succulent when set in the field. They should be planted in well-drained, adequately aerated soils. Peppers planted early in cold soil are likely to develop BER on their first fruits, with the severity of the disease often subsiding on fruits set later. Thus, planting peppers in warmer soils helps to alleviate the problem. Irrigation must be sufficient to maintain a steady even growth rate of the plants.

Mulching of the soil is often helpful in maintaining adequate supplies of soil water in times of moisture stress. When cultivation is necessary, it should not be too near the plants nor too deep, so that valuable feeder roots remain uninjured and viable. In home gardens, shading the plants is often helpful when hot, dry winds are blowing and soil moisture is low. Use of fertilizers low in nitrogen but high in superphosphate, such as 4-12-4 or 5-20-5, will do much to alleviate the problem of BER. In emergency situations, foliage can be sprayed with calcium chloride solutions. However, extreme caution must be exercised since calcium chloride can be phytotoxic if applied too frequently or in excessive amounts. Foliar treatment is not a substitute for proper treatment of the soil to maintain adequate supplies of water and calcium.

6.8.3   SUNSCALD

Sunscald, a noninfectious disease of pepper, is caused by sudden exposure of the fruit to intense direct sunlight and is most serious during periods of extreme heat. Sunscald is common on plants that have suffered premature loss of foliage from leaf spot diseases. Severely pruned plants or those attacked by Verticillium or Fusarium wilts are also likely to suffer from sunscald due to the loss of the lower foliage.

Fruit of healthy plants may be damaged by sunscald if the vines are disturbed or broken and the fruit is exposed to the sun during hot, dry weather. Tomato vines can be disturbed by pruning, natural spreading of the plant caused by the developing fruit load, or upsetting of the plants by a picker. Insufficient nitrogen after fruit set may further aggravate this problem.

6.8.4   SYMPTOMS

Irregular, light-colored, scalded areas appear on any part of the fruit exposed to direct sunlight. Affected areas soon become slightly sunken or wrinkled and creamy white on older fruit (Figure 6.12). On young fruits the spots are light brown. Scalded tissue dries out rapidly in hot weather,

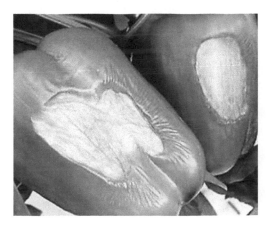

FIGURE 6.12   Sunscald pepper.

becoming thin and papery. Many secondary fungi, such as Alternaria, Brachysporium, and Cladosporium, commonly grow over the scalded surface, giving it a black, gray, or green moldy appearance. Such fungi often gain entrance through the scald, causing the entire fruit to decay.

6.8.5 MANAGEMENT

To prevent sunscald, practice good management techniques that include crop rotation, sanitation, and mulching to reduce the likelihood of defoliation by septoria leaf spot and early blight. Also, if staking, don't remove too much of the foliage. Furthermore, to prevent sunscald on pepper fruit, control foliar diseases and avoid heavy pruning or shoot removal.

6.8.6 MISSHAPEN FRUIT

Misshapen fruits (Figure 6.13).

6.8.6.1 Cause
- Physical barriers encountered by pepper fruit during expansion (Figure 6.13).
- Irregular, uneven growth during expansion.

6.8.6.2 Prevention
None. A low percentage of fruit will always become wedged between stems, stakes, and/or twine.

6.9 FUSARIUM STEM AND FRUIT ROT OF PEPPER

The causal agent of a stem and fruit rot disease of greenhouse sweet peppers are found to be Nectria haematococca (anamorph Fusarium solani). It infects wounds caused by salt damage at the stem base and scars left by careless leaf removal and fruit picking.

6.9.1 CAUSAL ORGANISM

Species	Associated Disease Phase	Economic Importance
F. solani	Stem Rot and Fruit Rot	Minor

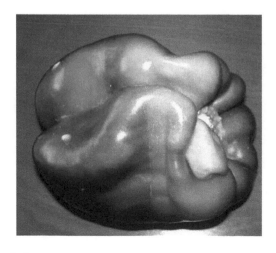

FIGURE 6.13   Misshapen fruit.

6.9.2 SYMPTOMS

Soft, deep brown or black cankers are formed on the stem, usually at nodes or wound sites (Figure 6.14). These may girdle the stem in later stages of disease development. Wounds or openings in the stem provide entry sites for the fungus. The lesions increase in length and width until they girdle the stem, causing the plant parts above the lesion to wilt and die. Leaves below the lesion do not wilt. The lesions may eventually develop cinnamon or light-orange colored, very small, flask-shaped fruiting structures known as perithecia, which are the fruiting bodies of the fungus. White cottony-like growth representing the imperfect stage of the fungus and known as fungus mycelium may also be present on the surface of stem cankers in late stages of disease development. Stem cankers restrict the upward flow of water, which results in wilting and death of the plant (Figure 6.15). Pepper fruits may also develop black, water-soaked lesions beginning around the calyx. The lesions grow, coalesce, and spread down the sides of the fruit. Copious mycelial growth of the pathogen occurs under humid conditions particularly when temperatures exceed 25°C. All these symptoms

FIGURE 6.14   Black cankers formed on stem.

FIGURE 6.15   Wilting and death of the plant.

have generally been reported in greenhouse and glasshouse-grown peppers. Fusarium stem rot has not been reported as a problem in field grown peppers.

6.9.3 CAUSE AND DISEASE DEVELOPMENT

Spore release at night is more favorable for disease development since periods of high relative humidity and even dew occur during this time. The other spores, called conidia, are produced asexually and in large numbers during the imperfect stage (F. solani). They are not ejected but transmitted passively, and therefore, are not as important in the natural dissemination in the greenhouse. These spores may be dispersed by water splash, on pruning knives and other tools, on clothing or on workers' hands.

6.9.4 MANAGEMENT

Sanitation measures offer the best control now. Removal and disposal of infected branches or plants during the cropping season reduce the inoculum in the greenhouse and spread of the disease. Using a sharp knife for pruning or harvesting is reported to reduce disease spread by promoting rapid healing of wounds. Between crops, the greenhouse should be thoroughly cleaned. All crop debris should be removed and all surfaces should be cleaned with a 1% bleach spray. Torn plastic covering the soil should be replaced. The fungus may be carried into the greenhouse on the rock-wool cubes of the transplants so maintaining a clean transplant production area is also important.

No fungicides have been labeled for control of Fusarium stem rot of pepper in the greenhouse. Controlled inoculation of cultivars "Cubico", "Kelvin", and "Triple 4", and observation of cultivars "Kelvin", "Cubico", and "Grizzly" in a commercial greenhouse did not suggest differences in susceptibility of cultivars to the fungus. Controlled temperature studies show that the rate of disease development is greater at higher temperatures. Keeping the greenhouse cooler could reduce losses due to the disease. High relative humidity also encourages the spread of the disease so increasing air movement in the greenhouse will slow the spread of the disease.

6.10 SOUTHERN BLIGHT ON PEPPER

Southern blight, also known as "southern wilt" and "southern stem rot" is a serious and frequent disease. It is caused by the soilborne fungus Sclerotium rolfsii and attacks many vegetable crops. The disease usually appears in "hot spots" in fields in early to mid-summer and continues until cooler, drier weather prevails. Losses may vary from light and sporadic to almost destruction of the crop.

6.10.1 CAUSAL ORGANISM

Species	Associated Disease Phase	Economic Importance
S. rolfsii	Lesions on Stem Causing Wilting of the Plant	Minor

6.10.2 SYMPTOMS

Southern blight is 1 of the most common causes of a sudden wilting and death of a plant. Mild yellowing of the leaves may occur prior to wilting (Figure 6.16). Under humid conditions, a thin, white, fan-shaped mold forms on affected stem tissues and adjoining surface soil. Even under dry conditions, at least a trace of the white mold should be evident on the stem surface. Soon after mold formation, seed-like bodies (sclerotia) develop in the mold (Figure 6.17). The sclerotia begin white, turning tan, then bronze. When the plant is pulled up, a brown, dry rot of the lower stem and upper roots are apparent.

FIGURE 6.16   Sudden wilting and death of a plant.

FIGURE 6.17   Seed-like bodies (sclerotia) developed in the mold.

6.10.3   FAVORABLE CONDITIONS OF DISEASE DEVELOPMENT

Southern blight is not a common disease of tomatoes. The disease is favored by high humidity and soil moisture and warm to hot temperatures (29°C–35°C).

6.10.4   DISEASE CYCLE

The fungus overwinters as sclerotia in the soil and in plant debris. A characteristic of the fungus is that it is generally restricted to the upper two or three inches of soil and will not survive at greater depths. The fungus is more active in hot, wet weather, and it requires the presence of undecomposed plant residue to initiate infection. S. rolfsii is more active under acidic soil conditions. The fungus does not have an airborne spore, so all infections result from contact of the plant tissue with soil. It is spread when infested soil particles are moved, as with cultivation. The fungal body is so strong that it can grow across the soil surface to reach a plant if old plant debris is available.

6.10.5   MANAGEMENT

6.10.5.1   Integrated Management

Southern blight is difficult to control because the fungus has a broad host range that includes over 500 plant species, and sclerotia can survive for several years in soil (Aycock, 1966). However, some integration of management methods may help to reduce the impact of southern blight during vegetable production.

- Use of pathogen-free transplants and resistant cultivars: Transplant nurseries should be located far from vegetable production fields and avoid excess watering and high temperatures. Vegetable transplants should always be inspected for overall health prior to planting. The use of resistant cultivars is always a preferred method of disease management. Unfortunately, resistance to S. rolfsii has not been identified or is limited for many host plant species, and it is currently not a viable option for most vegetables. Resistance has been identified for some hosts. Six tomato breeding lines—5635M, 5707M, 5719M, 5737M, 5876M, and 5913M—were released jointly from Texas A&M University Research Center, Coastal Plain Experiment Station, and the University of Georgia (Leeper et al. 1992). Recent attempts to reassess resistance to S. rolfsii in pepper confirmed some useful levels of resistance in several pepper species, including the bell-type cultivar "Golden California Wonder", which is conferred by a single recessive gene (Dukes and Fery, 1984; Fery and Dukes, 2005).
- Crop rotation: Although crop rotation is a traditional and preferred method to control disease, it is not very effective in controlling southern blight because of the broad host range of S. rolfsii and the survivability of sclerotia in the soil. Yet rotating with non-susceptible crops, such as corn or wheat, may help decrease disease incidence in following years by lowering initial inoculum (Mullen, 2001).
- Soil solarization: Sclerotia can be killed in four to six hours at 50°C (122°F), and in three hours at 55°C (131°F) (Ferreira and Boley, 1992). Covering moistened soil with clear polyethylene sheets during the summer season can reduce the number of viable sclerotia if the soil temperature under the sheet remains high enough for an appropriate length of time. While effective for smaller areas, solarization is generally impractical for larger commercial operations. Refer to http://edis.ifas.ufl.edu/in824 for more information about solarization.
- Deep plowing: Deep plowing in the fall or in the spring before bed preparation is another effective method. The ability of sclerotia to germinate is reduced with soil depths greater than 2.5 cm (1 inches). Soil depths of 8 cm (3.1 inches.) or greater prevent germination completely due to the mechanical stress created by the soil over the sclerotia (Punja, 1985). However, growers need to be aware that studies of other soilborne pathogens have shown that deep plowing can spread the pathogen, changing the distribution of future disease outbreaks (Subbarao et al. 1996).
- Soil amendments: Amending soils with organic fertilizers, biological control agents, and organic amendments—such as compost, oat, corn straw, and cotton gin trash—may help control southern blight. For example, the use of organic amendments, cotton gin trash, and swine manure was found to control southern blight through the improved colonization of soil by antagonistic Trichoderma spp. (Bulluck and Ristaino, 2002). Deep plowing the soil combined with applications of certain inorganic fertilizers—like calcium nitrate, urea, or ammonium bicarbonate—was also shown to control southern blight on processing carrots (Punja, 1986). Studies have proposed that the increased nitrogen inhibits sclerotia germination, whereas the increased calcium might alter host susceptibility. However, these approaches have not been tested in the sandy soils of Florida, and earlier studies indicated that the addition of inorganic fertilizers may be less effective in soils prone to leaching.

6.10.5.2 Chemical Control

The use of soil fumigants, such as methyl bromide, chloropicrin, and metam sodium, are the most practical means to treat seed beds and fields for a number of soilborne pathogens, including S. rolfsii (Mullen, 2001). They must be applied days to weeks before planting. However, the availability of methyl bromide is limited due to its status as an ozone-depleting material.

Preplant fungicides, such as Captan and pentachloronitrobenzene (PCNB), are effective in reducing disease severity. PCNB can effectively limit disease incidence when applied prior to infection and is registered for use on a limited number of vegetable crops. Some commercially available stro-bilurin fungicides (azoxystrobin, pyraclostrobin, and fluoxastrobin) are also labeled for the control of southern blight on certain vegetables and were found to provide some control of southern blight in peanut production (Culbreath, Brenneman, and Kemerait, 2009; Woodward, Brenneman, and Kemerait, 2007).

6.10.5.3 Biological Control

Some biological agents, such as Trichoderma harzianum, Gliocladium virens, Trichoderma viride, Bacillus subtilis, and Penicillium spp., were found to antagonize S. rolfsii and could suppress dis-ease. G. virens was found to reduce the number of sclerotia in soil to a depth of 30 cm, result-ing in a decreased incidence of southern blight on tomato (Ristaino, Perry, and Lumsden, 1991). Trichoderma koningii also reduced the number of sclerotia and the plant-to-plant spread of southern blight in tomato fields (Latunde-Dada, 1993). However, there is evidence from a greenhouse study that G. virens has better biocontrol capability against S. rolfsii than Trichoderma spp. (Papavizas and Lewis 1989).

> The information for Management ID.s taken from "PP272, one of a series of the Plant Pathology Department, Florida Cooperative Extension Service, Institute of Food and Agricultural Sciences, University of Florida. Original publication date March 2010".

6.11   POWDERY MILDEW OF PEPPER

Powdery mildew, caused by Leveillula taurica, is one of the most damaging diseases that affect greenhouse bell peppers. 1% mildew infection on the leaves would result in 1% yield loss. Studies show that the higher the level of powdery mildew infection, the higher the loss of production. An early, heavy infection with mildew had about 30% loss of production compared to a later, lighter infection. Powdery mildew generally has caused 10–15% yield loss in greenhouse pepper crops. Greenhouse pepper growers need to follow an intensive disease prevention plan because it is very important that powdery mildew never gets out of hand. Once pepper leaves are infected with pow-dery mildew it is difficult to control; if left unchecked the crop can be destroyed.

6.11.1   CAUSAL ORGANISM

Species	Associated Disease Phase	Economic Importance
L. taurica	Chlorosis. Necrotic, Brown Spots on Upper Leaf surface. Leaves Curl upward. Premature Defoliation. Sun Scald Because of Leaf Drop.	Severe

6.11.2   SYMPTOMS

The initial symptom on pepper is the appearance of small white to light gray colored spots on the underside of the leaf. In early infections, these spots often appear first on the lower leaves inside the canopy of the plant. The fungus produces asexual spores on conidiophores that emerge through stomatal openings. Profuse sporulation on the undersurface of the leaves appears (Figure 6.18). The mycelium is inside the leaf, so it does not have a white powdery appearance like other powdery mil-dews. Later infections may cause yellowing or appear as raised pimply areas. Infected leaves may curl resulting in the exposure of the fruit to sunburn. Fruits are not infected.

FIGURE 6.18   Profuse sporulation on the undersurface of pepper leaves.

6.11.3   CAUSE AND DISEASE DEVELOPMENT

These fungi have a wide host range. Airborne conidia from previous crops or weeds can be carried long distances by wind and act as initial sources of inoculum.

6.11.4   FAVORABLE CONDITIONS OF DISEASE DEVELOPMENT

High relative humidity is not required for infection. Warm temperature and low light conditions favor disease development. The fungus can cause disease in a wide range of environmental conditions (4°C–35°C, 0–100% relative humidity). However, optimal conditions for infection and disease development occur when the temperature is between 15.5°C and 27°C with humidity greater than 85%. Under favorable conditions, the fungus reproduces rapidly, and spores can germinate and infect a plant in less than 48 hours. Wind-disseminated spores cause secondary infections, which help spread the disease.

6.11.5   DISEASE CYCLE

The powdery mildew disease cycle starts when spores (known as conidia) land on a pepper leaf. Spores germinate much like a seed and begin to grow into the leaf. Pepper powdery mildew parasitizes the plant using it as a food source. The fungus initially grows unseen within the leaf for a latency period of 18–21 days. Then the fungus grows out of the leaf openings (stomata) on the undersurface of the leaf, producing conidiospores, which are borne singly on numerous, fine strands or stalks called conidiophores. These fungal strands become visible as white patches, i.e. mildew colonies, on the undersurface of the leaf. Air currents carry these microscopic, infectious spores to other plants. Spores are dispersed further through the greenhouse vents. In addition to dispersal by air currents or wind, powdery mildew can spread on ornamental plants and weeds, and by workers on their clothing. Repeated generations of powdery mildew can lead to severe outbreaks of the disease that economically damage the crop.

6.11.6 MANAGEMENT

Apply protectant fungicides before an epidemic or immediately after the first symptoms are observed. Provide for air circulation around plants and light penetration through the canopy. Excessive fertilization is reported to increase the severity of powdery mildew epidemics.

Disease is difficult to control once leaves become infected, therefore, preventive measures are necessary in the early stages of disease. Wettable sulfur, dusting sulfur, Rally (Mylobutanil) and Flint (Trifloxystrobin) are registered for use on peppers and give good control if applied preventively in both field and greenhouse peppers. Serenade, a biological, has some efficacy at low levels of disease in the greenhouse. There are no resistant varieties of pepper, but hot chilies seem to have less disease than bells, long greens, or sweet jalapenos.

6.12 PHYTOPHTHORA BLIGHT

Phytophthora blight of pepper is caused by the fungus Phytophthora capsici. Other names applied to this disease of peppers are Phytophthora root rot, crown rot, and stem and fruit rot. These names can apply since all parts of the pepper plant are affected.

Root rot, caused by the soilborne organism P. capsici, is most severe in heavy wet soils that favor growth and spread of P. capsici. Disease usually occurs in low spots in the field and is confined to discrete areas of poor drainage.

6.12.1 CAUSAL ORGANISM

Species	Associated Disease Phase	Economic Importance
P. capsici	Root Rot, Crown Rot, Stem Rot, and Fruit Rot	Severe

6.12.2 SYMPTOMS

Phytophthora blight of peppers can attack the roots, stems, leaves, and fruit, depending upon which stage plants are infected. A grower not knowing what to expect might first encounter the disease at mid-season when sudden wilting and death occur as plants reach the fruiting stage. Early infected plants are quickly killed, while later-infected plants show irreversible wilt. Often a number of plants in a row or in a roughly circular pattern will show these symptoms at the same time. Taproots and smaller lateral roots show water-soaked, very dark brown discoloration of cortical and xylem tissue. Very few lateral roots remain on diseased plants and the tap roots may also be shorter compared to those of healthy plants. The most striking difference between healthy and diseased plants is the total amount of root tissue.

Fungus-infected seedlings will damp off at the soil line, but relatively few plants die when temperatures are cool. Far more commonly, the disease will strike older plants which then exhibit early wilting. Stem lesions can occur at the soil line and at any level on the stem. Stems discolor internally, collapse, and may become woody in time (Figure 6.19). Lesions may girdle the stem, leading to wilting above the lesion, or plants may wilt and die because the fungus has invaded the top branches before the stem lesions are severe enough to cause collapse (Figure 6.20).

Leaves first show small dark green spots that enlarge and become bleached as though scalded. If the plant stems are infected, an irreversible wilt of the foliage occurs.

Infected fruits initially develop dark, water-soaked patches that become coated with white mold and spores of the fungus. Fruits wither but remain attached to the plant. Seeds will be shriveled and infested by the fungus (Figure 6.21).

FIGURE 6.19   Phytophthora symptoms within the stem.

FIGURE 6.20   Wilting of pepper plant.

FIGURE 6.21   Phytophthora symptoms on fruit.

6.12.3   CAUSE AND DISEASE DEVELOPMENT

P. capsici has motile asexual spores that move in free water in the soil and infect roots; these spores also move in irrigation water. Sexual spores are produced in plant tissue and can survive in residue or in the soil for prolonged periods of time.

6.12.4   FAVORABLE CONDITIONS OF DISEASE DEVELOPMENT

Phytophthora spreads rapidly during warm, wet weather. Ideal conditions for infection are moist soils above 18°C (65°F) and air temperatures of 24°C–29°C (75°F–85°F).

Phytophthora problems often follow field drainage patterns. They are most likely to occur in low-lying or poorly drained areas of the field.

6.12.5   DISEASE CYCLE

The fungus occurs naturally in most soils and can infect pepper and other crops at most stages of growth when there is excess soil moisture and warm, wet weather. The fungus overwinters in soil as thick-walled oospores. For pepper, the fungus also survives on and in infected seed, but this is not a factor with commercially purchased seed. Infected plants produce irregularly branched thread-like structures (sporangiophores), which in turn produce variously shaped sporangia spores. With adequate moisture the sporangia give rise to biflagellate (two-tailed) motile zoospores. These spores produce germ tubes that penetrate plant tissue. The cycle is repeated with the production of more sporangiophores and sporangia.

When the humidity is high, the sporangia can survive for long periods. During such times wind-borne sporangia can be carried long distances, causing widespread dissemination and a rapid increase in the disease. Zoospores are readily spread by splashing rain and by flowing irrigation and surface water. The disease develops first in low areas after heavy rains and can quickly spread throughout the field. Favorable conditions for the fungus include wet soils above 18°C (65°F) and prolonged wet periods with air temperatures in the 24°C–29°C (75°F–85°F) range (Figure 6.22).

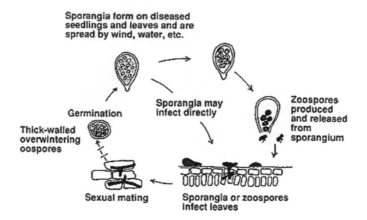

FIGURE 6.22   Disease cycle of P. capsica.

6.12.6   MANAGEMENT

Because Phytophthora blight is soilborne and more prevalent in poorly drained soils, careful attention must be given to cultural practices, especially on fields with a history of the disease.

- Practice crop rotation with crops other than tomato, eggplant, and cucurbits for at least three years.
- Avoid poorly drained fields for growing these crops.
- Plant the crop on a ridge, or better yet on raised, dome-shaped beds to provide better soil drainage. Un-mulched, low profile beds will deteriorate during the season and may not provide sufficient drainage in July and August when disease spread can occur. Maintaining the uniform soil moisture necessary to prevent BER of peppers is difficult with raised beds unless trickle irrigation is used. Overhead irrigation, like rainfall, will encourage disease spread and should be discontinued if the disease is present.
- Fungicide use will vary depending upon the crop grown, and in some cases, the particular disease phase to be controlled. Soil fumigation, although useful in greenhouse situations, is not practical for field use because the fungus quickly reinvades treated soil.

6.13   ROOT KNOT NEMATODE

Root knot nematode (RKN) is a serious malady in pepper. The functional root system is modified into galls and it impairs uptake of water and nutrients. Poor development of the root system makes the plant highly susceptible to drought. In addition, RKN in association with pseudomonas leads to bacterial wilt, which in association with soilborne rhizoctonia reduce seed germination and increases the root rot problem.

RKN are found worldwide and are known to affect over 2000 species of plants including various vegetable and crop species. There are several distinct species of RKN including Meloidogyne javonica, M. arenaria, M. incognita (southern RKN), M. chitwoodi (the Columbia RKN), and M. hapla (the northern RKN). Quarantines have been placed upon some countries and states known to have root knot to protect vegetable seed and crop production. Root knot is associated with other diseases such as crown gall and Fusarium diseases because the RKNs provide entryways for Agrobacterium sp. and Fusarium sp. to infect plants as well.

6.13.1   SYMPTOMS AND DAMAGE

The northern RKN produces small, discrete galls while the southern RKNs produce large galls and massive root swellings. Infected plants are stunted, appear yellow or pale green in color, and wilt easily, even when soil moisture is adequate. Severe infestations can dramatically reduce yields and eventually kill plants. Damage from RKN feeding may also increase the incidence of other soilborne diseases such as Fusarium wilt and cause Fusarium wilt-resistant varieties to become susceptible.

6.13.2 LIFE CYCLE

The juveniles hatch from eggs, move through the soil and invade roots near the root tip. Occasionally they develop into males but usually become spherical-shaped females.

The presence of developing nematodes in the root stimulates the surrounding tissues to enlarge and produce the galls typical of infection by this nematode. Mature female nematodes then lay 100s of eggs on the root surface, which hatch in warm, moist soil to continue the life cycle.

Continued infection of galled tissue by second and later generations of nematodes causes the massive galls sometimes seen on plants such as pepper at the end of the growing season. The length of the life cycle depends on temperature and varies from four to six weeks in summer to ten to fifteen weeks in winter. Consequently, nematode multiplication and the degree of damage are greatest on crops grown from September to May.

Nematodes are basically aquatic animals and require a water film around soil particles before they can move. Also, nematode eggs will not hatch unless there is sufficient moisture in the soil. Thus, soil moisture conditions that are optimum for plant growth are also ideal for the development of RKN (Figure 6.23).

6.13.3 MANAGEMENT

Management of root knot should focus on sanitation measures for preventing contamination of soils, reducing populations below damaging levels where infestations already exist, and variety selection. Sanitation measures include planting nematode-free pepper transplants and avoiding the introduction of nematodes on any other type of transplant stock or with soil. This is difficult in reality because soil clinging to plant roots may contain nematodes without obvious plant symptoms. Equipment and boots should be washed free of soil before working clean ground when moving from areas suspected of harboring nematodes. Strategies for reducing nematode populations include starving nematodes by two-year crop rotations with resistant crops like corn, milo, and nematode-resistant soybean varieties; or with clean (weed-free) fallow. Soil solarization may be effective in some situations, but soil fumigation provides more consistent control of nematode populations. Soil fumigants are restricted-use pesticides that can only be applied by certified applicators. Incorporation of cruciferous green manures such as cabbage, mustard, and rape into soil may also help reduce populations, particularly when combined with solarization. Many root knot resistant pepper varieties are available.

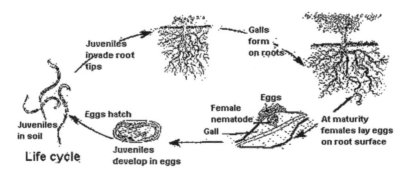

FIGURE 6.23 Life cycle of RKN.

Fertilizers like calcium cyanide, sodium cyanide, and urea cyanamide can release ammonia and NH4, which is poisonous for nematodes. Seed treatment with Aldicarb or Carbofuran (2gm/Kg) or nursery treatment (2gm/m²) or seedling root dip (1000 mg/L) for 30 minutes or main field application of chemicals (2Kg/ha) controls the nematode efficiently.

VAM fungus is proved to reduce the number of nematodes that develop into adults. Fungus Paecilomyces lilacinus can attack the eggs of nematodes when applied at 8 gm/plant.

6.13.4 STING NEMATODE ON PEPPER

Sting nematodes are among the most destructive plant-parasitic nematodes on a wide range of plants. Adults can reach lengths greater than 3 mm, making them one of the largest plant-parasitic nematodes. While there are several species of sting nematodes described, only Belonolaimus longicaudatus Rau is known to cause widespread crop damage.

6.13.5 SYMPTOMS

Plants damaged by sting nematodes often wilt, may be stunted, and may show symptoms of nutrient deficiency. Seedlings may sprout from the soil and then cease growing altogether. Plant death may occur with high population densities of sting nematodes. These symptoms may be caused by a number of plant diseases and disorders. Therefore, the only way to be certain whether sting nematodes are a problem is to have a soil nematode assay conducted by a credible diagnostic facility.

6.14 VERTICILLIUM WILT OF PEPPER

This includes a wide variety of vegetables and herbaceous ornamentals. Tomatoes, eggplants, peppers, potatoes, dahlia, impatiens, and snapdragon are among the hosts of this disease. Plants weakened by root damage from drought, waterlogged soils, and other environmental stresses are thought to be more prone to infection.

6.14.1 CAUSAL ORGANISM

Species	Associated Disease Phase	Economic Importance
2 Closely Related Soilborne Fungi, Verticillium dahlia and V. albo-atrum	Wilting of Plant	Minor

6.14.2 SYMPTOMS

Verticillium dahlia or V. albo-atrum can infect pepper plants at any growth stage. Symptoms include yellowing and drooping of leaves on a few branches or on the entire plant. The edges of the leaves roll inward on infected plants, and foliar wilting ensues. The foliage of severely infected plants turns brown and dry. Growth of pepper plants inoculated with aggressive strains of V. dahliae in greenhouse or of pepper plants infected early in the season under field conditions is severely stunted with small leaves that turn yellow-green. Subsequently, the dried leaves and shriveled fruits remain attached to plants that die. Brown discoloration of the vascular tissue is visible when the roots and lower stem of a wilted plant are cut longitudinally (Figure 6.24). The roots of V. dahliae-infected pepper plants show no external discoloration or decay.[6]

6.14.3 CAUSE AND DISEASE DEVELOPMENT

This is a soilborne disease which can survive in the soil for prolonged periods of time without host plants present.

FIGURE 6.24   Verticillium infected pepper plants.

6.14.4   FAVORABLE CONDITIONS OF DISEASE DEVELOPMENT

Symptoms can occur all season but are generally more severe after fruit set or during dry spells. Soil temperatures of 15°C–21°C (59°F–70°F) and air temperatures of 20°C–24°C (68°F–75°F) promote the development of symptoms. The disease is more severe where plant-parasitic nematodes are present.

6.14.5   DISEASE CYCLE

Both Verticillium albo-atrum and V. dahliae produce one-celled, colorless conidia that are short-lived. V. dahliae also produces minute, black, resting structures called microsclerotia, while V. albo-atrum produces microsclerotial-like dark, thick-walled mycelium but not microsclerotia. Optimum growth of V. albo-atrum occurs at 20°C to 25°C (68°F to 77°F), while V. dahliae prefers slightly higher temperatures (77°F to 81°F or 25°C to 28°C) and is somewhat more common in warmer regions. Different strains within each species differ considerably in virulence and other characteristics. Although some Verticillium strains show host specialization, most of them attack a wide range of host plants. Agricultural soils may contain up to 100 or more microsclerotia per gram. 6 to 50 microsclerotia per gram are sufficient to generate 100% infection in such susceptible crops as eggplant, pepper, potato, and tomato.

6.14.6   MANAGEMENT

Management of this disease is difficult since the pathogen survives in the soil and can infect many species of plants. As with many diseases, no single management strategy will solve the problem. Rather, a combination of methods should be used to decrease its effects. When a positive diagnosis has been made, the following recommendations may be followed:

*   Whenever possible, plant resistant varieties. There are many verticillium-resistant varieties of pepper. These are labeled "V" for verticillium-resistance. Note: Verticillium-resistant plants may still develop verticillium wilt if there is a high population of nematodes in the soil.
*   Remove and destroy any infested plant material to prevent the fungi from overwintering in the debris and creating new infections.
*   Keep plants healthy by watering and fertilizing as needed.
*   Fields should be kept weed-free since many weeds are hosts for the pathogen.
*   Soil fumigants are effective in reducing disease severity but are not recommended for use in gardening.

- Susceptible crops can be rotated with non-hosts such as cereals and grasses, although four to six years may be required since the fungi can survive for long periods in the soil.
- Avoidance of known infested fields and rotations of several years are recommended, although the microsclerotia of the fungus are known to survive for many years. Rotation to small grains and alfalfa, which are not hosts, may result in reduction of disease. Cotton rotations should be avoided since cotton is very susceptible to some strains of V. dahliae.

BIBLIOGRAPHY

1. Information referenced from Pessl Instruments *"Anthracnose Fruit Rot, Colletotrichum"*. http://docs. metos.at/Anthracnose+Fruit+Rot%2C+Colletotrichum?structure=Disease+model_en
2. Ritchie, D.F. 2000. "Bacterial spot of pepper and tomato", The Plant Health Instructor. https://www. apsnet.org/edcenter/intropp/lessons/prokaryotes/Pages/Bacterialspot.aspx
3. Sun X., Nielsen, M. C. and Miller, J. W., 2002, *"Bacterial Spot of Tomato and Pepper"*, Plant Pathology Circular No. 129 (Revised).
4. Cerkauskas R., 2004, *"Cercospora Leaf Spot"*, AVRDC – The World Vegetable Center.
5. Xie C. and Vallad, G., 2009, "Integrated Management of Southern Blight in Vegetable Production", University of Florida IFAS Extension. http://edis.ifas.ufl.edu/pp272
6. Bhat, R. G., Smith, R. F., Koike, S. T., Wu, B. M., and Subbarao, K. V. 2003. "Characterization of Verticillium dahliae isolates and wilt epidemics of pepper", *APS Journal*. https://apsjournals.apsnet.org/ doi/pdf/10.1094/PDIS.2003.87.7.789

7 Potato

Potato is a staple food in many parts of the world and an integral part of much of the world's food supply. It is the world's fourth-largest economical food crop following maize, wheat, and rice. The word "potato" may refer either to the plant itself or to the edible tuber. Wild potato species can be found throughout the Americas from the United States to southern Chile. In 2014, world production of potatoes was 382 million tons, an increase of 4% over 2013 amounts, and led by China with 25% of the world total. Other major producers were India, Russia, Ukraine, and the United States.

Being such a highly produced crop, potato suffers multiple diseases. The historically significant late blight remains an ongoing problem in Europe and the United States. Other potato diseases include Rhizoctonia Canker, powdery mildew, powdery scab, and leafroll virus. In this chapter, thorough information about diseases and its unique protective methods are given in relation to the potato (Figures 7.1 through 7.3).

7.1 BACTERIAL SOFT ROT AND BLACKLEG

Blackleg and tuber soft rot are all similar diseases caused by several types of soft rot bacteria. Blackleg and tuber soft rot occur wherever potatoes are grown.

Bacterial soft rot is the most serious causes of potato losses in storages. Soft rot bacteria infect potato tubers that have been damaged by mechanical injury or the presence of other diseases. Bacterial soft rot develops much faster when potatoes are wet. Tubers may be wet when put into storage, may become wet by the excessive application of a storage fungicide, or may become wet because of excessive respiration and water loss in storage. Wet conditions in storage allow the soft rot to spread from one tuber to another. Soft rot can be kept to a minimum if potatoes are kept dry.

The bacterium that causes the blackleg disease of potato is one of the pathogens that are tuberborne. The blackleg disease can cause severe economic losses to the potato crop. However, the occurrence of blackleg depends very much on the growing conditions, particularly temperature and rainfall after planting.

7.1.1 CAUSAL ORGANISM

Species	Associated Disease Phase	Economic Importance
Soft Rot	Soft Rot of Tuber	Severe
Pectobacterium carotovorum var. *carotovora, P. chrysanthemi*	Black Stem Decay	
Blackleg:		
Pectobacterium arotovorum var. *atrosepticum*		

7.1.2 SYMPTOMS

7.1.2.1 Soft Rot

When soft rot develops, tuber flesh decays and becomes cream to tan color. Often there is a black border between lenticel spot symptoms that may also be observed in the field, but usually are most noticeable four to ten days after harvest and packaging. Symptoms are characterized by tan to dark brown, circular, water-soaked spots or small lesions surrounding the lenticels on the tuber surface. Infected tissue usually does not extend deeper than about 4 mm into the tuber. Adjacent lesions may coalesce to form larger, irregularly shaped, sunken lesions. Under moist conditions, the lesions enlarge, and can

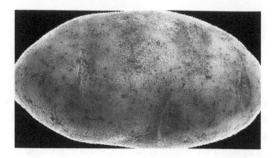

FIGURE 7.1   Potato.

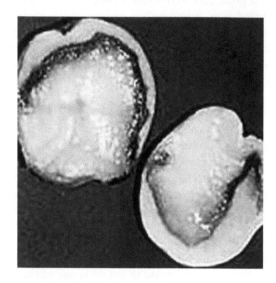

FIGURE 7.2   Symptoms of soft rot on potato tuber.

FIGURE 7.3   Symptoms of blackleg on potato foliar and tuber.

rapidly take on a puffy appearance due to the production of gases by respiring bacteria in the lenticels. Contamination of potato tubers occurs anytime they meet the bacterium, most commonly during the harvest through handling or washing. The bacterium invades the potato tuber chiefly through wounds. Most of the soft rot infections are in tissues that have been weakened, invaded, or killed by pathogens or by mechanical means. Soft rot in tubers is favored by immaturity, wounding, invasion by other pathogens, warm tuber and storage temperatures, free water, and low oxygen conditions. Tubers harvested at temperatures above 27°C can be predisposed to soft rot. Decay can be retarded by temperatures less than 10°C; the lower the temperature, the better. Immature tubers are susceptible to harvester-related injury and bacterial infection. Suberizing seed and treatment with fungicide is a tactic to reduce the risk of other seed infections that could lead to soft rot breakdown of the seed. Usually, the infected tubers remain firm and marketable, although of reduced quality and appearance. Infections via the lenticels can contribute to eventual soft rot of the entire tuber. However, if conditions remain dry, lenticel spot lesions remain limited, sunken, dry, and hard.[1]

7.1.2.2 Blackleg

7.1.2.2.1 Pre-emergence and Early Growing Season

Blackleg may occur early in the season causing the seed pieces to rot before emergence. When this happens, skips appear within the row and poor stand establishment is observed. After emergence, the infected plants are characterized by stunted, yellowish foliage with an upright habit. The lower stem typically blackens and decays giving the "blackleg" designation for this disease. Plants affected at this stage typically die.

7.1.2.2.2 Later Growing Season

More mature plants may also develop blackleg symptoms including plants that have begun flowering. Typically, the disease appears on healthy plants as a black discoloration of stems and rapid wilting and sometimes leaf yellowing. Black discoloration starts belowground and moves up the stem until the entire stem turns black and the plant wilts. Potato tubers infected with blackleg exhibit a soft rot and are watery. In later stages, the tubers often have a foul odor. Diseased tuber tissue is creamy or tan colored, often with a black border between it and healthy tissue.

7.1.3 FAVORABLE CONDITIONS OF DISEASE DEVELOPMENT

Wet fields and warm temperatures before harvest, plus a film of moisture on the tuber surface either in storage or transit, greatly favor development of both diseases. Generally, tuber soft rot caused by *Pectobacterium* occur at temperatures above 10°C (50°F) with 10°C to 30°C (77°F to 86°F) optimal.

Blackleg is most severe under cool, wet conditions at planting time followed by high soil temperatures (higher than 24°C) after plant emergence.

7.1.4 DISEASE CYCLE

7.1.4.1 Soft Rot

There are many ways in which a plant can become infected by a bacterial soft rot. They can be host to the bacteria either by being infected as seed or from direct inoculation into wounds or natural openings (stomata or lenticels) in mature plants, which is most common. But, when a plant is infected, and the conditions are favorable, the bacteria immediately begin feeding on liquids released from injured cells and start replicating. As they replicate they release more and more pectolytic enzymes that degrade and break down cell walls. And, because of the high turgor pressure within the cells, this maceration effectively causes the cells to explode and die, providing more food for the bacteria.[3]

As they gorge on intracellular fluid, the bacteria continue to multiply and move into the intercellular spaces, with their cell-wall-degrading enzymes ahead of them preparing the plant tissues for digestion. Often the epidermis is left unscathed, keeping the rotten flesh contained within until a crack allows the ooze to leak out and infect others around it.[3]

When the plant organs are harvested and placed into storage, those that are infected will automatically infect the others placed with it. When certain insects are present, the eggs laid over the stored vegetables will be invaded by the bacteria, becoming host and transporter, able to infect others as they grow. The bacteria then overwinter within the plant tissues, insect hosts, or in the soil and lay dormant until the conditions are right again to reproduce. If the infected storage organs are being used to propagate the plant, or if infected seed was produced, then when spring comes, the bacteria will begin to grow just as its host does. Also, in the spring, the contaminated insect eggs hatch into larvae and begin to cause infection within the host plant. The larvae then become adults, leave the infected host, and move on to unknowingly inoculate more plants to start the cycle over again.[3]

7.1.4.2 Blackleg

Blackleg is spread in contaminated seed pieces. The bacteria survive in or on seed tubers. All evidence suggests that the bacteria do not survive well apart from its association with potato plant tissue. Thus, the seed tuber is the most important source of inoculum and the key target for disease prevention. However, the bacteria may also overwinter in soil inside infected tubers or other plant debris. The disease is most severe under cool, wet conditions at planting time followed by high soil temperatures (higher than 24°C) after plant emergence.[3]

7.1.5 MANAGEMENT

- Plant only certified, disease-free seed tubers. If possible, use whole seed tubers that do not have to be cut.
- When receiving seed tubers in bags, do not stack more than five bags high. With bulk or bagged seed, store at 4°C–7°C until two to three weeks before planting, then warm to 13°C–15.5°C prior to cutting.
- Clean all equipment used for cutting seed tubers thoroughly and then sanitize with an appropriate disinfectant. Clean cutting equipment again periodically and definitely before cutting a new lot of seed tubers.
- Treat cut seed pieces with recommended fungicide dressings immediately after cutting.
- Plant treated cut seed pieces immediately if soil temperatures are 13°C–18°C at planting depth. Seed pieces can be held one or two weeks at 13°C–15.5°C and 95–99% relative humidity to hasten healing of cut surfaces. Condensation on surfaces of seed pieces must be avoided.
- Do not irrigate fields until plants are well emerged. Avoid using surface water for irrigation.
- During crop growth, monitor irrigation and nitrogen fertility to minimize excessive vine growth that will promote leaf wetness within the plant canopy.
- Harvest tubers only after the vines are completely dead to ensure skin maturity. Low spots in the field should be left unharvested if significant waterlogging has occurred.
- Take all precautions to minimize cuts and bruises when harvesting and handling tubers.
- Hold newly harvested potatoes at 13°C–15.5°C with 90–95% relative humidity for the first one or two weeks to promote wound healing. After this curing period, lower the temperature of table-stock to 3°C–4°C for long-term storage. Never wash tubers prior to storage.

7.2 BACTERIAL WILT OF POTATO

Bacterial wilt (also known as brown rot) limits production of potatoes, especially seed potatoes, worldwide. Appropriate chemical control measures that are practical and effective do not exist. Other integrated control components, however, may be effective. A requirement for propelling disease management is knowledge about the causal organism and its action. There are three races of bacterial wilt, *Ralstonia solanacearum*, which have been distinguished based on pathogenicity. Race three of the pathogen mainly affects potatoes.

7.2.1 CAUSAL ORGANISM

Species	Associated Disease Phase	Economic Importance
R. solanacearum	Wilting, Yellowing and Some Stunting of the Plants, Which Finally Die Right Back	Severe

7.2.2   Symptoms

R. solanacearum causes symptoms in both above and belowground organs of the potato plant.

7.2.2.1   Aboveground

Aboveground symptoms are wilting, stunting, and yellowing of the foliage. Wilting caused by *R. solanacearun* resembles that which is caused by lack of water, other pathogens such as *Fusarium* or *Verticilliurn* spp., as well as insect or mechanical damage at the base of a stem. The initial wilting of only part of the stems of a plant, or even 1 side only of a leaf or stem is characteristic of *R. solanacearun.* If disease development is rapid, entire plants wilt quickly (Figure 7.4). A cross-section through a young diseased potato stem reveals brown discoloration of the vascular system. Upon slight pressure, a milky slime may exude. In a longitudinal section, the vascular system may show dark, narrow stripes beneath the epidermis. The milky slime exuding from stems indicates the activity of the bacteria within the vascular system.

7.2.2.2   Belowground

External symptoms are not always visible on infected tubers. In case of severe infections, however, bacterial ooze collects at tuber eyes or stolon end, causing soil to adhere. A cut tuber often shows brownish discoloration of the vascular ring. Slight squeezing forces the typical pus-like slime out of the ring, or it oozes out naturally. The vascular ring, or the whole tuber, may disintegrate completely at more advanced stages of disease development (Figure 7.5). Not all tubers of a wilted plant may be affected, and not all affected tubers may show symptoms. Latent infections can be detected by incubating tubers at 30°C and high humidity. After two to three weeks, the typical tuber symptoms can be observed. Symptoms caused by *R. solanacearun* may be confused with those caused by *Corynebacteriurn sepedonicum* (ring rot). A major difference is that *R. solanacearun* causes direct collapse of green plants, whereas ring rot wilting is usually associated with chlorosis, yellowing, and necrosis of foliage.[4]

FIGURE 7.4   Aboveground symptoms of bacterial wilt.

FIGURE 7.5   Belowground symptoms of bacterial wilt.

7.2.3   CAUSE AND DISEASE DEVELOPMENT

Bacterial wilt is a devastating disease that affects potatoes when the soil becomes contaminated by disease-causing bacteria. The disease is spread when diseased potato seed is planted or when farmers plant potatoes on infected soil. Consequently, the affected plant stems die while potato tubers also rot. Although bacterial wilt has no known cure, it can be controlled to some extent if farmers can follow simple rules to manage it; the safest method is crop rotation.

Bacterial wilt can be spread through several ways: infected crop residues, contaminated surface run-off water, or even water used for irrigation. Infestation by root knot nematodes aggravates the disease. Farmers can spread the disease through farming tools such as hoes or forks when contaminated soil attaches itself to the tools. Insects or soilborne pests such as nematodes spread bacterial wilt into potatoes' plant roots.

7.2.4   FAVORABLE CONDITIONS OF DISEASE DEVELOPMENT

Bacterial wilt of potato is generally favored by temperatures between 25°C and 37°C. It usually does not cause problems in areas where mean soil temperature is below 15°C.

Under conditions of optimum temperature, infection is favored by wetness of soil. However, once infection has occurred, severity of symptoms is increased with hot and dry conditions, which facilitate wilting.

7.2.5   DISEASE CYCLE

R. solanacearum can overwinter in plant debris or diseased plants, wild hosts, seeds, or vegetative propagative organs like tubers. The bacteria can survive for a long time in water (up to 40 years at 20°C–25°C in pure water) and the bacterial population is reduced in extreme conditions (e.g. temperature, pH, salts). Infected land sometimes cannot be used again for susceptible crops for several years. *R. solanacearum* can also survive in cool weather and enter a state of being viable but nonculturable. In most cases, this stage is not an agricultural threat because the bacteria usually become avirulent after recovering.

R. solanacearum usually enters the plant via a wound. Natural wounds (created by excision of flowers, genesis of lateral roots) as well as unnatural ones (by agricultural practices or nematodes and xylem-feeding bugs' attack) would become entry sites for *R. solanacearum*. The bacteria get access to the wounds partially by flagellar-mediated swimming motility and chemotaxic attraction toward root exudates. Unlike many phytopathogenic bacteria, *R. solanacearum* potentially requires only 1 entry site to establish a systemic infection that results in bacterial wilt.

After invading a susceptible host, *R. solanacearum* multiplies and moves systematically within the plant before bacterial wilt symptoms occur (wilting should be considered as the most visible side effect that usually occurs after extensive colonization of the pathogen). When the pathogen gets into the xylems through natural openings or wounds, tyloses may form to block the axial migration of bacteria within the plant. In susceptible plants, this sometimes happens slowly and infrequently to prevent pathogen migration and may instead lead to vascular dysfunction by unspecifically obstructing uncolonized vessels.

Wilting occurs at a high level of bacterial population in the xylem and is partially due to vascular dysfunction in which water cannot reach the leaves sufficiently. At this time, extracellular polysaccharide (EPS1) content is about 10 μg/g tissue in the taproot, hypocotyl, and midstem; EPS1 concentration is higher later on at more than 100 μg/g tissue in a fully wilted plant. Wilting is due to vascular dysfunction that prevents water from reaching the leaves. *Ralstonia's* systemic toxin also causes loss of stomatal control but there is no evidence for excessive transpiration as its consequence. The primary factor contributing to wilting is probably blocking of pit membranes in the petioles and leaves by the high molecular mass EPS1. High bacterial densities, by-products of plant cell wall degradation; tyloses and gums produced by the plant itself are other contributing factors to wilting.

7.2.6 MANAGEMENT

Wilting and yellowing of the leaves as well as overall stunting of the plant are typical symptoms. The leaves may also take on a bronze cast along with stems becoming streaked and tuber eyes becoming discolored. Tubers will also start to rot if left in the ground. General sanitation practices are recommended to prevent the spread of the disease as chemical control is ineffective. Crop rotation with resistant crops is useful as well as altering the pH of the soil keeping it low in the summer and higher in the fall.

Crop rotation is not effective as the pathogen can survive for a long period (several years) in the soil and also attack a wide range of crops and solanaceous weeds.

Use plant varieties that are tolerant/resistant to bacterial wilt. Do not grow crops in soil where bacterial wilt has occurred. Remove wilted plants from the field to reduce the spread of the disease from plant to plant. Control root knot nematodes since they could facilitate infection and the spread of bacterial wilt. Where feasible, extended flooding (for at least 6 months) of the infected fields can reduce disease levels in the soil. Soil amendments (organic manures) can suppress the bacterial wilt pathogen in the soil.

7.3 FUSARIUM DRY ROT

Dry rot is probably the greatest cause of postharvest potato losses worldwide. Dry rot is caused by several fungal species in the genus *Fusarium*—thus, the name fusarium dry rot. The most important dry rot pathogen is *Fusarium sambucinum*, although *Fusarium solani* is also present.

7.3.1 CAUSAL ORGANISM

Species	Associated Disease Phase	Economic Importance
Fusarium spp.	Dry Rot of Tubers	Severe

7.3.2 SYMPTOMS

Fusarium dry rot is characterized by an internal light to dark brown or black rot of the potato tuber, and it is usually dry (Figure 7.6). The rot may develop at an injury such as a bruise or cut. The pathogen penetrates the tuber, often rotting out the center. Extensive rotting causes the tissue to shrink and collapse, usually leaving a dark sunken area on the outside of the tuber (Figure 7.7) and internal cavities. Yellow, white, or pink mold may be present.

Diagnosis of dry rot can be complicated by the presence of soft rot bacteria, which often invade dry rot lesions, particularly if the tubers have been stored at high humidity and condensation has occurred on the surfaces. Soft rot bacteria cause a wet rot that can very quickly encompass the entire tuber and mask the initial dry rot symptoms.

Several other diseases and physiological disorders, including Pythium leak, pink rot, late blight, and sub oxygenation (blackheart), also cause brown to black internal discoloration of tubers. Leak and pink rot are wet rots; tubers exude a clear fluid when squeezed. Late blight is a less aggressive rot that generally does not penetrate the center of the tuber and causes reddish-brown lesions. Poor air circulation or extremes in temperature that result in low internal oxygen concentrations can cause a smoky gray to black discoloration of tissue, but the tissue is never brown and is very firm.

7.3.3 CAUSE AND DISEASE DEVELOPMENT

Fusarium dry rot is caused by several species of the soilborne fungus *Fusarium*. These fungi are common in most soils where potatoes are grown and survive as resistant spores free in the soil or within decayed plant tissues. Although some infections may develop on tubers before harvest, most

FIGURE 7.6   Internal black rot of the potato tuber.

FIGURE 7.7   Symptoms of dry rot outside the tuber.

infections occur as the fungus enters tubers through harvest wounds. Small, brown lesions appear at wound sites three to four weeks after harvest and continue to enlarge during storage, taking several months to develop fully.

7.3.4 Favorable Conditions of Disease Development

The disease develops fairly rapidly at temperatures above 10°C, but lesions will cease enlarging below 4°C. The fungus is only dormant at these low temperatures, however, and will resume growth when tubers are warmed.

7.3.5 Disease Cycle

Fusarium dry rot is an important postharvest disease of potato worldwide. Fusarium dry rot can be caused by several different *Fusarium* spp., including *F. solani*, *F. sambucinum*, *F. avenaceum*, *F. culmorum*, and *F. oxysporum*, but *F. solani* appears to be the most aggressive and important. Dry rot *Fusarium* spp. originate from contaminated seed or infested soils, infecting tubers through wounds in the periderm that are common after potato cutting and handling practices. *Fusarium* spp. can be introduced into soils by contaminated seed can persist for years. Soilborne inoculum can infect tubers through wounds caused by other pathogens, insects, or during harvest and handling. The fungus can be seed- or soilborne. It enters tubers through wounds or bruises incurred during harvesting or handling. The disease can spread quickly if potatoes are improperly cured. Infected seed tubers result in low-quality seed that causes poor crop stands. The fungus can survive as resistant spores or mycelium in decayed plant debris in the soil.

7.3.6 Management

7.3.6.1 Cultural Control

Plant high-quality seed free from fusarium dry rot pathogens into soils without a history of fusarium dry rot. Varieties vary in their reaction to dry rot, and highly susceptible varieties should be avoided. Harvest tubers at least 14 days after vine kill to promote good skin set and reduce skinning injury that can increase storage dry rot. Avoid harvesting cold tubers that are more susceptible to injury. Provide conditions that promote rapid wound healing early in storage, including high humidity, good aeration, and temperatures of 13°C to 18°C for 14 to 21 days. Since fusarium dry rot increases with length in storage, short-term storage is advisable for fields where severe infection is expected.

7.3.6.2 Chemical Control

Post-harvest fungicide applications can reduce fusarium dry rot losses, but strains of *F. solani* and *F. sambucinum* resistant to benzimidazole fungicides are widespread (Table 7.1).

7.3.6.3 Biological Control

Dry rot caused by various species of *Fusarium* is a disease of significant importance in potatoes. Field trials were conducted in 2005 and 2006 in New Brunswick, Canada to assess the efficacy of *Pseudomonas fluorescens* and *Enterobacter cloacae* applied as a seed treatment in suppressing fusarium dry rot of potato (*Solanum tuberosum* L.), 'Russet Burbank' under field conditions. In 2005, the trial consisted of 5 treatments, namely, (1) non-treated, non-inoculated control; (2) non-treated control inoculated with *F. sambucinum*; (3) seed inoculated with *F. sambucinum* and treated with *P. fluorescens*; (4) seed inoculated with *F. sambucinum* and treated with *E. cloacae*; and (5) seed inoculated with *F. sambucinum* and treated with the fungicide fludioxonil. In 2006, a mustard meal treatment was added. After harvest, tubers were assessed for disease severity of dry rot in addition to other tuber diseases, including silver scurf (*Helminthosporium solani*), and

TABLE 7.1

Chemical Control

Trade Name	Target Pathogen	Rate	Remarks
		Chloroine	
Agclor 310	Decay causing organisms	10–15 fl oz. per 100-gal water	For treating potatoes in a pit system use a concentration of 100 to 15 ppm Cl2.
		Thiabendazole	
Decco Salt No. 19	Fusarium Tuber rot	5.67 per 2000 lb. of potatoes	Seed potatoes should be treated before cutting.
Mertect 340-F	Fusarium Tuber rot	Conveyer line: 0.42 fl oz. per 2000 lb. of potatoes; Dipping: 0.42 fl oz. per 1 gal water, dip tubers for 20 sec.	Seed potatoes should be treated before

common scab (*Streptomyces scabiei*). In addition, tubers were graded and assessed for total yield, tuber size, tuber number, and tuber weight.

Significant reduction in dry rot severity was obtained with all treatments compared to the non-treated control inoculated with *F. sambucinum*. The highest dry rot reduction averaged over the 2 years of the study was for the fludioxonil treatment (55.7%) followed by the treatment with mustard meal (47.5%; 2006 only), *P. fluorescens* (35%) and *E. cloacae* (26.5%). All treatments significantly reduced the severity of common scab and silver scurf compared to the non-treated, non-inoculated control. In both years, on average, seed treated with *P. fluorescens* and *E. cloacae* produced a higher total number of tubers. Both total and marketable tuber yields were significantly higher for the *E. cloacae* treatment compared to the non-treated, inoculated control. The results of this study suggest that *P. fluorescens*, *E. cloacae*, and mustard meal are viable options for controlling potato tuber diseases along with fludioxonil. This is the first study to investigate the effect of these bacteria on potato diseases under field settings.

7.4   PINK ROT OF POTATO

Pink rot is a fungal disease that occurs sporadically in many soils worldwide wherever potatoes are grown. The causal agent of pink rot is the soilborne pathogen *Phytophthora erythroseptica*, and it is 1 of the most damaging tuber diseases as apparently healthy tubers at harvest can deteriorate within a few weeks of being placed into store.

7.4.1   CAUSAL ORGANISM

Species	Associated Disease Phase	Economic Importance
P. erythroseptica	Rot of Potato Tubers and Also In Severe Cases, Plant Death	Severe

7.4.2   SYMPTOMS

Plants affected with pink rot disease often show signs of stunting or wilting during the later stages of the growing season. Wilting starts from the base of the plant's stem and progresses up causing a yellowing of the leaves, and the roots of affected plants may turn brown to black. Tuber symptoms are much easier to spot and include rubbery (but not discolored) tissue in the early stages with a tough, leathery surface on affected tubers. Cutting rotten tubers open will initially reveal odorless, cream colored tissue which, when left exposed to air for around 20–30 minutes, changes to pink (Figure 7.8). After around an hour, the tissue color changes to brown, before finally going black. Tubers infected with pink rot often have a characteristic smell of vinegar.

7.4.3 CAUSE AND DISEASE DEVELOPMENT

Pink rot is a major problem of potatoes. The pathogen survives for long periods in the soil and becomes active when the soil is saturated with water. The disease is usually associated with high soil moisture when tubers are approaching maturity and is a serious problem in poorly drained soils. Although the disease is found predominantly in wet fields, it can also develop in sandy soil without excessive moisture.

7.4.4 FAVORABLE CONDITIONS OF DISEASE DEVELOPMENT

Tuber infection occurs when soilborne fungal spores invade potato stolons, eyes, and lenticels. The disease is most severe in warm, wet soils (20°C–30°C) especially if these conditions occur late in the growing season. However, the optimal temperature for infection is 25°C. The disease is soilborne and there are many known alternate hosts of the fungus, including wheat and rye.

7.4.5 DISEASE CYCLE

Pink rot caused by *P. erythroseptica* can cause significant potato storage losses. Pink rot can often be found in the field before harvest. Infected tubers will appear dark and may be wet. When tubers are sliced open, infected tissues often turn pink or salmon-colored after exposure to air for 20 to 30 minutes. Pink rot usually begins from the stem end of the tuber and then progresses through the tuber in a uniform manner, often with a nearly straight line between the healthy and the diseased

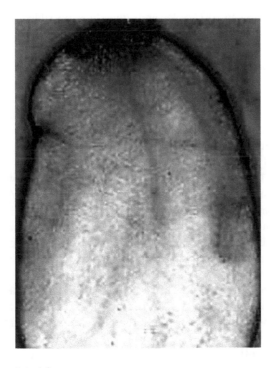

FIGURE 7.8   Pink rot on potato tuber.

portions of the tuber. Infected tissues retain some degree of firmness and have a texture like that of boiled potatoes. Pink rot by itself is not a slimy soft rot, but infected tissues are easily invaded by soft rot bacteria, which cause tubers to break down.

The pink rot pathogen survives in the soil by producing oospores. In the presence of potatoes, oospores germinate to produce mycelia and sporangia. When the soil becomes wet, sporangia can germinate directly, or release swimming spores called zoospores. Zoospores respond to chemicals released by the potato plant and swim toward potato roots. Once roots or stolons are infected, the pink rot pathogen can grow into the tuber. Severe plant infection can result in wilting symptoms in aboveground stems. Tubers can also become infected directly through eyes or lenticels during prolonged periods of high soil moisture. The pink rot pathogen readily produces oospores in the potato plant, and these oospores are returned to the soil when the potato vines are killed.[5]

Tubers can also become exposed to the pink rot pathogen at harvest. Tubers infected with pink rot may break apart during the harvesting process and meet healthy tubers. As these healthy tubers are wounded in the handling process, an avenue of entry is opened for the pink rot pathogen to invade. As a result, the pink rot pathogen may infect tubers at harvest, increasing the incidence of infected tubers.

7.4.6 MANAGEMENT

Managing pink rot of potato requires an integrated strategy. Several management practices need to be employed for optimal disease control. Reliance on a single control measure may not be effective. Control measures include:

1. Appropriate fungicide applications.
2. Establish a good skin set prior to harvest.
3. Avoid wounding tubers at harvest.
4. Appropriate water management.
5. Avoid harvesting when pulp temperatures reach 20°C.
6. Reduce pulp temperatures to 10°C or lower as quickly as possible.
7. Avoiding harvesting tubers from poorly drained areas.
8. Store tubers from suspect areas near storage doors.
9. Sort out infected tubers at harvest, prior to piling.

7.4.6.1 Fungicide Application

The first 3 recommendations improve the resistance of the tuber to disease. Recommendations 4–6 reduce the favorability of the environment for disease development. The last 3 recommendations are cultural practices that can reduce the threat of disease loss in storage.

Most studies have involved the use of Mefenoxam, which is the active ingredient in Ridomil Gold or Ultra Flourish and are applied as a 6- to 8-inch band directly over the seed piece prior to row closure at the rate of 0.42 fl oz. 4E/1000 ft of row or 0.84 fl oz 2E/1,000 ft of row, respectively. Alternatively, Ridomil Gold Bravo or Flouronil (mefenoxam/chlorothalonil) at the rate of 2 lb 76WP/A (or Ridomil Gold MZ at 2.5 lb. 68WP/A) can be used as a foliar spray with as much gallonage as possible. The first spray is applied at flowering (nickel size of setting tubers) with a second and third spray applied 14 and 28 days later. The third spray would be necessary only for fields with a history of storage rots. When applying these products as foliar sprays, it's important to achieve coverage of the soil surrounding the plants. Research has shown that mefenoxam moves systemically within a plant, and up to 96% enters through the stolon's. This implies that not much moves down the stem and therefore foliar applications are better if they are washed into the soil. The downside of using foliar application and allowing irrigation to move the product into the soil is that the entire fungal population is exposed to the product's active ingredient.

The activity of mefenoxam depends upon its availability in soils and this can be diminished in soils with higher organic matter, sand or sandy loam soils, and from frequent use of irrigation, since mefenoxam is very soluble. Microbial breakdown also proceeds faster at low pH, which can occur when fertilizers are mixed with mefenoxam. The half-life of metalaxyl in soil is generally considered to be 36 days, but some accounts report a half-life approaching 72 days for mefenoxam. This longer half-life may explain the good control obtained when mefenoxam is applied in the furrow.

7.5 BACTERIAL RING ROT OF POTATO

Bacterial ring rot is an important disease of potatoes and is 1 of the main reasons for rejection of seed potatoes from certification programs. This disease is particularly serious because it has the potential to spread quickly throughout a farm and may lead to severe losses if left unchecked. Ring rot was originally found in Germany in the late 1800s. The causal bacteria were introduced into the United States in the early 1930s, and by 1940, were found throughout the country. On an annual basis, economic losses due to ring rot are low, however, it is the constant threat of severe and devastating losses that warrants the continued vigilance of the potato industry. Although the bacterium that causes ring rot is primarily disseminated by seed potatoes, the bacterium is also capable of surviving outside its host for extended periods.

7.5.1 CAUSAL ORGANISM

Species	Associated Disease Phase	Economic Importance
Clavibacter michiganense subsp. *sepedonicus*	Yellowing and Wilting of Leaves, Surface Cracks on Potato Tuber, which Causes Other Diseases like Soft Rot	Severe

7.5.2 SYMPTOMS

Tuber symptoms have been seen when tubers are cut across the heel end (where the tuber was attached to the stolon). In the early stages, the tissues around the vascular ring appear glassy and water-soaked (Figure 7.9). As infection progresses, the vascular ring becomes discolored and a soft cheese-like rot develops around the vascular ring (hence the name "ring" rot). If a cut tuber is squeezed a cheese-like ooze emerges. External symptoms are not common, but in severe cases, the skin of the potato may crack, and ooze can emerge from the heel end and eyes, causing soil to adhere.

Wilting symptoms may occur late in the season and are often masked by the natural senescence of the crop (Figure 7.10). Symptom expression occurs at different rates in different varieties and is affected by temperature and other environmental conditions. Some varieties hardly ever express symptoms.

7.5.3 FAVORABLE CONDITIONS OF DISEASE DEVELOPMENT

Disease spread is largely via infected seed potato tubers. Ring rot can pass through 1 or more field generations without causing symptoms and latently infected tubers are an important means of spreading the disease. Ring-rot bacteria survive between seasons mainly in infected seed tubers. They are also capable of surviving 2–5 years in dried slime on surfaces of crates, bins, burlap sacks, or harvesting and grading machinery, even if exposed to temperatures well below freezing. Survival is longest under cool, dry conditions. Ring-rot bacteria do not survive in soil in the absence of potato debris but can survive from season to season in volunteer potato plants. Wounds are necessary for penetration of the bacteria into seed pieces. Disease development is favored by temperatures ranging from 18°C to 22°C.

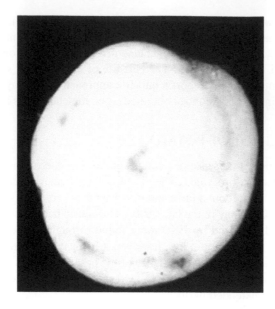

FIGURE 7.9   Symptom of ring rot on tuber.

FIGURE 7.10   Yellowing and wilting of the potato leaves.

7.5.4 DISEASE CYCLE

The bacterium *Clavibacter michiganensis* subsp. *sepedonicus* survives mainly in infected tubers but can also survive up to 5 years in dried slime on machinery, sacks, and other equipment even if exposed to temperatures below freezing. The ring-rot bacterium does not overwinter in fields unless infected volunteer tubers survive. The bacterium enters tubers through wounds. This is important when cutting seed pieces since a contaminated knife can result in the inoculation of a large number of potatoes. Disease development is favored by temperatures ranging from 18°C to 22°C.

7.5.5 MANAGEMENT

- Plant only classified seed.
- Control groundkeepers.

- Practice good hygiene—regularly clean and disinfect all machinery, equipment, containers, vehicles, and storage facilities used during potato production.
- Don't spread disease with waste—discarded potatoes and potato processing waste could harbor the disease.

7.6 SCLEROTINIA STEM ROT OF POTATO

Sclerotinia stem rot of potato, also called white mold, is caused by the fungus *Sclerotinia sclerotiorum*. This disease is a problem during very moist conditions and is especially common under center pivot irrigation. This disease is favored by high relative humidity and free moisture for long time periods and is especially severe where production practices include sprinkler irrigation and high nitrogen fertility which promote lush, dense foliage.

7.6.1 CAUSAL ORGANISM

Species	Associated Disease Phase	Economic Importance
S. sclerotiorum	Water-soaked Lesions with Development of White Mold	Major

7.6.2 SYMPTOMS

Sclerotinia stem rot first appears as water-soaked spots usually at the point where stems attach to branches or on branches or stems in contract with the soil. A white cottony growth of fungal mycelium develops on the lesions, and the infected tissue becomes soft and watery (Figure 7.11). The fungus may spread rapidly to nearby stems and leaves. Lesions may then expand and girdle the stem which causes the foliage to wilt. During dry conditions, lesions become dry and will turn beige, tan, or bleached white in color and papery in texture. Hard, irregularly shaped resting bodies of the fungus called sclerotia form in and on decaying plant tissues. Sclerotia are generally 2 to 3 inches in diameter, initially white to cream in color but become black with age and are frequently found in hollowed-out centers of infected stems. Sclerotia will eventually fall to the ground and enable the fungus to survive until the next growing season.

7.6.3 FAVORABLE CONDITIONS OF DISEASE DEVELOPMENT

S. rolfsii attacks many field and vegetable crops in warm regions. The fungus persists in the soil between crops. Germination and infection by the sclerotia are favored by hot temperatures

FIGURE 7.11 Lesion and white mycelial growth on potato stem.

(27°C to 32°C) and moist soil surfaces. Once infection has occurred, the disease develops most rapidly at low to moderate temperatures (15.5°C–25°C).

7.6.4 DISEASE CYCLE

Sclerotia are very durable and can survive in soil for at least 3 years. They require a conditioning period of cool temperatures before germination. During the growing season, sclerotia within 1–2 inches of the soil surface usually germinate when the canopy of the growing crop shades the ground and soil moisture remains high for several days. Sclerotia either germinate directly as mycelium, which may infect stems near the soil surface, or they produce fruiting bodies called apothecia (singular is apothecium). Apothecia are cup-shaped on their upper surface, about 0.5 inch in height, fleshy in texture and pale orange and pink or light tan in color. Millions of ascospores are formed in each apothecium and are ejected into the air. The ascospores are carried by air currents up to several miles in the distance and colonize dead or dying plant tissue when moisture is present. Ascospores external to potato fields are a major source of inoculums. Yellow leaves and blossoms lying on the ground serve as an energy base for the fungus to colonize green plant tissue that is in direct contact with the growing mycelium. Once infection has occurred, the disease develops most rapidly at low to moderate temperatures (15.5°C–25°C).

7.6.5 MANAGEMENT

Severity of sclerotinia stem rot can be reduced with a combination of practices such as limiting potato vine growth through nitrogen fertilizer management, avoiding over-irrigation, and excessive foliar applications of fungicides. Cultural practices need to be employed before stem rot begins developing in fields. High nitrogen fertility promotes lush dense crop canopies that provide a favorable environment of high relative humidity and prolong wet periods for disease development. Irrigation should be restricted during rainy weather, and on cool, cloudy days, whenever possible. Practices that promote long periods of leaf wetness or high relative humidity within the crop canopy should be avoided.

Protectant fungicides may be needed in areas with severe disease pressure. Ascospores are usually discharged over a period of two to eight weeks beginning before row closure. The microclimate within the crop canopy generally favors stem rot development after row closure, and latently infected blossoms at blossom drop are a major substrate for stem infections. A single application of fungicide after row closure and when latently infected blossoms begin to drop may be sufficient in areas where spores are discharged over a short period, and repeated application, beginning at blossom drop, will be needed in areas where ascospores are discharged for extended time periods. Fields should be monitored closely for the white, cottony mycelial growth of *S. sclerotiorum* on detached plant material on the soil surface because this indicates the presence of inoculum. A protectant fungicide should be applied at first signs of this mold on the ground.

7.7 PINK EYE OF POTATO

Pink eye is characterized by pink to brown blotches on the skin, usually around the eyes at the apical (bud) end of tubers. When the disease is severe, a shallow, reddish brown rot occurs beneath the discolored areas. Pink eye is often prevalent on varieties such as Kennebec and Superior, which are highly susceptible to verticillium wilt, and commonly occurs during and after a wet harvest season. As a rule, the disease is not commercially serious in table-stock or seed potatoes stored under cool, relatively dry conditions. Tuber appearance may be somewhat marred, but the affected skin and superficial rot usually dry up. However, in the case of chip-stock held at warm temperatures and high relative humidity, pink eye often opens the door to secondary soft-rotting bacteria, which frequently cause extremely heavy losses.

7.7.1 Causal Organism

Species	Associated Disease Phase	Economic Importance
Pseudomonas marginalia or *P. fluorescens*	Pinkish Brown Blotching on Moist Freshly Dug Tubers	Major

7.7.2 Symptoms

The pinkish brown blotching mentioned above shows up readily on moist, freshly dug tubers, but is usually difficult to notice on dry, unwashed potatoes (Figure 7.12). In severe cases, a reddish-brown decay extends a few mms (2–3) into the flesh. The rot may be confused with that caused by late blight, but it is not granular and is more superficial than the latter. If tubers are kept cool and dry, the only symptom commonly encountered is scaly, flaky skin over the affected areas. However, tissue weakened by pink eye is often invaded by soft-rotting bacteria if potatoes are held at warm (above 7°C) temperatures and high relative humidity. The result is often the slimy, foul-smelling decay typical of bacterial soft rot.

7.7.3 Favorable Conditions of Disease Development

A strong relationship exists between crop stress that occurs early in the season (high temperatures) and wet soil conditions later in the season. In general, older potato land with many potato crops in a rotation can also be associated with pink eye.

FIGURE 7.12   Pink eye on potato.

7.7.4 DISEASE CYCLE

The organism is common in soils and commercial composts and usually only causes a problem in susceptible cultivars. *P. fluorescens* requires wet soil conditions to be able to enter tubers through lenticels or via wounds. The disease can spread to adjacent tubers in storage under moist conditions when air temperatures exceed 7°C. Pink eye is sometimes found with verticillium wilt, Rhizoctonia stem canker, and Late blight, though it can occur independently of these diseases too.

7.7.5 MANAGEMENT

Unexpectedly wet weather obviously cannot be circumvented in humid areas. But selection of well-drained soils will help to reduce the effects of excessive rains. In irrigated regions, restriction of soil moisture at the end of the growing season is feasible and to be recommended. Because of the frequent correlation between pink eye and verticillium wilt, it is wise to avoid planting highly pink eye susceptible varieties in fields that have a history of serious wilt problems. Wilt-resistant varieties should be resorted to if land availability prohibits such a choice. If possible, tubers should be kept cool and dry after harvest to prevent further development of pink eye and secondary soft rot problems.

7.8 BLACK DOT DISEASE OF POTATO

Black dot disease of potato caused by the fungus *Colletotrichum coccodes* is generally considered to be a weak root pathogen of potato. Recent studies have revealed, however, that this disease must be considered as part of the total disease complex affecting potato.

Although not as serious a tuber- or soilborne pathogen as black scurf (*Rhizoctonia solani*), silver scurf (*H. solani*), or common scab (*S. scabies*), *Colletotrichum* can cause severe rotting of belowground plant parts and early plant decline leading to discolored tubers and reduced yields. The same black dot organism causes anthracnose or ripe-fruit disease of tomato and can occur on other solanaceous crops and weed species.

7.8.1 CAUSAL ORGANISM

Species	Associated Disease Phase	Economic Importance
C. coccodes	Small, Black, Dot-like Sclerotia on the Surface of Infected Stem, Stolon, and Tubers	Severe

7.8.2 SYMPTOMS

Black dot symptoms are first visible in the field in mid to late summer as yellowing and wilting of foliage in the tops of plants. These symptoms may go unrecognized because of their similarity to those caused by *Fusarium* and *Verticillium* spp. As the disease progresses, infected plants turn brown and die. Numerous small (1/50 inch) black "dots" (sclerotia, dormant fungal masses) may appear on infected stem tissue. The most striking symptoms appear on the belowground stem tissue. The cortical tissue scales away, exposing the woody vascular tissue that turns an amethyst color. The fungus produces black sclerotia on the internal and external stem surfaces (Figure 7.13). The amethyst coloration, sclerotia, and cortical sloughing are also found on infected stolons. Remnants of infected stolons may remain attached to tubers and aid in the identification of the disease in storage. Infected tubers exhibit grayish lesions on the surface that resemble silver scurf (Figure 7.14). Tuber injury is more severe on thin-skinned than on netted-skinned cultivars. Black dot may be more of a problem than previously considered since its foliar symptoms during the season are like early blight and several other diseases. This could explain the occasional failure of chemical

FIGURE 7.13 Black spots on stem of potato plant

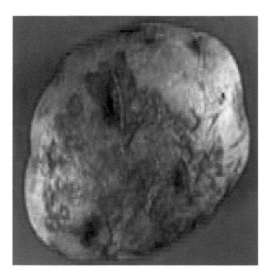

FIGURE 7.14 Grayish black spot symptoms on potato

controls for early blight, since the disease present may be black dot. There are no chemical means to control black dot.

7.8.3 Cause and Disease Development

Black dot is introduced into soil by infected seed pieces or tubers. Once introduced, it will remain there for years.

7.8.4 Favorable Conditions of Disease Development

Spores of *C. coccodes* are produced on potato plants and plant debris between 7°C and 35°C. However, in greenhouse studies, limited infection occurred below 15°C. As with most *Colletotrichum* species, *C. coccodes* favors temperatures above 20°C, and free moisture (from rain, irrigation, fog, or dew) is required for spore germination and infection of plant tissues.

7.8.5 DISEASE CYCLE

Sclerotium surviving on the surface of infected tubers is the source of inoculum that spreads the disease into new fields. Once the disease is established, the fungus sclerotia survive on infected plant residue in the soil for long periods of time. Potato plants growing in infested soil are exposed to inoculum arising from the sclerotia. When conditions are favorable the fungus invades underground stem tissue and moves upward in the plant. Airborne spores also infect the foliage, especially when tissue is injured by windblown sand, and the disease progresses downward into the stem and roots. In late stages of disease development, new sclerotia are produced, and the fungus population again builds to a high level in the soil. The fungus also invades several vegetable species in the potato family and a few weed species.

7.8.6 MANAGEMENT

7.8.6.1 Seed

Use certified seed, since seed is the only way known to infest a clean field. There are no resistant potato cultivars. Thin-skinned cultivars are more susceptible. Do not sterilize seed pieces.

7.8.6.2 Field

Rotate crops with grains, preferably 5 years before replanting potatoes on infected ground. Besides potatoes, other solanaceous crops and weeds can be hosts to black dot. Examples are tomatoes, peppers, eggplant, and nightshades. Keep potato fields free of nightshades. Fields should be clean of debris where black dot can overwinter. Keep soil adequately fertilized; petiole sampling will help. Keep fields irrigated but avoid excess watering, especially in low spots or poorly drained soils. Windbreaks may be useful on sandy soils in high wind areas. Avoid skinning or bruising tubers at harvest. Note: there are no chemical control measures for black dot.

7.8.6.3 Storage

Keep relative humidity at or above 90%. If possible, store at 4°C. If the field is infected, wash the tubers going into storage. Use good sanitary practices.

7.9 COMMON SCAB OF POTATO

Common scab of potatoes is caused by *Streptomyces scabies*, a very prevalent, soil-inhabiting bacterium. This serious disease can be found in all potato-growing areas throughout the world. The scab organism sometimes occurs in soils where potatoes have never been grown. In most potato soils, however, scab was probably introduced with infected seed tubers. The major loss from common scab is lower market quality because tubers are unsightly or disfigured and have poor customer appeal. While superficial scab lesions do not greatly affect the marketability of processing potatoes, deep-pitted lesions, however, do increase peeling losses and detract from the appearance of the processed product. The occurrence of scab and its severity varies by season and from field to field. Cropping history, soil moisture, and soil texture are largely responsible for this variability. The disease is often confused with powdery scab which causes somewhat similar symptoms.

7.9.1 CAUSAL ORGANISM

Species	Associated Disease Phase	Economic Importance
S. scabies	Scab on Tuber	Major

FIGURE 7.15   Symptoms of potato scab on potato tuber.

7.9.2 SYMPTOMS

Potato scab lesions are quite variable, and distinctions have been made between russet (superficial corky tissue), erumpent (a raised corky area), and pitted (a shallow-to-deep hole) scab (Figure 7.15). These can be caused by the same pathogen, *S. scabies*; however, the type of lesion probably is determined by host resistance, aggressiveness of the pathogen strain, time of infection, and environmental conditions.

Individual scab lesions are circular but may coalesce into large scabby areas. Insects may be involved in creating deep-pitted lesions. The term "common scab" generally refers to the response of the disease to soil pH. Common scab is controlled or greatly suppressed at soil pH levels of 5.2 or lower. Common scab is widespread and is caused by *S. scabies*. "Acid scab" seems to have a more limited distribution but has been found in several states in the Northeast. This disease occurs in soils below pH 5.2, as well as at higher levels. The causal agent *S. acidiscabies* is closely related to the common scab pathogen and can grow in soils as low as pH 4.0. Acid scab is controlled by crop rotation but can be a problem when seed is produced in contaminated soils. Acid scab lesions are similar, if not identical, to those caused by *S. scabies*.

7.9.3 CAUSE AND DISEASE DEVELOPMENT

Scab is caused by a group of filamentous bacteria called actinomycetes which occur commonly in soil. In soils with a pH above 5.5, *S. scabies* is usually responsible for common scab and can cause all the types of scab lesions described above. It is commonly introduced into fields on seed potatoes and will survive indefinitely on decaying plant debris once the soil is contaminated. The organism can also survive passage through the digestive tract of animals and be distributed.

7.9.4 DISEASE CYCLE

Most if not all potato soils have a resident population of *S. scabies* which will increase with successive potato or other host crops. Scab-susceptible potato varieties appear to increase soil populations faster than scab-resistant varieties. Rotation with grains or other non-hosts eventually reduces but does not eliminate the *S. scabies* population. This pathogen is a good saprophyte and probably reproduces to some extent on organic material in the soil. Given the right environmental conditions and a scab-susceptible potato variety, scab can occur in a field that has been out of potatoes for several years.

S. scabies infects young developing tubers through the lenticels and occasionally through wounds. Initial infections result in superficial reddish-brown spots on the surface of tubers. As the tubers grow, lesions expand becoming corky and necrotic. The pathogen sporulates in the lesion, and some of these spores are shed into the soil or re-infest soil when cull potatoes are left in the field. The pathogen survives in lesions on tubers in storage, but the disease does not spread or increase in severity. Inoculum from infected seed tubers can produce disease on progeny tubers the next season.

The disease cycle of S. *acidiscabies* is similar to that of S. *scabies*, but the acid scab pathogen does not survive in soil as well as common scab. Inoculum on seed tubers, even those without visible lesions, seems to be important in disease outbreaks.

7.9.5 MANAGEMENT

- Soil pH and fertilizer choice are very important. Keeping the soil pH at or below 5.2 will suppress scab. Sulfur can be applied to the soil to lower the pH and make it more acidic. Acid-forming nitrogen fertilizers, such as ammonium sulfate and diammonium phosphate, are more effective in reducing scab than ammonium nitrate, while calcium and potassium nitrate can raise soil pH and favor scab development. Fresh barnyard manure applied to the soil can create a scab problem and the pathogen may persist in the soil for many years afterward.
- Follow a crop rotation schedule when scab is a problem. Plant at least 3 years of non-susceptible crops between potato crops. Scab infection builds up following frequent crops of potatoes, beets, radishes, turnips, carrots, rutabagas, and parsnips. Rotations including rye, alfalfa, or soybeans may reduce scab severity.
- Keep soil moist during early tuber development (for about two weeks after the plants emerge from the soil). Avoid overwatering, as it may cause rotting or poor plant growth.
- Varietal resistance: Plant certified seed potatoes of the russet-skinned varieties, which are more resistant to scab. The variety Nooksack is highly resistant, while Russet Burbank, Netted Gem, and Norgold have moderate resistance. A red variety Red Norland is also moderately resistant.
- Fungicide Seed Treatment: In situations where growers are planting in the ground not planted previously with potatoes or where the field is known to be scab-free, treat seed tubers with a fungicide seed treatment to reduce scab introduction through the seed pieces.

7.10 *DICKEYA SOLANI* ON POTATO

- Previously called *Erwinia chrysanthemi*
- At least 6 species and probably more
- Closely related to *Pectobacterium atrosepticum*
- Causes blackleg, soft rot, and wilt of potato
- More pathogenic
- Wider host range than *P. atrosepticum*

This new pathogen causes blackleg-like symptoms, leaf wilts, and tuber soft rots. It has yet to be formally named but the name "*Dickeya solani*" has been proposed. It is a close relative of *Dickeya dianthicola*, previously known as *E. chrysanthemi*, which has caused problems sporadically in potato crops across the world.

7.10.1 CAUSAL ORGANISM

Species	Associated Disease Phase	Economic Importance
D. dianthicola	Blackleg-like Symptoms, Leaf Wilts, Tuber Soft Rots	Major

FIGURE 7.16   Wilting symptoms caused by *Dickeya solani*.

7.10.2   SYMPTOMS

Symptoms caused by "*Dickeya solani*" on the growing plant closely resemble blackleg in many cases. Wilting can be rapid with black soft rotting extending internally up the vascular system of the stem from the infected seed tuber (Figure 7.16) Symptoms may vary depending on variety, and in some, wilting can occur with no obvious sign of blackleg. High incidences of wilting (as much as 20%) have been observed. "*Dickeya solani*" is adapted to warmer temperatures.

7.10.3   MANAGEMENT

- In all cases, disposal of all waste from the crop (including soil and brock) will be controlled to prevent further spread of the pathogen. Care should be taken when using the crop as stock feed.
- It will be a requirement to control groundkeepers in the affected field for the 2 years following the crop.
- The field cannot be used to grow potatoes during that time.
- All machinery and boxes which have been in contact with the stock will be required to be cleaned and disinfected, i.e. planting, harvesting, grading, or other equipment which has been in contact with the stock.
- Other specific control measures will depend on the exact circumstances of the finding in each case but may include a ban on irrigation of the crop to prevent buildup of disease within the crop and controls on movement of machinery in and out of the crop to prevent spread of the disease.
- Any potential dangerous contact stocks will be traced and monitored to verify the disease has not spread to these or to control the disease if infection has occurred.
- High-risk crops and dangerous contacts will also be post-harvest tuber tested. Normally a 600-tuber sample per stock will be required but larger samples may be required in some cases.
- Water courses in the immediate vicinity of affected fields will be monitored for *Dickeya* infestation. In addition, all water courses used to irrigate potatoes will be monitored at least once every 3 years to ensure they remain free of *Dickeya*. Action on infested water courses may also be taken, including imposing irrigation bans for potatoes.

7.11   DISORDERS IN POTATO

7.11.1   BROWN CENTER AND HOLLOW HEART

Brown center and hollow heart are common, non-infectious, internal physiological disorders of Potato. Brown center is characterized by a region of cell death in the pith of the tuber that results

FIGURE 7.17   Hollow heart of potato.

in brown tissue. Hollow heart is characterized by a star- or lens-shaped hollow in the center of the tuber (Figure 7.17). Brown center is also called incipient hollow heart, brown heart, or sugar center.

Although brown center frequently precedes the development of hollow heart, both disorders often occur separately. Hollow heart may occur without being preceded by brown center. The probability of brown center incidence that results in hollow heart is based on the rate of tuber growth following a period of stress. The larger the tuber and the faster it grows, the greater the susceptibility of the tuber to incidence of hollow heart. Neither disorder is initiated by a disease organism.

Hollow heart and brown center negatively impact tuber quality. The disorders make cut fresh market tubers unattractive and can reduce repeat sales. Severe hollow heart negatively impacts the quality of chip-processing potatoes and can result in shipments not making grade. However, neither disorder is reported as harmful nor affects the tuber's taste or nutrition. Hollow heart is associated with rapid tuber growth after cool temperatures and moisture stress (too much or too little water). Large tubers usually, but not always, are more susceptible than small tubers.

Brown center and hollow heart arise at a higher incidence when growing conditions abruptly change during the season, such as when potato plants recover too quickly after a period of environmental or nutritional stress. When the tubers begin to grow rapidly, the tuber pith can die and/or pull apart, leaving a void in the center. Brown center and hollow heart effects likely form during tuber initiation but could also form during tuber bulking. If the disorder occurs during the early part of the season, then it is most often preceded by brown center and forms in the stem-end of the tuber, while late-forming hollow heart usually occurs near the bud-end with no brown center symptoms occurring.

Damage to cells signaling the onset of brown center can occur under conditions such as when soil temperatures are less than 13°C for five to eight straight days, or when available soil moisture is greater than 80%. Incidence of brown center and hollow heart also increases with periods of stress because of high or low moisture levels, especially if heavy water applications follow a period of stress because of low moisture levels. Since large tubers are more prone to develop the disorder, and wide interplant spacing produces larger tubers, this situation can result in higher incidence of brown center and hollow heart.

Grower decisions and management practices can be utilized to reduce the incidence of these disorders. Selecting potato varieties that are known to be less susceptible and delaying planting until soil temperature reaches adequate levels can lessen the occurrence of brown center and hollow heart. Planting with larger seed pieces that are less aged can also reduce brown center and hollow heart risk because of increased stem number per seed piece. Achieving recommended stand establishment, avoiding planting skips, and applying multiple small or split fertilizer applications (especially nitrogen) are practices that can reduce incidence of brown center and hollow heart. Maintaining consistent soil moisture through uniform irrigation applications is also critical for avoiding the disorders.[6]

7.11.1.1 To Reduce Hollow Heart

- Plant susceptible varieties at closer spacings.
- Prepare seed well and avoid apical dominance.
- Maintain uniform soil moisture throughout the entire growing season.
- Ensure adequate and balanced fertilization.

7.11.2 GROWTH CRACKS

It's a physiological disorder of the potato tuber in which the tuber splits while growing. The split heals but leaves a fissure in the tuber. Growth cracks generally start at the bud or apical end of the potato and can extend lengthwise. Growth cracks can vary in severity from appearing as a surface abrasion to a split through the tuber. The severity depends on the stage of growth the initial cracking occurred. Growth crack incidence increases when growing conditions are uneven or sudden environmental changes occur. Critical environmental conditions include both soil moisture and temperature. Growth cracks increase when relatively poor growing conditions are followed rapidly by relatively good growing conditions. There are differences in the susceptibility of potato varieties to growth cracks.

The potato above has a growth crack (Figure 7.18). The crack develops while the potato is in the ground, growing. Due to heavy rains or too much fertilizer the potato has a sudden growth spurt, leaving a furrow or crack in the potato. The potato continues to grow in the ground, and the growth crack heals over. In addition to potatoes, you will find growth cracks affecting tomatoes, carrots, and celery, just to name a few.

To reduce the incidence of growth cracks, maintain proper soil moisture during the season. This is especially important during the bulking stage when the plants are large, and tubers are rapidly expanding. Large plants and expanding tubers require more water to maintain good growth. Other recommendations to reduce the incidence of growth cracking include spacing plants uniformly, applying irrigation evenly to keep soil moisture levels consistent, and ensuring accurate application amounts and fertilizer placement.

The only recourse when the incidence of severe growth cracks is high is to select out tubers with severe growth cracks prior to packing and/or shipping.

7.11.3 POTATO GREENING

Potato greening refers to the development of a green pigment on the potato skins while they are stored or when they are in the field and some parts of the vegetables are exposed to light.

FIGURE 7.18   Formation of crack on tuber.

FIGURE 7.19   Greening of potato tuber.

The greening is caused by the presence of chlorophyll. The skin of the tuber turns green (Figure 7.19). Greening is strongly affected by 3 light factors: quality, duration, and intensity. Chlorophyll is green because it reflects green light while absorbing red-yellow and blue light.

In potato tubers, the greening is a sign that there may be an increase in the presence of glycoalkaloids, especially the substance solanine. Unlike chlorophyll, light is not needed for solanine formation, but, with light, glycoalkaloid formation is increased. Of critical importance is the potato category which produces toxic levels of the glycoalkaloid solanine during the greening process. Glycoalkaloids are toxic to humans; the lethal dose is 3–6 mg per kg of body mass. Improper display and lighting of potatoes contributes to accelerated greening and represents a substantial food safety and economic risk to food retailers and consumers.

A key factor is temperature during light exposure. This is important because greening is an enzymatic response and enzyme activity is increased with increasing temperature. There is no greening when the temperature is less than 4°C, refrigeration temperature, and is most rapid at 20°C, room temperature. The difference in greening at 10°C versus 20°C is how long it takes to fully green.

When the potato greens, solanine increases to potentially dangerous levels. Increased solanine levels are responsible for the bitter taste in potatoes after being cooked. Solanine biosynthesis occurs parallel but independent of chlorophyll biosynthesis; each can occur without the other. Unlike chlorophyll, light is not needed for solanine formation but is substantially promoted by it.

To prevent greening, inspect the potatoes before purchasing and store them in a cool dark place with good air circulation.

7.12   EARLY BLIGHT OF POTATO

Early blight (EB) is a disease of potato caused by the fungus *Alternaria solani*. It is found wherever potatoes are grown. The disease primarily affects leaves and stems, but under favorable weather conditions, and if left uncontrolled, can result in considerable defoliation and enhance the chance for tuber infection. Premature defoliation may lead to considerable reduction in yield. The disease can also be severe on tomatoes and can occur on other solanaceous crops and weeds.

7.12.1   CAUSAL ORGANISM

Species	Associated Disease Phase	Economic Importance
A. solani	Infection on Stem, Leaves, and Tuber	Severe

7.12.2 Symptoms

The first symptoms usually appear on older leaves and consist of small, irregular, dark brown to black, dead spots ranging in size from a pinpoint to 1/2 inch in diameter. As the spots enlarge, concentric rings may form because of irregular growth patterns by the organism in the leaf tissue. This gives the lesion a characteristic "target-spot" or "bull's eye" appearance. There is often a narrow, yellow halo around each spot and lesions are usually bordered by veins (Figure 7.20). When spots are numerous, they may grow together, causing infected leaves to turn yellow and die. Usually the oldest leaves become infected first and they dry up and drop from the plant as the disease progresses up the main stem.

On potato tubers, early blight results in surface lesions that appear a little darker than adjacent healthy skin. Lesions are usually slightly sunken, circular, or irregular, and vary in size up to 3/4 inch in diameter. There is usually a well-defined and sometimes slightly raised margin between healthy and diseased tissue. Internally, the tissue shows a brown to black corky, dry rot, usually not more than 1/4 to 3/8 inch deep. Deep cracks may form in older lesions (Figure 7.21).

7.12.3 Cause and Disease Development

This fungus is universally present in fields where these crops have been grown. It can also be carried in potato tubers. Spores form on infested plant debris at the soil surface or on active lesions over a wide temperature range, especially under alternating wet and dry conditions. They are easily carried by air currents, windblown soil, splashing rain, and irrigation water.

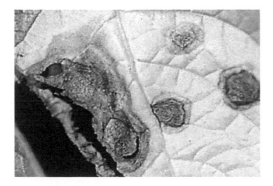

FIGURE 7.20 Symptoms of early blight on potato leaf.

FIGURE 7.21 Symptoms of early blight on potato tuber.

7.12.4 Favorable Conditions of Disease Development

Infection of susceptible leaf or stem tissues occurs in warm, humid weather with heavy dews or rain. Early blight can develop quite rapidly in mid to late season and is more severe when plants are stressed by poor nutrition, drought, or other pests. Infection of potato tubers occurs through natural openings on the skin or through injuries. Tubers may meet spores during harvest and lesions may continue to develop in storage. Warm, humid (24°C–29°C/ 75°F–84°F) environmental conditions are conducive to infection. In the presence of free moisture and at an optimum of 28°C–30°C (82°F–86°F), conidia will germinate in approximately 40 minutes.

7.12.5 Disease Cycle

The fungus overwinters either on potato tubers or in dead, infected plant debris either in the soil or on the soil surface. The concentration of initial or primary inoculum from these reservoirs is usually low. Therefore, primary infection is difficult to predict since EB is less dependent upon specific weather conditions than late blight. Environmental factors and plant vigor also help to determine when the first EB lesions are found. Infection can occur from early to mid-July when frequent rains or dews occur, and daytime temperatures remain near 24°C–29°C. The fungus can penetrate the leaf surface directly through the epidermis and spots begin appearing in two to three days. Lesions are most numerous and pronounced on lower, older, and less vigorous leaves and on early maturing varieties. The lesions are dark brown and appear leathery with faint, concentric rings giving a "target-spot" effect. At first the spots are small (1/8 inch in diameter) and oval or angular in shape, but later the spots can enlarge to about 1/2 inch. In many cases, they are bounded by the larger leaf veins. More spores are produced on the EB spots and lesions may coalesce, greatly increasing the secondary spread of wind-borne conidia between plants and between fields. At harvest time, spores from blighted vines may be deposited onto tubers. These spores germinate during wet and warm weather and invade the tissue, primarily through cuts, bruises, or wounded surfaces.

Tuber infections appear as generally small, irregular, brownish-black spots, which are usually slightly sunken (approx. 1/16 inch). Externally the spots resemble those caused by late blight, but internally they are shallower and darker in color. The rotted tuber tissue is firm, hard, and somewhat corky. EB tuber rot develops slowly and may not be severe until quite late into the storage period. This decay may allow the entry of secondary organisms such as *Fusarium* fungi and soft rot bacteria. (Figure 7.22).

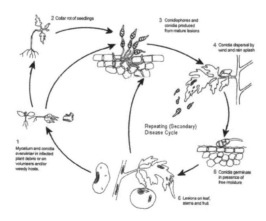

FIGURE 7.22   Disease cycle of *A. solani*.

7.12.6   Management

7.12.6.1   Cultural Practices

In many cases, employing sound cultural practices that maintain potato plants in good health will keep early blight losses below economic levels. Because the pathogen overwinters on infected crop debris, in-field sanitation practices that reduce initial inoculum in subsequent crops are beneficial. Consideration should be given to removing infected material such as decaying vines and fruits from the vicinity of production fields. Ensuring that seed or transplants are pathogen-free and rotating fields to a non-susceptible crop will also help to reduce buildup of soilborne inoculum.

Optimal tuber maturity is the most important factor for control of tuber infection. Tubers harvested before maturing are susceptible to wounding and infection. Tuber infection can be reduced by careful handling during harvest to minimize wounding as well as avoiding harvesting during wet conditions if possible. Tubers should be stored at 10°C to 13°C (50°F to 55°F), at high relative humidity and with plenty of aeration to promote wound healing, which will reduce the amount and severity of tuber infections that develop in storage.

7.12.6.2   Chemical Practices

The most common and effective method for the control of early blight is through the application of foliar fungicides. Protectant fungicides recommended for late blight control (e.g. maneb, mancozeb, chlorothalonil, and triphenyl tin hydroxide) are also effective against early blight when applied at approximately seven- to ten-day intervals. Resistance to the strobilurin group of fungicides has been reported. The geographical spread of this resistance is not known but applications of strobilurins should be made in combination with tank-mixtures of the fungicides listed above. Other products that have shown efficacy against early blight include azoxystrobin, trifloxystrobin, famoxodone, pyrethamil, fenamidone, and boscalid.

The application of foliar fungicides is not necessary for plants at the vegetative stage when they are relatively resistant. Accordingly, spraying should commence at the first sign of disease or immediately after bloom. The frequency of subsequent sprays should be determined according to the genotype and age-related resistance of the cultivar. Protecting fungicides should be applied initially at relatively long intervals and subsequently at shorter intervals as the crop ages.

7.12.6.3   Resistant Cultivars

Complete resistance to early blight does not exist in commercial potato cultivars. Apparent levels of resistance are often correlated with plant age. Immature potato plants are relatively resistant to early blight, but, following tuber initiation, susceptibility increases gradually, and mature plants are very susceptible.

7.13   LATE BLIGHT OF POTATO

Late blight is one of the most devastating diseases of potato worldwide. It was responsible for the devastating Irish potato famine of the 1840s and has continued to be important to the present. If left unmanaged, this disease can result in complete destruction of potato crops. The fungus attacks both tubers and foliage at any stage of development and is capable of rapid development and spread. Soft rot of tubers often occurs in storage following tuber infections. Consequently, the tolerance for late blight is usually very low.

7.13.1   Causal Organism

Species	Associated Disease Phase	Economic Importance
Phytophthora infestans	Lesions on Leaves and Tubers	Severe

7.13.2 SYMPTOMS

Late blight is visible as pale-green, water-soaked spots on the leaf's edges or tips. These circular to irregular lesions are often surrounded by a yellowish border that merges with healthy tissue. Lesions expand rapidly, taking on a purplish, brownish, or blackish color. During high moisture periods, lesions are often ringed with grayish-white fungal growth. Foliar destruction can be severe, with large crop areas defoliated in as little as 1 week from the time of initial symptom development.

Late blight-infected tubers have brown, dry, sunken lesions that cover granulated brown or tan tissue up to 12 inches deep. Secondary infections, caused by bacteria or other fungi, can result in the slimy breakdown of additional tuber tissue in storage (Figures 7.23 and 7.24).

7.13.3 CAUSE AND DISEASE DEVELOPMENT

Late blight is caused by the fungus *P. infestans*. Unlike most pathogenic fungi, the late blight fungus cannot survive in soil or dead plant debris. For an epidemic to begin in any 1 area, the fungus must survive the winter in potato tubers (culls, volunteers), be reintroduced on seed potatoes or tomato transplants, or live spores must blow in with rainstorms.

7.13.4 FAVORABLE CONDITIONS OF DISEASE DEVELOPMENT

Disease development is favored by cool, moist weather. Rain, fog, or heavy dew are ideal. Under these conditions, lesions may appear on leaves within three to five days of infection, followed by the

FIGURE 7.23   Symptoms of late blight on potato leaves.

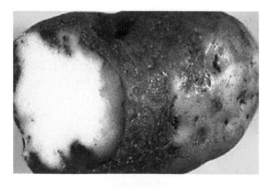

FIGURE 7.24   Symptoms of late blight on potato tubers.

white mold growth soon thereafter. Spores formed on the mold are spread readily by irrigation, rain, and equipment. They are easily dislodged by wind and rain and can be blown into neighboring fields within 5–10 miles or more, thus, beginning another cycle of the disease. They begin to germinate when conditions are damp (>90% relative humidity) and relatively cool (10°C–15°C, below 25°C). These conditions can easily occur under the leafy canopy of a growing potato field.[7]

7.13.5 DISEASE CYCLE

The pathogen overwinters as mycelium in infected potato tubers. These tubers may be in the ground as volunteer potatoes, in cull piles, or planted as seed. The mycelium will spread up the new sprouts, particularly in the cortical tissue. When the mycelium reaches the aerial plant parts, it will produce sporangiophores that emerge through stomata of the stems and leaves. The sporangia produced on these sporangiophores can become airborne or rain dispersed. When the sporangia land on wet potato leaves or stems, they can germinate and cause new infections. Mycelium emerging from this infection site will penetrate new tissue leading to the formation of lesions. New sporangiophores will be produced and protrude through stomata. They will appear as white fungal growth on the underside of leaves where lesions are apparent. These sporangiophores will produce more sporangia, which can be spread by wind and rain. The sporangia can not only be dispersed to leaves and stems, but also to the soil or tubers at the soil surface. The sporangia can infect directly or through the formation of zoospores. The zoospores produced from the sporangia can move in water and penetrate the tubers through lenticels or wounds. At harvest time tubers can be contaminated, and consequently, infected by sporangia on the soil surface or plant tissue. The infection of tubers may not be apparent at harvest, but it will develop in storage.

The development of late blight epidemics will depend greatly on temperature and humidity. Sporangia lose their viability in 3 to 6 hours at relative humidities below 80%. Sporangia germinate only when free moisture or dew is present on leaves. The fungus will sporulate most abundantly at humidities near 100% (Figure 7.25).

7.13.6 MANAGEMENT

A combination of several management practices is necessary to achieve consistently good control of late blight. The first practice is to avoid introducing late blight into a field. Thus, disease-free seed tubers should be planted, and cull and volunteer potatoes should be destroyed. Second, resistant varieties should be planted. No cultivar is immune to late blight, but several are moderately resistant

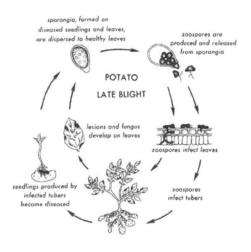

FIGURE 7.25   Disease cycle of *P. infestans* causing late blight of potato.

TABLE 7.2
Management of Blight

Moderately Resistant	Slightly Resistant	Susceptible
Abnaki	Atlantic	Kennebec
Belchip	Bake-King	Rosa
BelRus	Frito Lay 657	Sebago
Chieftain	Green Mountain	
Chippewa	Katahdin	
Hudson	Russett Burbank	
Monona		
Norchip		
Russet Rural		
Superior		
Wauseon		

and could be planted if blight was expected to be a problem (Table 7.2). The third practice is to apply fungicides as needed throughout the growing season.

Several forecast techniques have been developed that predict when a spray will be necessary based on environmental conditions favorable to the development of the fungus. Fourth, hilling and vine killing reduce the incidence of tuber infection. Infected tubers should be removed before potatoes are stored.

7.14   POTATO BLACK RINGSPOT VIRUS (PBRSV)

Potato black ringspot is caused by potato black ringspot nepovirus, which is classified as

Family: Comoviridae
Genus: *Nepovirus*
Species: Potato black ringspot nepovirus

7.14.1   SYMPTOMS

Several potato cultivars develop calico-like symptoms. A bright yellow area on the margins of middle and upper leaves gradually increase in size to form large patches (Figure 7.26). Most of the plant foliage may eventually turn yellow without stunting or leaf deformations. Primarily infected plants show local and systemic necrotic spots and ringspots and sometimes systemic necrosis.

7.14.2   MEANS OF MOVEMENT AND TRANSMISSION

Local spread is by contact between plants and possibly insect vectors. PBRSV is readily transmitted through tubers. In international trade, it could be carried by potato tubers or by true seed of potato.

7.14.3   PREVENTION AND CONTROL

As with all potato viruses, control depends on the production of high-quality seed potatoes from virus-free nuclear stock.

FIGURE 7.26   Symptoms of potato black ringspot nepovirus on potato leaves.

7.15   POTATO LEAFROLL VIRUS

Potato leafroll virus (PLRV) occurs worldwide where potatoes are grown and causes significant yield and quality losses. It is the major aphid-borne virus in potato production in many countries. The disease may be introduced to a field either by planting infected seed or by an aphid vector of the virus. PLRV is one of the most important potato viruses worldwide but particularly devastating in countries with limited resources and management. It can be responsible for individual plant yield losses of over 50%. One estimate suggests that PLRV is responsible for an annual global yield loss of 20 million tons.

7.15.1   CAUSAL ORGANISM

Potato leafroll virus (PLRV) is a member of the

Family: Luteoviridae
Genus: *Polerovirus*

7.15.2   SYMPTOMS

PLRV infects members of the Solanaceae family. The most economically important host is the potato, *S. tuberosum* spp. In potato, symptoms of primary infection, infection in the growing season, occurs in the youngest leaves. Leaf margins become necrotic, turning brown and purplish and curl inward toward the center of the leaf. Secondary infection, which starts from infected potato culls, produces more severe symptoms. Leaf rolling (Figure 7.27) is more apparent and the entire leaf can become chlorotic and sometimes also has a purple discoloration. Necrosis of the phloem tissue particularly in the haulm is observed after the onset of symptoms. Plants infected with PLRV experience stunted growth and produce smaller tubers. Infected tubers retain normal shape but experience necrosis of the vascular tissue. Symptoms on tubers do not always appear, and when they do, it is after several weeks in storage. Closest to the stem end, the inner tissue will show strands of discoloration called net necrosis. This usually appears as small brown spots scattered

FIGURE 7.27   Leaf roll symptom of PLRV.

throughout the tissue. Necrosis of the tuber may not be apparent at harvest and can develop in storage. Net necrosis of potato is the result of infection by PLRV. This symptom is caused by the selective death and damage to cells in the vascular tissues of the tuber. The fact that only specific cells within the tuber are affected by this problem while others remain normal causes the characteristic net symptom. Infection by the virus may directly cause the damage to and death of the vascular tissues, or the presence of the virus may make these sensitive tissues more susceptible to damage from other stresses. There is a strong resemblance between PLRV net necrosis and another tuber defect known as stem end discoloration (SED). Unlike PLRV, SED is believed to be a physiological disorder.

7.15.3   MEANS OF MOVEMENT AND TRANSMISSION

Potato leafroll virus can be introduced to a potato field by infected seed tubers or by aphids that have fed on infected potato plants. The most efficient vector of the virus is the green peach aphid. Several minutes to hours are required for the aphid vector to acquire the virus, but once the virus has been acquired, the aphid carries the virus for life. Winged aphids carried in air currents spread the virus for long distances between fields, and non-winged aphids are important in plant-to-plant spread. Aphid feeding introduces potato leafroll virus into the phloem tissue where the virus multiplies, spreads, and initiates disease. Potato leafroll virus is not transmitted mechanically by machinery or contact with leaves.

7.15.4   PREVENTION AND CONTROL

Since PLRV is persistently transmitted it makes for easier means of control. Studies have shown a minimum of 12 hours is required for the virus to be transmissible by an aphid. Therefore, PLRV can be controlled effectively by reducing aphid populations. Systemic and foliar insecticides can be used to prevent aphid feeding. Since the virus takes several hours to be transmitted by aphids, systemic insecticides are utilized and the aphid dies before it can transmit the virus. Foliar pesticides are utilized when colonizing aphid populations get too high and are useful as a knockdown method, to rapidly reduce aphid populations. Imidacloprid, Methamidophos, and Endosulfan are commonly used in aphid control. One study found that ethyl-methyl parathion pesticides are less effective at controlling the green peach aphid at lower temperatures. At 25°C aphid mortality was 95%. At 17°C mortality was 90%, at 10°C mortality was 80%, and at 7°C mortality was 67%. This suggests that some chemical controls are more effective at higher temperatures. Other management strategies include sanitation and seed certification. Cleaning any volunteer tubers from a field reduces any reservoirs that may persist. Seed certification programs test seed lots utilizing ELISA for the detection of multiple potato viruses. Seed lots can then be rated by the amount of infection. Highly infected seed lots are rejected and not used the next season.

7.16   POTATO VIRUS Y

Potato virus Y (PVY) has become a serious problem for the seed potato industry, with increased incidence worldwide. PVY infection of potato plants results in a variety of symptoms depending on the viral strain.

7.16.1   CAUSAL ORGANISM

PVY is a plant pathogenic virus of the family Potyviridae, and one of the most important plant viruses affecting potato production.

Family: Potyviridae
Genus: *Potyvirus*
Species: Potato virus Y

7.16.2   SYMPTOMS

The appearance of symptoms on a crop depends on the strain of virus Y (Yo: ordinary Y or Yn: necrotic), the variety, the weather conditions, and the type of infection (primary or secondary).

In the case of strains of type Yo, contamination during the current year (primary infection) emerges through the appearance of black necrotic blemishes, on the veins and the undersides of leaves. The leaves become brittle and dry while remaining attached to the plant. It can also produce a distorting mosaic (crinkle), often located on a stem or part of the plant.

On the other hand, contamination during the previous year (secondary infection) produces much more pronounced symptoms, which are highly variable according to the variety. They comprise 3 types:

Crinkle deformation of the leaves with mosaic, or in an embossed pattern together with the appearance of a glossy and mosaic effect,
Mottle dwarfism with large necrotic blemishes on the foliage ribs and with substantial distortion of the plants,
Mosaic (alternating pale green and dark green areas) non-distorting, but more or less obvious according to the variety and more visible in cloudy weather (Figure 7.28).

7.16.3   MEANS OF MOVEMENT AND TRANSMISSION

PVY may be transmitted to potato plants through grafting, plant sap inoculation, and through aphid transmission. The most common manner of PVY infection of plant material in the field is through

FIGURE 7.28   Symptoms of potato virus Y on potato leaves.

the aphid, and although aphids on their own can directly damage potato plants, it is their role as viral vectors that has the greatest economic impact. In cold climates aphids spend the winter either as wingless aphids giving birth to live young (viviparae) or as eggs. Hosts such as weeds and other crops serve as breeding grounds for these aphids and form a temporary area of colonization before the aphids migrate to the potato fields. In moderate climates, such as in South Africa, aphids are thought to reproduce asexually on weeds, other crops, indigenous plants, and garden plants. This means that there are a number of aphids present year-round. Wingless aphids have not yet been linked to the spread of PVY in potato fields.

The green peach aphid (*Myzus persicae*) has been found to be most effective in its role as viral vector, but others such as *Aphis fabae*, *Aphis gossypii*, *Aphis nasturtii*, *Macrosiphum euphorbiae*, *Myzus (Nectarosiphon) certus*, *Myzus (Phorodon) humuli*, and *Rhopalosiphum insertum* are also strongly associated with viral transmission. The efficiencies of some of these aphids to function as PVY vectors were also established and were found to vary between the different species. Apart from being classed according to efficiency as vectors, aphids can also be divided into two subgroups, namely colonizing and non-colonizing species. Colonizing aphids are aphids that reproduce and establish themselves on potato plants, specifically, while non-colonizing aphids do not reproduce nor establish colonies on potato plants. Colonizing aphids are better adapted to life on potato plants and are thus generally considered as better PVY vectors than non-colonizing aphids. Non-colonizing aphids do not primarily feed on potato plants but do occasionally feed on them while searching for a more suitable host. Their lower efficiency as PVY vector is canceled out by the sheer numbers in which they occur. Because of this, all aphids present in and around potato fields must be considered as possible vectors and their numbers carefully monitored.

7.16.4 PREVENTION AND CONTROL

Virus diseases such as PVY cannot be cured, but measures can be taken to manage them. The single most important control strategy is the use of certified seed. If accompanied with an appropriate seed handling practice, weed control in and around crops, and good control of volunteer potatoes, it is possible to have excellent control of this virus.

7.17 ROOT KNOT NEMATODE ON POTATO

Meloidogune chitowoodi is the root knot nematode which damages the potatoes economically worldwide. The root knot nematode is commonly known as "eelworm".

7.17.1 SYMPTOMS

The infected potato plant may be stunted, yellowing of leaves may be a general problem, and the plant may wilt. Wilting occurs under moisture stress. The roots have a common generation of beads or knots (hence the name of the nematode) after the initiation of swelling. The tuber might swell or develop blisters.

Symptoms are most severe when crops are grown on sandy soils and warm climates above 25°C. The root knot nematode reduces the quality, size, and number of tubers. After the potato is infested by the root knot nematode, it also gets susceptible to various other diseases like verticillium wilt, bacterial wilt, and diseases caused by *Rhizoctonia* (Figure 7.29).

7.17.2 SPREAD

The spread is generally by planting infested tubers.

FIGURE 7.29   Knot formation on the roots of potato.

7.17.3   MANAGEMENT

- Since symptoms do not appear immediately, it is necessary that the crops are monitored regularly for the symptoms of the root knot infestations.
- Crop rotation can help to keep the nematode away. Crops that are resistant or immune like rye, wheat, etc., or non-host crops such as grasses can be planted.
- Maintaining hygiene also helps control the emergence of the nematode.
- Planting of susceptible crops should be avoided.

7.17.4   ROOT LESION NEMATODE

Pratylenchus penetrans is the root lesion nematode that lives in soil and plant roots that causes root lesions on potato roots.

7.17.5   SYMPTOMS

The infected potato plant may show reduced tuber yield and quality, reduced growth of roots and foliar.

As with root knot nematode incidents, the plants infected with root lesion nematodes also become prone to fungal invasion causing severe fungal diseases. Also, the plant becomes less tolerable to moisture stress. As much as a 40% reduction in tuber yield is observed due to infection with root lesion nematode. Tubers also tend to become prone to becoming misshapen.

7.17.6   MANAGEMENT

- Soil fumigation.
- Crop rotation can help to keep the nematode away. Crops that are resistant to the nematode, like Ryegrass and Marigold, can be planted.
- Avoid soybeans and red clover as rotation crops as they are good host of *P. penetrans*.
- Planting resistant cultivars of potato may help, however, no fully resistant cultivar is available to date.

7.18   RHIZOCTONIA CANKER

Rhizoctonia canker (also known as black scurf) is caused by the soilborne fungus *R. solani*. Disease causes lesions or cankers on sprouts and young stems and development of persistent dark sclerotia (resting structures of the fungus) on the surface of tubers, hence the disease common name black scurf. Disease is most destructive when young stems are weakened or girdled and killed, resulting in delayed development and reduced yields. Severe infections may cause malformation, cracking, or the pitting of tubers.[8]

7.18.1 CAUSAL ORGANISM

Species	Associated Disease Phase	Economic Importance
R. solani	Lesions on Stem, Stolon, and Dark Sclerotia on Tuber	Severe

7.18.2 SYMPTOMS

The two types of symptoms are caused by the same fungus *R. solani*, which can be soilborne or introduced into pathogen-free fields by planting contaminated seed tubers. The phase of the disease called black scurf is common on tubers produced commercially and in home gardens. The irregular, black to brown hard masses on the surface of the tuber are sclerotia, or resting bodies, of the fungus (Figure 7.30). Although these structures adhere tightly to the tuber skin, they are superficial and do not cause damage, even in storage. They do perpetuate the disease and inhibit the establishment of a plant from the tuber if it is used as seed.

Black scurf is the most noticeable sign of *Rhizoctonia*. But the most damaging phase of the disease occurs underground and often goes unnoticed. The fungus attacks underground sprouts before they emerge from the soil. Stolons that grow later in the season can also be attacked. The damage varies. The fungal lesion, or canker, can be limited to a superficial brown area that has no discernible effect on plant growth. Severe lesions are large and sunken, as well as necrotic. They interfere with the normal functioning of stems and stolons in translocating starch from leaves to storage in tubers. If the fungal lesion expands quickly, relative to the growth of the plant, the stolon or stem can be girdled and killed.

If *Rhizoctonia* damage is severe and lesions partially or completely girdle the shoots, sprouts may be stunted or not emerge above the soil. Stolon cankers reduce tuber numbers and size and are identical to shoot cankers in appearance. *Rhizoctonia* lesions are always dry and usually sunken.

Late season damage to plants is a direct result of cankers on stolons and stems causing problems with starch translocation. Tubers forming on diseased stolons may be deformed. If stolons and underground stems are severely infected with Rhizoctonia canker, they cannot carry the starch produced in the leaves to the developing tubers. In this case, small, green tubers, called aerial tubers,

FIGURE 7.30   Black scurf on potato tuber.

may form on the stem above the soil. Formation of aerial tubers may indicate that the plant has no tubers of marketable quality belowground.

At the end of the growing season, the fungus produces its sexual state *Thanetephorus cucumeris* on stems just above the soil line. It appears as a superficial delicate white mat which is easily removed. The fungus does not damage the tissue beneath this mycelium. *R. solani* is a specialized pathogen. Only a subset of the isolates of this fungal species can cause cankers on potato. Isolates are grouped by the ability of their hyphae to fuse; isolates that can fuse, or anastomose, are in the same anastomosis group (AG). Isolates that are pathogens of potato are in AG-3. Rarely, isolates in other AG groups can form sclerotia on tubers and mycelial mats on stems. Though not damaging to potato, other AGs of *R. solani* cause diseases on sugar beet, beans, crucifers, and rice. In the absence of host plants, *R. solani* can exist by deriving its nutrients as a soil saprophyte from organic debris (Figures 7.31).

7.18.3 CAUSE AND DISEASE DEVELOPMENT

Damage is most severe at cold temperatures when emergence and growth of stems and stolons from the tuber are slow relative to the growth of the pathogen. Wet soils also contribute to damage because they warm up more slowly than dry soils and excessive soil moisture slows plant development and favors fungal growth.

7.18.4 FAVORABLE CONDITIONS OF DISEASE DEVELOPMENT

The fungus only infects juvenile tissue. Disease development is favored by relatively wet, cool (13°C to 15.5°C) soils. Sclerotia form on the surfaces of mature tubers under cool, moist conditions, generally after the vines have begun to die. The temperature range for AG-3 is 5 to 25°C, so plants will be most susceptible to infection when the soil temperatures are within this critical range. Cool temperature, high soil moisture fertility, and a neutral to acid soil (pH 7 or less) are thought to favor the development of *Rhizoctonia* disease in potato.

7.18.5 DISEASE CYCLE

The disease cycle is very straightforward. Inoculum usually is introduced into fields on potato seed tubers, although it may be introduced via contaminated soil. Sclerotia in soil or on seed tubers germinate, and the resulting mycelium colonizes plant surfaces where nutrients are available. Seed inoculum is particularly effective in causing stolon damage because it is so close to developing

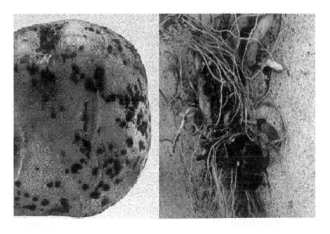

FIGURE 7.31 Rhizoctonia canker on stem and stolon.

sprouts. The fungus penetrates young, susceptible tissue, causing cankers that slow or stop the expansion of the infected stem or stolon. Cankers can sever the stolon or shoot from the plant or kill the growing point. The plant's resistance to stolon infection increases after emergence, eventually limiting the expansion of the lesions. Sclerotia form on tubers and in the soil, providing inoculum for other growing seasons.

7.18.6 MANAGEMENT

Getting potato plants to emerge quickly in the spring is the key to minimizing damage to shoot and stolon cankers because plants are more susceptible before emergence. Planting seed tubers in warm soil and covering them with as little soil as possible will speed the emergence of the shoots and increase resistance to canker infection. Plant fields with coarse-textured soils first because they are less likely to become waterlogged and will warm up faster.

Crop rotation reduces inoculum that can cause cankers because those *R. solani* isolates are specific to potato. Sclerotia are relatively resistant to degradation in the soil, however, and may survive for several years in the absence of potato. The fungus can also exist as a saprophyte in soil by colonizing organic debris. The longevity of the population is determined by the initial density of sclerotia at the start of the rotation period, the soil conditions, and the amount of microbial activity in the soil. Planting sclerotia-free seed is an excellent management strategy. Fungicide treatments applied to tubers may help suppress tuber-borne inoculum but are not a replacement for clean seed. Several products have been developed specifically for control of seedborne potato diseases. Some of them are Tops MZ, Maxim MZ (and other Maxim formulations + Mancozeb), Quadris, Nubark Maxim, and MonCoat MZ.

Black scurf, or sclerotia, can be minimized by harvesting soon after vines are killed. Sclerotia begin to form on tubers as vines senesce and become larger and more numerous over time. Therefore, harvesting tubers as soon as possible after skin set, reduces tuber scurf significantly. Sclerotia do not form and grow in storage, and there is no increase in tuber storage rot.

7.19   SILVER SCURF OF POTATO

Silver scurf, caused by the fungus *H. solani*, is a common disease of potato and is present in all major production areas of the world. The brown blemishes that develop on the tuber surface lower the market value of the crop. These losses are greatest when the disease occurs on tubers with white or red skin intended for the table-stock market. The increased water loss from infected tubers during storage results in shrinkage that can be economically significant. Silver scurf-infected tubers may be more susceptible to secondary infection by other pathogens. Silver scurf infection has been associated with decreased seed-tuber vigor, but the importance of these effects is not well documented.

7.19.1   CAUSAL ORGANISM

Species	Associated Disease Phase	Economic Importance
H. solani	Silvery Appearance of Dead Skin on the Surface of Infected Tubers	Severe

7.19.2   SYMPTOMS

H. solani infects only the periderm (skin) of the potato tuber. Experimental attempts to infect stems, stolons, and roots have not been successful. Tubers are infected during the growing season, and lesions become visible in three to five weeks. Symptoms usually appear at the stolon end of the tuber as small pale brown spots. Lesions may be difficult to detect at harvest, particularly if the tubers are not washed. Tubers that appear to be disease-free at harvest, may develop symptoms in storage. Silver scurf symptoms can be confused with black dot.

The appearance of silver scurf lesions often changes during the storage period. Severe browning of the surface layers of tubers may occur, followed by sloughing-off of the outer layers of the periderm so that the tuber is protected only by the inner periderm. Lesions have definite margins and are circular, but individual lesions may coalesce as the disease progresses. The silvery appearance of older lesions, for which the disease is named, is most obvious when the tubers are wet and results from air pockets in dead periderm cells. After some time in storage, the surface of the infected tubers may become shriveled and wrinkled due to excessive water loss from the silver scurf lesions. Cell death, however, never extends beyond the periderm into the tuber flesh.

The appearance of lesions also varies with skin type. Brown blemishes are obvious on red varieties because the red pigment in the tuber skin is destroyed. In contrast, infection of russet-skinned tubers is masked by the dark thick periderm. Russet varieties are susceptible, however, and sporulation on the tuber surface is sometimes apparent. Under conditions favorable for the disease (relative humidity >90% and temperature >4°C), masses of dark conidia are formed on the surface of the tuber. The obclavate, straight or curved conidia, are arranged in whorls on the sides of a black, multiseptated conidiophore. Conidia have two to eight pseudosepta and are attached to the conidiophore from its broad end (Figure 7.32).

7.19.3 CAUSE AND DISEASE DEVELOPMENT

Silver scurf is almost always derived from infected seed (it is very difficult to spot signs of the disease in seed) and once planted, is encouraged by warm and humid conditions. The longer the tubers are left in the soil, the more severe the problem can become. Smooth-skin types are more susceptible than russet types, and the scurf is especially visible on their tubers.

7.19.4 FAVORABLE CONDITIONS OF DISEASE DEVELOPMENT

H. solani infects tubers through lenticels or directly through the skin and remains confined to the periderm and outer layers of cortex cells. The incidence and severity of the disease can increase over time if tubers are left in infested soil. Once tubers are placed into storage, silver scurf lesions can continue to develop, and secondary spread of the disease to other tubes in storage can continue when temperatures are above 3°C and humidity is above 90%. Disease incidence increases the longer tubers are held in storage. With time, all tubers within storage can be severely infected by this pathogen.

7.19.5 DISEASE CYCLE

H. solani is seedborne and infected seed tubers are probably the main source of inoculum for infection of daughter tubers. The pathogen produces reproductive structures, called conidia, on the

FIGURE 7.32 Symptoms of silver scurf on potato tuber.

surface of the seed tuber. These conidia are washed off the seed tuber and through the soil by rain or irrigation; some of these are deposited near or on the surfaces of the daughter tubers. Conidia germinate in response to free moisture and infect tubers directly through the periderm or through lenticels. The pathogen then colonizes the periderm cells. Infection can occur as soon as tubers are formed and may continue throughout the growing season. *H. solani* does not infect or grow along roots, stems, or stolons.

Research completed in Europe and Canada suggests that the soil is not a significant source of overwintering inoculum; the survival of spores in the soil is thought to be poor. Experiments in Canada showed that fallowing and treatment of the soil with fungicides prior to planting did not affect disease incidence in a subsequent crop. Since *H. solani* can certainly survive on tubers left in the soil from previous crops, volunteer potatoes can be a source of inoculum for silver scurf and many other diseases.

The incidence and severity of silver scurf can increase significantly during storage. Temperature and humidity conditions commonly found in potato storages are favorable for disease spread and symptom development. The fungus sporulates on the surface of infected tubers at high relative humidity (>90%) and temperatures of 3°C (38°F) or higher. Conidia carried by air currents in storage land on healthy tubers where free moisture, caused by fluctuating temperatures at high relative humidity, allows germination of the conidia, infection of tubers, and lesion development and expansion. Disease incidence and severity can increase greatly on tubers stored for a few months.

7.19.6 MANAGEMENT

Control practices recommended for silver scurf include:

1. Early harvest of tubers
2. Manipulation of storage conditions
3. Use of uninfected seed
4. Control of volunteers
5. Crop rotation

Early harvest shortens the time daughter tubers are exposed to inoculum and thereby limits disease incidence. The soil environment is often favorable for disease development in late summer and early fall. Late-season infection of mature tubers, even after the vines are dead, can greatly increase the level of silver scurf in the crop.

Fluctuation of temperatures, and cold spots that allow moisture condensation or dripping on tubers, will stimulate disease spread in storage. Although the pathogen continues to grow at temperatures as low as 3°C (38°F), disease progress is slow. As storage temperatures rise, disease severity increases. If tubers are kept at a lower relative humidity (<90%), spore production and germination are halted, preventing tuber-to-tuber spread.

Using clean seed, rotating crops, and controlling volunteers will reduce the amount of initial inoculum, and therefore, help to control disease. Unfortunately, it is sometimes difficult to obtain seed that is free of silver scurf. Crop rotation undoubtedly reduces problems with silver scurf but little is known about the effect of specific rotation crops. Alternate hosts for the fungus have not been reported and further research is needed. Volunteer potato plants provide a reservoir of inoculum in the field since the pathogen can overwinter in the periderm of the potato tuber. Very little information is available on the relative susceptibility of potato varieties to *H. solani*. Red potato cultivars have been categorized as susceptible because lesions are very noticeable. Varieties with russet skins are considered resistant because symptoms are not obvious. Current recommendations for potato disease control should be consulted for appropriate fungicide seed treatments and resistant varieties.

REFERENCES

1. Inglis, D. A., Schroeder, B. K., 2011, *"Bacterial Soft Rot and Lenticel Spot on Potato Tubers"*, Washington State University Extension.
2. Elphinstone, J. G., 1987, *"Soft Rot and Blackleg of Potato"*, Technical Information Bulletin 21, International Potato Center (CIP).
3. Martin, C. and French, E. R., *"Bacterial Wilt of Potato"*, Technical Information Bulletin 13, International Potato Center (CIP).
4. Syngenta group company, 2007, *"Management of Pink Rot (Phytophthora erythroseptica)"*.
5. Zotarelli, L., Hutchinson, C., Byrd, S., Gergela, D., and Rowland, D. L., 2003, *"Potato Physiological Disorders - Brown Center and Hollow Heart"*, UF/IFAS Extension.
6. *"Late Blight of Potato and Tomato"*, Information referenced from Ohio State University as seen on: http://www.ncbuy.com/flowers/articles/01_10359.html
7. Wharton, P., Kirk, W., Berry, D., and Snapp, S., *"Potato Diseases: Rhizoctonia Stem Canker and Black Scurf"*, Michigan State University Extension. http://msue.anr.msu.edu/resources/potato_diseases_rhizoctonia_stem_canker_and_black_scurf_e2994

8 Onion

Onion, also known as the bulb onion or common onion, is a vegetable and is the most widely cultivated species of the genus *Allium*. It was first officially described by Carl Linnaeus in his 1753 work *Species Plantarum*.

Onion is cultivated and used around the world. The onion plant is characterized by a fan of hollow, bluish-green leaves, and its bulb at the base of the plant begins to swell when a certain day length is reached. The bulbs are composed of shortened, compressed, underground stems. The onion is most frequently a biennial or a perennial plant but is usually treated as an annual and harvested in its first growing season.

A variety of diseases and disorders affect onion crop. Most of the diseases are caused by fungi or bacteria, whereas disorders may be caused by adverse weather, air pollutants, soil conditions, nutritional imbalances, and pest control products. Sometimes several diseases and/or disorders can be present at the same time. Accurate disease diagnosis is very important. This chapter helps to identify the disease and gives accurate control measures for it (Figure 8.1).

8.1 BASAL PLATE ROT

This disease is caused by the *Fusarium oxysporum* f. sp. *Cepae*. It is caused mainly in the subtropical or warmer regions thus it has caused huge losses of yield around the world.

8.1.1 PATHOGEN

F. oxysporum is a fungal soilborne necrotroph and an imperfect state of ascomycete fungus. The disease severity is mainly dependent upon the pectolytic isoenzymes produced by the pathogen.

8.1.2 SYMPTOMS

Yellowing can be seen in the leaves along with a twist. In the later stages of infection, the leaves die off starting at the tip. The entire plant has the tendency to wilt in the early stages of the infection. Infected roots show dark-brown discoloration and culminate into rotten bulbs during storage. There is a development of white fungus on the basal plate of the crop without any sclerotia. The lower part of the bulb and the scales turn watery and grayish.

8.1.3 DISEASE DEVELOPMENT AND EPIDEMIOLOGY

The outbreak of this disease is at high temperatures. Maximum disease growth is seen at a temperature of 25°C. Soilborne fungus fusarium is reported to survive for many years in the form of thick-walled resting spores, chlamydospores, macroconidia, and less often, microconidia. The propagules or the chlamydospores, produce mycelium on the surface. The following shows the hyphal penetration (Figure 8.2).

It may take place through or without wounds. Appresoria is not always formed. Throughout the crop development, there is a risk of infection to basal plate rot. Primary damage to the crops at the basal part of the bulb, for instance, physical damage through flies, increases the chances of fusarium basal plate rot. This disease spread in the lot while storage even if the pathogen is latent in the plant and the bulbs.

FIGURE 8.1   Onion.

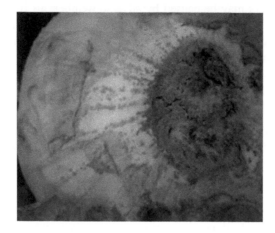

FIGURE 8.2   Bulb showing fusarium rot.

8.1.4 CONTROL MEASURES

- For the fields with the previous history of infection, varieties of high resistance should only be cultivated. Wilt-resistant cultivars should be planted.
- Crop rotation of four years or more is helpful.
- Infected lots should be discarded immediately.
- Temperature should be preferably being kept below 4°C.

8.2 FUSARIUM ROT DISEASE (YELLOWS)

Fusarium rot is a very serious disease because *F. oxysporum* f. spp. *zingiberi* can grow saprophytically in the absence of the host plant. It is more widespread than bacterial wilt and it can devastate the crop field entirely. In South Africa, *F. oxysporum* f. spp. *zingiberi* causes ginger wilt. Yellow was first reported in Queensland, then in Hawaii, followed by India and Asia.

8.2.1 SYMPTOMS

The first symptom that is seen is the yellowing of the margins of the older leaves followed by the younger ones. Plants also show drooping, wilting, drying in patches or wholly and yellowing. Plants infected by this fungus do not wilt rapidly as in bacterial wilt.

The plants do not fall on the ground. Wilting of ginger leaves, in this case, is very slow and growth of the plants is stunted. The lower leaves dry out over an extended period. It is not difficult to notice the presences of stunted and yellowed shoots aboveground in between the green shoots.

The fungus invades the entire vascular system of the plant, and hence, the plant dries out eventually.

The basal portion of the plant becomes water-soaked and soft and the plants can be pulled out easily from the rhizome. The rhizomes show creamy to brown discoloration of vascular tissues and a very prominent darkening of the cortical tissues. The central portion of the rhizome shows prominent rotting. Later, only the fibrous tissue is seen in the rhizome.

On storage, a white cottony fungal growth may be seen.

8.2.2 Pathogen Detection

Trujillo (1963) reported the causal organism to be *F. oxysporum* f. spp. *zingiberi*. Sharma and Dohroo (1990) found *F. oxysporum* as the major cause of yellows, which was present in all the ginger growing areas of Himachal Pradesh, India. Moreover, one out of the three isolates of this causal fungus caused 100% mortality of inoculated ginger seedlings while the other to isolates showed 60% and 80% mortality respectively. The other 4 species isolated were *F. solani*, *F. moniliforme* (*Gibbrella fujikuroi*), *F. graminearum* (*G. zeae*), and *F. equiseti*.

Spread of disease and the growth of the pathogen was seen when the moisture of the soil was 25% to 30% and the soil temperature was 24°C to 25°C. Galled root (galls of *Meloidogyne incognita*) incorporated *extract* of ginger showed better growth of the pathogen.

8.2.3 Pathogen Life

The pathogen survives in the soil and spreads through the infected soil, water, and rhizomes. The pathogen survives as chlamydospores which remain viable for many years. Microconidia, macroconidia, and chlamydospores are three types of asexual spores produced by *F. oxysporum*. The plants are invaded either through its root tips, wounded roots or formation point of the lateral roots and then either via a sporangial germ tube or mycelium and grows through the root cortex. Gradually the mycelium reaches the vessels and invades the xylem through the pits. Eventually, the mycelium grows and produces microconidia. These small characters are pulled upward by the plant's sap stream. The infection can also spread laterally as via the adjacent xylem vessels and pits.

8.2.4 Control Measures

1. *Chemical control*: Good quality rhizome germination acquired when the seeds were drenched in 0.1% HgCl2 followed by thiram (0.5%), ceresan wet (0.5%), and Dithane Z-78 (0.2%). Seed treatment with mercurial fungicides at 2 lb/40 gals of water, trizone as a fumigant, Arasan as a seed treatment and Benzimidazole-type fungicides is proved to be beneficial in the control of the disease.
2. *Seed selection*: It is obvious to select the healthy non-infected rhizome to protect the future crops from the pathogen and loss of yield.
3. *Intercropping*: Intercropping will help in the eradication of the pathogen from the soil. Intercropping with capsicum is advised; it is observed that intercropping with capsicum results in 76% control on yellows disease.
4. *Biological control*: In-vitro studies have shown that *T. harzanium* and *T. virens* inhibit the action of *F. oxysporum* f. sp. *Zingiberi*. In addition to this, *Bacillus stubtilis* also showed its antagonistic effect on the pathogen in the pot culture study. When *T. harzanium*, *T. hamatum*, and *T. virens* are used with chemicals like mancozeb and carbendazim as seed treatments, soil application proved to reduce the disease. The action of *F. oxysporum*

in the presence of six different *Streptomyces* sp. (SSC-MB-01 to SSC-MB-06) was studied. The bacteria are successful in suppressing the fungal activity of the pathogen and were found to protect the ginger plant against *F. oxysporum* by both dual culture method as well as agar culture method. A study showed that concentrated cow urine, prepared by dry heating in an oven at 50°C for four days, and 10% aqueous extract of leaves of *Parthenium hysterophorus* inhibited the growth of the fungi.

5. *Biotechnological approach*: Shoot buds of the rhizome encapsulated in 4% sodium alginate gel (in-vitro and in the field) were studied. The results were found to be as ginger siblings that were free of yellows. Yellows is scientifically known as Trujillo of ginger (Yellows is a disease caused by Fusarium oxysporum f. sp. zingiberi on Zingiber officinale).

6. *Resistant cultivars*: Using cultured rhizomes, which are resistant or less susceptible to the disease, provides excellent quality yield. SG 666 is a cultivar, which was proved resistant to the pathogen filtrate.

8.3 ONION LEAF ROT AND LEAF BLIGHT

Onion leaf rot, leaf blight, and blast are caused by the fungi *Botrytis squamosa*. Unlike *B. aclada*, both the asexual and sexual state of this fungi may be responsible for the cause of this disease.

8.3.1 PATHOGEN

B. squamosa is the asexual state of the ascomycete *Sclerotinia squamos*. Conidia (asexual spores) and ascospores (sexual spores) can both cause the disease.

8.3.2 SYMPTOMS

During the early stages of bulbing, onions are highly susceptible to leaf blight. It causes a substantial loss in the yield. There are grayish, whitish, or yellowish lesions on the leaf (leaf spots), which can be 1 to 10 mm long and 1 to 4 mm wide from circular to elliptical in shape. They are slightly impaired and outlined with a silver halo. On careful observation, the lesion penetrates the leaf blade. Chronic infections cause the death of onion leaves from its tips backward. The disease level can be demonstrated from the level of infection spread on the leaf surface. This fungus sporulates only on the dead tissue.

Ozone air pollution during hazy, calm, hot humid weather may exaggerate the problem of leaf blightening. Ozone of 1 ppm for some hours is sufficient for severity of the disease and proliferation of the disease. The bulbs in the leaf blightened field are small due to loss of leaves (Figures 8.3 through 8.5).

The disease is more prevalent in the densely populated crops of onion in the United Kingdom but more destructive in the temperate regions of the world.

Primary inoculum that is the sclerotia and secondary inoculum that is the conidia, dispersal of spores, survival, and infection to other plants and colonization of *B. squamosa* is the source of leaf blight (Figure 8.6).[1]

8.3.3 EPIDEMIOLOGY

The primary inoculum produces conidia along with temperature and humidity regulation. The most favorable temperature was found to be 5°C to 10°C and soil water potential to be -2Mpa for production of the conidias. The wetness should be for more than 12 hours for the pathogen to produce the spores. *B. squamosa* produces conidia on the onion leaf tip at night when the temperature is as low as 8°C. Temperature and leaf wetness are the main factors that are considered. Leaf penetration is direct and localized infection is seen on the leaf. The lesions increase with the increase in the

FIGURE 8.3 Leaf blight.

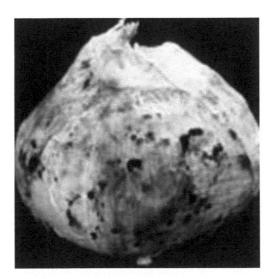

FIGURE 8.4 Development of black sclerotia on the onion bulb.

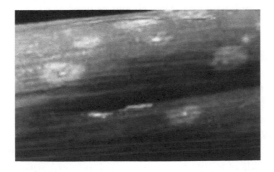

FIGURE 8.5 White spots surrounded by a greenish halo.

BOTRYTIS LEAF BLIGHT DISEASE CYCLE

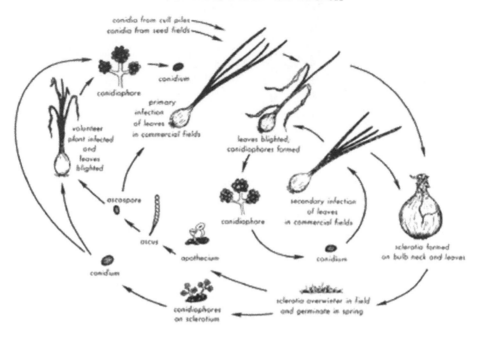

FIGURE 8.6   Botrytis leaf blight disease cycle.

number of hours of leaf wetness (duration up to 48 hours). This is followed by leaf blightening if the leaf wetness persists. In 1992, Lobeer discovered that the greatest number of conidias are produced in the necrotic leaf tissues rather than on the leaf spots, lesions or blightened leaf.

8.3.4   Control Measures

Plantation: Plant only healthy and non-contaminated seedlings of onions in a well-drained soil after confirming about the good air circulation by placing maximum distance between the two rows. Increasing the airflow would eventually decrease the wetness that is fruitful for the spread of the disease incidence. Deep plowing would be useful to decrease the number of sclerotia in the soil.

Fertilizers: Use nitrogen in split applications for more productivity and efficiency. Use dolomitic limestone to maintain a pH of 5.4 or higher in organic soils and at least 5.6 in other soils.

Crop rotation: Practice crop rotation for more than two years after the disease incident.

Fungicides: Fungicides like cyprodinil, fluazinam, iprodione, mancozeb, penthiopyrad, pyrimethanil, difenoconazole, etc. are used individually and in combination against the leaf blight disease before seven to fourteen days of harvesting. They should be used as per the instructions (Table 8.1).

Disease Forecasting: It can limit the use of fungicides used thus helping in the production of organic crops. BOTCAST and SIV were the methods used in the Netherlands, which were found to be effective in reducing the chemical sprays by 54% without any loss in yield or severity of the disease.

Integrated pest management programs in the United States have been found to be effective in delaying the use of fungicides until a critical disease level (CDL). Microclimate measurements are taken so that the severity of the disease can be calculated on a daily basis so that the time to spray the fungicides can be computed accurately. BLIGHT-ALERT is the weather-based forecasting system available commercially, like BOTCAST.

Biological Control: Applications of *Gliocladium roseum* were found to be effective only half the time compared to the fungicide chlorothalonil. However, it was effective in controlling the disease without any loss in the yield. This was approved using the BOTCAST.

TABLE 8.1

Fungicides Used Commercially for Leaf Blight

Disease	Active Ingredient	Rate/a of Commercial Product	Days to Harvest	Remarks and Suggestions
Leaf blight and purple blotch (*contd.*)	Cyprodinil	10.0 oz. Vangard WG	7	Do not apply more than 28.0 oz/a per plot per year of Vangard. Do not exceed 2 sequential applications of Vangard before alternating to a labeled fungicide with a different mode of action.
	Cyprodinil + Fludioxonil	11.0–14.0 oz Switch 62.5WG	7	Do not exceed 56.0 oz product per year per acre. Do not plant rotational crops other than strawberries or onions for 12 months following the last application.
	Difenoconazole + Cyprodinil	16.0–20.0 fl oz Inspire Super	7	For bulb onions, do not apply within 7 days of harvest. For green onions, do not apply within 14 days of harvest.
	Fluazinam	1.0 Pt Omega	7	Do not make more than 6 applications/a per year. Do not use with an adjuvant.
	Iprodione	1.5 pt Iprodione 4L AG, Meteor, Nevado 4F, Rovral 4F	7	Spray every 7 days during the growing season (more often during periods favoring the disease).
	Mancozeb	1.6–2.4 qt Penncozeb 4FL. 2.4 at Dithane F-45, Manzate Flowable.	7	Consult labels for seasonal product limits. Do not apply to exposed bulbs. Spray every 7–10 days depending on weather conditions. Mancozeb will also control downy mildew.
		2.0–3.0 lb Penncozeb 75DF, Penncozeb 80WP.	7	
		3.0 lb Dithane DF Rainshield, Manzate Pro-Stick.	7	
	Penthiopyrad	16.0–24.0 fl oz Fontelis	3	Do not apply more than 72.0 fl oz/a per season. Make no more than 2 sequential applications before alternating to a fungicide with a different mode of action.
	Pyrimethanil	18.0 fl oz Scala SC alone 9.0–18.0 fl oz Scala SC in tank mix	7	Scala belongs to the Group 9 fungicide category. Use the 9.0 fl oz rate of Scala in a tank mix with a broad-spectrum fungicide. Alternating the tank mix combination with a broad-spectrum fungicide is a resistance management strategy.
	Zoxamide + Mancozeb	1.5–2.0 lb. Gavel 75 DF	7	Do not apply to exposed bulbs. Do not make more than 8 applications/a per year.

(*Continued*)

TABLE 8.1 (CONTINUED)
Fungicides Used Commercially for Leaf Blight

Disease	Active Ingredient	Rate/a of Commercial Product	Days to Harvest	Remarks and Suggestions
Leaf blight (Botrytis) and purple blotch (Alternaria)	Azoxystrobin+Chlorothalonil	1.6–32 pt Quadris Opti	14	Do not apply more than one foliar application without alternating to a fungicide with a different mode of action. See label for specifications on tank mixing.
	Azoxystrobin+Difenoconazole	12.0–14.0 fl oz Quadris Top	7	Quadris Top and Cabrio belong to the Group 11 (strobilurin) fungicide category. Do not exceed 1 application of Group 11 products before alternating with a fungicide having a different mode of action. Do not exceed 4 applications of strobilurin fungicides per year. Do not exceed 56.0 fl oz Quadris Top (dry bulb) and 72.0 oz/a Cabrio per season.
	Pyraclostrobin	8.0–12.0 oz Cabrio EG	7	
	Boscalid	6.8 oz Endura WDG	7	Endura belongs to the Group 7 (anilide) fungicide category. Do not exceed 2 sequential applications of Endura before alternating to a labeled fungicide with a different mode of action. Do not exceed 6 applications per season. Do not exceed 41.0 oz/a Endura per season. Note comments on Pristine below.
	Boscalid+Pyracbstrobin	10.5–18.5 oz Pristine WDG	7	Pristine belongs to Groups 7 and 11 fungicide categories. Do not exceed 2 sequential applications of Pristine before alternating to a labeled fungicide with a different mode of action. Do not exceed 4 applications of Pristine or other Group 11 fungicide per season. Do not exceed 111.0 oz/a Pristine per season.
	Chlorothalonil	1.0–3.0 pt Bravo Weather Stik, Echo 720, Equus 720	7	Do not exceed 15.0 lb ai/a chlorothalonil per season. Spray every 7–10 days during the growing season (more frequently during periods favoring the disease). You can alternate chlorothalonil with mancozeb. Do not use on Sweet Spanish onions. Excessive use can reduce yields.
		1.5–425 pt Bravo Zn, Echo Zn, Equus 500 Zn	7	
		0.875–1.625 lb Echo 90DF	7	
		0.9–2.7 lb Bravo Ultrex 82.5WDG, Equus DF	7	

Post-Harvest Care: Piles of onion debris and bulbs should be destroyed to prevent the spread of the inoculum.

8.4   ONION NECK ROT

Onion neck rot caused by a necrotrophic fungus named *Botrytis aclada* is a post-harvest disease. Cool, moist conditions near the harvest favor the spread of the disease.

In some more years, botrytis neck rot could spoil the bulb to an extent of 50 to 70%. It is an ana-morph that is an asexual form of an ascomycete fungus. The sexual stage is absent. *B. aclada*, *B. byssoidea*, and *B. allii* are the three possible isolates studied by Yohalem et al. (2003). Preliminary molecular examination of the bulbs gave a possibility of the presence of all the three isolates. Even *B. cinerea* is found to be the causative organism (Figure 8.7).

8.4.1   *B. ACLADA*: THE PATHOGEN

Fresenius used the adjusted spore measurements to conclude *B. acalada* as a pathogen in the early 1850s. This is the storage disease of the bulb generally in a temperate part of the world. Munn in 1917 found this pathogen to cause small sclerotical neck rot in onions and as confirmed by Walker in 1926. Researchers in the United States have reported that *Botrytis* survives in the soil for more than three years and is transmitted to the emerging seedling.

8.4.2   SYMPTOMS

The pathogen enters the onion bulb through the succulent necks when the tops are removed or through wounds and matured leaves. After eight to ten weeks of storage of the onion bulbs, the pathogen infection is visible. The upper part of the onion becomes soft and water-soaked, and on opening the wrapper of leaves a sclerotical black mass encompassing the neck region can be seen. The infection spreads generally on the first one or two layers; down into the scales, with the perished sour color and having a brown cooked appearance. Each sclerotia are spread up to 5mm in diameter but sometimes the infection is severe and proliferated into the deeper layers of the bulb. However, the fungus might have partially spoiled the bulb before the external injury is visible (Figure 8.8).

FIGURE 8.7   Onion neck rot.

FIGURE 8.8   Botrytis sporulation observed.

The fungus sporulates on the dead onion bulb and is then dispersed to other onion plants. Secondary soft rot is when the bulb becomes watery.

This disease is practically symptomless. The disease spreads along the conidias by aerial dispersion. To know the presence of this disease, it is necessary to observe incubation of the bulbs for the presence of conidia and conidiophores.

Many scientists have studied the percentage of seed infection to the percentage of bulbs infected. However, the combined result showed that in some parts of the world the ratio was higher while in certain areas no relationship was evident. Moreover, in the seven-year study of the United Kingdom, it was found that the increase in rainfall along with the increase in environmental relative humidity (RH) resulted in the increased number of bulbar infections of the neck (Figure 8.9).

8.4.3   CONTROL MEASURES AND DISEASE MANAGEMENT

1. *Biological controllers*: Many mycofloral antagonists on the dead onion leaf tissue were demonstrated that might critically affect the sporulation of *B. aclada*. Parasitism was not observed. The mode of action causing the anti-fungal effects were as follows:
   a. Hyphal plasmolysis
   b. Antibiosis
   c. Exclusion of rapid colonization of the necrotic leaf tissue

These all could be used as a potential biocontrol fungus.

2. *Crop rotation*: If there had been a disease incidence, then there should be three to four crop rotations before replanting onion.

FIGURE 8.9   Botrytis neck rot infected onion bulbs.

3. *Disposal of the culled onions and sanitation*: The pathogen is disseminated from the culled onions; hence, such dead culled onions either should be buried into the deep layers of soil or should be covered in them. No culls should be exposed to the fields where the onions are to be grown. Proper disposal of the rotted onions is important in preventing the spread of the disease, otherwise, the pathogen will be dispersed in the field.

4. *Seed selection and planting*: High-quality onion seeds which are free from contamination must be transplanted. Avoid late or excess application of nitrogen as it prevents bulb maturity. Application of nitrogen in parts is recommended. Follow good weed planting techniques and management practices. Fungal spores can spread in the field or storage shed by air, water, farming equipment, and insects or workers.

5. *Chemical treatment*: Labeled fungicides, such as chlorothalonil, mancozeb, metalaxyl, Acrobat, pristine, Cabrio, Rovral, should be applied late season to protect the plants from foliage destruction and reduce neck contamination.

6. *Storage*: Select the proper good-quality bulbs from the rotten, culled, dried, scallions. If additional curing or drying is required, then pass 2 ft^2 of warm air per minute for 10 or more minutes. To prevent neck rot, proper care should be taken during lifting and harvesting. At -0.7°C, onions freeze. Store the dried onions at 0.5°C to 4°C and 70 to 75% humidity. Monitor storage temperature regularly. Poor ventilation, high humidity, and temperatures greater than 40°C stimulate the growth of the pathogen and the rot. Direct sunlight should be avoided. Gaps should be left in between crates to promote air circulation.

8.5   ONION WHITE ROOT ROT

Onion white rot root is caused by the soilbourne fungus *Sclerotium cepivorum*. It can infect all plants in the *Allium* family (including leeks and chives) but garlic and onions are the most susceptible. It is the most important and widespread destructive fungal disease. It spreads very quickly, 1 plant infected with the disease immediately spreads within and between paddocks and lead to crop loss. The organism remains dormant in soil for years until next crop is planted, hence, making it difficult to manage. In Australia, it is a major threat. Currently, there are no available resistant varieties.[2]

8.5.1   SYMPTOMS

Aboveground, the first symptoms appear on the leaf. Leaves turn yellow and wilting of foliage occurs. Leaf decay initiates from the base, with the older leaves being the first to collapse. Underground, the pathogens infect and cause root to rot and invade the bulb. A semi-watery decay of the bulb scales results. Under the wetter conditions, the plants may not wilt; however, they will become loose in the soil. Roots rot, and the plant can be easily pulled from the ground. The rot is associated with a fluffy white growth, the fungal mycelium, which develops around the base of the bulb. As the disease progresses, the mycelium becomes more compacted, less conspicuous, with numerous small spherical black bodies (sclerotia) forming on this mycelial mat. These sclerotia, the resting bodies of the pathogen, are approximately the size of a pin head or poppy seed. Plants can be infected at any stage of growth, but as observed in California, symptoms usually appear from mid-season to harvest.

8.5.2   DISEASE CYCLE

The pathogens survive as small, dormant structures, known as sclerotia, in the soil. Sclerotia has a capability to survive for over 20 years, even in the absence of a host plant. The sclerotia are highly resistant to adverse temperatures and conditions, and therefore, can remain alive in the soil for 20 or more years, even in the absence of a host.

Disease severity is dependent on sclerotia levels in the soil at planting. As few as one sclerotium per 10 kgs of soil can initiate and spread the disease. Sclerotia can be spread throughout a field or from one field to another field by floodwater, equipment, or on plant material, including windblown scales. Sclerotia remain dormant in the absence of onion or other *Allium* crops, making it difficult to control. Their germination is stimulated by *Allium* root extracts and exudates that extend into the soil about 0.5 inches from the root.

Disease development is favored by cool, moist soil conditions. The soil temperature range for infection is 10°C to 24°C, with the optimum temperature being 15.5°C to 18°C. At soil temperatures above 25.5°C, the disease is markedly inhibited. Soil moisture conditions that are favorable for onion and garlic growth are also ideal for white rot development (Figures 8.10 and 8.11).

8.5.3 CONTROL MEASURES

8.5.3.1 Sanitation

Strictly use clean seed or transplant stock. Infected planting stock is the cause of spreading the disease over large areas. With the infected soil, sclerotia are moved as well. They can survive and move in water, on equipment, on shoes, and in the wind. Regularly wash tractors, spray rigs, harvesting equipment, etc. especially when moving between infected and uninfected fields. Wash equipment with water only and make sure all soil is washed off. If muddy, wash off boots between fields.

FIGURE 8.10   Symptoms seen in onion bulb.

FIGURE 8.11   Aboveground symptoms.

8.5.3.2   Cultural Controls

White rot grows under the same conditions that are ideal for onion and garlic growth (cool weather and moist soil), so it is not possible to prevent the pathogen by changing planting dates. In areas with hot summers, planting in the Spring and harvesting in the Fall will reduce potential disease. This strategy is not effective in areas with cool summers. Reducing irrigation slows down the spread of the disease.

Do not move cull bulbs, litter, and soil from infested to non-infested fields. Onion seed does not carry sclerotia, but transplants and sets can. The fungus is vulnerable at temperatures above 46°C, thus dipping in hot water will greatly reduce the amount of pathogen and is a good preventative measure, although it may not completely eradicate the fungus. If the disease is observed, cessation of irrigation will minimize damage but not stop the disease. In addition, follow a long-term rotation schedule and do not follow *Allium* crops with other *Allium* crops. Rotation alone will not control white rot, but it does help prevent a build-up of the pathogen.

8.5.3.3   Sclerotia Germination Stimulants

Sclerotia germination stimulants are highly effective in reducing the numbers (and initial inoculum) of sclerotia in the soil. They reduce numbers of sclerotia in the soil by over 90%. The active ingredient is the chemical diallyl disulfide (DADS), a chemical from *Allium* roots. This triggers sclerotia to germinate. When DADS is applied artificially in the field in the absence of *Alliums*, sclerotia germinate. Since they cannot find the *Allium* host, this causes them to germinate and die, rather than lying dormant. DADS are available as a commercial product, called Alli-up. There can be no *Alliums* planted in the treated area for at least a year after application of DADS. However, it's fine to grow other crops, during the one-year *Allium*-free period. After this period, *Allium* crops can be grown again. It is, however, still very important to apply a fungicide at planting, even after DADS treatment.

8.5.3.4   Chemical Controls

Currently, three fungicides are registered for white rot control, namely, tebuconazole, fludioxonil, and boscalid. They are needed to be applied in a four to six inch bandwidth at planting and in the seed furrow. Later fungicide applications are ineffective in controlling the disease. Tebuconazole is the most effective fungicide out of the registered anti-fungal, for white rot control. Tebuconazole is phytotoxic on onions if it is applied at a higher concentration than the recommended 20.5 fl. oz./ A. Phytotoxicity can also occur if the bandwidth is narrowed. If a narrower bandwidth is applied, then the tebuconazole concentration per acre must also be reduced. Fludioxonil and boscalid are found to be effective in reducing white rot.

8.5.3.5   Organically Acceptable Practices / Biological Control

Organic growers should follow sanitation and cultural control practices. Sclerotia germination stimulants from natural sources (such as garlic oil/extract and powder) are acceptable for organic use. Currently, there are no other organically acceptable controls which have proven to be effective.

8.6   DOWNY MILDEW

Downy Mildew is a foliar disease caused by the obligate biotrophic phycomycete fungus *Peronospora destructor Peronosporacea*. It affects mainly during the cool, wintry, humid weather. It greatly affects the quality and quantity of the harvest. It causes loss of approximately 65% to 70% loss in onions and almost 100% loss in salad onions. The estimated field losses in Colorado, USA is found to be 25% to 50%. It is the major disease recorded in onions.

8.6.1   Pathogen

The pathogen survives in persistent dew, humid climates, and rain. The optimal temperature for the growth and spread of the disease is 25°C to 29°C. The pathogen *P. destructor* grows systematically in the plant tissue. The pathogen persists as mycelium in infected onion buds and seeds. This is the primary source of infection. The pathogen is found to persist in the soil for many years in the form of oospores. Under wet conditions, the pathogen sporulates on the infected tissues and disperses to other onion plants.

8.6.2 SYMPTOMS

The primary/initial symptoms that appear on the infected crop are the pale green oval patches or lesions, which on further infection become covered with violet gray patterns. The rest of the leaf is darker in color as compared to the elongated patch. The infection is mainly caused in older leaves. The foliage which have been affected with this pathogen become more susceptible to other fungi, for instance, *Alternaria* and this combined infection turn the entire foliage black. The affected leaves gradually become paler and finally turn yellow.

The stalk collapses and germination of the seeds is reduced, thus, hampering the yield. The bulbs become soft, watery, and shriveled. The other source of the pathogen may be the infected mother seeds (Figure 8.12).

8.6.3 EPIDEMIOLOGY

Conidia-bearing sporophores are susceptible to low humidity and produce conidia when exposed to high humidity and wetness. The temperature plays a secondary role in the growth of the conidia. The sporulation occurs only when there is a dark period followed by a light period with favorable relative humidity. Continuous dark or light periods prevent the sporulation. Therefore, darkness followed by the two hours of a humid period supports sporulation and the spread of the disease along with the temperature of the area being 10°C to 22°C. Moreover, if the humidity period is delayed by 4 to 5 hours, the temperature for sporulation needs to be 14°C to 18°C.

A greenhouse study by V. Buloviene and E. Surviliene concluded that shortest period of spore germination and sporulation was four days on the onion leaves inoculated with 10^6 spores per ml distilled water suspension (Figures 8.13 and 8.14).

8.6.4 CONTROL AND MANAGEMENT OF DISEASE

8.6.4.1 Cultural Management

Various practices are followed that are proven to have aided in the reduction in the chances of crops getting infected and to slow down the disease development as well.

FIGURE 8.12   Pale patches caused by fungus P. destructor.

FIGURE 8.13   Collapse of the leaf.

1. Disease-free seed sets or transplants are to be chosen for plantation to avoid infection from entering the field.
2. Following three to four year of crop rotation.
3. Prevent overwintering onion, infected debris, and destroy the culls. Infected plants in the fall after harvest will hasten decomposition.
4. Well-drained soils with good air circulation to promote rapid drying of foliage after rain, dew, or irrigation.

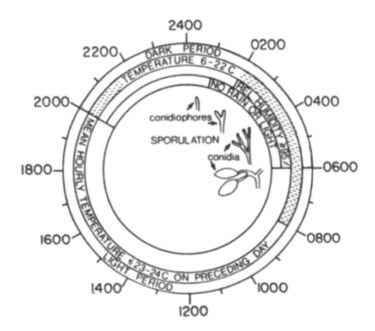

FIGURE 8.14   Diagrammatic representation of factors responsible for sporulation viz. humidity, temperature, and dark and light periods.

5. Prevent overhead irrigation to avoid wetting of foliage. Use of drip and sub-surface irrigation systems provide the plant with sufficient water to meet crop requirement without wetting foliage.

8.6.4.2 Chemical Treatment

Treatment of foliar applied protectant fungicides is necessary to protect the plant in cool, humid seasons. The dithiocarbamate fungicides such as maneb and mancozeb are used as protectants. Another fungicide widely used is chlorothalonil. Raziq et al. demonstrated the efficacy of 10 fungicides, namely, fungicides, namely Aliette, Antracol, Benlate, Cobox, Daconil, Derosal, Dithane, Polyram, Ridomil, and Topsin M.

8.6.4.3 Disease Forecast

The forecasting method "DOWNCAST" is used to speculate the initiation of the sporulation-infection period. It predicts sporulation and infection; however, it cannot detect the quantity if sporulation. Moreover, ONIMIL is the device which can detect both the infection as well as the quantity of sporulation. In addition, a lateral flow device is an immunological test that has been invented to detect the conidia of the pathogen well before the symptoms appear on the crops. This type of forecasting also helps in limiting the use of fungicides.

8.7 ONION SMUT

Onion smut is a fungal disease caused by the pathogen *Urocystis cepulae*. The disease was first reported in the United States in 1850. This disease is 1 of the most destructive of all onion diseases and hampers the crop greatly. The crops are killed in just three to four weeks after initial infection, that is, by the time of the development of the first true leaf. Later, the plants become resistant this. This disease affects only the seedlings and younger crops; hence, this disease should be looked after in the initial stages of crop development.

8.7.1 Pathogen

The pathogen is an obligate biotroph, basidiomycete fungus specific to *Allium* spp. Teliospores or chlamydospores play the main part in spreading the disease and hampering the crops. They are a thick-walled central cell enclosed in a hyaline cell-line. They produce uninucleate non-pathogenic thalli and binucleate pathogenic mycelium. The binucleate mycelium infects the new cotyledon leaf part situated underground.

8.7.2 Symptoms

Immediately after infection, dark-blackish gray lesions or black bands appear on the cotyledons of the onion leaves. A few scientists have described the symptoms as blisters with a silver sheen. The infected leaves are thick with dark pustules, curled backward. Later it may also affect the bulb, and the infection to the bulb may give entry to the secondary infections that cause rot.

Lesions rupture releasing the teliospores that are like black powder and can be transplanted into the field by various means, for instance, hands and feet of the farmers, farming equipment, running water, etc. The plants are killed soon after emergence and the ones that are not killed showed stunted growth, lesioned leaves, and brittle texture. Moreover, the disease often becomes systemic in this case (Figures 8.15 and 8.16).

Moreover, the smut does not cause any rot during storage, but the bulbs shrink rapidly.

FIGURE 8.15   Lesioned leaves of onion.

8.7.3   FACTORS AFFECTING THE DISEASE

The fungus can stay alive in the soil for more than 15 years.

   Temperature: The infection is at its peak when the soil temperature is between 10°C to 12°C. There is a moderate chance of infection if the temperature is up to 25°C. The pathogen loses its viability at 29°C. At the temperature above 25°C, the hyphal growth and teliospore germination are totally restricted; hence, this is the optimum temperature for the growth of the seedlings.

8.7.4   CONTROL MEASURES

   1. Use healthy transplants in the smut infected soil to minimize the infection.
   2. If possible, plant the crops which have already developed the first true leaf.
   3. Crop rotation must be practiced.
   4. Proper sanitation care should be taken. If possible, the tops should be burned after harvest.

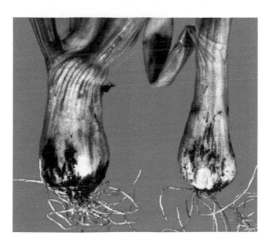

FIGURE 8.16   Plants that survived the disease, showing stunted growth of the bulbs.

5. Furrow and seed treatment with fungicides has been to be useful. Treatment of Thiram (45m for .45kg of seed) and spraying formaldehyde solution in the furrow is found to be useful.

Seed treatment with Captan 2.5g/kg and the seedbed when treated with methyl bromide (1 kg/25m²) is also helpful in controlling the disease.

8.8   PINK ROOT

Pink root is caused by weak necrotroph *Pyrenochaeta terrestris* also called as *Phoma terrestris*. It causes great yield losses and devastates fields of onions. It is mainly found in the onions grown in the tropical and subtropical regions.

8.8.1   PATHOGEN

The soilborne pathogen is an ascomycete fungus in its pycnidia asexual state. There is no teleomorph. The isolates of the fungus were found to be weakly associated with the roots of the onions. The pathogen is very active in the uppermost 15 cm of the soil.

8.8.2   DISEASE DEVELOPMENT AND SYMPTOMS

Roots are the most susceptible to infection, while infection cannot occur in plant wounds. Hyphal penetration is through the roots. This turns them pale pink first, that is; in the initial stages of infection. In later stages, the roots become deep pink in color, finally turning purple and then when they shrivel and die, turn black. The bulbs show stunted growth and roots become water-soaked, dried, and gradually disintegrate. Whereas the plants show drought-like symptoms. The mycelium is hyaline, septate, and anastomosing (Figure 8.17).

The pink discoloration can penetrate the scales of the bulbs too. However, the discoloration is only confined to the roots and the outer scales of the bulbs.

8.8.3   EPIDEMIOLOGY

The fungus most often is found to survive in the soil in the form of chlamydospores, pycnidia, or pycnidiospores. They are also found in the roots or plant debris of the infected plants. If onions are present in the soil, the population of the pathogen is found to increase at a higher rate[3].

FIGURE 8.17   Pink root by P. Terrestris.

Temperature if the main factor for the pathogen to survive and replicate in the soil. Reduced infection is seen at a temperature of 16°C to 20°C. On the contrary, the activity of the fungus is at its peak from 24°C to 28°C. The humidity or the moisture of the soil is equally important. However, in the recent studies, *P. terrestris* is found to reduce the number of plants as well as the weights of the onion bulbs.

8.8.4 CONTROL MEASURES

8.8.4.1 Crop Rotation

It is not an effective measure but Crop rotation of four to six years with a resistant variety should be done in the field with a history of this disease. In addition, longer the onion grows in the field, the incidence of the disease increases. Do not plant onions after cereals because the potential of the inoculum is greater.

8.8.4.2 Resistant Cultivars

Yellow Bermuda was found to be the most commercially resistant cultivar. Cultivars with a strong root system were infected less than the cultivars with a weak system according to the studies carried out in the United States and Mexico.

8.8.4.3 Chemical Methods for Curing

Fumigants like metam sodium and chloropicrin can be used if the other strategies for the prevention of the disease fail. Fumigation is an expensive treatment. TerraClean 5.0 application on a timely basis was also found to be effective.

8.8.4.4 Solarization

Soil solarization includes mulching the soil and then covering it with tarp (generally transparent polyethylene cover) to trap the solar energy for decontaminating the soil, and it is an effective means of doing so. The trapped heat arms the soil, thereby, increasing the soil temperature and killing pathogens like bacteria, fungi, etc. This method should be done depending on the month of the plantation and the location of the field.

8.8.5 PREVENTIVE MEASURE

Maintain healthy plants by optimum fertility and keeping the pests and rodents away from the field.

8.9 PURPLE BLOTCH

Purple blotch is caused by necrotrophic fungus *Alternaria porri (Ellis) Ciferri*. It is the asexual conidial-producing stage of ascomycete fungi. This fungus is known to cause harm to the species of *Allium*, but its effect is most seen on onion worldwide. This is the most virulent in onions growing in hot and humid climatic conditions.

8.9.1 SYMPTOMS

Older leaves are the most vulnerable to this disease. After infection, the primary symptom recorded is the formation of the small white water-soaked lesions on the leaf which turn brownish purple in color where the spores are seen. Secondary infection causes the edges of the flecks to turn reddish purple surrounded by a yellow zone or yellow concentric circles. When the flecks unite with each other and the disease progress, it affects the entire health of the leaf and causes it to turn yellow-brown, lose its erectness, and wilt or die. This is seen within four weeks of initial infection and result in the loss of the onion field (Figures 8.18a and 8.18b).[3]

FIGURE 8.18a   Purple blotch lesion.

FIGURE 8.18b   Wilting of leaves from purple blotch.

8.9.2   EPIDEMIOLOGY

The major factor to be considered is the soil wetness, temperature, and aerial humidity. The pathogen is active over a wide range of temperatures with the most favorable being 25°C. However, the fungal activity is found to be minimum or null at a temperature below 13°C. The spores or the conidia are mainly formed in the night damp conditions when the humidity is 90% or more. The spores are fully mature when they are exposed to 12 to 15 hours of dew. The sporulation occurs in the night under favorable moist conditions and high relative humidity. The conidial concentration in the air increases with the increase in windy air, irrigation, and spraying. The sporulation also occurs from the dead leaf debris. It affects the bulbs in major proportions.

In case of large onion dumps, (Thomas, 1951) wind played a key role in the dissemination of the spores. Moreover, Ajrekar (1922) found thrips to play an important role in the dispersal of the primary inoculum in Mumbai, India. Spore dispersal in *A. porri (Ellis)* Neerg. on onions in Nebraska.

8.9.3   EPIDEMIC

Environmental factors like hail storm, wind, and pollution favor injuries. Moreover, infections of *Botrytis* causing injuries when followed *A. porri*, results in epidemics and loss of onion yield on a very large scale. In Colorado, the most probable month of infection is mid-July. In addition to this, spraying of fungicides has also contributed to the detachment of the conidia from the conidiophore and the spread of a number of small conidia, eventually resulting in an outbreak of the epidemic.

8.9.4   CONTROL MEASURES

8.9.4.1   Crop Rotation
3 to 4 years of crop rotation is advisable in the field with the history of purple blotch disease.

8.9.4.2   Sanitation and Crop Debris
Crop debris is found to be the most important means of the spread of the disease to the succeeding crops. Proper disposal of the onion culls and debris after harvest and from the shed should be done to prevent the spread of conidia to the field. Pandotra (1965a) concluded from his findings that the fungus remains viable for eight months on the crop debris left in the field. Contaminated crops should be immediately buried 5 to 15cm deep in the soil. It was found by Pandotra (1965a) that the pathogen lost its viability after two months when buried in the soil.

8.9.4.3   Plantation and Crop Handling
Summer plowing and plowing of fields three times before sowing reduced the occurrence and severity of purple blotch and resulted in a high yield of bulbs. Late application of nitrogen fertilizers should be avoided. A split application of nitrogen is recommended. Only high-quality seedlings should be sowed and transplanted. A good weed management system should be followed. Do not irrigate within 10 to 14 days of lifting. Minimum bruising should be done. Maintain the temperature and humidity in the shed after harvesting is done.

8.9.4.4   Chemical Control
Fungicides are found to be effective against *A. porri*. Godfrey (1945) and Pandotra (1965b) found that the Bordeaux mixture was found to be effective against the pathogen. Weekly applications of mancozeb (Dithane M-45) and dichloran (Allisan) were found by Bock (1964) and Gupta et al. (1986). Husain (1960) found that anilazine was the most effective fungicide. Seedling dipped in and followed by the foliar spray with 0.2% copper oxychloride gave the best results in controlling the disease. Dithiane and difolatan have also been found to be effective against *A. porri*. A mixture of fentinacetate/maneb, which is called Brestan and iprodione (Rovral), are the two registered fungicides in South Africa to control purple blotch. Rovral is used as the preventive measure.[4]

8.9.4.5   Resistant Cultivars
Behera et al. (2013) found that out of the 22 varieties of the cultivars only VG-18 is resistant, 12 varieties are moderately resistant, and nine are moderately susceptible. Moreover, VG-18 succeeded in giving the highest yield (288.18q/ha) in comparison to others.

REFERENCES

1. Jones, D. G., 1998, *"The Epidemiology of Plant Diseases"*, pp. 410–415, Springer Science Business Media, B. V.
2. Ehn, B., Ferry, A., Turini, T., and Crowe, F., 2012, *"White Rot of Onion and Garlic: Symptoms and Controls".*
3. *"Pink Root on Onions"*, 2012, *New Mexico State University Extension on Plant Pathology, U.S. Department of Agriculture cooperating.*
4. Michailides, T. J., *"Pest, disease and physiological disorders management: Above Ground Fungal Diseases"*, University of California Davis Fruit and Nut Information, pp 214–232.
5. Yohalem, DS, Nielsen, K, Nicolaisen, M., "Taxonomic and Nomenclatural Classification of the Onion Neck Rotting Botrytis Species", *Mycotaxon* 85: 175–182, 2003.
6. Trujillo, E.C., 1963. Fusarium yellows and rhizome rot of common ginger. Phytopath. 53: 1370–1371.
7. Sharma, S.K. and Dohroo, N.P., 1990. Occurrence and distribution of fungi causing ginger yellows in Himachal Pradesh. Pl. Dis. Res., 5(2): 200–202.
8. Prachi, T.R., Sharma, T., and Singh, B.M., 2001. In vitro and in vivo phytotoxic effect of culture filtrates of Fusarium oxysporum f. sp. zingiberi on Zingiber officinale. Advances in Horticultural Sciences 14: 52–58.
9. Behera, S., Santra, S., Chattopadhyay, S., Das, S., and Maity, T. K., 2013. "Variation In Onion Varieties For Reaction To Natural Infection Of Alternaria Porri (Ellis) Ciff. And Stemphylium Vesicarium (Wallr.)". *The Bioscan.* 8(3): 759–761.

9 Chili

The name chili pepper (also chile pepper, chilli pepper, or simply chili) is from Nahuatl. Chili is the fruit of plants from the genus *Capsicum* and members of the nightshade family, *Solanaceae*. Chilis are widely used in many cuisines throughout the world to produce a spicy flavor in food. The substances present in chili that give chili peppers their intensity when ingested or applied topically are capsaicin and related compounds known as capsaicinoids.

Chili peppers are believed to be originated in Mexico. After the Columbian Exchange, many cultivars of chili pepper spread across the world. Chili is used for both food and traditional medicine.

As recorded in 2014, 32.3 million tons of green chili peppers and 3.8 million tons of dried chili peppers were produced worldwide. China is the world's largest producer of green chilies, providing half of the global total.

This chapter provides information about diseases which are affecting chili produce and discusses possible effective remedies for those diseases as well.

9.1 ANTHRACNOSE OF CHILI

Anthracnose disease is 1 of the major economic constraints to chili production worldwide, especially in tropical and subtropical regions. Accurate taxonomic information is necessary for effective disease control management. Little information is known concerning the interactions of the species associated with the chili anthracnose, although several *Colletotrichum* species have been reported as the causal agents of chili anthracnose disease worldwide.[1]

Anthracnose is an economically important disease of chili affecting both fruit and seed quality. The disease is more severe in all southern states.

9.1.1 CAUSAL ORGANISM

Anthracnose is caused by different species of *Colletotrichum*, mentioned in the table below.

Species	Associated Disease Phase	Economic Importance
Colletotrichum sps.	Fruit Rot	High

9.1.2 SYMPTOMS

Anthracnose causes extensive pre- and post-harvest damage to chili fruits through anthracnose lesions. Even small anthracnose lesions on chili fruits reduce their marketable value.

The infected leaves and fruits have small or large lesions, or purplish or brown patches without the formation of definite lesions. The stems and petioles have girdles and the inflorescences turn yellow, causing dieback and shriveling. The fruit usually develops lesions during the ripening process, but lesions may develop on fruit of any size and possibly on the foliage and stems at later stages of infection. As the fruit ripens, its susceptibility to infection increases. The lesions on the fruit are circular and may reach 3 cm in diameter or bigger on the larger fruit. The concentric rings at the center of the lesion may be tan or orange to black. Initial infections are undefined tanned colored lesions that may appear in a matter of days after infection.

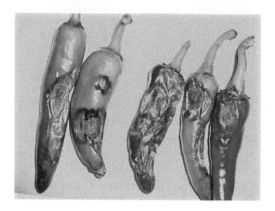

FIGURE 9.1   Anthracnose symptoms on chili fruits.[1]

Symptoms are seen on the leaf, stem, and fruit. Small, circular spots appear on the skin of the fruit and expand in the direction of the long axis of the fruit. The fruits with many spots drop off prematurely resulting in a heavy loss of yield. Fungus may also attack the fruit stalk and spread along the stem causing dieback symptoms (Figures 9.1 and 9.2).

9.1.3   CAUSE AND DISEASE DEVELOPMENT

In the *Colletotrichum* patho-system, different *Colletotrichum* species can be associated with anthracnose of the same host. Different species cause diseases of different organs of the chili plant; for example, *C. acutatum* and *C. gloeosporioides* infect chili fruits at all developmental stages, but usually not the leaves or stems, which are mostly damaged by *C. coccodes* and *C. dematium*. Different *Colletotrichum* species may also play an important role in different diseases of mature stages of chili fruit as well. For example, *C. capsici* is widespread in red chili fruits, whereas *C. acutatum* and *C. gloeosporioides* have been reported to be more prevalent in both young and mature green fruits.

9.1.4   FAVORABLE CONDITIONS OF DISEASE

The ideal temperature for the disease is 28°C–32°C. The pathogen infects the host at a minimum temperature of 22°C–25° C. Relative humidity of above 80% is ideal. Heavy prolonged dew

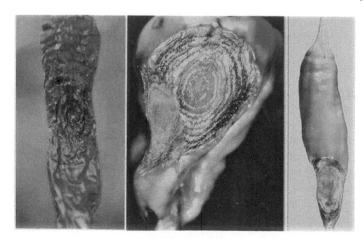

FIGURE 9.2   Anthracnose symptoms on chili fruits.

deposition is important for dieback development. Wet periods of about 12 hours or more favors the occurrence of infection.

9.1.5 DISEASE CYCLE

Environmental factors play a major role in the development of disease epidemics. The relationships among rainfall intensity, duration and crop geometry, and the dispersal of inoculum possibly lead to different levels of disease severity.

Colletotrichum species produce a series of specialized infection structures such as germ tubes, appressoria, intracellular hyphae, and secondary necrotrophic hyphae. These pathogens infect plants by either colonizing subcuticular tissues intramurally or being established intracellularly. The preinfectional stages of both are very similar. Conidia adhere to and germinate on the plant surface producing germ tubes that form appressoria, which in turn penetrate the cuticle directly. Following penetration, the pathogens that colonize the intramural region beneath the cuticle invade in a necrotrophic manner and spread rapidly throughout the tissues. There is no detectable biotrophic stage in this form of parasitism. In contrast, most anthracnose pathogens exhibit a bio tropic infection strategy initially by colonizing the plasmalemma and cell wall intracellularly. After the biotrophic state, intracellular hyphae colonize one or two cells and subsequently produce secondary necrotrophic hypha. These pathogens are therefore regarded as hemibiotrophs or facultative biotrophs. For example, *C. gloeosporioides* on avocado, chili, and citrus can produce both types of colonizations: intracellular biotrophy at an early stage and intramural necrotrophy later.[2]

Although the mechanisms developed by *Colletotrichum* species appear similar in prepenetration events, there are differences between species in the later mechanisms such as spore adhesion, melanization, and cutinization in penetration of the plant cuticle by the appressoria.

9.1.6 MANAGEMENT

Diseases can be managed using the following strategies:

- Chemical control
- Biological control
- Cultural control
- Integrated pest management

9.1.6.1 Chemical Control

Chemicals are the most common and practical method to control anthracnose diseases. However, fungicide tolerance often arises quickly, if a single compound is relied upon too heavily. The fungicide traditionally recommended for anthracnose management in chili is Manganese ethylenebisdithiocarbamate (maneb), although it does not consistently control the severe form of anthracnose on chili fruit. The strobilurin fungicides azoxystrobin (Quadris), trifloxystrobin (Flint), and pyraclostrobin (Cabrio) have recently been labeled for the control of anthracnose of chili but only preliminary reports are available on the efficacy of these fungicides against the severe form of the disease. The disease can be controlled under normal weather conditions with a reasonable spray program. However, there are numerous reports of negative effects of using chemicals on farmers' income and health, and toxic contamination to the environment, particularly in developing countries.

9.1.6.2 Biological Control

So far, biological control methods for chili anthracnose disease have not received much attention. The potential for biological control of *Colletotrichum* species had been suggested as *Pseudomonas fluorescens*. Antagonistic bacterial strains (DGg13 and BB133) were found to effectively control *C. capsica*—the major anthracnose pathogen in Thailand. It is also believed that *Trichoderma* species are able to effectively compete for surface area, thereby, reducing pathogen infection success. *Trichoderma* species have been applied to control *Colletotrichum* species in chili, strawberries, and citrus with concomitant disease reduction. Other biological control agents that have been tested for efficacy against *C. acutatum* include *Bacillus subtilis* and *Candida oleophila*.

9.1.6.3 Cultural Control

Pathogen-free chili seed should be planted, and weeds eliminated. Crops should be rotated every 2–3 years with crops that are not alternative hosts of *Colletotrichum*. Transplants should be kept clean by controlling weeds and solanaceous volunteers around the transplant houses. The field should have good drainage and be free from infected plant debris. If disease was previously present, crops should be rotated away from solanaceous plants for at least 2 years. Sanitation practices in the field include the control of weeds and volunteer chili plants. Choosing cultivars that bear fruit with a shorter ripening period may allow the fruit to escape infection by the fungus. Wounds in fruit from insects or other means should be reduced to the extent possibly because wounds provide entry points for *Colletotrichum* spp. and other pathogens such as bacteria that cause soft rot. At the end of the season, infected plant debris from the field must be removed or deep plowed to completely cover crop diseases.

9.2 BACTERIAL LEAF SPOT OF CHILI

Bacterial leaf spot (BLS) is the most common, and 1 of the 2 most destructive diseases of peppers in the eastern United States. Leaf spots are water-soaked initially, and then turn brown and irregularly shaped. Affected leaves tend to turn yellow and drop. Yield is reduced because raised, scab-like spots may develop on fruit and because affected leaves drop off plants, thereby reducing plant productivity and exposing fruit to potential sunscald.

BLS is a serious disease because it has a high rate of spread, especially during periods with wind-driven rains because adequate control measures are not available and because fruit symptoms reduce marketable fruit.

9.2.1 Causal Organism

BLS is mostly caused by the fungi mentioned in the table below.

Species	Associated Disease Phase	Economic Importance
Xanthomonas campestris pv. *vesiccatoria*	Leaves and Fruits	High

9.2.2 Symptoms

Leaves, fruit, and stems are affected. Lesions on the leaf begin as circular, water-soaked spots that become necrotic with a brown center and thin chlorotic borders. Slowly the spots may enlarge and develop straw-colored centers. Generally, they are raised on the lower surface of leaf. Severely affected leaves turn yellow and drop, and defoliation is very common. Fruit symptoms occur as raised, brown lesions that are wart-like in appearance. Narrow, elongated lesions or streaks may develop on stems, petioles, and fruits.

Early symptoms appear on the lower side of leaves. Initially water-soaked circular or irregular lesions occur that become necrotic with brown centers and thin chlorotic borders. Lesions on leaves grow up to 10 mm in diameter. They are sunken on the top surface and slightly raised on the bottom. Under favorable conditions these lesions coalesce and appear blighted and later turn yellow and drop prematurely. The lesions become light brown and rough in appearance. Fruit lesions initially begin as pale green spots, which enlarge and become brown in color. These spots are raised with a cracked, rough watery appearance up to 5 mm diameter. Wounds caused by wind, insects, mechanical injuries, and sand particles on leaves and fruits are vulnerable sites of infection.

The leaves exhibit small, circular or irregular, dark brown or black greasy spots. As the spots enlarge in size, the center becomes lighter surrounded by a dark band of tissue. The spots coalesce to form irregular lesions. Severely affected leaves become chlorotic and fall off. Petioles and stems are also affected. On the fruits, round, raised water-soaked spots with a pale yellow border are produced (Figures 9.3 through 9.5).[3]

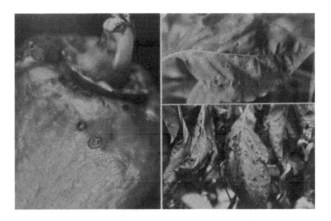

FIGURE 9.3   Bacterial leaf spot symptoms.

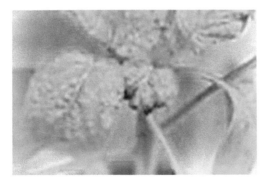

FIGURE 9.4   Small, yellow-green lesions on young leaves.

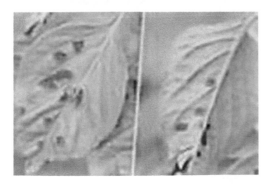

FIGURE 9.5   Dark, water-soaked, greasy-appearing lesions on older foliage.

9.2.3 Cause and Disease Development

Bacterial spot was first identified in South Africa as early as 1914. For almost half a century, a single bacterial species, classified as *Xanthomonas vesicatoria*, and later as *X. campestris* pv. *vesicatoria*, was considered the cause of bacterial spot of both pepper and tomato. In the early 1990s, it was shown that two distinct genetic groups (possibly species) existed within strains of pv. *vesicatoria*. In 1995, Vauterin et al. restructured the classification of the genus *Xanthomonas* by proposing species status for these groups: *X. vesicatoria* and *X. axonopodis* (syn. *campestris*) pv. *vesicatoria*. More recently, it has been shown that bacteria belonging to 4 distinct groups (previously designated A, B, C, and D) cause bacterial spot.

9.2.4 Favorable Conditions

The bacterium needs a temperature of 22°C to 34°C with high humidity for maximum infection. Infection takes place at a wide range of temperatures from 15°C to 35°C but the optimum is 24°C to 30°C. 40 to 50-day old plants are most susceptible to this disease.

9.2.5 Disease Cycle

The bacteria have a very limited survival period of days to weeks in the soil, and thus their survival is almost always in association with debris from infected or diseased plants. The pathogens have been reported to persist in association with roots of wheat as well as a few weed species; weeds, however, are considered to play only a minor role in pathogen survival (Figure 9.6).

9.2.6 Management

The disease can be managed using the following strategies:

9.2.6.1 Chemical Control

Seed can be treated with hot water or Clorox® bleach (calcium hypochlorite) to kill the pathogen. Hot-water treatment is more thorough than Clorox; however, high temperatures can adversely affect germination if proper precautions are not taken. It is best to have seed custom treated, which some seed companies will do. If you decide to do it yourself, treat at 50°C for 25 minutes. The best way to control temperature while treating seed is to use a stirring hot plate and a precision laboratory

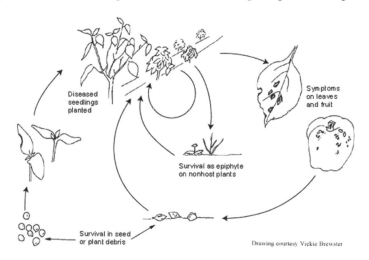

FIGURE 9.6   Disease cycle of bacterial leaf spot.

thermometer. Hot plates and thermometers can be purchased from a laboratory supply company such as Fischer Scientific (800-766-7000). Hot-water treatment can be done successfully using a large pot on a stovetop and a precision laboratory thermometer.

Clorox® Commercial Solutions Ultra Clorox germicidal bleach is labeled for pepper seed treatment (EPA Reg. No 67619-8). There is less chance of seed being damaged with bleach than with hot water; however, chemical controls such as Clorox are effective for pathogens on the seed surface only; hot-water treatment can kill bacteria inside as well as on the outside of seed. To Clorox treat seed, mix 24 oz product with 1 gallon of water to obtain a solution with 10,000 ppm available chlorine. Use one gallon of this solution per pound of seed. Put up to 1 pound of seed in a cheesecloth bag, submerge in this solution and provide continuous agitation for 40 minutes, rinse seed under running tap water for 5 minutes, then dry seed thoroughly on a paper towel in a location free of mice. Prepare a fresh batch of the dilute Clorox solution for each 1-pound batch of seed. The soak can stimulate germination, so if the seed is dried and held too long, germination will be reduced. To legally make this treatment, a label with this use must first be obtained from the Clorox company (800-446-4686) or by going to the following web site https://www.clorox.com/contact-us/.

Label instructions: Spray 3 g. of copper oxychloride + 200ppm Plantomycin

Mixed application of copper and mancozeb as spray in field along with 20ppm streptocycline

9.2.6.2 Biological Control

The second disease control strategy is to reduce the rate at which the disease develops in a planting. This can be accomplished by selecting resistant varieties, applying bactericides if necessary, and/or avoiding conditions that enable the pathogen to spread and multiply rapidly. An integrated management program for BLS is recommended to ensure effective control as bactericides or resistant varieties alone will not be sufficient when conditions are very favorable for BLS (hot and wet) or when less susceptible BLS races are present.

BLS resistant pepper varieties usually provide effective control of BLS and excellent yield. Resistant varieties have performed well in several experiments. For example, susceptible Merlin and North Star were severely infected by BLS and produced only 0.7–1.4 ton/A of marketable fruit, while resistant Boynton Bell produced 20.5 ton/A. All 7 resistant varieties tested on Long Island provided better control of BLS than a weekly preventive spray program of the copper fungicide/bactericide Kocide 2000 + Maneb 75DF (which was applied with a tractor-mounted boom sprayer delivering 100 gpa at 250 psi). These varieties yielded as well as the susceptible standard Camelot treated preventively. Applying Kocide + Maneb to the resistant varieties did not improve disease control or increase yield in these experiments. However, in Florida, where conditions can be very favorable for BLS, growers feel they need to use foliar sprays on resistant peppers.[4]

9.2.6.3 Cultural Control

Field sanitation is important. Also, seeds must be obtained from disease-free plants.

Warm, wet conditions are favorable for diseases caused by bacteria. Therefore, the irrigation method is an important consideration in managing BLS in the field. Overhead irrigation provides both a means for the pathogen to spread and favorable conditions for disease development, therefore trickle irrigation is recommended. In addition to movement by splashing water drops, the pathogen can be spread through any mechanical means imaginable when plants are wet, including on worker's hands and on machinery such as cultivators. If possible, avoid working fields when the plants are wet and work infested areas last. Disinfect machinery used in infected sections of the field after the job is completed.

Low nitrogen or potassium, and extra high magnesium and calcium levels have been associated with increased crop susceptibility to BLS. Pepper crops that show visible signs of nitrogen deficiency (light-colored leaves) have been severely affected by BLS in Connecticut. Researchers have also found that BLS is more severe on pepper plants grown in soils adjusted with dolomitic lime, which is high in magnesium, than plants grown in soils adjusted with Cal limestone ($CaCO_3$). Maintain nutrients at the proper levels (moderate to high) to help plants resist infection.

9.3 CERCOSPORA LEAF SPOT OF CHILI

The leaf spot disease is seen both in nursery and main fields. The disease causes loss due to defoliation and reduction in photosynthesis.

9.3.1 CAUSAL ORGANISM

Cercospora leaf spot is mostly caused by the fungi mentioned in the table below.

Species	Associated Disease Phase	Economic Importance
Cercospora capsici	Leaves, Fruits, and Flowers	High

9.3.2 SYMPTOMS

The disease first manifests as small brownish spots on the leaves and gradually develops into the big, circular grayish spots with a whitish center. Later they form large lesions due to the coalescing of the spots. Infection on fruit stalks and calyxes is also very common in severe cases.

Small, circular water-soaked brown spots are observed on the leaves, and these spots later develop a light center surrounded by dark band. The spots frequently coalesce affecting large portions of leaves which wither and drop off. Defoliation of the leaves may often be due to this disease. The spots may be formed on the petiole, branches, and peduncle (Figures 9.7 through 9.10).

9.3.3 CAUSE AND DISEASE DEVELOPMENT

Cercospora leaf spot is caused by *C. capsici*. Stromata are well developed. Conidiophores are 30 to 60 x 4.5 to 5.5 μm. Conidia are subhyaline to colored, acicular to obculate and 75 and 125 x 4 to 5 μm.

9.3.4 FAVORABLE CONDITIONS

Mean temperature of 22.5°C–23.5°C, relative humidity of 77–85% and more sunshine hours (>5 hours per day) favors the disease. The number of rainy days favors the disease. Intermittent rains were also found to be favorable for disease development.

FIGURE 9.7   Initially brownish spots with yellow margin.

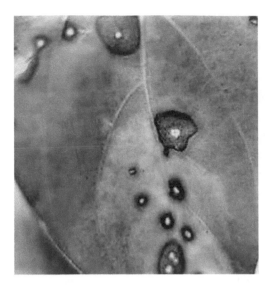

FIGURE 9.8   Brownish spots with whitish center.

FIGURE 9.9   Coalescing of spots.

FIGURE 9.10   Necrosis of leaves.

9.3.5 MANAGEMENT

The disease can be managed using the following strategies:

9.3.5.1 Chemical Control

Seed treatment with Captan 3g/kg of seed (UAS, Dharwad).

Spraying Copper hydroxide 77 WP @ 2g/l of water was found effective for the management of Cercospora leaf spot (UHS, Bagalkot).

Spraying with 1g Carbendezim or 2g Mancozeb/1lit (UAS, Dharwad).

Spraying with Mancozeb 2g/lit or Copper oxychloride @ 2.5g/lit (TNAU).

Spray Carbendazim @ 1g/lit or Mancozeb @ 2.5g/lit of water 2–3 times at 1-week intervals (POP, Andhra Pradesh).

9.4 PEPPER MOTTLE MOSAIC VIRUS (PMMV)

Pepper veinal mottle virus (PVMV) (genus *Potyvirus*, family *Potyviridae*, PPV) has been a major constraint to cultivation of the bell pepper (*Capsicum annuum* L.) in parts of Nigeria. The virus is efficiently transmitted non-persistently by aphids, which themselves are often difficult to control. The distribution of the aphid vectors was studied by trapping the aphids weekly on bell pepper between 2003 and 2005 from March through November within pepper producing areas of 6 agro-ecological zones in Nigeria using green water pan traps. The potential aphid vectors were identified, and the seasonal dynamics of the aphid species were described within the agroecological zones.

9.4.1 CAUSAL ORGANISM

Pepper Mottle Mosaic Virus (PMM) is mostly caused by the fungi mentioned in the table below.

Species	Associated Disease Phase	Economic Importance
PVMV (genus *Potyvirus*, family *Potyviridae*, PPV)	Leaves, Fruits, and Flowers	High

9.4.2 SYMPTOMS

Symptoms of infection on pepper include mottle and puckering of leaves and misshapen fruit. Heavily infected plants may be stunted, and fruit yields reduced. Foliar symptoms of pepper mild mottle virus (PMMV) consist of mottling and yellow/green mosaic, while fruit may be small, malformed, and mottled with sunken or raised necrotic spots. Yield loss is considerable when young plants become infected.

Symptoms of PVMV on pepper include mild mottle, mosaic, vein banding, ring spots, various types of necrosis, leaf discoloration, deformation, blistering, and severe stunting of the whole plant. Mild chlorosis and stunting occurs, especially if plants are infected when young. Fruits are small, malformed, mottled, and some have necrotic depressions.

Peppers infected by PVMV present varied symptom expression on leaves, stems, flowers, or fruit and can kill plants. The virus is efficiently transmitted by aphids, which themselves are difficult to control. About 5% of infections are due to primary spread by aphids acquiring the virus from a reservoir host. The other 95% of infections are believed to occur during the secondary cycle where the virus is spread by aphids that acquired the virus within plants. This study was undertaken to examine the occurrence and distribution of aphid pests of pepper that are responsible for transmission of PMMV on bell peppers (Figure 9.11).

 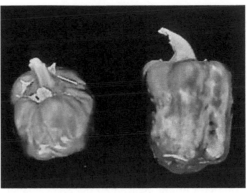

FIGURE 9.11   (a) Leaves develop a mild systemic mosaic and at times show crinkling. (b) Fruit develop severe symptoms such as distortion, rings, lines, and necrotic spots.

9.4.3   CAUSE AND DISEASE DEVELOPMENT

In the past the virus has been introduced into pepper growing areas on infected transplants, then spreads to nearby weeds which act as reservoirs for the virus. The virus is not seedborne. The virus is quite stable and highly infectious and is easily spread from plant to plant during normal crop maintenance. Also, the virus can persist in the previous crop in infected pepper debris such as leaves, stems, or roots in soil for several months.

PVMV are flexuous filaments containing a single molecule of linear, positive-sense, single-stranded ribonucleic acid (ssRNA), about 9.7–10 Kb in size, which has a poly (Λ) tract at the 3_ end and is covalently linked at the 5_ end to the virus-encoded VPg protein. PVMV share similar properties with other potyviruses, including induction of viral inclusion bodies and "pinwheels" in the cytoplasm of infected cells, but is serologically unrelated to several pepper potyviruses.

9.4.4   MANAGEMENT

The disease can be managed using the following strategies.

Use seed from healthy plants. If seed is suspected of infection with PMMV, then do the following: soak seeds in 10% trisodium phosphate (TSP) for 2.5 hours while stirring the seeds in the solution. Change the TSP once after 30 minutes. Rinse seeds thoroughly in tap water after the treatment to remove residues of TSP and spread seeds out to dry. The above procedure will significantly reduce levels of PMMV in external and internal portions of the pepper seed. Do not re-contaminate seed by placing them in used containers. Use a 1-year rotation to avoid continuous pepper cultivation. Keep the production area and seedbeds free of volunteer peppers that can serve as over-wintering hosts for the virus.

If growing seedlings and transplants in a greenhouse, then use steam-pasteurized soil in which plant debris has been allowed to thoroughly decompose since PMMV may be protected in thick pieces of root and stem refuse.

Avoid touching or handling pepper plants prior to setting them in the field. Remove diseased pepper seedlings that show mild mottle or mosaic symptoms on the foliage. Remove 1 or 2 plants adjacent to those plants that show symptoms. Do not touch other seedlings while discarding them. Dip hands in milk while handling plants every 5 minutes (more often if different lots of plants are handled). Rubber gloves will protect hands. Do not clip or damage young seedlings since this increases the possibility of mechanical transmission of the virus from contaminated tools or hands.

Remove diseased plants from the field as soon as virus symptoms are noticed. This will reduce the spread of the virus by direct contact between plants. Work in diseased portions of fields last, after working in healthy portions of fields. Cultural practices should be used that minimize contact with plants by workers and equipment or tools.

Disinfect tools, stakes, and equipment before moving from diseased areas to healthy areas. This can be done by: 1) soaking 10 minutes in a 1:10 dilution of a 5.25% sodium hypochlorite, do not rinse; or 2) by washing (enough to clean) in detergent at the concentrations recommended for washing clothes or dishes. Keep all solutions fresh. Hands and tools may be washed with soap or milk. Work in diseased areas after working in unaffected parts of a field. Wash clothing that comes into contact with ToMV/TMV-infected plants with hot water and a detergent.

9.5 PHYTOPHTHORA BLIGHT OF PEPPER/CHILI

Phytophthora blight, caused by the oomycete *Phytophthora capsici*, is a serious threat to the production of peppers worldwide. Phytophthora blight is a devastating disease on both bell and non-bell peppers. *P. capsici* was first described by Leonin in 1922 on chili pepper. The disease was subsequently reported in many pepper growing areas in the world. Phytophthora blight causes yield losses up to 100% in pepper fields in Illinois. *P. capsici* has a broad host range, among which cucurbits, eggplants, and tomatoes are severely affected in Illinois.

Other names applicable to this pepper disease are damping off and phytophthora root rot, crown rot, and stem/fruit rot. All of these names can apply since all parts of the pepper plant are affected. The disease has occurred sporadically in New York for more than 40 years, and has been responsible for serious losses in New Jersey, California, New Mexico, and Florida.

9.5.1 CAUSAL ORGANISM

Phytophthora blight is mostly caused by the fungi mentioned in the table below.

Species	Associated Disease Phase	Economic Importance
P. capsici	Leaves, Fruits, and Stems	High

9.5.2 SYMPTOMS

Damping off is primarily caused by *Pythium* spp. Seedlings affected by damping off fail to emerge or fall over and die soon after emergence. Stems usually have a dark, shriveled portion at the soil line. Damping off is generally limited to areas where drainage is poor or where soil is compacted, but whole fields can be affected, especially in early plantings exposed to rain.

Root and crown rot is primarily caused by *P. capsici*. Symptoms on affected pepper plants include rapid wilting and the death of pepper plants. Close examination of the roots and stems is necessary to confirm the cause of disease. The disease can develop at any stage of pepper plant growth. Taproots and smaller lateral roots show water-soaked, very dark brown discoloration of surface, cortical, and vascular tissues. Very few lateral roots remain on diseased plants and the tap roots may also be shorter compared to those of healthy plants. The most striking difference between healthy and diseased plants is the total amount of root tissue. Stems are usually infected at the soil line. Stem lesions first become dark green and water-soaked, followed by drying and turning brown. A lesion can girdle a stem, resulting in wilting of plants above the lesion and subsequent death.

Phytophthora blight of peppers can attack the roots, stems, leaves, and fruit depending upon which stage plants are infected. A grower not knowing what to expect might first encounter the disease at mid-season when sudden wilting and death occur as plants reach the fruiting stage. Early infected plants are quickly killed, while later-infected plants show irreversible wilt. Often a number of plants in a row or in a roughly circular pattern will show these symptoms at the same time.

Fungus-infected seedlings will damp off at the soil line, but relatively few plants die when temperatures are cool. Far more commonly, the disease will strike older plants, which then exhibit early wilting. Stem lesions can occur at the soil line and at any level on the stem. Stems discolor internally, collapse, and may become woody in time. Lesions may girdle the stem, leading to wilt above the lesion, or plants may wilt and die because the fungus has invaded the top branches before the stem lesions are severe enough to cause collapse.

Leaves first show small dark green spots that enlarge and become bleached, as though scalded. If the plant stems are infected, an irreversible wilt of the foliage occurs. Infected fruits initially develop dark, water-soaked patches that become coated with white mold and spores of the fungus. Fruits wither but remain attached to the plant. Seeds will be shriveled and infested by the fungus.

Because of the wide host range and the various phases at which plants can be infected, refer to the table for clarification of the crops affected and the Phytophthora species involved.

9.5.3 Cause and Disease Development

P. capsici is a heterothallic oomycete. The fungus occurs naturally in most soils and can infect pepper and other crops at most stages of growth when there is excess soil moisture and warm, wet weather. The fungus overwinters in soil as thick-walled oospores. For pepper, the fungus also survives on and in infested seed, but this is not a factor with commercially purchased seed.

Phytophthora species are soil-inhabiting pathogens that are favored by wet conditions. Although previously considered fungi, *Phytophthora* species are now considered to be in a separate classification called oomycetes. Species of *Phytophthora* produce resting spores that survive for years in moist soil in the absence of a suitable host.

9.5.4 Favorable Conditions

Disease initially occurs in low areas of fields where water accumulates, often leading growers to believe that stunting and death of the cultivar is due to waterlogging. *P. capsici* grows best at 27°C. It rapidly spreads in warm wet conditions. The asexual spore-bearing structures called sporangia are spread by irrigation water, drainage water, and rain. These indirectly germinate and release zoospores.

9.5.5 Disease Cycle

P. capsici is a soilborne pathogen. The pathogen produces several types of spores which enable it to spread throughout the field, and to persist in the field between crops. *P. capsici* survives between crops as oospores or mycelium in infected tissue. An oospore is a thick-walled sexual spore and is formed when mycelia of two opposite mating types (similar to male and female) grow together. Oospores are resistant to desiccation, cold temperatures, and other extreme environmental conditions, and can survive in the soil in the absence of the host plant for many years. Once pepper plants are transplanted into a field, and the environmental conditions are favorable, oospores germinate and produce sporangia and zoospores (asexual spores). Rainfall, soil saturation, and temperatures between 24°C–29°C (75°F–85°F) are necessary for development of Phytophthora blight. Zoospores, released in water, swim, and upon contact with host tissue, initiate infection. Following infection, a girdling lesion is formed at the base of the plant near the soil line. Sporangia are produced on the lesion surface and are spread by splashing rain. Upon landing on a pepper plant, sporangia release zoospores which initiate infection. Production and spread of sporangia are repeated throughout the season. Plants eventually die, and oospores formed within the lesions are released into the soil as the plant decomposes. Oospores will persist in the soil until another susceptible crop is planted.[5]

9.5.6 Management

The disease can be managed using the following strategies:

9.5.6.1 Chemical Control

The most effective way of preventing Phytophthora rot diseases is to provide good drainage and to practice good water management. Along with the appropriate cultural controls, the fungicide fosetyl-al (Aliette) may be used on many ornamental plant species to help prevent *Phytophthora* infections. When applied as a foliar spray it is absorbed by foliage and moves into roots. The fungicide mefenoxam has historically been used to control *P. capsici* on pepper. However, resistance of *P. capsici* to mefenoxam has been documented in the United States and in Michigan, which has necessitated the use of other fungicides.

9.5.6.2 Biological Control

Plant resistant varieties, whenever it is possible. 'Emerald Isle', 'Paladin', and 'Rainger' have moderate to high resistance against *P. capsici* isolates of Illinois. Also 'Arda' has been reported to be resistant to *P. capsici*. When growing resistant pepper varieties, implement the cultural practices.

9.5.6.3 Cultural Control

It may be possible to slow the spread of *Phytophthora* within an orchard by avoiding movement of infested soil, water, and plant parts from an area where Phytophthora rot has developed. Surface and subsurface drainage water and anything that can move moist soil can carry the pathogen to a new area, including boots, car tires, and tools. If the physical setting allows drainage water to flow from infested to uninfected areas within the garden during wet weather, consider putting in drains to channel the water away from healthy plants.

Select fields with no history of Phytophthora blight, if possible. Select fields that did not have peppers, cucurbits, eggplants, or tomatoes for at least 3 years. No rotation period has been established for effective management of Phytophthora blight in pepper fields in Illinois. Select fields that are well isolated from infested fields with *P. capsici*. Select well-drained fields. Do not plant the crop in the low areas or the areas that do not drain well. Clean farm equipment of soil between fields. Plant peppers on raised beds (a minimum of 9 inches high). Beds should be prepared prior to transplanting—do not transplant on flat culture and make the beds by cultivation. The beds must be made with a central crown for water to run-off during rainfall. Equipment must be adjusted properly to avoid making beds with depressions. In fields where low areas exist, beds should be broken, and drainage areas established. Maintain dome-shaped status of beds throughout the season. Do not leave a depression around the base of a transplant. Fill soil depression around the plants. Avoid excessive irrigation. Do not irrigate from a pond that contains water drained from an infested field. Scout the field for the Phytophthora symptoms, especially after major rainfall, and particularly in low areas. As Phytophthora symptoms become obvious, remove infected plants to reduce the number of spores produced. Do not save seed from a field where Phytophthora blight occurred.

9.6 POWDERY MILDEW OF CHILI

Powdery mildew disease commonly infects hot pepper plants and harms their growth. The disease looks exactly as the name describes it—like a powdery white coating on the plant's leaves. It is caused by the *Leveillula taurica* fungus, which is possible to control with fungicides. If left untreated, the disease can significantly slow plant growth and reduce pepper output. Gardeners should check pepper plants for mildew regularly, because immediate treatment helps remedy the disease much better than delayed treatment.

Powdery mildew causing heavy yield loss ranging from 14 to 20% is due to severe defoliation and reduction in photosynthesis, size and number of fruits per plant. This disease is very common from November to February.

9.6.1   CAUSAL ORGANISM

Powdery Mildew is mostly caused by the fungi mentioned in the table below.

Species	Associated Disease Phase	Economic Importance
L. taurica	Leaves, Fruits, and Flowers	High

9.6.2   SYMPTOMS

Powdery mildew usually appears on the lower surfaces of older plant leaves, just after peppers set. The white and powdery growth under the leaves often has a patchy appearance. The leaves might also turn yellowish or brown on the topsides above areas where the undersides have the white mildew. When an infection becomes severe, the powdery white mildew might also show up on the topsides of leaves. The leaves may begin to curl upward and eventually fall off. With fewer leaves, the plant will not shade peppers from the sun very well, resulting in sunburn damage to the pepper crop.

White powdery coating appears mostly on the lower surface. Sometimes the powdery coating can also be seen on the upper surface. Correspondingly on the upper surface yellow patches are seen. Severe infection results in the drying and shedding of affected leaves. Powdery growth can also be seen on young fruits, and branches. Diseased fruits do not grow further and may drop down (Figures 9.12 through 9.15).

FIGURE 9.12   Powdery growth on lower surface of leaves.

FIGURE 9.13   Powdery growth on upper surface of leaves.

FIGURE 9.14   Drying of leaves due to severe infection.

9.6.3   CAUSE AND DISEASE DEVELOPMENT

L. taurica is unique among powdery mildews because of its internal growth habit. The fungal hyphae grow inside the host tissue, rather than on the surface of the leaf tissue, as is common with powdery mildews that infect most other plants. The fungus typically infects the older leaves first and can be seen with the naked eye when masses of the spore-bearing structures (conid-iophores) extend out of the leaf tissue and appear as typical white, fuzzy powdery mildew growth.

9.6.4   FAVORABLE CONDITIONS

Powdery mildew grows best when temperatures are warm, and humidity climbs above 85%. Wind can spread the spores from one plant to another, as can rain splash and insects, to a lesser extent. Low relative humidity of a minimum of 50% and temperatures of 20°C–30°C are required.

FIGURE 9.15   Powdery growth on stalk of plant.

9.6.5   DISEASE CYCLE

The conidiophores are most often seen on the under sides of pepper leaves; however, when infection is severe, they can also be observed on the upper sides of the leaves. The spores of pepper powdery mildew, unlike most other powdery mildew types, contain sufficient water for growth, and as a result can germinate at relatively low humidities. For instance, they can germinate at relative humidities of 40% to 90%, with the optimum being 90%. The additional water that the spores contain makes them susceptible to bursting in the presence of free water in the environment from sources such as rain or sprinkler irrigation.

Pepper powdery mildew is typically observed at first harvest. However, during years of severe infection, it may be seen earlier in the growth cycle. If left uncontrolled, powdery mildew can affect the majority of the leaves on the plant. Severe infection causes yellowing of the leaves and subsequent defoliation of the plant, exposing pepper fruit to sunburn damage. Losses from sunburned fruit can range from 50% to 60%.

9.6.6   MANAGEMENT

The disease can be managed using the following strategies:

9.6.6.1   Chemical Control

Many gardeners choose to control powdery mildew disease on hot peppers with fungicides. Fungicides are most effective when applied as soon as the gardener notices an infection. Some pepper growers even use sulfur as a preventative fungicide in areas that have previously had problems with powdery mildew. Sulfur and potassium bicarbonate are three fungicides that organic gardeners can use to get rid of the disease. Azoxystrobin, Myclobutanil, Quinoxyfen, and Trifloxystrobin are non-organic fungicides that work against powdery mildew. As always, it is important to carefully follow fungicide manufacturer instructions to avoid water pollution, damage to the plants, or health risks from the chemicals.

Sulfur is a better preventive fungicide. Five pounds of sulfur per acre, applied on a 10-day spray application schedule provided significantly greater control of pepper powdery mildew. Triadimefon (Bayleton) and myclobutanil (Rally) provided excellent control of powdery mildew. Bicarbonate materials are active on several species of powdery mildew. However, ammonium and potassium bicarbonates had limited efficacy on pepper powdery mildew and also caused phytotoxic effects such as brittleness of the leaves and marginal leaf burn.

9.6.6.2   Biological Control

AQ-10 is a fungus (*Ampelomyces quisqualis*) that parasitizes and kills powdery mildew. Applications of this material provided limited control early in the season, but efficacy quickly declined. HV-12 is a long-podded sweet pepper from France that is resistant to powdery mildew. Variety 94-128, a hybrid cross between HV-12 and a commercial bell pepper line, is slightly susceptible to pepper powdery mildew.

The National Gardening Association has endorsed milk as a natural powdery mildew remedy. Although not as traditional as fungicides, research shows that a solution of one part milk and nine parts water reduces powdery mildew on a variety of crops. Natural horticultural oils, such as vegetable oil and neem oil, also reduce problems with powdery mildew and insect infestations. Natural horticultural oil sprays are often available at gardening stores, and gardeners can also rub oil directly on plant leaves (Table 9.1).

9.7   VERTICILLIUM WILT OF CHILI

Verticillium wilt is a serious disease of a large number of diverse plants. The causal agents *Verticillium albo-atrum* (Reinke and Berthold) and *V. dahliae* (Kelb) are ubiquitous, soilborne pathogens. The disease incidence and severity vary from year to year and from one location to another. The disease significance also varies with host susceptibility, pathogen virulence, soil type, and environmental conditions.

9.7.1   CAUSAL ORGANISM

Verticillium wilt is mostly caused by the fungi mentioned in the table below.

Species	Associated Disease Phase	Economic Importance
V. albo-atrum (Reinke and Berthold) and *V. dahliae* (Kelb)	Leaves and Stems	High

9.7.2   SYMPTOMS

The first symptoms on chili peppers are stunting and a slight yellowing of the lower foliage. As the disease progresses, excessive yellowing and shedding of leaves may occur. Symptom severity depends highly on soil and air temperatures and nutrient availability. The fungus invades the xylem elements and disrupts water transport. As the disease develops, varying degrees of vascular discoloration may occur, and the plant begins to wilt because of water stress. Infected plants may recover at night for a few days before permanent wilting and death occur.

Verticillium dahliae can infect pepper plants at any growth stage. Symptoms include yellowing and drooping of leaves on a few branches or on the entire plant. The edges of the leaves roll inward on infected plants, and foliar wilting ensues. The foliage of severely infected plants turns brown and dry. Growth of pepper plants inoculated with aggressive strains of *V. dahliae* in greenhouse or of pepper plants infected early in the season under field conditions is severely stunted with small leaves that turn yellow-green. Subsequently, the dried leaves and shriveled fruits remain attached to plants that die. Brown discoloration of the vascular tissue is visible when the roots and lower stem of a wilted plant are cut longitudinally. Another important soilborne disease of pepper in California phytophthora root rot, causes similar foliar symptoms; however, phytophthora root rot causes extensive browning and rotting of the root cortex, while the roots of *V. dahliae*-infected pepper plants show no external discoloration or decay.

Symptoms of verticillium wilt vary by host and environmental conditions. In many cases, symptoms do not develop until the plant is bearing flowers or fruit or after periods of stressful hot, dry weather. Older leaves are usually the first to develop symptoms, which include yellowing, wilting, and eventually dying and dropping from the plant. Infected leaves can also develop pale yellow blotches on the lower leaves and necrotic, V-shaped lesions at the tips of the leaves (Figures 9.16 through 9.18).

TABLE 9.1

A Summary of Registered Fungicides and Label Information (Please Adhere to Product Label Instructions When Using Each Chemical)

Product	Chemical/Biological Ingredient	Chemical Group	Mode of Action	REI[a]	PHI[b]	Application
Actinovate SP	*Streptomyces lydicus*	Biological	Preventative/Suppressive	1 hr	0 days	Use preventatively; Apply at 7–14 days interval; Use the product within 4 hrs of preparation
Bartlett microscopic sulphur	sulfur	M	Suppressive	24 hrs	NA	Use preventatively; Do not exceed 10 applications per crop cycle; Apply at 14 days interval
MilStop	Potassium bicarbonate	NC	Preventative, non-systemic	4 hrs	0 days	Use preventatively; Apply at 7 days intervals; Treated produce cannot be exported to the USA
Nova 40W	myclobutanil	3	Preventative/Some curative action, Locally systemic	12 hrs	3 days	Use preventatively; Do not exceed 3 applications per crop cycle; Apply at 10–14 days interval
Pristine	boscalid + pyraclostrobin	7 & 11	Preventative/Some curative action, locally systemic	12 hrs	1 day	Use preventatively; Do not exceed 1 application per crop cycle, Hence, use in rotation, after 7 days, with other fungicides

NA – information is not available (please refer product label & contact the manufacturer).

[a] PHI - pre-harvest interval.

[b] REI - re-entry interval.

FIGURE 9.16   Early symptoms of verticillium wilt in chili peppers.

FIGURE 9.17   Defoliation caused by verticillium.

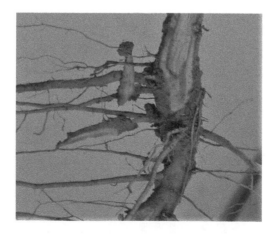

FIGURE 9.18   Vascular discoloration exhibited by chili pepper infected with Verticillium.

9.7.3 Cause and Disease Development

V. dahliae and *V. albo-atrum* are incredibly versatile fungi in their ability to cause disease on a wide range of diverse plant species over a large geographic area. However, these fungi exist in different races or strains, which vary in virulence and host range. Most isolates of both species can infect a number of different crop plants and weeds but a few isolates of *V. dahliae*, including the isolates from chili peppers, are largely host-specific or have unique host ranges. Conversely, isolates from bell pepper generally can infect many different hosts. The only *Verticillium* isolates that are consistently unable to infect chili pepper are those from cotton and cabbage, but chili pepper isolates can infect cotton. Strains isolated from the same host may vary in their abilities to cause disease in the host (pathogenicity). For example, some isolates from tomato infect peppers, while others do not. Additionally, isolates from the same host may vary in pathogenicity on the originating host.

9.7.4 Favorable Conditions

Environmental conditions that favor disease are similar for both *Verticillium* spp., although *V. dahliae* is a somewhat warmer-temperature pathogen (optimum 25°C) than *V. albo-atrum* (optimum 21°C). The chili pepper isolate, specifically, is favored by soil temperatures of 29°C to 35°C. Both pathogens require moisture for growth and development, but *V. dahliae* appears to tolerate dry conditions better than *V. albo-atrum*.

9.7.5 Disease Cycle

When temperature and moisture are favorable for pathogen growth, root exudates of susceptible plants stimulate microsclerotia to germinate. When roots of a host crop come near the resting structure (about 2 mm), root exudate promotes germination and the fungi grows out of the structure and toward the plant. Being a vascular wilt, it will try to get to the vascular system on the inside of the plant, and therefore, must enter the plant. Natural root wounds are the easiest way to enter, and these wounds occur naturally, even in healthy plants because of soil abrasion on roots. *Verticillium* has also been observed entering roots directly, but these infections rarely make it to the vascular system, especially those that enter through root hairs.

Once the pathogen enters the host, it makes its way to the vascular system, and specifically the xylem. The fungi can spread as hyphae through the plant, but can also spread as spores. *Verticillium* produce conidia on conidiophores and once conidia are released in the xylem, they can quickly colonize the plant. Conidia have been observed traveling to the top of cotton plants, 115 cm, 24 hours after initial conidia inoculation, so the spread throughout the plant can occur very quickly. Sometimes the flow of conidia will be stopped by cross-sections of the xylem, and here the conidia will spawn, and the fungal hyphae can overcome the barrier, and then produce more conidia on the other side.

A heavily infected plant can succumb to the disease and die. As this occurs, the *Verticillium* will form its survival structures and when the plant dies, its survival structures will be where the plant falls, releasing inoculates into the environment. The survival structures will then wait for a host plant to grow nearby and will start the cycle all over again.

Besides being long lasting in the soil, *Verticillium* can spread in many ways. The most common way of spreading short distances is through root-to-root contact within the soil. Roots in natural conditions often have small damages or openings in them that are easily colonized by *Verticillium* from an infected root nearby. Airborne conidia have been detected and some colonies observed, but mostly the conidia have difficulty developing aboveground on healthy plants. In open channel irrigation, *V. dahliae* have been found in the irrigation ditches up to a mile from the infected crop.

Without fungicidal seed treatments, infected seeds are easily transported and the disease spread, and *Verticillium* has been observed remaining viable for at least 13 months on some seeds. Planting infected seed potatoes can also be a source of inoculum to a new field. Finally, insects have also been shown to transmit the disease. Many insects including potato leaf hopper, leaf cutter bees, and aphids have been observed transmitting conidia of *Verticillium* and because these insects can cause damage to the plant creating an entry for the *Verticillium*, they can help transmit the disease.

9.7.6 MANAGEMENT

The disease can be managed by using the strategies mentioned below.

The verticillium wilt fungi are difficult to control. Their ability to survive in the soil for long periods with or without a host plant and the colonization of the water-conducting tissues within a plant limit any scheme to eradicate the pathogens. The first effort to manage *Verticillium* starts with proper diagnosis. Only laboratory culturing of infected plant material can positively identify *Verticillium* as the causal agent. Similar symptoms are produced by other pathogens. When *Verticillium* has been identified, several measures can be taken to reduce the effects of the disease in nurseries, fields, and landscape plantings. While various fungicides have been tested for application directly to plants, none have been found practical for continued use. An exception is the use of benomyl (Benlate 50 WP) fungicide as a root dip for transplanted seedlings. Benomyl, however, is a temporary measure and will not protect the plant after new roots emerge and colonize the untreated soil.

Chemical control of verticillium wilt has been shown to be economically practical in strawberry beds, in small vegetable or flower beds, and in soil in greenhouse benches. A common procedure is to treat the soil with a soil fumigant. These chemicals also will control weeds, insects, and nematodes in the soil. Fumigation is usually done not by the grower but by commercial applicators who are licensed to handle restricted chemicals. Prevention of the disease and the use of resistant varieties or cultivars are perhaps the best methods for controlling verticillium wilt.

The following are some suggested recommendations:

1. Steam the soil used for potted plants or for bench crops in the greenhouse and nursery at 82°C (180°F) for 30 minutes or 71°C (160°F) for 1 hour.
2. Do not grow susceptible plants on land where crops previously have been killed by verticillium wilt. For vegetables, flowers, and field crops, rotations of 5 years or more may help to reduce the amount of infection. Only non-host crops should be used in the crop rotation cycle.
3. Control weeds that can act as inoculum reservoirs in and around planting sites. Common weed hosts include ground cherries, lamb's-quarter, pigweed, horse nettles, and velvet leaf.
4. Fertilize to promote vigorous growth and maintain a balance of nitrogen, phosphorus, and potassium. Fertilizing can help reduce symptoms in nursery, field, and landscape plantings. Apply a fertilizer containing ammonium sulfate following the suggestions in a soil test report. Affected trees and shrubs should be fertilized and watered as soon as possible after initial wilt symptoms are exhibited. For a quick response, the fertilizer should either be injected into the soil in liquid form or be applied to the soil surface and watered in. Ammonium sulfate can be applied at the rate of 29 pounds per 1,000 ft². Water well immediately after application.
5. Water trees and shrubs that show symptoms every 10 to 14 days during dry periods of the growing season, applying 1 to 2 inches (600 to 1,200 gallons per 1,000 ft²) each time.
6. Destroy dead plants in nurseries or flower beds, removing as much of the root system as possible.
7. Branches or entire trees with recent wilt symptoms should not be removed immediately. They may recover in response to watering and fertilizing (see 4 and 5 above). Dead branches on trees should be removed. Cut well below the area of internal discoloration. This wood should not be chipped and used as a much as it may spread the fungus to other

plantings. Pruning tools should be disinfected by swabbing them with 70% rubbing alcohol after working on an infected plant.

8. Plant only resistant species, varieties, or cultivars where verticillium wilt is a problem.[6]

REFERENCES

1. Than, P. P., Prihastuti, H., Phoulivong S., 2008, *"Chilli anthracnose disease caused by Colletotrichum species"*, *Journal of Zhejiang University Science B* pp. 764–78.
2. Sarah, E., Perfect, H., Hughes, R., 1999, *"Colletotrichum: A Model Genus for Studies on Pathology and Fungal–Plant Interactions"*, School of Biological Sciences, University of Birmingham, Birmingham, United Kingdom.
3. TNAU Agritech Portal, *"Bacterial leaf spot: Xanthomonas campestris pv. vesicatoria"* http://agritech. tnau.ac.in/crop_protection/chilli_diseases_4.html
4. Boucher T. J., 2012, *"Managing Bacterial Leaf Spot in Pepper"*, IPM, University of Connecticut. http://ipm.uconn.edu/documents/raw2/Managing%20Bacterial%20Leaf%20Spot%20in%20Pepper/ Managing%20Bacterial%20Leaf%20Spot%20in%20Pepper.php?display=print
5. Report on Plant Disease, 2001, *"Phytophthora Blight Of Pepper"*, Department of Crop Sciences, University of Illinois at Urbana Champaign.
6. Report on Plant Disease, 1997, *"Verticillium Wilt Disease"*, Department of Crop Sciences, University of Illinois at Urbana Champaign. *International Journal of Systematic Bacteriology*, Vol. 45, No. 3, pp. 472–489, 1995.

10 Cucurbits

Cucurbits belong to a plant group with the most species used as human food in the Cucurbitaceae family. Within this family, the genus *Cucurbita* stands out as 1 of the most important. The cucurbits, also called Cucurbitaceae and the gourd family, are a plant family consisting of about 965 species in around 95 genera mainly in tropical and subtropical regions. The plants in this family are grown around the tropics and in temperate areas, where those with edible fruits were among the earliest cultivated plants both in the Old and New Worlds. The Cucurbitaceae family ranks among the highest of plant families for number and percentage of species used as human food.

Most of the plants in this family are annual vines, but some are woody lianas, thorny shrubs, or trees (Dendrosicyos). Many species have large, yellow or white flowers. The stems are hairy and pentangular. Tendrils are present at 32°C to the leaf petioles at nodes. Leaves are exstipulate alternate simple palmately lobed or palmately compound. The flowers are unisexual, with male and female flowers on different plants (dioecious) or on the same plant (monoecious). The female flowers have inferior ovaries. The fruit is often a kind of modified berry called a pepo. Despite the current marginalization of some of these species, all of the plants in this family have contributed essential food products to the diet of rural and some urban communities on the American continent and in many other parts of the world.

In this chapter, I have discussed some commonly occurring diseases affecting cucurbits and their management.

10.1 ANTHRACNOSE OF CUCURBITS

Anthracnose, caused by the fungus *Colletotrichum lagenarium*, can be a destructive disease of cucurbits during warm, wet growing seasons. The disease occurs worldwide. At least 3 races of *Colletotrichum* have been reported. The disease attacks watermelon, cantaloupe, cucumber, and gourds. Squash and pumpkin are almost immune.

This disease has been known since 1867 when it was first described in Italy. The bacterium causes symptoms on the leaves, stems, blossoms, and fruit. It is favored by warm, humid climates. It can cause fruit to drop or become spotted.[1]

10.1.1 CAUSAL ORGANISM

Anthracnose is mostly caused by the fungi mentioned in the table below.

Species	Associated Disease Phase	Economic Importance
C. lagenarium	Leaves, Fruits, and Stems	High

10.1.2 SYMPTOMS

Symptoms typically are much more common on watermelon than on cucumber or muskmelon. On leaves, the lesions are typically irregular and jagged in appearance. The centers of larger, older leaf lesions may fall out, which gives the leaf a "shot-hole" appearance. On cucumber and muskmelon, leaf lesions are less angular than those on watermelon leaves. Stem lesions are light brown and appear spindle-shaped. Muskmelon and watermelon fruit also may have anthracnose lesions that appear sunken and round, and may be orange or salmon-colored. Such lesions often start on the

FIGURE 10.1   Anthracnose lesions on muskmelon leaves.

lower surface of the fruit where moisture accumulates. It can cause leaf, fruit, and/or stem lesions. Foliage lesions are tan to brown except on watermelon foliage where they are dark brown to black. Fruit may develop brown to black, sunken, water-soaked spots. Tiny, black fruiting structures called acervuli form within the lesion. In wet weather, pink or orange spores ooze from these fruiting bodies. Symptoms often become severe when the plant canopy has developed sufficiently to provide a favorable environment for the fungus to infect (Figures 10.1, 10.2, 10.3 and 10.4).

10.1.3   CAUSE AND DISEASE DEVELOPMENT

Anthracnose is caused by the fungus *C. lagenarium.* The anthracnose fungus overwinters on diseased residue from the previous vine crop. The pathogen may also be carried on cucurbit seed.

FIGURE 10.2   Cucumber leaves with light brown lesions and a "shot-hole" appearance.

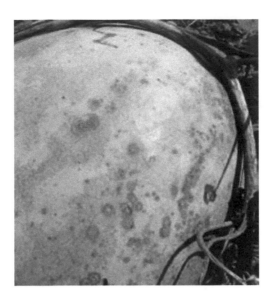

FIGURE 10.3 Watermelon fruit with severe foliar symptoms of salmon-colored anthracnose lesions.

FIGURE 10.4 Anthracnose lesions on watermelon stems tend to be light brown.

10.1.4 FAVORABLE CONDITIONS

The fungus depends on wetness and fairly high temperatures, 24°C (75°F) being considered optimum. Conidia do not germinate below 4.4°C (40°F) or above 30°C (86°F) or if they are not supplied with a film of moisture. Symptoms can appear within six days after infection has taken place.

10.1.5 DISEASE CYCLE

The *Colletotrichum* fungus overwinters in refuse from a previous vine crop (possibly up to 5 years) or in weeds of the cucurbit family. It may be seedborne and is also transmitted by cucumber beetles. Frequent rains accompanied by surface drainage water and temperatures around 23°C (75°F) favor the spread and buildup of the fungus. It may also be carried by workers. The fungus can penetrate leaves directly and does not require natural openings (e.g., stomates) or wounds. Initial infection requires a period of 100% relative humidity for 24 hours and a temperature of 20°C–23°C (68°F–75°F). Symptoms can appear within six days after infection has taken place. Spore-bearing structures (acervuli) break through the host surface and produce tremendous numbers of 1-celled, colorless spores (conidia), which exude from the acervuli in moist weather. The pinkish masses of conidia are formed on new lesions and serve as inoculum for secondary disease cycles.

Most damage from anthracnose generally occurs late in the season after the fruits are well formed. Post-harvest infections occur when fruit are wounded and washed with contaminated water. Several distinct races of the *Colletotrichum* fungus are known. Different kinds of host plants and crop cultivars may react differently to these races.

10.1.6 MANAGEMENT

The disease can be managed using the following strategies:

- Chemical control
- Biological control
- Cultural control
- Integrated pest management

10.1.6.1 Chemical Control

Chemical control can be obtained through a regular spray program of eradicant or protective fungicides. Coverage of leaf undersides and fruit is crucial to success.

Fungicides are listed here by active ingredient with examples of brand names in parentheses. (These brands are usually available to commercial growers.) On fungicide labels, the active ingredient is usually listed below the brand name. Chlorothalonil (Bravo, Evade, Echo); Mancozeb (Manzate 200 DF, Dithane DF, F- 45, Manex II); Maneb (Maneb 80, Manex); Thiophanate-methyl (Topsin M).

There are several fungicides labeled for the control of cucumber anthracnose in Florida. For conventional producers some of these include chlorothalonil, potassium bicarbonate, copper, pyraclostrobin, mancozeb, azoxystrobin, thiophanate-methyl, and *Bacillus subtilis* strain QST 713. For organic producers, some permitted fungicides include potassium bicarbonate, coppers, *B. subtilis* strain QST 713, and some horticultural oils. It is important for both conventional and organic producers to rotate fungicide chemistries to avoid the development of pathogen resistance.[2]

10.1.6.2 Biological Control

The use of resistant cultivars should be the first step in managing any plant disease and can greatly reduce yield losses due to anthracnose. Several seed companies offer cultivars with varying levels of resistance to this disease, which may be listed as anthracnose, *Colletotrichum orbiculare*, or "Co" on the seed package, and/or catalogue. Some anthracnose-resistant slicing cucumber cultivars include "Diamante", "Stonewall", and "Greensleeves". Pickling cucumber types that have resistance to anthracnose include "Cross Country", "Eclipse", "Feisty", "Fortune", "Spunky", and "Treasure".

Numerous plant growth-promoting rhizobacteria have been shown to effectively reduce the severity of anthracnose when applied to cucumber seeds before or at planting. This reduction is due to the induction of systemic acquired resistance (SAR) in the cucumber.

10.1.6.3 Cultural Control

1. Plant only certified, disease-free seed grown in a semiarid area in the western United States.
2. Rotate with crops other than cucurbits for 3 years or longer. Plant in well-drained soil, free from surface run-off water.
3. Avoid cultivating or handling plants when they are wet with dew or rain.
4. Control all weeds, especially wild and volunteer cucurbits.
5. Where feasible, collect and burn or plow down cleanly all infected plant debris after harvest.
6. Follow a rigorous, weekly spray program starting at the first true leaf stage. Thorough coverage of all plant parts must be accomplished if desired control is to be achieved. Try to time sprays just before rainy periods when infections occur. Follow the manufacturer's directions regarding amounts to use, the interval between the last spray and harvest, and compatibility between fungicides and insecticides.

TABLE 10.1

Fungicides and their Groups

Common Name (Trade Name)	Fungicide Group	Amount/Acre	Remarks
Chlorothalonil Bravo Ultrex, etc.	M5	1.4–1.8 lb	Do not apply more than 19.1 lb/acre/season.
Mancozeb (Dithane DF) (Dithane F-45) (Dithane M-45) (Penncozeb) 75DF	M3	1–2 lb 0.8–1.6 qt 1–2 lb Label rates	Labeled for cucumbers, melons, watermelon, and summer squash only.
Maneb (Maneb 75 DF) (Maneb 80) (Manex)	M3	1.5–2 lb 1.5–2 lb 1.2–1.6 qt	

7. To prevent post-harvest losses, avoid wounding (bruising, scratching, or puncturing) fruits. Immersing fruits in clean and fresh wash water containing 120 ppm of chlorine, aids in preventing new infections but does not eradicate previous infections.
8. Where practical, grow watermelon and cucumber cultivars with resistance to common races of the anthracnose fungus. (Table 10.1)

Group numbers are assigned by the Fungicide Resistance Action Committee (FRAC) according to different modes of actions (for more information, see http://www.frac.info/). Fungicides with a different group number are suitable to alternate in a resistance management program. In California, make no more than one application of fungicides with mode of action group numbers 1, 4, 9, 11, or 17 before rotating to a fungicide with a different mode of action group number; for fungicides with other group numbers, make no more than 2 consecutive applications before rotating to a fungicide with a different mode of action group number.

10.2 ALTERNARIA LEAF BLIGHT (SPOT) OF CUCURBITS

Alternaria leaf spot or blight of cucurbits or vine crops is caused by the fungus *Alternaria cucumerina*. The disease is widespread and often damaging after wet weather with temperatures between 20°C–32°C (68°F–90°F). Alternaria leaf spot is most severe on muskmelon and cantaloupe on sandy soils. The causal fungus also attacks summer and winter squashes, cucumber, pumpkin, vegetable marrow, watermelon, citron, and bur gherkin. Alternaria blight causes damage by defoliating the vines and reducing fruit yield, size, and quality. Even with only partial defoliation, the fruit may sunscald and ripen prematurely. This disease is usually first observed about the time of early fruit development. This disease was first reported in Italy in 1893 and has been reported in the United States since the early 1900s.

10.2.1 Causal Organism

Alternaria leaf spot or blight is mostly caused by the following fungal organism.

Species	Associated Disease Phase	Economic Importance
A. cucumerina	Leaves	Serious

10.2.2 Symptoms

Plants usually develop circular spots or lesions on the oldest crown leaves (Figure 10.5). The number of spots increases rapidly in warm, humid weather, later spreading to the younger leaves toward the tips of the vines (Figure 10.6). At first, the lesions are small, circular, and somewhat water-soaked or transparent. They enlarge rapidly until they are 1/2 inch or more in diameter, turning light brown on muskmelons (Figure 10.7), cucumber, and squash, and dark brown or black on watermelon when mature. Definite concentric rings may often be seen in the older, round to irregular spots, giving them a target-like appearance (Figure 10.8). Spots may merge, blighting large areas of the leaf. Muskmelons and cantaloupes are more susceptible than other cucurbits. The leaves commonly curl, wither, and fall prematurely. Vines may be partly or completely defoliated by harvest time. Often

FIGURE 10.5   Alternaria leaf blight on older leaves of muskmelon plants.

FIGURE 10.6   Alternaria leaf blight of melons.

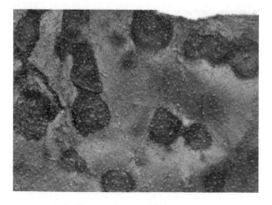

FIGURE 10.7   Close-up of Alternaria leaf blight of target-like spots.

FIGURE 10.8   Alternaria rot spots on 2 cucumber fruits. The black lesions are covered with the Alternaria fungus.

the spots become covered with a dark olive to black mold, which are the spores (conidia) of the *Alternaria* fungus.

Affected fruit may rot. On summer squash, rot starts at the blossom end. Fruit turn brown and shrink, later becoming black and mummified. The rot on muskmelon and cucumber fruit is often associated with sunscald injury or over-ripeness.

10.2.3   FAVORABLE CONDITIONS OF DISEASE DEVELOPMENT

Disease severity increases with the duration of leaf wetness periods from 2 to 24 hours over a range of temperatures from 12°C–30°C (54°F–86°F). Disease development is favored by frequent rainfall, which increases the relative humidity within the canopy, causes splash-dispersal of conidia, and increases the duration and frequency of leaf wetness periods.

10.2.4   DISEASE CYCLE

The mycelium of *A. cucumerina* is dark. In older diseased tissue it produces large numbers of short, simple, erect conidiophores usually with several prominent conidial scars. The large, multicellular conidia with 6–9 transverse cross walls, several longitudinal cross walls, and a long beak are usually borne singly on the conidiophores. The conidia are easily detached and can be found in the air and dust everywhere around cucurbit fields.

The *Alternaria* fungus overwinters as dormant mycelium in diseased and partly decayed crop refuse, in weeds of the cucurbit family and possibly in the soil. Fungus conidia can survive under warm, dry conditions for several months. Conidia produced on diseased plants or crop refuse may be blown by the wind for long distances. Clothing, tools, and other equipment and running and splashing water are other means of spread. The germinating spores penetrate susceptible tissue directly or through wounds and soon produce a new crop of conidia that are further spread by wind, splashing rain, tools, or workers.

At least 18 hours of high relative humidity, producing leaf wetness, is required before infection can occur. The period between infection and the appearance of symptoms varies from three to twelve days. Young plants less than a month old and plants that are bearing fruit and 70–75 days old appear to be more susceptible than plants 45–60 days of age. The *Alternaria* fungus is normally a vigorous pathogen only on cucurbit plants weakened by malnutrition, drought, insects, other diseases, a heavy fruit set, or other kinds of stress (Figure 10.9).

10.2.5   MANAGEMENT

The disease can be managed using the following strategies:

- Chemical control
- Biological control

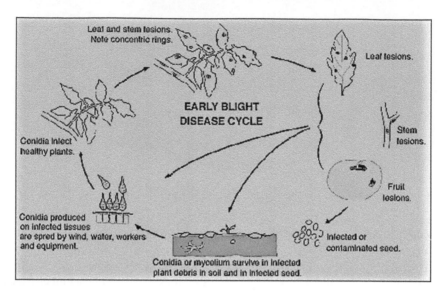

FIGURE 10.9   Disease cycle of alternaria leaf blight.

- Cultural control
- Integrated pest management

10.2.5.1   Chemical Control

Fungicide application is needed to limit disease development and spread.

Put Bravo Ultrex at 1.8–2.7 lb/A in water. This may be applied through sprinkler irrigation. It may also be applied on the day of harvest. There is a 12-hour re-entry.

Carboxamide (Group 7) formulations are registered for use. Do not make more than 2 sequential applications before alternating to a labeled fungicide with a different mode of action.

- Endura at 6.5 oz/A on 7- to 14-day intervals. Preharvest interval is 0 days. 12-hour re-entry.
- Fontelis at 12–16 fl oz/A on 7- to 14-day intervals. Preharvest interval is 1 day. 12-hour re-entry.

Copper products are not recommended as stand-alone materials.

- Cueva at 0.5–2 gal/100-gal water on 7- to 14-day intervals. This can be applied on the day of harvest. 4-hour re-entry.
- Cuprofix Ultra 40D at 1.25–2 lb/A on 5- to 7-day intervals. 12-hour re-entry.
- Liqui-Cop at 3 to 4 teaspoons/gal water.

Inspire Super at 16–20 fl oz/A on 7- to 10-day intervals. Do not make more than 1 application before alternating to a labeled fungicide with a different mode of action (non-Group 3 or 9). Preharvest interval is seven days. 12-hour re-entry.

JMS Stylet-Oil at 3–6 quarts/100-gal water. Do not spray if temperature is below 10°C or above 32°C or when plants are wet or under heat or moisture stress. 4-hour re-entry.

Manzate 75 DF at 2–3 lb/A. Do not apply within five days of harvest. 24-hour re-entry.

Strobilurin fungicides (Group 11) are labeled for use. Do not make more than 1 application of a Group 11 fungicide before alternating to a labeled fungicide with a different mode of action.

- Cabrio EG at 12–16 oz/A on 7- to 14-day intervals. Preharvest interval is 0 days. 12-hour re-entry.

TABLE 10.2

Fungicides Labeled for the Control of Alternaria Leaf Spot

Fungicide	Typical Application Interval	Examples of Trade Names
Azoxystrobin	7–14 days,	Quadris
Chlorothalonil	7–14 days,	Daconil, Terranil, Echo, Bravo, others
Copper products	7–14 days,	Basic Copper Sulfate, copper hydroxide (Kocide 2000 and others), Tenn-Cop 5E, copper resinate, Others
		Copper products may cause plant injury and reduce yields of watermelon and muskmelon
		Homeowner: Copper Fungicide, Bordeaux
Mancozeb	7–14 days	Dithane, Penncozeb, Manex

- Pristine at 12.5–18.5 oz/A on 7- to 14-day intervals. Preharvest interval is 0 days. 12-hour re-entry.
- Quadris Flowable at 11.0–15.4 fl oz/A or Quadris Opti at 3.2 pints/A or Quadris Top at 12–14 fl oz/A on 7- to 14-day intervals. Preharvest interval is 1 day. 4-hour re-entry for Quadris Flowable; 12-hour re-entry for Quadris Opti and Quadris Top.
- Reason 500 SC at 5.5 fl oz/A on 5- to 10-day intervals. Do not apply within 14 days of harvest. 12-hour re-entry.
- Trilogy at 0.5%–2%. Do not use above 32°C or when plants are under heat or moisture stress. Do not use when foliage is wet as good coverage is essential. 4-hour re-entry.

10.2.5.2 Biological Control
No cucurbit cultivar resistant to Alternaria leaf blight has been reported.

10.2.5.3 Cultural Control
- Rotate vegetables so that 3 or more years go by before planting any member of the squash family in the same location.
- Use drip irrigation instead of overhead sprinklers if possible.
- Do not work in plants when wet.
- Remove and destroy infected plants at the end of the season in small gardens (Table 10.2).

10.3 GUMMY STEM BLIGHT OF CUCURBITS

Gummy stem blight is an important disease of cucurbits in many parts of the United States. Gummy stem blight, caused by the fungus *Didymella bryoniae* previously known as *Mycosphaerella melonis*, is a common disease of cantaloupes, watermelons, and cucumbers.

Under conditions favorable to disease development, commercial growers and home gardeners may experience heavy losses. This disease can occur at any point in plant growth, from seedling stage to fruit in storage. Gummy stem blight is the name given to the disease when leaves and stems are infected. Muskmelon (cantaloupe), cucumber, and watermelon are most commonly affected by this phase of the disease. Black rot refers to the same disease on fruit; it is seen less often than the foliar phase. The fungus attacks all parts of the plant, and under favorable conditions, causes severe economic losses.

10.3.1 CAUSAL ORGANISM

Gummy stem blight is majorly caused by the fungi mentioned below.

Species	Associated Disease Phase	Economic Importance
D. bryoniae, previously named *M. melonis* and *M. citrullina*	Leaves, Fruits, and Stems	High

10.3.2 SYMPTOMS

10.3.2.1 Stems

Early symptoms consist of pale brown lesions at wounds created by removing leaves, fruits, and lateral shoots. These lesions become dotted and almost entirely covered with tiny black spore-producing fruiting bodies. They may also crack and exude a gummy amber-colored sap (Figure 10.10). Such lesions often occur at the base of the main stem, and if girdling by the lesions is complete, wilting and plant death will result. If seeds are contaminated with this pathogen, damping off will occur.

10.3.2.2 Fruits

Fruits can be affected internally and externally. Internal fruit rot is not externally visible and is characterized by a tapering of the blossom end (Figure 10.11) and discolored centrally located tissues (Figure 10.12). The brownish-black internal discoloration often extends for 1–2 cm along the length of the fruit. If the disease is more severe, the blossom end tapers even more directly to a point and becomes black due to the profusion of fruiting bodies externally (Figure 10.13). External fruit lesions appear as irregular circular spots that are initially yellow, then gray to brown. These lesions are soft, wet, sunken, and often contain spots of gummy exudate at the center. Symptoms become visible mostly during storage.

10.3.2.3 Leaves

Leaf symptoms are usually visible at the tips (Figure 10.14) as pale yellow or brown dead tissue, often with a yellow halo, extending backward in a V-shape (Figure 10.15). Sometimes the entire margin is affected, creating a brown edge and a downward cupping of the leaf. Lesions may also consist of circular spots on the leaves.

FIGURE 10.10 Stem cracks, amber-colored gummy droplets, and tiny black spore-producing bodies at base of cucumber stem.

FIGURE 10.11   Narrowing of blossom end of fruit due to gummy stem blight infection.

FIGURE 10.12   Internal discoloration and rotting of fruit is due to gummy stem blight.

FIGURE 10.13   Advanced external symptoms of gummy stem blight.

FIGURE 10.14   Early symptoms of gummy stem blight infection on leaf beginning at leaf tip.

10.3.3 Cause and Disease Development

The causal fungus (*D. bryoniae*) survives from season to season on infected crop debris and weeds. The causal agent is also seed- and transplant-borne. Infections occur when spores are carried by wind (ascospores) or splashing water (conidia) to susceptible tissues during moist weather.

10.3.4 Favorable Conditions

The ideal temperature range for disease development is between 16°C and 24°C. However, moisture is the most important factor for infection by the pathogen and subsequent spread. Leaf wetness is required for germination of spores, infection, and expansion of lesions. Frequent rains favor spore production and short-distance spread of spores.

10.3.5 Disease Cycle

D. bryoniae is highly resistant to dry conditions and it can survive as dormant mycelia or hardy structures called chlamydospores in undecomposed debris for up to 2 years. The fungus essentially

FIGURE 10.15   Advanced v-shaped symptoms of gummy stem blight infection on mature leaf.

produces 2 types of spores: conidia that may be dispersed by splashing water, and ascospores that are spread by air current. Both types of spores are also spread on fingers, knives, wet hands, and clothing.

Gummy stem blight develops in humid conditions and in free moisture on leaf surfaces. The most significant contributor to establishing the infection is how long plant surfaces remain wet. 1 hour of free water on leaves is sufficient for initial infection; however, continuous leaf wetness is required for subsequent expansion of lesions. Germination and spore production, development of symptoms on stems, and infection of cucumber leaves, petioles and flowers can occur over a wide range of temperatures (5°C–35°C), but optimum temperatures are 24°C–25°C.

Wounding is another factor that facilitates infection, particularly in older plant parts. In young plant tissue, however, wounding is not necessary for infection to occur. Wounding may result from damage due to crop activities (e.g. deleafing, fruit harvest) or guttation (exudation of plant sap from the ends of leaf veins and its evaporation, which can cause a localized buildup of salts) (Figure 10.16). Guttation is often seen in very humid greenhouses in the early morning.

However, water shortage before and after flowering increase internal fruit rot. Removing flowers from the fruit reduces infection but is time-consuming. Thus, a cultivar in which the flowers fall off early in fruit development would be advantageous. In addition, rapidly developing fruit are more capable of limiting the fruit rot process, compared to slowly developing fruit (i.e. less than 10 cm growth in 14 days).

10.3.6 MANAGEMENT

The disease can be managed using the following strategies:

- Chemical control
- Cultural control
- Integrated pest management

10.3.6.1 Chemical Control

Fungicides are listed here by active ingredient with examples of the brand names in parentheses. (These brands are usually available to commercial growers.) On fungicide labels, the active ingredient is usually listed below the brand name. Chlorothalonil (Bravo, Equus, Echo); Mancozeb

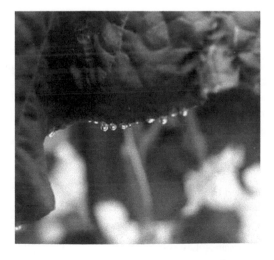

FIGURE 10.16   Guttation or exudation of droplets from ends of leaf veins.

(Manzate 75 DF, Dithane DF, F-45, Manex II); Maneb (Maneb 80, Manex); Thiophanate-methyl (Topsin M). Apply fungicides mancozeb and benomyl and use recommended rate. Apply fungicides at 7–14-day intervals. A minimum of 6 fungicide applications are required to manage gummy stem blight adequately on watermelon grown in disease-conducive environments.

10.3.6.2 Cultural Control

Implement good sanitation measures to reduce sources of the fungus.

- During and after crop production, remove all plant debris from the greenhouse area, and bury or compost it. *D. bryoniae* persists longer in debris left on the soil surface compared to those that are buried. This pathogen can survive for at least 10 months in cucumber stems buried in dry, nonsterile soil in a greenhouse, and for 18 months in dried, infected cucumber stems left on the soil surface. Such undecomposed crop residue becomes a source of airborne spores when it is wetted. Ascospores can be released 3 hours after wetting of infected plant material.
- In between crops, remove all plant debris, including plant tendrils on the wire, prior to washing and disinfecting the greenhouse interior and superstructures.
- Prevent or minimize periods of leaf wetness.
- Maintain a low relative humidity or moderate vapor pressure deficit (VPD) to decrease water condensation on fruit and leaves.
- Always cut fruit flush with the stem surface. Pulling damages both the stem and the fruit, giving easy access to the fungus.
- In a heavily infested crop, disinfect pruning knives frequently.
- Prune crops regularly. Remove wilted leaves and shoots which, if left on the plant, can provide nutrients to the fungus, leading to increased spore production and more infections.

10.3.6.3 Fungicides

Apply preventatively before the onset of disease or at the first sign of symptoms or under conditions (high RH, low light, and high inoculum levels) favorable for gummy stem blight (GSB). Use fungicides (Table 10.3) in rotation to avoid pathogen from developing resistance to a fungicide and prolong fungicide efficiency.

10.4 ANGULAR LEAF SPOT OF CUCURBITS

Angular leaf spot of cucurbits is caused by the bacterium *Pseudomonas syringae* pv. *lachrymans* and is the most common bacterial disease of cucurbits in the world.

The disease is more severe on cucumber, Zucchini squash, and honeydew melon but it also can infect muskmelon, cantaloupe, watermelon, other squashes, pumpkins, various gourds, vegetable marrow, West Indian gherkin, and bryonopsis. Losses in processing cucumbers can exceed 50% in wet seasons where control measures are not practiced. The bacterium causes symptoms on the leaves, stems, blossoms, and fruit. It is favored by warm, humid climates. It can cause fruit to drop or become spotted.

10.4.1 CAUSAL ORGANISM

Angular leaf spot is mostly caused by the fungi mentioned below.

Species	Associated Disease Phase	Economic Importance
P. syringae pv. *lachrymans*	Leaves and Fruits	High

TABLE 10.3

Fungicides to Avoid Pathogen from Developing Resistance

Product	Chemical/Biocontrol Ingredient	Chemical Group	Mode of Action	REI	PHI	Application
Nova 40W	myclobutanil	3	preventative, locally systemic	until dry	2 days	use preventatively; apply at 14-day intervals; do not exceed 6 applications per year or 2–3 applications per crop cycle
Pristine	boscalid + pyraclostrobin	7 & 11	preventative, locally systemic	until dry	0 days	do not exceed 1 application per crop cycle; use in rotation with other fungicides at 7–14 day intervals.
Rovral	iprodione	2	preventative, non-systemic	12 hrs	2 days	use preventatively; apply at 7 day intervals. Treated produce cannot be exported to the USA.
Rhapsody ASO	*Bacillus subtilis QST 713*	44	suppressive	NA	0 days	biofungicide; use preventatively before onset of disease or under low disease pressure at 7-10-day intervals.

10.4.2 Symptoms

Angular leaf spot affects aerial parts of cucurbits including leaves, petioles, stems, and fruits. Small, round to irregular, water-soaked spots appear on infected leaves. The spots expand until they are limited by larger veins, which give the spots an angular appearance. The spots on the upper leaf surfaces turn whitish gray to brown and die (Figure 10.17). On the lower leaf surfaces, the lesions are gummy and shiny. As these spots dry they shrink and commonly tear away from the healthy portions of the leaf, leaving large irregular holes (Figure 10.18). Therefore, infected foliage appears ragged and yellowish. Young, fully expanded leaves are more susceptible than older leaves. Under humid conditions, water-soaked spots are covered with a white exudate. The exudate dries to form a thin white crust on or adjacent to the spots under the leaf surface. On squash the brown leaf spots vary in size and are surrounded by a yellow halo. The tissue next to the halo may be water-soaked, especially on the lower leaf surfaces following damp weather. On watermelon the leaf spots begin as small, usually circular, dark lesions surrounded by a yellow halo. The center of the lesions may be white. The enlarging spots become angular and can involve entire lobes or even larger areas of a leaf.

The nearly circular, water-soaked spots on ripening cucumber fruit are much smaller than those on the foliage (Figure 10.19). When the infected tissue dies, the centers of the lesions become chalky white and may crack open. Infected fruit is frequently invaded by secondary fungi and bacteria.

FIGURE 10.17   Angular leaf spot symptoms on a squash leaf.

FIGURE 10.18   Close up of cucumber leaf infected by angular leaf spot.

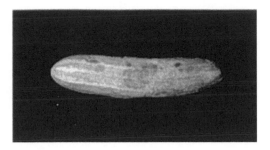

FIGURE 10.19   Early symptoms of angular root spot on cucumber fruit.

This fruit breaks down to form a slimy, foul-smelling rot. Lesions may develop after harvest when the fruit is in transit or storage. Droplets of bacterial ooze frequently appear at infection sites on leaves, stems, petioles, and fruit in very humid weather. The droplets dry to form a whitish crust. Fruits infected when they are young can become deformed and curved.

10.4.3   CAUSE AND DISEASE DEVELOPMENT

Angular leaf spot of cucurbits is caused by the bacterium *P. syrinage* pv. *lachrymans*, and is the most common bacterial disease of cucurbits in the world. The disease cycle begins when seedborne inoculum or bacterial cells in infested crop debris colonize cotyledons upon germination. The bacterium infects hosts through natural openings and wounds, multiplying in and on leaves. Windblown soil containing infested crop debris is an effective means of spreading the bacterium and disease. The pathogen can be disseminated within and among fields by irrigation water, splashing rain, insects, workers, contaminated equipment, and by wind as aerosols. *P. syrinage* pv. *lachrymans* survives between cucurbit crops in contaminated seed and infested crop debris.[3]

10.4.4   FAVORABLE CONDITIONS

Angular leaf spot is a warm weather disease. It is most active at temperatures of 24°C–27°C (75°F–80°F). Seedborne infections will usually appear shortly after emergence in the spring. Soilborne infections may appear at any time.

Infection and disease development are promoted by water-soaking of leaves which follows extended periods of rainfall, high relative humidity (95% or above), and a combination of warm, moist soil and cool nights followed by warm days. Two weeks of dry weather will stop disease development but a high temperature of 36°C (98°F) for five days will not. High nitrogen levels result in more severe disease.

10.4.5   DISEASE CYCLE

Little is known about either the life cycle of the pathogen or the disease cycle. The *Pseudomonas* bacterium is a seedborne pathogen. The pathogen can overwinter in infested crop residues. Upon germination of seed, the cotyledons become infected. The bacterium multiplies in the intercellular space of leaves and colonizes leaf surfaces. During warm rainy weather or sprinkler irrigation, the bacteria are splashed from infected seedlings to the foliage of healthy plants and later to the fruit. The bacteria are carried from plant to plant by splashing rain, by insects, on the hands and arms of pickers, and on farm machinery. Windblown sandy soil containing infested debris and irrigation water contaminated with the bacterium are effective in spreading the disease. If bacteria reach the developing seed, they can infect the seed coat.

10.4.6   Management

The disease can be managed using the following strategies:

- Chemical control
- Biological control
- Cultural control
- Integrated pest management

10.4.6.1   Chemical Control

Copper-based bactericides are often necessary to reduce the severity of angular leaf spot in warm, humid production regions. A 4- to 7-day spray interval is often necessary when conditions are highly favorable for disease development. Chemical controls are most effective when integrated with sound cultural control practices.

At the first sign of disease, apply fixed copper + maneb at label rates. Repeat sprays every 7 days.

10.4.6.2   Biological Control

Angular leaf spot can be managed by using disease-free seed. Some resistant varieties of cucumber are available. Crop rotation helps to lower inoculum levels.

10.4.6.3   Cultural Control

Plant pathogen-free seed or varieties that have tolerance. Plow under residue to lower inoculum levels. Rotate to non-host crops for 3 or more years. Limit overhead irrigation, and handling of wet foliage. Avoid excessive nitrogen fertilization. Schedule preventive weekly treatment with fixed copper compounds when conditions are favorable for disease development. Apply insecticides to protect from insect wounds. Minimize fruit wounds at harvest.

Common Name (Trade Name)	Fungicide Group	Amount/A	Remarks
COPPER HYDROXIDE 37.5%	M1	1–1.33 pt	Do Not Apply More Than 19.1 lb/A/season

Preharvest Interval (P.H.I.) (Table 10.4).

10.5   CERCOSPORA LEAF SPOT OF CUCURBITS

Cercospora leaf spot is caused by the fungus *Cercospora citrullina*. The disease is most damaging to watermelon, other melons, and cucumber.

10.5.1   Causal Organism

Powdery mildew is mostly caused by the fungi below.

Species	Associated Disease Phase	Economic Importance
C. citrullina	Leaves	High

10.5.2   Symptoms

Cercospora leaf spot is a foliar disease most commonly found on watermelons. Characteristic symptoms are small circular spots having dark green to purple margins, becoming white to light tan in

TABLE 10.4

Product List for Angular Leaf Spot

Pesticide	Product per Acre	Application Frequency (Days)	Remarks
		Copper Fungicides	
Champ Dry Prill	1.33 lb	5–7 days	May cause injury
Champ Formula 2	1.33 pt	5–7 days	May cause injury
Copper-Count-N	4-6 pt	7 days	May cause injury
Kocide 101	1.5–3 lbs	5–7 days	May cause injury
Kocide DF	1.5–3 lbs	5–7 days	May cause injury
Kocide 4.5LF	1–2 pts	5–7 days	May cause injury
Kocide 101	1.5–3 lbs	5–7 days	May cause injury
Kocide 3000	0.5–1.25 lb	5–7 days	May cause injury
Nordox	1.5–2.0 lb	5–7 days	May cause injury
Tri Basic Copper	2–4 pt	5–7 days	May cause injury
		Copper/EBDC	
Cuprofix MZ Disperss	4-7.25 lb	3–10 days	Maximum of 63.1 pounds per season; 5 days PHI
ManKocide	2.0-2.5 lb	7-10 days	Maximum of 128 pounds per season; 5 days PHI

the center. The leaf lamina around the spots may become chlorotic and eventually the entire leaf may turn yellow and fall off.

Cercospora leaf spot symptoms occur primarily on foliage, but petiole and stem lesions can develop when conditions are highly favorable for disease development. Fruit lesions are not known to occur. On older leaves, small, circular to irregular circular spots with tan to light brown lesions appear. The number and size of the lesions increase, eventually coalescing and causing entire leaves to become diseased. On cucumber, squash, and melon the centers of lesions may grow thin and fall out. Lesion margins may appear dark purple or black, and may have yellow halos surrounding them. Severely infected leaves turn yellow, senesce, and fall off. On watermelon, lesions often form on younger rather than older foliage. Cercospora leaf spot can reduce fruit size and quality, but economic losses are rarely severe (Figures 10.20 and 10.21).

FIGURE 10.20 Cercospora leaf spot causes brown necrotic lesions that may have gray centers on watermelon leaves.

FIGURE 10.21   On pumpkin leaves, Cercospora leaf spot causes light gray necrotic lesions.

10.5.3   Cause and Disease Development

Cercospora leaf spot is caused by the fungus *C. citrullina*. The fungus overwinters in crop debris and on weeds in the cucurbit family. The spores can be windblown or carried in splashing water. Free water on leaf surfaces is required for infection, which is favored by temperatures of 26°C–32°C. The disease progresses rapidly at these temperatures and infections of new leaves can occur every seven to ten days.

10.5.4   Favorable Conditions

The spores are airborne and may be carried great distances on moist winds. Infection requires free water and is favored by 26°C–32°C (80°F–90°F) temperatures. The disease develops quickly at these temperatures.

10.5.5   Disease Cycle

Cercospora leaf spot is caused by the fungus *C. citrullina*. The disease is most damaging to watermelon, other melons, and cucumber. The disease cycle begins when spores (conidia) are deposited onto leaves and petioles by wind or splashing water. Conidia germinate during moderate to warm (25.5°C–32°C) temperatures in the presence of free moisture. New cycles of infection and sporulation occur every seven to ten days during warm, wet weather. The pathogen is readily disseminated within and among fields by wind and splashing rain and irrigation water, and survives between cucurbit crops as a pathogen on weeds and in infested crop debris.

10.5.6   Management

The disease can be managed using the following strategies:

- Chemical control
- Biological control
- Cultural control
- Integrated pest management

10.5.6.1   Chemical Control

Fungicide sprays are necessary for disease control in wet, humid weather but are not required most years in the High Plains, USA. Spray sulfur, copper, or neem at first sign of the disease. (Cucurbits are very copper sensitive; test your varieties before spraying; use copper sprays very sparingly, and never on bright, sunny days with temperatures above 27°C–29°C).

Several fungicides are labeled for the control of Cercospora leaf spot including products with the active ingredient chlorothalonil (e.g., Bravo®, Echo®, Equus®), Cabrio®, Quadris®, and Inspire Super®.

10.5.6.2 Biological Control

No biological control practices have been developed for Cercospora leaf spot.

10.5.6.3 Cultural Control

Elimination of crop debris and cucurbit weed hosts is essential for Cercospora leaf spot management. Practice a 2- to 3-year rotation to no hosts. No resistance to Cercospora leaf spot resistance has been identified in commercial cucurbit varieties.

Grow resistant varieties. Soak seed in 50°C water for 25 minutes before sowing. Use a 3-year rotation between each susceptible vegetable crop (i.e. tomatoes, carrots, lettuce). Keep water off leaves. Increase air movement by staking if possible.

To encourage new leaf growth, use foliar fish fertilizer each time you irrigate (Table 10.5).

10.6   CHARCOAL ROT OF CUCURBITS

Charcoal rot is a soilborne root and stem disease of cucurbits that develops in the mid to late summer when plants are under stress, especially heat and drought stress. Charcoal rot is a root disease caused by the soilborne fungus *Macrophomina phaseolina*. Infected plants may die prematurely and are often wilted and stunted. Significant yield losses can occur. The disease is common in the southern states of America and occurs in the Midwest in seasons with hot, dry conditions. Charcoal rot was first confirmed in Minnesota in 1999 and North Dakota in 2002 and may be an expanding cucurbits disease in the northern Midwest of the United States.

10.6.1   CAUSAL ORGANISM

Species	Associated Disease Phase	Economic Importance
M. phaseolina	Stems and Roots	Medium

10.6.2   SYMPTOMS

Charcoal rot affects all cucurbits. First symptoms are the yellowing and death of crown leaves and water-soaked lesions on the stem at the soil line. As the disease progresses, the stem of infected plants ooze amber-colored gum, and the stem eventually becomes dry and tan-to-brown in color. The stem may be girdled by the lesion, resulting in plant death. Numerous microsclerotia, visible as black specks, are embedded in the dead plant tissue.

The charcoal rot fungus attacks roots, stems, and fruits. Stems develop basal cankers that girdle the stem, resulting in yellowing of foliage and eventual wilting and collapse of the entire plant. Initially the lesions are brown and may have amber-colored droplets on them (resembling gummy stem blight), but they later become light tan in color and are dotted with small, spherical dark colored fruiting bodies (microsclerotia). Fruit develop large sunken areas that are dark gray to black in color (Figure 10.22).

10.6.3   CAUSE AND DISEASE DEVELOPMENT

M. phaseolina is a soilborne fungus occurring in most soils in California. The fungus persists in soil as microsclerotia for 3–12 years and can infect 500 plant species. The pathogen most commonly

TABLE 10.5

Pesticides and their Quantity of Application

Pesticide	Product per Acre	Application Frequency (Days)	Remarks
Chlorothalonil and Chlorothalonil Mixtures			
Bravo 720	1.5-2 pt	7 days	Do not graze or feed debris to livestock; 7 day PHI
Bravo Ultrex	1.4-1.8 lb	7–10 days	Maximum of 16.5 pounds per season; 0 day PHI
Bravo WeatherStik	1.5-2.0 pt	7–10 days	Maximum of 20 pints per season; 0 day PHI
Echo 720	1.5-2.0 pt	7–10 days	Maximum of 2.5 gallons per season; 7 day PHI
Echo 90DF	1.2-1.6 lb	7–10 days	Maximum of 16.67 pounds per season; 7 day PHI
Echo Zn	2.2–2.8 pt	7–10 days	Maximum of 3.6 gallons per season; 7 day PHI
Ridomil/Bravo	1–2 lb	7–14 days	7 day PHI
EBDC, Copper/EBDC, and EBDC/Zoxamide Mixtures			
Cuprofix MZ Disperss	5–7.25 lb	3–10 days	Maximum of 63.1 pounds per season; 5 day PHI
Dithane	2–3 lb	7–10 days	Maximum of 25.6 pounds per season; 5 day PHI; use a nonionic surfactant to improve performance
Gavel 75DF 1.5	2.0 lb 7	10 days	Maximum 16 pounds per season; 5 day PHI; include a nonionic surfactant to improve performance
Maneb 75 DF	1.5–2.0 lb	7–10 days	Maximum of 17.1 pounds per season; 5 day PHI
Manex 80W	1.5–2.0 lb	7–10 days	Maximum of 16.0 pounds per season; 5 day PHI
Manex	2.4-3.2 pt	7–10 days	Maximum of 25 pints per season; 5 day PHI
ManKocide	2.0–2.5 lb	7–10 days	Maximum of 128 pounds per season; 5 day PHI
Penncozeb 80W	1.5–3.0 lb	7–10 days	Maximum of 24.0 pounds per season; 5 day PHI
Penncozeb 75DF	1.5–3.0 lb	7–10 days	Maximum of 25.6 pounds per season; 5 day PHI
Strobilurins and Strobilurin Mixtures			
Cabrio	12-16 oz	7–14 days	Maximum of 4 applications or 64 oz per season; Alternate with different modes of action; 0 day PHI
Quadris	11.0–15.4 fl oz	5–14 days	Maximum of 4 applications or 2.88 quarts per season; Alternate Quadris with fungicides with different modes of action; 1 day PHI

FIGURE 10.22 (a) Early stage of symptom. Water-soaked lesion at the base of the crown caused by *M. phaseolina*. (b) Advanced stage. Water soaking progresses and older parts of lesions become necrotic.

infects melon stems at the soil line within one to two weeks after planting, but the first disease symptoms occur late in the growing season, usually within one to two weeks of harvest.

The fungus is a stress pathogen and disease incidence increases with increases in water stress, a heavy fruit load, and high temperatures. Although severe charcoal rot is relatively uncommon in furrow-irrigated fields, ironically, its prevalence has increased with the use of buried drip irrigation systems. This may have occurred as a result of increased salt levels (stress) in beds, particularly at the soil surface. Additionally, disease incidence and severity is most common in fields cropped multiple times to melons.

10.6.4 FAVORABLE CONDITIONS OF DISEASE DEVELOPMENT

Disease development is favored by high temperatures and wet conditions. Plants can be infected at any time during the growing season. Much infection may occur early in the season, but symptoms typically do not develop until after flowering when plants become stressed. Hot, dry weather favors disease development. Disease is most severe where plants have been growing under conditions of stress or injury.

10.6.5 MANAGEMENT

No fungicides are labeled for control of charcoal rot on cucurbits.

Practice a 3- or 4-year crop rotation. Earlier planting will result in plants growing a larger shading canopy, thus, reducing soil temperatures and conditions conducive to the fungus. Avoid overcrowding. Provide sufficient levels of phosphorus and potassium. Maintain moisture, apply a mulch if necessary.

Start looking for charcoal rot during the vegetative growth stage, and note infections to make management decisions for the next crop. Rotation to a non-host crop for 2–3 years can be an effective disease management strategy in some crop production systems. However, avoidance of drought stress throughout the growing season is paramount to disease management. Leaching soil to reduce salinity levels, particularly at soil surface layers, may help reduce the incidence of disease in drip-irrigated fields. Furthermore, destruction of infected plant tissue before the pathogen reproduces at the end of the growing season will prevent a buildup of soil inoculum. The use of grafted transplants (i.e. susceptible scions grafted onto resistant cucurbit rootstock) has been proposed as an effective management strategy for the control of charcoal rot as well as many other soilborne root-infecting pathogens where the use of chemicals is not feasible. No preplant or postplant chemical control measures have been reported. Solarization is not promising for diseases favored by heat like charcoal rot.[4]

10.7 CHOANEPHORA WET ROT OF CUCURBITS

Choanephora Rot is otherwise called as the blossom end rot. This wet rot disease is very common in plants like squash and cucumbers. This disease occasionally occurs in cucurbits. This will include pumpkins and muskmelons.

10.7.1 CAUSAL ORGANISM

Powdery Mildew is mostly caused by the fungi mentioned below.

Species	Associated Disease Phase	Economic Importance
Choanephora cucurbitarum	Fruits	Low

10.7.2 Symptoms

Symptoms begin as a soft, wet rot of flowers and the blossom end of fruit. Infected fruits decay rapidly, becoming soft and watery. A profuse, fuzzy fungal growth with large masses of black spores forms on infected tissues. The pathogen's distinctive appearance (like numerous small black-headed pins sticking out of a pincushion) is diagnostic for this disease.

Fruits rot rapidly and fungal mold appears on the infected area. The fruit resembles a pin cushion with numerous small, black-headed pins stuck into it. Initially, the heads are white to brown but turn purplish black within a few days. Affected flowers, pedicels and immature fruit become water-soaked, and a soft wet rot develops. An entire fruit can rot in a 24- to 48-hour period. Symptoms begin usually on the blossom end of the fruit (Figures 10.23 and 10.24).

10.7.3 Cause and Disease Development

Choanephora rot is caused by the fungus *C. cucurbitarum*. The fungus survives from season to season in crop debris and is spread to new flowers by insects, splashing water, or wind. Infection most commonly occurs on flowers, although the fungi can also infect through wounds on the fruit. Infected flowers are soft, rotted, and quickly become covered with first white then purplish black fungal growth. In female flowers, the infection progresses into the fruit and results in soft water rot of the blossom end of the squash. The fungus thrives in wet conditions.

10.7.4 Favorable Conditions

Development of wet rot is favored by high relative humidity and excessive rainfall, warm (>25°C) and wet weather. Both blossoms and fruit are affected and fruit nearest the ground are more likely to

FIGURE 10.23   Choanephora fruit rot on yellow straight neck squash.

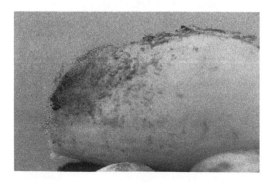

FIGURE 10.24   Pin-cushion symptom of wet rot on squash.

become diseased. It is not unusual to find 30–40% of blossoms and/or fruit infected with the fungus. While the disease is destructive, it is also as short-lived as the conditions that promote it. Subsequent fruit sets are usually not affected unless conducive conditions reoccur.

10.7.5 Disease Cycle

The fungus overseasons as a saprophyte (living on dead plant tissue) or in a dormant spore form. In spring, fungal spores are spread to squash flowers by wind and by insects such as bees and cucumber beetles; infection occurs through the blossom and spreads into the fruit and stem. Development of wet rot is favored by high relative humidity and excessive rainfall.

10.7.6 Management

Since this disease is caused by the fungus, 1 of the most sensible and the easiest way to destroy the disease is to use the fungicides. These fungicides need not be the broad-spectrum ones. They need to exclusively attack the fungus that grows on these plants. But there are 2 viewpoints to this method. 1 group considers this as an effective method and the other group feels that it is not an effective method as new flowers that are susceptible to this disease keep coming out each day. Apart from all these traditional methods, 1 of the best ways to prevent this disease is by following the practice of crop rotation. When you are trying to irrigate the plants, overhead irrigation is not needed to be done. When the plants are grown in an overcrowded state, this will give the disease a chance to progress. If you wish to avoid the disease and provide the plants with a healthy environment to thrive, then it is good to grow them apart. There must be proper spacing between the plants. A planning planting must be done to avoid the disease in the crop fields. When you plant them with a proper space, then there will be enough air circulation. The flowers and fruits will always be in dried condition. This will not give the fungus a chance to grow.

No effective control practices are available for wet rot. Fungicide sprays are impractical because new blossoms open daily and need to be protected soon after. Drip irrigation (beneath the foliage) watering may reduce development and spread of the disease during dry seasons.

10.8   CUCUMBER MOSAIC VIRUS DISEASE

Cucumber mosaic virus disease (CMV) attacks more than 40 families of plants worldwide, including all vine crops. Strains of CMV differ in their host range, symptoms, and method of transmission. Cucurbits are susceptible at any stage of growth.

This virus was first found in cucumbers (*Cucumis sativus*) showing mosaic symptoms in 1934, hence the name cucumber mosaic. Since it was first recognized, it has been found to infect a great variety of other plants. These include other vegetables such as squash, melons, etc.

10.8.1 Causal Organism

CMV is mostly caused by the fungi mentioned below.

Species	Associated Disease Phase	Economic Importance
CMV	Leaves and Fruits	High

10.8.2 Symptoms

When plants are vigorously growing, symptoms appear on the youngest leaves. Leaves have small yellowish areas, are curled slightly downward at the edges, and become puckered or crinkled with the tissue between the small veins becoming raised. As the leaf expands, it becomes distinctly mottled yellow and green. All future leaves will grow this way. As the leaves age, the puckering becomes more distinct and the mottling less distinct. Leaves are smaller than normal. Plants are severely stunted. The plant will produce little fruit.

When older, less actively growing plants are infected, the symptoms are less distinctive. Older leaves may look healthy for some time. A few leaves near the growing tip may turn yellow and wilt and have brown, withered edges. Occasionally, all the leaves of a shoot may appear this way. The oldest leaves may gradually die. Usually, portions of the leaf turn yellow, often starting at a lobe of the leaf or a v-shaped section of the leaf, at the edge. The yellow portions, and usually the entire leaf, quickly turn brown and die. The leaves usually remain attached to the vine.

Cucumber fruit may show yellow and green mottling or have dark green "warts" on pale green fruit. Cucumber fruit produced in the later stages of the disease is sometimes smooth and pale whitish green (called "white pickle") and more blunted at the ends than fruit produced on healthy vines. Watermelon, muskmelon, and winter squash fruit may be mottled and warty with raised areas lighter in color than surrounding tissue (Figure 10.25).

10.8.3 CAUSE AND DISEASE DEVELOPMENT

CMV is a plant pathogenic virus in the family Bromoviridae. It is the type member of the plant virus genus *Cucumovirus*. This virus has a worldwide distribution and a very wide host range.

CMV is made up of nucleic acid (ribonucleic acid, RNA) surrounded by a protein coat. Once inside the plant cell, the protein coat falls away and the nucleic acid portion directs the plant cell to produce more virus nucleic acid and virus protein, disrupting the normal activity of the cell. CMV can multiply only inside a living cell and quickly dies if outside a cell or if the cell dies.

10.8.4 FAVORABLE CONDITIONS

Symptoms develop more rapidly at 26°C–32°C (79°F–89°F) than at 16°C–24°C (61°F–75°F). The severity of symptoms is at least partially related to the virus concentration.

10.8.5 DISEASE CYCLE

CMV survives over winter in reservoir hosts. CMV is usually introduced into cultivated vine crop fields and gardens by more than 60 species of aphids (especially the green peach aphid, *Myzus persicae*) after they pick up the virus by feeding on reservoir hosts for a few seconds to a minute

FIGURE 10.25   CMV disease on cucumber leaf.

(non-persistent transmission). Primary infection can also occur from mechanical inoculation, especially in greenhouses on workers' hands and pruning knives when plants are handled, from use of infected seed, and from the feeding of virus-infected, striped, and 12-spotted cucumber beetles. Secondary spread of disease within a crop is usually the result of aphid vectors, but may occur when workers handle healthy plants after handling infected plants, and by the feeding of cucumber beetles. The cycle is completed when reservoir hosts are infected, usually by the feeding of virus-contaminated aphids, but sometimes from infected weed seed (especially chick-weed) or very rarely when seed of cultivated cucurbits is planted.

10.8.6 MANAGEMENT

The disease can be managed using the following strategies:

- Chemical control
- Biological control
- Cultural control
- Integrated pest management

10.8.6.1 Chemical Control and Biological Control

There is no chemical biological control that cure a CMV-infected plant, nor any that protect plants from becoming infected.

10.8.6.2 Cultural Control

- Plant varieties with resistance to these viruses whenever possible. The availability of varieties with virus resistance varies depending on the virus and the vine crop in question. Major seed producers list resistant varieties in their seed catalogs.
- Control weeds that could harbor the viruses from areas around production fields. Pokeweed is a primary weed in the Ohio area that harbors many plant viruses. Weeds also serve as hosts for insects that transmit the viruses.
- Since the virus is moved about by insects, it would seem logical that insect control would be a primary method of virus control. However, in this case it is not that simple. Since aphid vectors need only probe the plant to transmit the virus, insect control is not really that effective in controlling the virus. Even if the aphid were to die immediately following probing, virus transmission would still take place. To control the insect effectively, insecticides would need to be applied on a daily basis. The economics of this may not be feasible. Aphids also move long distances with weather fronts making local control difficult.
- When planting in the same fields in successive years, do not plant varieties of plants that are susceptible to CMV. Since CMV can overwinter in perennial plants and weeds, the virus can enter the roots and present itself at the top of the plant in the springtime where it can be retransmitted by aphids.

10.9 ROOT KNOT NEMATODE OF CUCURBITS

Root knot disease is caused by various species of *Meloidogyne*. It has long been considered "the nematode" disease by farmers and other plant growers because of the severe yield reduction and obvious root-galling symptoms that are caused by these pests.

Root knot nematodes (*Meloidogyne* spp.) are major pathogens of vegetables throughout the United States and the world, impacting both the quantity and quality of marketable yields. In addition, root knot nematodes interact with other plant pathogens, resulting in increased damage caused by other diseases.

10.9.1 CAUSAL ORGANISM

Root knot disease is mostly caused by the fungi mentioned below.

Species	Associated Disease Phase	Economic Importance
Meloidogyne hapla	Roots	High

10.9.2 SYMPTOMS

Infected plants are stunted and produce little or no fruit. They wilt conspicuously in warm weather. Knots or galls of infected roots are usually much larger and more numerous on greenhouse-grown crops than on field-grown crops.

Infections by root knot nematode cause decline in the host, and under some conditions, may kill the plant. Infected plants may be stunted and chlorotic, usually wilt easily, and are not productive. However, the extent of damage caused by root knot nematode infections varies with host, timing of infection, and cultural conditions. Root knot nematode infection is often easy to identify because of the swellings in roots that look like "knots". The swellings become large and easy to see on some hosts such as squash, but may be smaller and less conspicuous on others such as chili pepper. Multiple infections on 1 root result in a swollen, rough appearance. Root knot nematodes are very small and can only be observed using a microscope.

Root symptoms induced by sting or root knot nematodes can oftentimes be as specific as aboveground symptoms. Sting nematode can be very injurious, causing infected plants to form a tight mat of short roots, oftentimes assuming a swollen appearance. New root initials generally are killed by heavy infestations of the sting nematode, a symptom reminiscent of fertilizer salt burn. Root symptoms induced by root knot cause swollen areas (galls) on the roots of infected plants. Gall size may range from a few spherical swellings to extensive areas of elongated, convoluted, tumorous swellings, which result from exposure to multiple and repeated infections. Symptoms of root galling can in most cases provide positive diagnostic confirmation of nematode presence, infection severity, and potential for crop damage (Figures 10.26, 10.27 and 10.28).

10.9.3 CAUSE AND DISEASE DEVELOPMENT

Nematodes are microscopic round worms found in many habitats. Most are beneficial members of their ecosystems, but a few are economic parasites of plants and animals. There are several plant

FIGURE 10.26   Root knot nematode induced stunting and galling of cucumber seedlings.

FIGURE 10.27   Close-up view of root knot nematode induced galling of plant roots.

FIGURE 10.28   Root knot nematode (*Meloidogyne* sp.) induced galling of watermelon roots.

parasitic nematodes that cause problems on landscape and garden plants in Arizona. The most widespread and economically important are the root knot nematodes.

10.9.4   FAVORABLE CONDITIONS

Nematodes are most active in warm weather in moist, but well aerated, sandy soils in the presence of host plants. They are most abundant in the upper foot of soils, but will follow roots several feet deep. Root knot nematodes are not indigenous to soils of the Southwest of America and are not found in native desert plants unless they have been introduced. In adverse conditions, the eggs can persist in the soil for long periods of time ranging from months to years.

10.9.5   DISEASE CYCLE

Root knot nematodes are sedentary endoparasites in that the female no longer is motile once she has established a feeding site in the root. The female deposits single-celled eggs in a gelatinous mass at or near the root surface. Embryonation of the egg begins immediately and continues until a worm-shaped larvae hatches. This larvae is about 1/60 inch long and is in the second stage, having passed through 1 molt within the egg. It then migrates either into the soil or to a different location in the root.

The larva in the soil penetrates a suitable root by repeatedly thrusting its feeding structure, the stylet, into cells at the surface. After forcing its way into the root, the larva moves between and through cells to the still-undifferentiated conductive tissues. Within two or three days, the larva becomes settled with its head embedded in the developing vascular cylinder and begins feeding. The nematode then begins to grow in diameter, loses its ability to move, and matures.

While the nematode is maturing, it goes through two additional larval stages interspersed by molts. The only significant growth is in diameter, so the mature female is not much longer than the second-stage larva. Her body is now spherical or pear shaped, with a diameter of about 1/40 inch and a narrow neck. The male develops in the same way as the female, except that he reverts to the worm shape at the last molt. Males are not always required for reproduction and usually appear in high numbers only under adverse conditions. The cycle is completed when the female begins laying eggs. Her egg masses normally contain 300 to 500 eggs but may range from almost none under unfavorable conditions to as many as 2,000 under highly favorable conditions.

Roots begin to swell within a day after infection. Cells around the nematode are stimulated to multiply and enlarge abnormally in response to its salivary secretions. Important microscopic changes begin to occur in the conductive tissues. Walls of cells around the head of the nematode dissolve and cell contents are incorporated into an ever-enlarging multinucleate syncytium or giant cell. The nematode feeds upon the giant cells throughout the rest of its life. Continued enlargement of these cells, rapid multiplication of other cells, and growth of the nematode contribute to the developing root gall, which protects the maturing nematode from the outside environment. The conductive tissues no longer function properly. Translocation of water and nutrients is impeded and, as a result, top growth is affected adversely. The heavier the infection burden, the more stunting and chlorosis occur aboveground.[5]

The length of the life cycle and rate of population increase depend upon several factors, the most important of which are soil temperature, host suitability, and soil type. At 27°C (80°F), which is about optimum for most *Meloidogyne* species, 1 generation on a good host requires approximately 21–25 days, whereas at 19°C (67°F) at least 87 days are necessary. Thus, 3–6 generations are possible out-of-doors in Illinois, depending on location. Many more generations can occur indoors. The life cycle is lengthened on a less-suitable host. Sandy, organic muck and peat soils are more favorable for population build-up than are heavier clay soils (Figures 10.29).

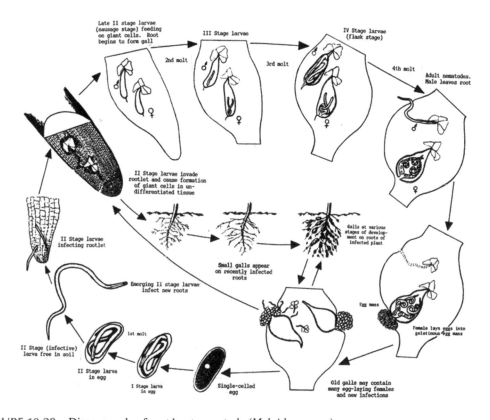

FIGURE 10.29   Disease cycle of root knot nematode (*Meloidogyne* sp.).

10.9.6 MANAGEMENT

The disease can be managed using the following strategies:

10.9.6.1 Chemical Control

Control of root knot nematodes through the use of chemicals is highly effective and practical, particularly on a field basis and where crops of relatively high value are involved. It may be the only alternative where crop rotation cannot be practiced or resistant varieties are unavailable. Both fumigant and nonfumigant chemicals are available for nematode control.

Chloropicrin has proved very effective against diseases but seldom nematodes or weeds. Telone (1,3-dichloropropene) is an excellent nematicide but generally performs poorly against weeds and diseases. Bacterial pathogens have not been satisfactorily controlled by any of the fumigants. Metam sodium and metam potassium can provide good control of weeds when placed properly in the bed; however, research to evaluate modification of rate, placement, and improved application technology have not resolved all problems of inconsistent pest control. Dimethyl disulfide (DMDS), the newest entry to registered fumigants in Florida, has demonstrated good to excellent control of nematodes, disease, and weeds when co-applied with chloropicrin.

All of the fumigants are phytotoxic to plants and as a precautionary measure should be applied at least three weeks before crops are planted. When applications are made in the spring during periods of low soil temperature, these products can remain in the soil for an extended period, thus delaying planting or possibly causing phytotoxicity to a newly planted crop. Mocap 15G at 13 lb/A in a band 12–15 inches wide on the row at planting. Do not use a seed furrow treatment or let seeds contact the product. 48-hour re-entry.

10.9.6.2 Biological Control

MeloCon WG at 2–4 lb/A at 4- to 6-week intervals for nematode suppression.

Use of resistant varieties is perhaps the best method of controlling root knot nematodes. However, these varieties are usually resistant to only 1 or 2 species of *Meloidogyne*.

Therefore, this method is limited to situations in which 1 or perhaps 2 *Meloidogyne* species are present. Resistance may not provide protection against even 1 species, since numerous intraspecific races and biotypes are known to exist in nature.

10.9.6.3 Cultural Control

The best way to prevent root knot nematode is to make sure that only clean materials are introduced into a planting site. Carefully examine all plants before they are planted. When adding soil and/or sand to a planting site, be sure it originated from a site that is not infested. If you are concerned that planting materials may be infested, have a sample analyzed by a professional laboratory or the nematology lab.

Starting plants directly from true seed also prevents introduction of root knot nematodes on plant material since they are not seedborne. Once soils are infested with root knot nematode, control is extremely difficult. Summer dry fallow will reduce the nematode populations in soils but will not eradicate them.

Soil treatment by solarization will also reduce nematodes. Solarization is a good option for small vegetable gardens and flower beds if treatments are applied according to recommendations. Addition of large amounts of compost give some control and enhances solarization. Soil removal and replacement in infested sites is a short-term option since it is usually impossible to remove all the nematodes. Avoidance, by planting only winter annuals and using a summer fallow, is an option for heavily infested sites.

10.10 DOWNY MILDEW OF CUCURBITS

Downy mildew caused by *Pseudoperonospora cubensis* is 1 of the most important foliar diseases of cucurbits. It occurs worldwide where conditions of temperature and humidity allow its establishment and can result in major losses to cucumber, melon, squash, pumpkin, watermelon, and other cucurbits. Downy mildew can begin to develop at any time during cucurbit crop development in the northeastern United States. Fortunately, it has occurred sporadically in this region, usually appearing late enough in the growing season that the yield is not impacted.

10.10.1 Causal Organism

Downy mildew is mostly caused by the fungi mentioned below.

Species	Associated Disease Phase	Economic Importance
P. cubensis	Leaves, Fruits, and Flowers	High

10.10.2 Symptoms

Downy mildew affects plants of all ages. Although the disease only infects foliage, a reduction in photosynthetic activity early in plant development results in stunted plants and yield reduction, especially in cucumber. Premature defoliation may also result in fruit sunscald due to overexposure to direct sunlight. Symptoms of downy mildew infection exhibit themselves differently on the various cucurbit crops.

Downy mildew symptoms first appear as small yellow spots or water-soaked lesions on the top-side of older leaves (Figure 10.30, left). The center of the lesion eventually turns tan or brown and dies (Figure 10.30, right). The yellow spots sometimes take on a "greasy" appearance and do not have a distinct border. During prolonged wet periods, the disease may move onto the upper crop canopy.

In cucumbers, the lesions are often confined by leaf veins and appear angular in shape (Figure 10.30). In cantaloupe crops, the lesions appear irregular shaped (Figure 10.31), whereas the lesions are smaller and rounder on infected watermelon leaves (Figure 10.32). As the disease progresses, the lesions expand and multiply, causing the field to take on a brown and "crispy" appearance.

Under humid conditions, the lesion often develops a downy growth on the underside of the light-yellow lesions observed on the top of the leaf. This downy growth is particularly noticeable in the mornings after a period of wet weather or when conditions favors dew formation. The downy growth on the underside of the lesions is frequently speckled with dark purple to black sporangia (spore sacks) that can be observed with a hand lens (Figure 10.33). The presence of the downy growth on the underside of the lesion is key to diagnosing this disease. Lesions are sometimes invaded by secondary pathogens such as soft rot bacteria or other fungi (Figure 10.34).

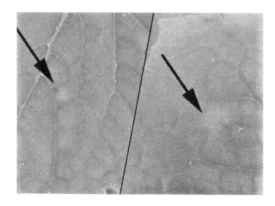

FIGURE 10.30 Small yellow "greasy" spots on the topside of leaves (left) are often the first symptom of downy mildew infection. The yellow spot eventually develops a tan brown color.

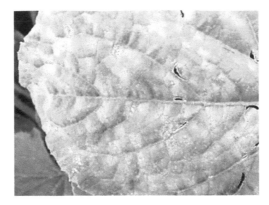

FIGURE 10.31   Expanding lesions on cucumber leaves are often restricted by leaf veins, giving the lesion an angular or square appearance.

FIGURE 10.32   Downy mildew lesions on the upper surface of melon leaves appear irregular shaped.

FIGURE 10.33   Downy mildew lesions on watermelon leaves appear smaller and rounder than on cucumber.

Due to the rapid spread of this disease and because symptoms often do not appear until 4–12 days after infection, a successful disease management program must be implemented prior to the appearance of the disease symptoms.

10.10.3   CAUSE AND DISEASE DEVELOPMENT

P. cubensis is a fungal-like organism that belongs in the Kingdom Straminipila and phylum oomycota. *P. cubensis* is a member of Peronosporaceae (the downy mildew family) in the order Peronosporales within the class oomycetes. Like other downy mildew organisms, *P. cubensis* is

FIGURE 10.34 Sporangia (spore sacks) in the lesions on the underside leaf surface appear as black specks.

a biotroph or obligate parasite, meaning that the organism requires living host tissue to grow and reproduce. The organism cannot be propagated on artificial media. *P. cubensis* overwinters on infected cucurbits, either wild or propagated, in areas that do not experience a hard frost, such as southern Florida in the eastern United States.

10.10.4 Favorable Conditions

Downy mildew is favored by cool, wet and humid conditions. The pathogen produces microscopic sac-like structures called sporangia over a wide range of temperatures (5°C–30°C). Optimum sporangia production occurs between 15°C and 20°C and requires at least 6 hours of high humidity. This disease may progress slowly or stop temporarily when temperatures rise above 30°C during the day. Nighttime temperatures of 12°C–23°C will promote disease development, especially when accompanied by heavy dews, fog, or precipitation. With nighttime temperatures around 15°C and daytime temperatures around 25°C, downy mildew infections on cucurbits produce more sporangia within four days.

10.10.5 Disease Cycle

P. cubensis causes a polycyclic disease. Sporangia are the source of primary inoculum. Sporangia are transported from infected plants via wind currents and travel to local or distant places. Optimal temperature for sporulation is 15°C with 6–12 hours of available moisture. Symptomatic plants with yellow lesions have the greatest sporulating capacity. The sporulating capacity of purely necrotic lesions is low and that of yellow-necrotic lesions is intermediate. Sporangia and sporangiophores are greatly affected by changes in temperature and humidity. Warming and drying of the atmosphere, typical of early morning hours, causes twisting of the sporangiophores, which may be of importance for the detachment of sporangia.

Once sporangia land on a susceptible host, free moisture is required for each sporangium to release 5–15 zoospores. Free moisture is also important for zoospore movement, germ tube development and penetration of host tissue by the germ tube. However, excess moisture may reduce the duration of sporangia viability. Zoospores can be released between temperatures of 5°C and 28°C. The temperature optimum for zoospore release depends on the duration of the leaf wetness period.

The optimum temperature for zoospores to form cysts is 25°C. High temperatures induce immediate cyst formation. Zoospores encyst on a stomatal opening and then form a germ tube that will enter the host via the stomate. Once the host tissue is infected, intercellular hyphae form haustoria within plant cells, which provide nutrients for survival and asexual reproduction.

New sporangiophores, differentiated from the mycelium, emerge singly or in groups from the epidermis, usually via the stomata. The new sporangia are produced 4–12 days after initial infection, depending on temperature and day length. Because of the higher frequency of hyphae infecting the spongy parenchyma, the vast majority of sporangiophores and sporangia are produced on the lower leaf surface. High humidity is required for the emergence of sporangiophores.

Symptoms appear 3–12 days after infection, depending on temperature, presence of free moisture, and inoculum dose. High temperatures (>35°C) are not favorable for disease development. However if cooler nighttime temperatures occur, disease development may progress (Figure 10.35).

10.10.6 MANAGEMENT

The disease can be managed using the following strategies:

- Chemical control
- Biological control
- Cultural control
- Integrated pest management

10.10.6.1 Chemical Control

Chemical control can be very effective at managing downy mildew on cucurbits. However, when chemical control is used in combination with cultural practices, host resistance, and disease forecasting, growers can reduce pesticide use and save money. Efficacious fungicides include fluopicolide, famoxadone + cymoxanil, cyazofamid, zoxamide, and propamocarb hydrochloride. *P. cubensis* is known to develop resistance to fungicides very rapidly. Reduced efficacy of mefenoxam, metalaxyl, and the strobilurin fungicides has been reported.

10.10.6.2 Biological Control

Most popularly used cultivars of cucumber and cantaloupe, and to a lesser extent squash and pumpkin, have some level of downy mildew resistance bred into them. Even though cultivars with downy mildew resistance may become diseased, disease onset may be delayed, less severe, or the pathogen may produce fewer sporangia than on cultivars without resistance. However, since a new, more virulent strain of *P. cubensis* arrived in the eastern United States in 2004, cucumber production cannot rely solely on downy mildew resistant cultivars for control.

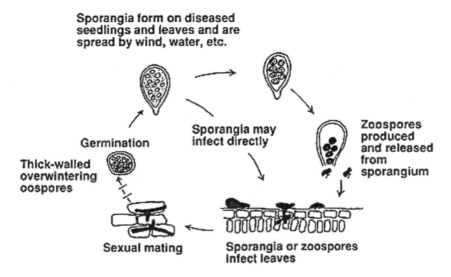

FIGURE 10.35 Disease cycle of downey mildew.

10.10.6.3 Cultural Control

Because this disease is carried to most fields on light winds, cultural practices like crop rotation and sanitation have a limited effect on the incidence of downy mildew. Still, there are several things that growers can do to suppress the disease. Growing vigorous plants, capable of withstanding or repelling disease onslaughts, is the first step.

Good soil fertility management can often be backed up with foliar fertilization, which some growers believe can assist in pest resistance.

Further cultural considerations include selecting growing sites with good air drainage, full sunlight, and low humidity. Using drip irrigation, or scheduling overhead irrigation to avoid excessive leaf wetness, will also reduce disease incidence. When detected early, disease spread might be slowed somewhat by removing and destroying infected plants and by taking care not to transport the disease by hand or on infected tools and equipment (Table 10.6).

10.11 POWDERY MILDEW OF CUCURBITS

Powdery mildew is a common and serious disease of cucurbit crops. This disease occurs in cucumbers, muskmelons, squash, gourds, and pumpkins grown both in field and greenhouse conditions. Previously, powdery mildew was an occasional problem for watermelons, but for the past 5 years, the incidence of powdery mildew outbreaks has increased (Roberts and Kucharek, 2005). A powdery mildew infection acts as a sink for plant photosynthesis causing reductions in plant growth, premature foliage loss, and consequently a reduction in yield. The yield loss is proportional to the severity of the disease and the length of time that plants have been infected (Mossler and Nesheim, 2005). For instance, in cucumber there is a negative linear relationship between disease severity and

TABLE 10.6
Fungicides and their Groups

Common Name (Trade Name)	Fungicide Group	Amount/Acre	Remarks
Cyazofamid Ranman 400SC	21	2.1–2.75 fl oz	Do not apply more than six applications of Ranman per growing season in cucurbits.
Flupicolide Presidio	43	3–4 fl oz	A labeled rate of another product with a different mode of action effective on the target pathogen must be mixed with Presidio fungicide
Propamocar Previcur Flex	U	1.2 pt or 0.6–1.2 pt in tank mix	Do not apply more than 6 pt/season in cucurbits
Mefenoxam/Chlorothalonil Ridomil Gold/Bravo	M5	1.5–2 lb	–
Famoxadone/Cymoxanil Tanos	11/27	8 oz	Do not make more than one application without alternating with a fungicide that has a mode of action other than QoI (Group 11).
Mancozeb Dithane DF Dithane F-45 Dithane M-45 Penncozeb 75 DF	M3	1–2 lb 0.8–1.6 qt 1–2 lb Label rates	Labeled for cucumbers, melons, watermelons, and summer squash only.
Chlorothalonil Bravo Ultrex, etc.	M5	1.4–1.8 lb	Do not apply more than 19.1 lb/acre/season
Mefenoxam/Chlorothalonil Ridomil Gold/Bravo	4/M5	1.5–2 lb	–

yield (Dik and Albajes, 1999). If this disease is not controlled in a timely manner, symptoms can be severe enough to cause extensive premature defoliation of older leaves and wipe out the crop.

10.11.1 CAUSAL ORGANISM

Powdery mildew is mostly caused by the 2 fungi below.

Species	Associated Disease Phase	Economic Importance
Podosphaera xanthii (Previously Known as *Sphaerotheca fuliginea*)	Leaves, Fruits, and Flowers	High
Golovinomyces cucurbitacearum (Previously Known as *Erysiphe cichoracearum*)		

10.11.2 SYMPTOMS

Symptoms of a powdery mildew are often easier to identify than symptoms of any other disease because powdery mildew forms obvious pads of whitish mycelium on upper and lower leaf surfaces, petioles, and stems.

During the crop growing season, the fungus produces hyphae and asexual spores called conidia on leaves. This disease can be first noted on older leaves, which develop reddish-brown, small, restrained round spots. At this point, a microscopic examination is necessary in order to discern if typical conidia of powdery mildew are present. Later on, those spots become white as hyphae and spores are produced in abundance. Infected areas enlarge and coalesce quickly, forming a white powdery mycelium that resembles talc. Severely infected leaves lose their normal dark green color, turn pale yellow and then brown, and finally shrivel, leaving cucurbit fruits exposed to sunburn (Figures 10.36 and 10.37).[6]

10.11.3 CAUSE AND DISEASE DEVELOPMENT

Powdery mildew of cucurbits is caused by 2 organisms, *S. fuliginea* (syn. *P. xanthii*) and *E. cichoracearum* (syn. *G. cichoracearum*). *S. fuliginea* is more commonly reported worldwide, and prefers warmer weather, while *E. cichoracearum* prefers cooler weather. The 2 organisms have similar conidia and can only be differentiated by the fibrosin bodies only present in conidia of *S. fuliginea*. Spores dispersed over a long distance from alternate hosts are the primary source of inoculum. Powdery mildew is diagnosed by white, powdery mold on plant tissues.

FIGURE 10.36   Abundant powdery mildew on pumpkin leaves.

FIGURE 10.37   (a) Severely powdery-mildew-infected muskmelon and (b) Beit Alpha cucumber leaves.

10.11.4 FAVORABLE CONDITIONS

Unlike most fungi, the spores of powdery mildew do not require free water for germination and are actually inhibited in its presence. High humidity is beneficial but not necessary for spore germination. Infection has been known to occur below a relative humidity of 50%, although the humidity at the surface of the leaf is undoubtedly higher. High humidity also increases the rate at which the fungus grows after infection occurs. Spores will germinate above 10°C with an optimum temperature of around 27°C and an upper limit of 32°C.

Temperatures between 24°C–29°C and elevated levels of relative humidity (80%–95%) in the absence of rainfall promote the development of this disease.

10.11.5 DISEASE CYCLE

Powdery mildew develops quickly under favorable conditions because the length of time between infection and symptom appearance is usually only 3–7 days and a large number of conidia can be produced in a short time. Favorable conditions include dense plant growth and low light intensity. High relative humidity is favorable for infection and conidial survival; however, infection can take place in as low as 50% relative humidity. Dryness is favorable for colonization, sporulation, and dispersal. Rain and free moisture on the plant surface are unfavorable. However, disease development occurs in the presence or absence of dew. Mean temperature of 20°C–27°C is favorable; infection can occur at 10°C–32°C. Powdery mildew development is arrested when daytime temperatures are at least 38°C. Plants in the field often do not become affected until after fruit initiation. Susceptibility of leaves is greatest 16–23 days after unfolding.

10.11.6 MANAGEMENT

The disease can be managed using the following strategies:

- Chemical control
- Biological control
- Cultural control
- Integrated pest management

10.11.6.1 Chemical Control

For susceptible cucurbit cultivars, fungicide is the most effective means of control. According to Konstantinidou-Doltsinis (1998), the need to control powdery mildew disease is 1 of the reasons for the increased use of fungicides in cucurbits. Most of the fungicides to control powdery mildew are primarily preventive, that is, to be effective they must be applied before the fungus infects the plant. In addition, powdery mildew fungi can develop resistance to specific fungicides (Brown, 2002).

The following fungicides are approved to control powdery mildew infections under Florida environmental conditions: Flint® (trioxystrobin), Nova® (myclobutanil), Quadris® (azoxystrobin), and Pristine® (boscalid and pyraclostrobin).

Strictly follow all label recommendations for handling, application and disposal of these fungicides. The misuse of these fungicides increases the risk of environmental contamination and also increases the possibility that powdery mildew fungi will develop resistance to these chemicals. Contact County Extension Agents for updated registration, recommendations, and supplementary assistance.

Other non-systemic fungicides, such as sulfur and copper, have some efficacy to control powdery mildew outbreaks. Sulfur is 1 of the oldest natural fungicides to control powdery mildews. Sulfur is active against a variety of targets in the fungus and resistance has not developed.

Sulfur may be applied as Microthiol Disperss®, which is a micronized, wettable sulfur compound that allows uniform dispersal over the plant surface and increases the anti-fungal activity. However, according to Mossler and Nesheim (2005), sulfur only provides a moderate level of control in squash, and this lack of control has been confirmed by extensive personnel. Moreover, several cucurbit species, mostly muskmelons and honeydews, are very sensitive to sulfur and phytotoxicity, as scorch occurs when sulfur is applied to the leaves.

10.11.6.2 Biological Control

There are several biofungicides that have been registered for the control of powdery mildew in cucurbits. 1 of these, AQ10 (Ecogen, Inc.) was developed specifically for powdery mildew. Its active ingredient is fungal spores of *Ampelomyces quisqualis* Ces., which act to parasitize and destroy the powdery mildew fungi.

Another biological control is Serenade® (AgraQuest, Inc.), which has a bacterium, *B. subtilis*, as an ingredient and prevents the powdery mildew from infecting the cucurbit plant.

Recently, a yeast-like fungus *Sporothrix flocculosa* (syn. *Pseudozyma flocculosa*) has been tested for control of powdery mildew in greenhouse-grown cucumbers with promising results. It has been formulated as a wettable powder (Sporodex®) for use against powdery mildew on greenhouse crops (Paulitz and Bélanger, 2001). A problem with many biological control agents is that they require a higher humidity for survival than the powdery mildews do.

Consequently, biological fungicides are not as effective at controlling powdery mildew as the biorationals and non-harmful chemicals or other fungicides.

10.11.6.3 Cultural Control

Plant cucurbits in a sunny location with good air circulation. Avoid planting new crops next to those that are already infected with powdery mildew. Remove old and heavily diseased leaves to improve air circulation and reduce inoculum (Table 10.7).

10.12 FUSARIUM WILT DISEASE OF CUCURBITS

Fusarium wilt is 1 of the most economically important diseases that affects most cucurbits. It is caused by *Fusarium oxysporum*, which may persist for long periods in the soil as durable spores (chlamydospores) or in association with plant debris. This disease was first described by Erwin F. Smith in 1894 in the southeastern region of the United States. Currently, it is severe in California, the Midwestern states, central Wisconsin, and lower Ontario. However, the disease is now widespread throughout most regions in the world.

TABLE 10.7

Fungicides and their Groups

Common Name (Trade Name)	Fungicide Group	Amount/ Acre	Remarks
Triflumizole Procure480SC	3	4-8 fl oz	Do not apply more than 40 fl oz of Procure 480SC/ acre/season.
Myclobutanil (Rally 40W)	3	2.5–5 oz	Do not apply more than 1.5 lb/acre/season.
Pyraclostrobin/Boscalid (Pristine)	11, 7	12.5–18.5 oz	Do not make more than one application before alternating to a fungicide with a different mode of action other than Group 11
Micronized Sulfur (Microthiol)	M 2	4–6 lb	Sulfur can injure plants, especially when temperatures reach 35°C. Do not use on sulfur-sensitive varieties
Quinoxyfen (Quintec)	13	4–6 fl oz	Registered for use on melons, including cantaloupe and watermelon. Do not apply more than 24 fl oz/ acre/season.
Azoxystrobin (Quadris)	11	11–15.4 fl oz	Do not apply more than one application before rotating to a fungicide with a different mode of action (i.e., group number)
Trifloxystrobin (Flint)	11	1.5–2 oz	Do not apply more than one application before rotating to a fungicide with a different mode of action (i.e., group number)
Pyraclostrobin (Cabrio)	11	16 oz	Do not apply more than one application before rotating to a fungicide with a different mode of action (i.e., group number).
Kresoxim-Methyl (Sovran)	11	3.2–4.8 oz	Do not apply more than one application before rotating to a fungicide with a different mode of action (i.e., group number).
Potassium Bicarbonate (Kaligreen)		2.5–5 lb	Use the higher rate when disease pressure is severe. Direct contact with the fungus is required for control. Conditionally allowed in an organically certified crop; check with your certifier.
Cinnamaldehyde (Cinnacure)		0.25–1 gal	Make no more than two consecutive applications before rotating to a fungicide with a different mode of action. May not provide good control under all conditions.

10.12.1 Causal Organism

Fusarium wilt is mostly caused by the fungi mentioned below.

Species	Associated Disease Phase	Economic Importance
F. oxysporum f. sp. *niveum*	Leaves, Stems, Fruits, and Roots	High

10.12.2 Symptoms

Symptoms of fusarium wilt are similar on all cucurbits and are dependent on several factors, including the amount of inoculum in the soil, environmental conditions, nutrients (particularly nitrogen), and susceptibility of the host. Fusarium wilt is characterized by loss of turgor pressure of the vines. Vines may recover during the evening, but eventually wilt permanently (Figure 10.38). Initial

FIGURE 10.38   Yellowing of leaves due to wilt.

symptoms often include a dull, gray green appearance of leaves that precedes a loss of turgor pressure and wilting. Wilting is followed by a yellowing of the leaves and finally necrosis. The wilting generally starts with the older leaves and progresses to the younger foliage. Initial symptoms often occur as the plant is beginning to vine and wilting may occur in only 1 runner leaving the rest of the plant apparently unaffected (Figure 10.39). Under conditions of sufficiently high inoculum density, or a very susceptible host, the entire plant may wilt and die within a short time. Affected plants that do not die are often stunted and have considerably reduced yields. Under high inoculum pressure, seedlings may damp off as they emerge from the soil.

The primary diagnostic symptom of fusarium wilt is a discoloration of the vascular system (xylem), which can be observed readily in longitudinal or cross-section of roots or stems (Figure 10.40). In watermelon and muskmelon, a brown necrotic streak may be visible externally on the lower

FIGURE 10.39   1 runner affected leaving the rest of the plant apparently unaffected.

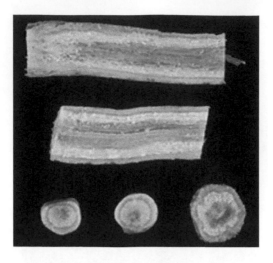

FIGURE 10.40   Discoloration of the vascular system (xylem) due to fusarium wilt.

stem and extending along the length of the vine. Since the pathogen is soilborne, symptomatic plants often occur in clusters corresponding to the distribution of high inoculum densities in the soil (Figure 10.41).

10.12.3   CAUSE AND DISEASE DEVELOPMENT

The fusarium wilt pathogen is very host-specific; it only infects watermelon. There are 3 pathogenic races of *F. oxysporum* f. sp. *niveum*. A race is a subgroup within a species that may be capable of attacking only particular cultivars of a host. The fungus can live for many years in the soil in the absence of a susceptible watermelon crop. It survives in crop debris and on old diseased vines. Morphologically, this species produces 3 types of spores: macroconidia, microconidia, and chlamydospores. Chlamydospores are the dormant structure that helps this species survive for a long time without a host or in a harsh environment. The spores in the soil begin to germinate when roots

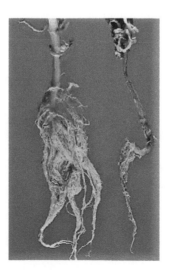

FIGURE 10.41   Healthy green tissue and whiter roots will become rotten and dark over time due to wilt.

of susceptible host plants are available. The fungus enters through the root tips or any openings in the root tissue.

10.12.4 FAVORABLE CONDITIONS

This disease may attack the plant at any stage, although, outbreaks often follow periods of crop stress, including hot, dry weather conditions. Soil temperatures of 18°C–25°C (64°F–77°F) and 50%–65% relative humidity support rapid disease progression.

10.12.5 DISEASE CYCLE

This disease is not spread aboveground from plant to plant but rather from fungal spores in the soil. The fungus can be introduced to the soil through contaminated seeds, in compost, from farming tools and vehicles, on the feet of humans and animals, and with drainage water. If the seeds are infected, the fungus can be transported directly into the xylem from infected inner tissues. If the fungus is in the soil, it germinates and penetrates the fibrous root system at the root tips or at any openings. From inside the root tissue, the fungus enters the xylem and produces conidia that block water flow in the host. three to four weeks after planting, symptoms of a 1-sided wilt will appear. In the field, this symptom often occurs in patches. When the plants are dead, the fungus remains on the dead vines or in the soil in the form of white mycelia, macroconidia, and chlamydospores until a new susceptible host is available. This disease can be quite severe in fields with light, sandy soils with pH 5.5–6.5, less than 25% soil moisture, and high nitrogen content.

10.12.6 MANAGEMENT

The disease can be managed using the following strategies:

- Chemical control
- Biological control
- Cultural control

10.12.6.1 Chemical Control

Moderate reductions in disease severity have been observed with commercial soil treatment products containing methyl bromide, chloropicrin, and metam sodium. However, fumigation with these chemicals is seldom completely successful, primarily because of the resistant nature of the chlamydospores, roots that grow from the fumigated portion of the field into non-fumigated, infested portions of the field (below the fumigated zone or between fumigated rows), re-colonization and/or re-infestation of the soil by the pathogen, or because of improper application of the fumigants. In addition, many compounds once approved for soil treatment, including methyl bromide, are being phased out around the world because of health and environmental concerns. Finally, if the pathogen is introduced into a previously fumigated field via infected seeds or transplants, the fumigation would have little effect on control of the disease. In some situations, soil fumigation may not be an economically viable option.

Seed disinfection: Use 96- "Tianda hymexazol" powder 3000×liquid+"Tianda 2116" 500×liquid soaking for 20 minutes, then rinse with water, then germination sowing will take place or sterilize 72 hours after seeding with a constant temperature of 70°C.

10.12.6.2 Biological Control

Numerous studies have examined the potential of various types of biological controls for fusarium wilts in a number of crop plants, including watermelon and muskmelons. While many demonstrate promise in laboratory and greenhouse trials, few have shown significant control at the field level. Examples of some of the approaches investigated are:

1. The use of antibiotic-producing soil fungi and bacteria, i.e. *Gliocladium* spp., *Trichoderma* spp., and *Pseudomonas* spp.
2. Use of non-pathogenic strains of *F. oxysporum* that compete with pathogenic forms for root colonization (i.e. *F.o.* strain 47).
3. The use of other formae speciales and races of *F. oxysporum* to induce resistance in plants to pathogenic forms.
4. Natural or induced soil suppressiveness. The use of cover crops, such as hairy vetch, as soil amendments ("green manures") may reduce fusarium wilt in some locations, but the magnitude of the wilt reduction may vary by location.

While none of these methods give adequate control in the field, they may be useful when used in combination with other management strategies.

10.12.6.3   Cultural Control

1. Plant clean, quality seed of resistant cultivars. Because there are races of wilt forms, it is necessary to know which races are present before choosing a resistance variety.
2. Plant on land not previously cropped with the cucurbit species.
3. Liming applications to bring the soil pH to 6.5–7.0 can reduce disease.
4. Do not move soil from infested fields.
5. Not replanting the same cucurbit species for 5–7 years can help manage watermelon wilt but is considered ineffective for melon and cucumber wilt.

There are no fungicides available for this disease. Management can be achieved through crop rotation, use of resistant varieties, and other cultural practices.

1. Crop Rotation: Controlling this disease is very difficult once the soil is infested. Crop rotation for 4–10 years might reduce the density of the fungus in the soil, and with it, disease incidence. Crop rotation should be done with non-cucurbit vegetation. Avoid planting watermelon in the same field for a minimum of 5–7 years.
2. Resistant varieties of watermelons: Growers should use resistant varieties available in the market such as "Afternoon Delight", "Crimson Sweet", "Indiana", etc.
3. Use disease-free transplants and seeds: When transplanting in the green house, the soil used must be pathogen-free, or disinfected by steam or soil fumigation. The seed also must be pathogen-free so that the fungus is not introduced into the field. Seed from infested fields should not be saved.
4. Biocontrol options: Biocontrol strategies for this disease are not commercially available at this time but may become a management option in the future. Greenhouse studies of organic fertilizers in combination with different antagonistic strains of non-pathogenic *F. oxysporum* and other microorganisms have shown promise but are insufficient for recommendations at this time.

10.13   PYTHIUM DISEASE OF CUCURBITS

Pythium is a genus of organisms that cause common crop diseases such as pythium root rot, Pythium fruit rot and in seedlings can cause damping off. These are commonly called water molds. Pythium damping off is a very common problem in fields and greenhouses where the organism kills newly emerged seedlings, usually as a result of overwatering.

Cottony leak, also referred to as pythium fruit rot, affects most cucurbits; however, it is most common on cucumber and squash. *Pythium* spp. are found in most soils in the world.[7]

10.13.1   CAUSAL ORGANISM

Pythium disease is mostly caused by the fungi below.

Species	Associated Disease Phase	Economic Importance
Pythium spp.	Fruits and Roots	High

10.13.2   SYMPTOMS

10.13.2.1   Fruit Rot

Certain species of *Pythium* are capable of causing fruit rot of numerous crops. Pythium fruit rot, described as cottony leak or watery rot, occurs during wet weather or in poorly drained areas of fields.

Fruit rot caused by *Pythium* generally starts as small, water-soaked lesions on immature or mature fruit near or in contact with the soil. The whole fruit is infected within 72 hours; the epidermis is ruptured, and the fruit collapses. Under high moisture, a white, cottony growth may be apparent on the lesion surface.

Fruit infection occurs through wounds or where fruit touches wet soil. The pathogens grow quickly through diseased tissues. The pythium fruit rot pathogens are readily disseminated within and among fields in irrigation water and on contaminated equipment. *Pythium* spp. survive between cucurbit crops as pathogens on many crops and weeds and as dormant resting structures in the soil (Figure 10.42).

10.13.2.2   Root Rot

In mature plants, *Pythium* causes crown and root rot, where plants suddenly wilt when weather turns warm and sunny and when plants have their first heavy fruit load. Often, upper leaves of infected plants wilt in the day and recover overnight but plants eventually die. In the root system, initial symptoms appear as brown to dark-brown lesions on root tips and feeder roots and, as the disease progresses, symptoms of soft, brown stubby roots, lacking feeder roots, become visible. In larger roots, the outer root tissue or cortex peels away leaving the string-like vascular bundles underneath. Pythium rot also occurs in the crown tissue at the stem base. In cucumber, diseased crown turns orange-brown in color, often with a soft rot at the base; brownish lesions extending 10 cm up the stem base may be seen (Figure 10.43).

10.13.3   CAUSE AND DISEASE DEVELOPMENT

Several species of *Pythium*, a fungus-like organism, have been involved in this disease. These soilborne pathogens can overwinter as dormant spore structures in the residue of many different crops

FIGURE 10.42   Pythium fruit rot (cottony leak disease) on cucumber.

FIGURE 10.43   (a) Pythium crown and root rot in cucumber showing orange discoloration of the crown area. (b) Pythium crown and root rot in cucumber showing orange discoloration of the root tips.

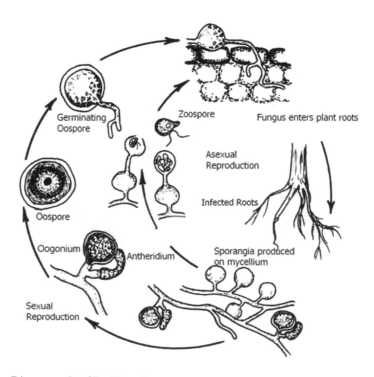

FIGURE 10.44   Disease cycle of Pythium disease.

and weeds. Infection occurs through wounds or where the fruit touches the wet ground. *Pythium* spp. is easily disseminated via water and soil particles. Wet conditions promote infection and decay.

10.13.4   Disease Cycle

Pythium can be introduced into soil and irrigation water. Insects such as fungus gnats (*Bradysia impatiens*) and shore flies (*Scatella stagnalis*) can also carry *Pythium*. *Pythium* spreads by forming sporangia, sack-like structures, each releasing hundreds of swimming zoospores (Figure 10.44). Zoospores that reach the plant root surface encyst, germinate, and colonize the root tissue by producing fine threadlike structures of hyphae, collectively called mycelium. These hyphae release hydrolytic enzymes to destroy the root tissue and absorb nutrients as a food source. *Pythium* forms oospores and chlamydospores on decaying plant roots which can survive prolonged adverse conditions in soil and water, leading to subsequent infections.

10.13.5   Management

The disease can be managed using the following strategies:

- Chemical control
- Biological control
- Cultural control
- Integrated pest management

10.13.5.1   Chemical Control

Chemical controls should be used in combination with cultural controls to be most effective.

10.13.5.2   Biological Control

No biological control strategies have been developed for pythium fruit and root rot.

10.13.5.3   Cultural Control

Sanitation: Field soil, debris, pond and stream water, and roots and plant refuse of previous crops can contain *Pythium*. Follow a strict sanitation program throughout the year and a thorough year-end clean up.

Irrigation water: Untreated water from rivers or streams poses a great risk for *Pythium* introduction, while treated, municipal water is considered safe from *Pythium*. Water storage and nutrient tanks need to be disinfected periodically and covered to prevent *Pythium* contamination.

Disease monitoring: Plants must be monitored for any signs of *Pythium* diseases throughout the cropping cycle. Remove and destroy severely infected plants and replant in new growing bags. Infected plant materials, including grow bags, must be safely disposed away from the greenhouse by deep-burying, incinerating, or composting.

Control fungus gnats (*B. impatiens*) and shore flies (*S. stagnalis*), which spread *Pythium* (Table 10.8).

REI—re-entry interval
PHI—pre-harvest interval
NA—information is not available

10.14   SCAB OR GUMMOSIS OF CUCURBITS

Scab of cucurbits or vine crops is caused by the fungus *Cladosporium cucumerinum*. Scab is primarily a disease of cucumber, cantaloupe, honeydew melon, muskmelon, summer and winter squash, true and other pumpkin types and gourds. Watermelon is very resistant to the disease and many varieties of cucumber that have resistance to scab are now available. In the absence of resistance, scab has resulted in 50% or greater losses in cucumbers during prolonged, cool and moist weather.

TABLE 10.8

Chemicals and their Use

Product	Chemical/Biocontrol name	Chemical Group	Mode of Action	REI	PHI	Application
Previcur	propamocarb	28	preventative, locally systemic	12 hrs	2 days for cucumber; 1 day for tomato & pepper	use preventatively; maximum 2 applications per crop cycle after transplanting, thereafter 7–10 days interval
Mycostop	*Streptomyces* Strain K61	biological	suppressive, non-systemic	NA	0 days	use preventatively; apply to growing medium soon after transplanting, thereafter every 3–6 weeks interval; store unopened product in a cool (≤8°C), dry place.
Prestop	*Gliocladium catenulatum*	biological	suppressive	4 hrs	0 days	use preventatively; apply to growing medium soon after transplanting, thereafter every 3–6 weeks interval; store unopened product in a cool (≤4°C), dry place
RootShield WP	Trichoderma harzianum Rifai, strain KRL–AG2	biological	suppressive	4 hrs	0 days	use preventatively; apply to growing medium soon after transplanting, repeat thereafter; store unopened product in a cool (2°C–5°C), dry place
Ridomil Gold 480EC or 480SL	metalaxyl-M, S isomers	4	preventative, systemic	12 hrs	21 days	use preventatively; one application per crop cycle; apply as drench immediately after transplanting

The disease was first described in the United States in New York in 1887. It has been reported in many cool, temperate parts of North America, Europe, and Asia.

10.14.1 Causal Organism

Scab or Gummosis is mostly caused by the fungi mentioned below.

Species	Associated Disease Phase	Economic Importance
C. cucumerinum	Leaves, Fruits, and Stems	Moderate

10.14.2 Symptoms

The scab fungus can attack any aboveground portion of the plant including leaves and petioles, stems and fruit.

On leaves and runners, pale-green water-soaked areas are the first sign of the disease. These spots gradually turn gray to white and become angular shaped. A chlorotic halo may appear around the lesion. If weather conditions are favorable, a scab can deform young leaves, and the apical runners of young plants like melons can be killed. Sporulation on leaves tends to be sparse.

On fruit, scab can produce the greatest damage, especially if they are infected when young. Spots first appear as small sunken areas, similar to insect stings, about 1/8 inch in diameter. A sticky substance may ooze from the infected area. The spots become darker with age and may create a cavity in summer squash fruit, which are very susceptible. The cavities may be lined with a dark olive green, velvety layer of spores. Secondary soft-rotting bacteria may also invade the cavities and lead to foul-smelling decay. On highly resistant cucurbit fruits, spores are more difficult to detect, and lesions may remain quite superficial. The time when fruits are infected may determine the relative severity of symptoms. Watermelon is considered to be highly resistant as shown in the superficial infection of fruit under severe disease pressure.

On highly resistant fruits, especially on certain squashes and pumpkins, irregular, knoblike formations may develop. Affected fruits are often invaded later by soft-rotting bacteria that produce a mushy, foul-smelling decay. Scab is a common transit and storage decay of muskmelon and occasionally of late-planted squash (Figures 10.45, 10.46, 10.47 and 10.48).

FIGURE 10.45 Scab lesions initially appear water-soaked and brown with a yellow halo surrounding the lesion. Lesions eventually crack, producing a "shot-holed" appearance not unlike bacterial angular leaf spot.

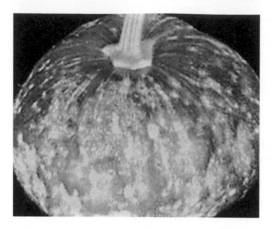

FIGURE 10.46   Numerous sites of scab infection on immature fruit of the susceptible cheese pumpkin.

FIGURE 10.47   Summer squash showing characteristic scab lesion on leaves and fruit. Fleshy fruit, like summer squash, are particularly susceptible to Gummosis and extensive fungal sporulation.

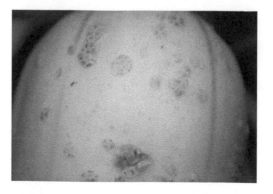

FIGURE 10.48   Scab infection on butternut squash showing superficial infection of the rind.

10.14.3 CAUSE AND DISEASE DEVELOPMENT

Scab is caused by *C. cucumerinum*. Extensive presence of the fungus is not evident on infected leaves and stems, but sporulation can be profuse on fleshy fruit. The fungus consists of septate and branching mycelium, which appear as hyaline when young and turn greenish to black with age. Conidia are oblong, colored, mostly continuous (or, in some cases, 1-septate), and borne terminally on short, branched, dark conidiophores. The 1-septate conidia measure 4.6–5.7 × 16.4–22.5 μm. Structures intermediate between conidiophores and conidia also become detached and germinate. These are larger than conidia, have thicker walls, and have 1 to many cells.[8]

10.14.4 FAVORABLE CONDITIONS

Temperatures around 17°C (63°F) and alternating between 12°C and 25°C (59°F and 77°F), accompanied by moist weather with frequent fogs, heavy dews, and light rains, are most favorable for scab development.

10.14.5 DISEASE CYCLE

The scab organism survives in soil on squash, melon, and pumpkin vines and reportedly may grow extensively as a saprophyte. The fungus may also be seedborne. It is disseminated on clothing and equipment and by insects. The conidia can survive long-distance spread in moist air. The most favorable weather conditions for disease development are wet weather (valley fogs, heavy dews, and light rains) and temperatures near or below 21°C, which usually occur after mid-season in the northern United States. At 17°C the growing tips of young plants are killed. Conidia germinate and enter susceptible tissue within 9 hours. A spot may appear on leaves within 3 days, and a new crop of spores is produced by the fourth day.

10.14.6 MANAGEMENT

The disease can be managed using the following strategies:

- Use only disease-free seed and treat with a seed fungicide to additionally control seed decay and damping off. Do not save your own seed if disease is present.
- Grow scab-resistant cucumber varieties so that scab fungicide sprays can be omitted.
- Select sites that have well-drained soils and are conducive to good air drainage to allow for rapid drying of foliage.
- Follow crop rotations of 2 or more years between cucurbit crops and non-host crops.
- Use fungicides to control scab, but specific points need to be made to increase their effective use. During cool, wet weather, fungicide sprays are not totally effective because of the short disease cycle. During these periods, sprays should be applied every 5 days rather than weekly. Early applications before fruit formation are important. For squash, apply when plants begin to bloom or, if conditions warrant, when first true leaves appear. For melons and pumpkins, apply when vines begin to run or earlier if conditions warrant.
- The following fungicides are presently registered for use, but consult the most recent Vegetable Recommends or the label for specific rates. Dithiocarbonates include maneb (flowable and WP) for melons (muskmelon and watermelon), pumpkin, and squash (summer and winter); mancozeb (WP) for use on melons and summer squash only; metiram (Polyram) for muskmelons; captafol (Difolatan) for melons; chlorothalonil (Bravo) for melons, squash, and pumpkin; and anilazine (Dyrene) for melons, squash, and pumpkin. Bravo Ultrex at 1.8–2.7 lb/A. This may be applied through sprinklers. 12-hour re-entry. Liqui-Cop at 3–4 teaspoons/gal water.

- Use resistant cultivars whenever available. Scab-resistant cultivars include "Dasher II", "Raider", "Encore", "Sprint", "Poinsett 76", "Turbo", "Regal", "Flurry", "Calypso", "Quest", "Gemini", "Marketmore", "Pioneer", "SMR-58", and "SMR-18". Commercial growers should consult processors for the resistant cultivars to grow.

List of Cucurbitaceae Plant:
winter melon (*Benincasa*)
watermelon (*Citrullus lanatus)*
muskmelon (*Cucumis melo*)
pumpkin *(Cucurbita)*
cucumber (*Cucumis sativus*)
bottle gourd (*Lagenaria siceraria*)
ridge gourd luffa (*Luffa acutangula*)
smooth gourd luffa (*Luffa aegyptiaca*)
snake gourd (*Trichosanthes cucumerina*)
bitter gourd (*Momordica charantia*)
round gourd (*Praecitrullus*)

Winter Melon (*Benincasa*):
powdery mildew
downy mildew
Alternaria leaf blight
squash vine borers
Cercospora leaf spots
charcoal rot of fruits
mosaic diseases
gummy stem blight

Watermelon (*Citrullus lanatus*):
powdery mildew
downy mildew
Alternaria leaf blight
squash vine borers
Anthracnose
Fusarium wilt
Cercospora leaf spots
charcoal rot of fruits
angular leaf spot
mosaic diseases
gummy stem blight

Bacterial Rind Necrosis
Only watermelon is affected by bacterial rind necrosis.

Muskmelon (*Cucumis melo*):
powdery mildew
downy mildew
angular leaf spot
Alternaria leaf blight
squash vine borers

Fusarium wilt
Fusarium root rot
Cercospora leaf spots
charcoal rot of fruits
mosaic diseases
gummy stem blight

Pumpkin (*Cucurbita*):
powdery mildew
downy mildew
angular leaf spot
Alternaria leaf blight
Cercospora leaf spots
charcoal rot of fruits
mosaic diseases
gummy stem blight
Choanephora wet rot

Cucumber (*C. sativus*):
powdery mildew
downy mildew
Ulocladium leaf spot
angular leaf spot
Alternaria leaf blight
Anthracnose
Cercospora leaf spots
charcoal rot of fruits
bacterial leaf spots same as angular leaf spot
Mosaic diseases
scab or gummosis
root knot nematode of cucumber

Bottle Gourd (*Lagenaria siceraria*):
powdery mildew
downy mildew
Anthracnose
Fusarium wilt
Cercospora leaf spots
charcoal rot of fruits
fruit rot or cottony lock disease
mosaic diseases
gummy stem blight
Choanephora wet rot

Ridge Gourd Luffa (*Luffa acutangula*):
powdery mildew
downy mildew
Cercospora leaf spots
charcoal rot of fruits
mosaic diseases
gummy stem blight

Smooth Gourd Luffa (*Luffa aegyptiaca*):
powdery mildew
downy mildew
Cercospora leaf spots
charcoal rot of fruits
fruit rot or cottony lock disease
mosaic diseases
gummy stem blight

Snake Gourd (*Trichosanthes cucumerina*):
powdery mildew
downy mildew
Anthracnose
Cercospora leaf spots
charcoal rot of fruits
fruit rot or cottony lock disease
mosaic diseases
gummy stem blight

Bitter Gourd (*Momordica charantia*):
powdery mildew
downy mildew
Fusarium root rot
Cercospora leaf spots
charcoal rot of fruits
fruit rot or cottony lock disease
mosaic diseases
gummy stem blight

Round Gourd (*Praecitrullus*):
powdery mildew
downy mildew
Cercospora leaf spots
charcoal rot of fruits
mosaic diseases
gummy stem blight

Common Diseases of All Cucurbits:
powdery mildew
downy mildew
Cercospora leaf spots
Anthracnose
angular leaf spot
Alternaria leaf blight
squash vine borers, not in use
Fusarium wilt
Pythium disease same as fruit rot or cottony lock disease
charcoal rot
mosaic diseases
gummy stem blight
Choanephora wet rot

Insect Pests of Cucurbits:
cucurbit fruit fly
red pumpkin beetle
epilachna beetle
root knot nematode

REFERENCES

1. Sikora, E. J., 2011. *Common Diseases Of Cucurbits*, Extension Plant Pathologist, Auburn University.
2. Palenchar, J., Treadwell, D. D., and Datnoff, L. E., 2009. *Cucumber Anthracnose in Florida*, University of Florida IFAS Extension.
3. Report on Plant Diseases, 2012. *Angular Leaf Spots of Cucurbits*, Department of Crop Sciences, University of Illinois at Urbana Champaign.
4. Gubler, W. D., 2009. *Cucurbits: Charcoal Rot*, Plant Pathology, UC Davis.
5. Report on Plant Diseases, 1993. *Root Knot Nematodes*, Department of Crop Sciences, University of Illinois at Urbana Champaign.
6. Nuñez-Palenius, H. G., Hopkins, D., and Cantliffe, D. J., *Powdery Mildew of Cucurbits in Florida*, University of Florida IFAS Extension.
7. Wick, R. L., 2013. *Root Diseases of Greenhouse Crops*, University of Massachusetts Amherst.
8. Zitter, T. A., *Scab*, Department of Plant Pathology, Cornell University.
10. Mossler, M. A., Nesheim, O. N. 2005. Florida Crop/Pest Management Profile: Squash. Electronic Data Information Source of UF/IFAS Extension (EDIS). CIR 1265. February, 3, 2005. http://edis.ifas.ufl.edu/.
11. Dik, A., Albajes. R., 1999. Principles of epidemiology, population biology, damage relationships and integrated control of diseases and pests. In Albajes R L. Gullino, J. van Lenteren, Y. Elad, (eds.), Integrated pest and disease management in greenhouse crops. Dordrecht, The Netherlands: Kluwer Academic Publishers: 69–81.
12. Konstantinidou-Doltsinis, S., Schmitt, A. 1998. Impact of treatment with plant extracts from Reynoutria sachalinensis (F Schmidt) Nakai on intensity of powdery mildew severity and yield in cucumber under high disease pressure. Crop Protection 17: 649–656.
13. Brown, J., 2002. Comparative genetics of avirulence and fungicide resistance in the powdery mildew fungi. In Bélanger, R., W. R. Bushnell, A. J., Dik, T. L. W. Carver, (ed.), The Powdery Mildews. A Comprehensive Treatise. St. Paul, MI: APS Press: 56-65.
14. Paulitz, T. C., Belanger, R. R., 2001. Biological control in greenhouse systems. *Annual Review of Phytopathology* 39: 103–133.

11 Ginger

Ginger, *Zingiber officinale*, is an erect, herbaceous perennial plant in the family Zingiberaceae grown for its edible rhizome (underground stem), which is widely used as a spice. The English origin of the word, "ginger", is somewhere from the mid-fourteenth century. The ginger rhizome is brown, with a corky outer layer and pale yellow, scented center. The aboveground shoot is erect and reed-like with linear leaves that are arranged alternately on the stem. The shoots originate from a multiple basis and wrap around one another. Typical leaves of ginger can reach 7 cm (2.75 inches) in length and 1.9 cm (0.7 inches) broad. Flowering heads are borne on shorter stems and the plant produces cone shaped, pale yellow flowers. The ginger plant can reach 0.6–1.2 m in height (2–4 ft) and is grown as an annual plant. Ginger is also referred as true ginger, stem ginger, garden ginger, or root ginger, and it is believed to have originated in the Southeast Asia. In 2014, with a global production of 2.2 million tons of raw ginger, India accounted for 30% of the world total, followed by China (19%), Nepal (13%), Indonesia (12%), and Thailand (7%).

This chapter provides information on the identification of diseases in ginger plant.

11.1 GINGER WILT

The rhizomes of edible ginger (*Z. officinale* Roscoe) are a valued spice or fresh herb ingredient in international cuisine. Bacterial wilt of ginger is a very destructive, parasitic disease. Ginger wilt, caused by a bacterium known as *Ralstonia solanacearum* (Smith, 1896) Yabuuchi (1995), is the most limiting factor in the production of culinary ginger (*Z. officinale* Roscoe). It is a complex and difficult disease to control, infecting the ginger crop through all phases of a production cycle. It is present systemically in seed rhizomes as both an active and latent infection that contaminates seed-pieces when they are cut and prepared for field planting.[1]

11.1.1 CAUSAL ORGANISM

Species	Associated Disease Phase	Economic Importance
R. solanacearum	Rhizome	Severe

11.1.2 SYMPTOMS

The first symptoms of wilt are a slight yellowing and wilting of the lower leaves. The wilt progresses upward, affecting the younger leaves, followed by a complete yellowing and browning of the entire shoot. Under conditions favorable for disease development, the entire shoot becomes flaccid and wilts with little or no visible yellowing. However, the plant dries very rapidly, and the foliage becomes yellow-brown in 3–4 days. Young succulent shoots frequently become soft and completely rotted and these diseased shoots break off easily from the underground rhizome at the soil line. The underground parts are also completely infected. Grayish-brown discoloration of the rhizomes may be localized if the disease is at an early stage of infection, or discoloration may be general if the disease is in an advanced stage. A water-soaked appearance of the central part of the rhizome is common. In advanced infections, the entire rhizome becomes soft and rots. Bacterial wilt of ginger can be distinguished from other rhizome rots of ginger by the condition of the rhizome and the foliage. A better diagnostic feature is the extensive bacterial ooze that shows as slimy, creamy exudate on the surface of a cut made in the rhizome or on the aboveground stem of an infected plant. (Figures 11.1 through 11.6).

FIGURE 11.1 A mature ginger plant with bacterial wilt.

11.1.3 DISEASE CYCLE

Disease "signs" refer to the observable presence of a plant pathogen in or on infected host tissues. Signs can be useful in diagnosing the pathogen and the disease. The signs of bacterial wilt of ginger will be discussed below.

- Bacterial streaming, i.e., large populations of bacteria that exude from the cut surface of infected plant tissue when observed with a microscope or observed macroscopically when a diseased ginger rhizome is suspended in a glass or beaker of water.
- Bacterial ooze from infected tissues, especially from infected rhizomes. Ooze is the emission of bacterial colonies from infected tissues, seen as moist, milky mounds collecting on the tissues surfaces.

R. solanacearum colonizes the xylem, the water-conducting elements of a ginger plant's vascular system, and causes wilt. The symptoms caused by this pathogen on ginger include the following:[1]

FIGURE 11.2 An immature edible ginger plant showing the typical symptoms of bacterial wilt: green wilt (green ginger leaves roll and curl due to water stress caused by bacteria blocking the water-conducting vascular system of the ginger stems), leaf yellowing, and necrosis.

FIGURE 11.3   A field of young ginger plants affected by bacterial wilt disease. The wilting, yellowing plants are infected with *R. solanacearum* race 4.

1. Green wilt: The diagnostic symptom of the disease. This occurs early in the disease cycle and precedes leaf yellowing. Infected green ginger leaves roll and curl due to water stress caused by bacteria blocking the water-conducting vascular system of the ginger stems.
2. Leaf yellowing and necrosis: Leaves of infected plants invariably turn yellow and then necrotic brown. The yellowing should not be confused with another disease of ginger causing similar symptoms, Fusarium yellows. (Figures 11.7 and 11.8).

11.1.4   CAUSE AND DISEASE DEVELOPMENT

Ginger *(Z. officinale* Roscoe) is a fibrous-rooted perennial plant with branched underground stems called rhizomes. The aboveground portion of the plant consists of several aerial shoots which are about 3 ft tall. The plant is indigenous to the tropics and susceptible to frost injury. The crop is planted in late winter or early spring and harvested in the winter months. Because ginger does not produce true seeds, portions of the underground rhizome weighing from 1 to 4 ounces known as "seed pieces" are used for planting material. The cost of planting material is high, and growers usually save some of the best portions of a crop.

FIGURE 11.4   An edible ginger rhizome infected with *R. solanacearum* race 4, showing a discolored, necrotic center.

FIGURE 11.5   Discolored ginger rhizome and stems affected by bacterial wilt.

Continuous re-cropping of the same land is a common problem. This, combined with the use of planting material saved from the previous year, has contributed to the buildup of a number of disease-producing organisms which have become destructive in many areas of the world.

The bacterium that causes this disease is soilborne and may be carried in infected rhizomes. There is also a possibility that some of the insects that feed on the diseased ginger foliage or rhizome could be vectors of the bacteria if they later feed on healthy plants. Commonly, bacterial wilt

FIGURE 11.6   Growers should control other pests that damage ginger plants such as the lesser corn stalk borer (*Elasmopalpus lignosellus*), which creates holes in the stems of ginger plants.

FIGURE 11.7   When an infected rhizome is suspended in water, the white, milky bacteria stream from the diseased tissue.

can be spread from diseased to healthy rhizomes with the knife used to cut the rhizomes into seed-pieces. New soil can be contaminated with the bacteria by use of contaminated farm equipment or by irrigation water flowing through infested soil.

11.1.5   FAVORABLE CONDITIONS OF DISEASE DEVELOPMENT

Plant diseases are of 2 general types, parasitic and non-parasitic. The parasitic diseases are caused by living organisms, which subsist in whole or in part on the plant, thereby, making the natural develop-ment of the plant impossible. These diseases are caused by fungi, bacteria, viruses, and nematodes. Non-parasitic diseases are caused by unfavorable environmental conditions such as drought, tem-peratures unfavorable for plant growth, and nutrient deficiencies or excesses in the soil.[2]

The bacterial wilt of ginger is caused by a strain of *Pseudomonas solanacearum*. The strains of the bacteria attacking ginger will weakly attack tomato, pepper, and eggplant. Although laboratory tests have shown that the strain that causes severe tomato wilt does not attack ginger, a relationship between the strains of the bacteria does exist. Although relationships may be of little importance to the grower, there is a possibility that under field conditions, the tomato strain could be severe on ginger and the ginger strain severe on crops belonging to the tomato family. Thus, it is highly

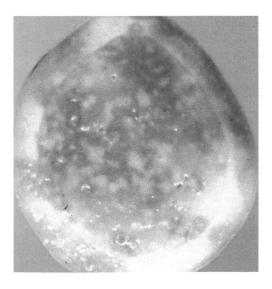

FIGURE 11.8   Bacterial ooze from an infected ginger rhizome.

FIGURE 11.9 Bacterial wilt of ginger showing foliar symptoms. Rapid wilting of the green foliage prior to yellowing is characteristic. 2 or 3 days later the leaves turn yellow-brown and dry.

undesirable to plant either crop in a field in which the other had been diseased. The bacterium that causes this disease is soilborne and may be carried in infected rhizomes. There is also a possibility that some of the insects that feed on the diseased ginger foliage or rhizome could be vectors of the bacteria if they later feed on healthy plants. Commonly, bacterial wilt can be spread from diseased to healthy rhizomes with the knife used to cut the rhizomes into seed pieces. New soil can be contaminated with the bacteria by use of contaminated farm equipment or by irrigation water flowing through infested soil (Figures 11.9).

11.1.6 Management and Control

When the life history of a disease-producing organism is known, it is usually possible to destroy it or to prevent its spread. The most serious diseases of ginger are seed- or soilborne. The seed phase of the diseases may be controlled by seed selection and by application of the seed treatments described below. The most effective means of eradicating these pathogens from the seed is by a process of seed certification involving "subculture" of apical shoots in sterile nutrient agar media and propagating such material in disease-free soil. However, this method of seed certification is a costly operation that cannot be handled by small growers.

The soilborne phase of these diseases may be controlled by soil sterilization or soil disinfection.[2]

11.1.7 Seed Selection

Because certified or inspected seed is not available to protect the grower from bacterial wilt, the grower must be careful about the source of seed that is planted. He must be familiar with the symptoms of each major disease and avoid saving seed from diseased plants.

The hot-water treatment is ineffective against *R. solanacearum*, the bacterial wilt organism, when these pathogens are inside the vascular tissues of the seed pieces. The seed can be surface-disinfected after cutting by dipping the ginger in a 10% Clorox solution for 10 minutes (use 1 part commercial Clorox to 9 parts water). Other materials such as panogen, PMA (phenyl mercuric acetate), etc., could be used if the seed pieces are discarded at harvest time; these materials have not been cleared by the United States Food and Drug Administration.

TABLE 11.1
Management

Chemical	Application
10% Clorox	Seed must be surface disinfected after cutting by dipping the ginger in the chemical. Use one part of chemical with nine parts of water.
Methyl Bromide and Chloropicrin	It is less reliable. Grower should avoid continuous re cropping of land.

Soil fumigation with a material composed of methyl bromide and chloropicrin also offers excellent weed control. Since the control of bacterial wilt with fumigation is less reliable, the grower should avoid the continuous re-cropping of the land. Bacterial wilt-infested fields should be placed on rotation using non-susceptible crops such as cucumbers or sweet corn. Fumigation after a year's rotation may prove more effective. Also, land that had a previous tomato family crop with bacterial wilt should not be used for ginger production without proper fumigation (Table 11.1).

11.2 BACTERIAL WILT OF GINGER

Ginger, *Z. officinale* Roscoe, is not only known as a flavoring agent but also known for its medicinal values. This rhizome is an herbaceous perennial plant used in the treatment of constipation, flatulence, motion sickness, etc. Bacterial wilt is 1 of the major diseases that is responsible for the damage and loss of better quality ginger on a large scale due to its lethality, wide host range, and geographic distribution.

R. solanacearum is an aerobic, non-sporing, gram-negative plant bacterium. *Ralstonia* was recently classified as *Pseudomonas* due to its similar characteristics, except that it does not produce fluorescent pigment like *Pseudomonas*. The bacteria is known to cause rapid and fatal wilting of ginger by colonizing the xylem. This pathogen holds the second position for its scientific and economic importance in plant diseases. The associated difficulties with this pathogen are because of endophytic growth, survival in water, and deep layers of soil.

11.2.1 CAUSAL ORGANISM

Species	Associated Disease Phase	Economic Importance
R. solanacearum	Bacterial Wilt	High

Being soil and water-borne, this pathogen may spread through any of these means:

1. *Water*: It may spread through the water for irrigation, contaminated flood water, or any other related means of water.
2. *Soil*: contaminated soil can be transported from 1 part of the field to the other through hands, boots, legs, vehicles, farming equipment, insect vectors, etc.
3. *Infected rhizome ginger*: The infected rhizome will spread the wilting to other healthy ginger plants. The infection is spread to a very high population ($10^8 - 10^{10}$ cfu/g tissue).

R. solanacearum hibernates in the host plant in winters. However, the bacterial population ceases to minimal at this time; the pathogen can still survive for over 40 years in cold conditions.

The bacterium invades the host cell by entering its roots, attaching to its epidermis, and eventually infecting the cortex, followed by late stage colonization in the xylem. Thus, it causes the plant to wilt and eventually die. The bacteria are then released into the environment and it seems to survive in soil and/or water, reservoir plants, biofilm formation, etc. until a new host plant is found.

Infection: The most visible side effect of immense colonization of the pathogen is wilting. The pathogen attacks the plant through the wound and moves forward systematically. It multiplies extensively inside the host plant thus causing the wilt. Tyloses may be developed in order to cease the further injury to the plant.

11.2.2 Symptoms

The leaf of the damaged ginger plant appears yellowish-brown and is easily detached from the stem. The stem generally shows a sign of water soaking and color darkening along with a white exudate spurting out of the stem, if pressed, due to the colonization of the pathogen in the xylem (the water transporter) (Figures 11.10 and 11.11).

The symptoms of the bacterial wilt and the disease cycle are as follows:

1. Most often the bacteria invade by the host by root colonization, xylem penetration, or cortical infection by its swimming motility, chemotaxis, and aerotaxis.
2. Green wilt: This is the early stage of the infection and the plant is green in color. The leaves are curled and rolled due to the water stress caused by vascular dysfunction.
3. Yellowing of leaves: The leaves are yellowed rapidly, unlike how the *fusarium* yellows. The wilt progresses from the lower leaves to the upper leaves.
4. Growth of the plant is stunted. The leaves break off easily and the underground stem can be separated from the rest of the plant without any strong force.
5. The plants die before harvesting.
6. Rhizomes are discolored and often water-soaked. If the disease is in the early stage, then the discoloration may be localized, otherwise, a general discoloration is seen throughout.

Under certain favorable conditions for disease development, there is no visible yellowing. The bacterial ooze or bacterial streaming, milky white exudate, can be observed under the microscope as the large population of the bacteria in the late stages of host colonization.

11.2.3 Pathogen Detection and Disease Diagnosis

Detection of a pathogen at an early stage will help in preventing the crops from deteriorating. There are various sources to check the presence of pathogen, namely, soil, water, and infected plant. This can be done by using various techniques. Some are listed below.

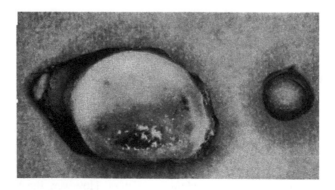

FIGURE 11.10   Bacterial wilt ginger plants.

FIGURE 11.11   Severe bacterial wilt in ginger.

11.2.3.1   Bioassay

Bioassay is a simple and reliable technique used by farmers to check the possible development of bacterial wilt. The tissue-cultured or natural plant, which are pathogen-free, are grown in pots using the field soil and are checked for wilting by raising the temperature to 27°C and through a constant water supply. This technique has the following advantages-

1. Reliable method
2. Easy to implement
3. After transplantation, if the soil is R. *solanacearum* positive, wilting is seen in one to two weeks. If the plants are tissue-cultured, then the growth of the bacteria is faster.

However, bacterial wilt symptoms are closely related to other disease. Hence, this technique is not an inference test. Signs of bacterial oozing or streaming should be checked for the presence of pathogen.

11.2.3.2   Immunoassay

This test is used for the plants exhibiting the pathogen symptoms and to test bacterial culture samples. This test is done in 30 minutes using the immunostrip and sample bag with BEB1 buffer. Sample selection should be appropriate for accurate results. The lower detection limit of the immunoassay test is 10^5 cfu/mL. Better results are seen if the experiment is carried out in temperatures between 15°C and 35°C.

11.2.3.3   Polymerase Chain Reaction and DNA Test

Polymerase chain reaction that amplifies the 16S ribosomal rRNA gene selectively and appears on the electrophoresis gel as a band and a DNA test such as LAMP (loop-mediated isothermal amplification) can confirm the presence of pathogen.

11.2.4   MANAGEMENT OF THE DISEASE AND CONTROL MEASURES

Edible ginger (Z. *officinale*) is used worldwide. The bacteria responsible for spreading the disease is also present on a very large scale and it is very difficult to stop the spread of the disease. Hence, integrated disease management is the best strategy to improve the yield. However, acceptable management of the disease is required to increase the quality and quantity of edible ginger. Here are some points of effective control measures and management techniques for the disease.

11.2.4.1 Chemical Procedures

Over time, there is a steep increase in the use of pesticides. It is seen that for every 1.8% increase in the use of pesticide per hectare, there is 1% increase in the crop output per hectare. Moreover, the use of herbicides is more as compared to fungicides and insecticides. The insecticide imidaclopride in the concentration of 42.8%, 22.6%, and 40.4% when used against *R. solanacearum* has proved to be effective in reducing the bacterial wilt.

Biofumigation: Use of essential oils from *Cymbopogon martini* (Palmarosa), *C. citratus* (lemongrass), and *Eucalyptus globulus* (Eucalyptus) on the crops is found to produce good-quality yield. The crops planted in essential oil treated field did not show any reduced growth. Experiments like epifluorescence microscopy showed bacterial cell degradation from 95%–100% at all concentrations (0.04, 0.07, and 0.14% vol/vol) of lemongrass and palmarosa oil, and at a concentration of 0.14%, eucalyptus oil showed its bactericidal activity. These anti-bacterial activity-possessing plants, if planted a few months before sowing the ginger seed is a very effective biofumigant.

Site selection: Before planting, the soil should be assayed to confirm pathogen-free soil. Downslope soil field should be avoided as the run-off water from the bacteria-contaminated ginger field above might bring the pathogen to your field. The sowing of seed pieces should be avoided in wet weather because the pathogens can be transmitted through various means like boots, hands, etc. causing more damage. Sowing, planting, and harvesting should be done in the drier season to avoid the spread of pathogen within fields. If the pathogen comes in contact with the soil, it persists for a long time.

Planting precautions: It is important to use pathogen-free tractors, farming equipment, etc. to avoid the spread of the pathogen. The seeds should be bacterial wilt-free. Such seeds can be prepared in the green house and multiplied. Bleaching in 10% bleach solution could be done to sterilize the surface of the ginger. Do not allow any outsider to enter the field as this may bring in the pathogen. Dirty boots should be dipped in the 10% bleach solution to prevent the dispersal of bacterial wilt pathogen from spreading. In addition to this, dirty equipment should not be brought into the field and the field should be properly fenced to keep away pigs and other animals. Organic amendments like compost and mulch help in reducing the *R. solanacearum* activity through antibiosis, competition, or both.

11.2.4.2 Crop Rotation and Inter-cropping

The bacteria *R. solanacearum* has a huge host of plants; hence, it has a limited number of crops with whom rotation could be done (Table 11.2).

According to the range of hosts, this phytopathogen is classified into 5 categories viz; race 1, race 2, race 3, race 4, and race 5.

Race 1 having the highest number of host plants, this race is present in the 5 continents. The following are the list of the host plants affected by *R. solanacearum* race 1:

> solanaceous crops like chili and sweet pepper, eggplant, potato, tobacco, and tomato; nonsolanaceous crops like bean, groundnut, and sunflower; ornamental plants like *Anthurium* spp., *Dahlia* spp., *Heliconia* spp., *Hibiscus* spp., *Lesianthus* spp., *Lilium* spp., marigold, palms, *Pothos* spp., *Strelitzia* spp., *Verbena* spp., and *Zinnia* spp.; trees like Eucalyptus and fruit trees like black sapote, custard apple, and neem.

TABLE 11.2
List of Some Host Plants

Common Name	Scientific Name
Potato	*S. tuberosum*
Tomato	*L. esculentum*
Aubergine (eggplant)	*S. melongena*
Banana	*Musa spp.*
Geranium	*Pelargonium spp.*
Tobacco	*N. tabacum*
Sweet pepper	*Capsicum spp.*
Olive	*O. europea*
Woody night shade	*S. dulcamara*
Groundnut	*A. hypogea*
Black night shade	*S. nigrum*

In addition to this, abaca, cowpea, cucurbits, hyacinth beans, jute, moringa, mulberry, nutmeg, patchouli, Perilla crispa, sesame, strawberry, water spinach, wax apple, and winged bean.

Race 2 typically appears in south America and some parts of Philippines; this has the following host plants namely wild *Heliconia* spp. and cooking and dessert bananas, plantain, other *Musa* spp.

Race 3 is listed as the race of agro terrorism. It is wide spread throughout the world. The host plants for this species are weeds like *Solanum dulcamara* and *S. nigrum* and crops like *Capsicum* spp., eggplant, geranium, potato, and tomato.

Race 4 occurs in Asia and affects the ginger plantation and its related species.

Race 5 is limited to China and it affects the *Morus* spp.

The rotation should be done with crops like sweet potato and taro.

11.2.5 STORAGE

Avoid delay in harvesting the crop because that leads to desiccation and shrinkage. Harvest the crop at correct maturity for minimal wounds on the tuberous stem, unlike immature crop. In addition, unlike overmature plants, it does not cause any development of cracks on the peel. Minimal bruising should occur. Air dry harvested tubers to wilt the succulent roots, so that they are easily removed and cause minimal injury to the ginger. Quality inspection of the stored ginger should be carried out.

11.3 FUNGAL SOFT ROT

Soft rot is also known as rhizome rot or *Pthyium* rot. The rot was first observed in India caused by *Pythium gracile* in 1907. Since 1907, numerous incidence of soft rot was recorded caused by various *Pythium* spp. The disease has been observed in all the ginger growing countries such as India, Japan, China, Nigeria, Fiji, Taiwan, Australia, Hawaii, Sri Lanka, and Korea (Table 11.3).

TABLE 11.3

The Pathogen

Sr. No.	Organism	Location
1	*P.aphanidermatum* (Edson) Fitz.	Kerela
		Hyderabad (Telangana)
		Nagpur (Maharashtra)
		Madhya Pradesh
		Pusa (Bihar)
2	*P. butleri* Subram	Kerela
		South Kanara (Karnataka)
		Ceylone
3	*P. delience* Meurs	Madhya Pradesh
4	*P. gracile*	Bengal, Gujrat
		Kerela
		Assam,
		Fiji
5	*P. completans* Braun	Ceylon
6	*P. myriotylum*	Kerela
		Pune
		Mumbai
		Nagpur
		Taiwan
		Ceylon
		Hong Kong
7	*P. myriotylum* and *Fusarium*	Rajasthan
8	*P. graminiolum*	Ceylon
9	*P. ultimum* Trow.	Himachal Pradesh
10	*P. pleroticum* T.	Solan (Himachal Pradesh)
11	*P. vexans*	Kerela
12	*P. zingiberum*	Korea
		Osaka (Japan)

11.3.1 Disease Cycle

Oospore is the primary source of soft rot in the diseased rhizome. It produces a germ tube that elon gate and either produces sporangium or penetrates the host directly. In another case, oospore germinate, further producing sporangium and zoospores. Infection can be by appresoria formed due to hyphal elements. The disease spreads by waterborne zoospores. This zoospore is attached to the ginger root and encyst, producing germ tubes that infect root. On the surface of lesions, sporangia are produced. Oospore is formed generally in the host tissue or soil. Oospores are dormant and aid in perennation.

Weather and soil are 2 vital factors affecting the Pythium infection and disease development. High water retention of soil, high relative humidity, and low temperature are favorable for the disease to develop and spread. The disease spreads to adjacent clumps through soil, water via zoospores, and hyphal fragments.

FIGURE 11.12   Aboveground rot symptoms; ginger affected by rot disease.

11.3.2   DISEASE MANAGEMENT

6 important primary preventive measures and disease management techniques are as follows:

Healthy rhizome selection: Rhizome being the source of perennation and spread of soft rot; to prevent disease, the rhizome selected should be disease-free.

Narrow ridge cultivation: Kim et al. have reported that the method of narrow ridge cultivation is as effective to reduce the disease as compared to unridged control in all the fields.

Mulching: The plots mulched with maha neem leaves were found to be free from rhizome rot.

Soil solarization: It is a disinfection practice for soil. The soil is covered with a transparent polythene film and high temperature and intense solar radiation is applied. It is a non-hazardous and economical method, leaving no harmful residues. The soil solarization leads to high soil temperature (37°C–52°C), sufficient to kill the pathogenic fungi. The thermal inactivation of Pythium species occurs in between 37°C–52°C.

Chemical control: Different chemicals have been experimented with to kill the rhizome-borne inoculum. Various treatments demonstrated are 0.1% mercuric chloride for 24 hours/90 minutes, Ceresan (0.25%), and Agrosan-GN (0.25%) for 30 minutes. Many workers have established a treatment of soil against rhizome to rot. Treatment of soil with Bordeaux mixture, perenox (0.35%) and Dithane Z-78 (0.15%), 0.1% HgCl$_2$, Ceresan wer (0.5%), Dithane (0.2%), methyl bromide, Dithane Z-78, Difolatan, Aliette, Dithane M-45, and Difolatan and Ridomil granular have proven effective.

A mixture of Ridomil and captafol increased the yield of ginger when used on the seed pieces 1 day before and in field soil 3 months after planting the rhizome rot. 4% formaldehyde along with Topsin M 0.1% and 0.1% Bavistin or 0.3% Dithane M-45 along with formaldehyde was proved to be an excellent fungicide against *F. oxysporum*, which causes rhizome rot. Metalaxyl formulations (Ridomil 5G and Apron 35 WS) were most effective against *P. aphanidermatum* among Fosctyl-Al, oxadixyl, propanocarb, and ethazole (epidiazole).

The disease is seedborne and soilborne in nature; disinfected seed and drenched soil are effective simultaneously. Rosenberg observed seed treatment with aresan to be effective with soil fumigation with trizone to control the disease. Echlomezol is observed to prevent spread of infection upon application to drench around the source of primary infection. Ghorpade and Ajri observed reduction in soft rot incidence utilizing amendments like oil seed cakes in the field. It also caused significant rise in yield.

Biological control: Thomas demonstrated biological control of Pythium species using *Trichoderma lignorum* as an antagonist. Antibiosis causes increased acidity of medium, which results in reduction of growth of *Pythium*. Bhardwaj and Gupta studied invitro antagonism studies of *Trichoderma* sp. to *P. aphanidermatum, F. equiseti, F. solani, Clasdosporium, and Mucor hiemalis.*

Sorghum grain treated infected soil improved the rhizome sprouting. *Trichoderma viride* is responsible for the inhibition of *P. myriotylum* and *F. solani* by the production of a non-volatile substance. Rhizomes steeped in the suspension of *T. viride* or *T. hamatum* were found to be effective in inhibiting the growth of *P. aphanidermatum* and *F. equiseti.* Rhizome rot along with sawdust was effectively reduced rhizome rot and increased the yield of edible ginger.

Rhizome treatment with *T. harzianum, T. aureoviride, Gliocladium virens,* and *T. viride* reduced ginger rhizome rot, a rhizome- and a seedborne disease caused by *F. solani* or *P. myriotylum* or both and significantly increased the yield (Ram et al., 2000). Shanmugam et al. (2000) indicated that *T. harzianum* and *T. viride* are potential antagonists against *P. aphanidermatum.* Bhat (2000) found that *Oxyspora paniculata* extracts gave highest inhibition of *P. aphanidermatum,* while *Macaranga denticulate* extracts gave complete inhibition of *Pythium* sp.

Resistant Cultivars: Cv. Maran has demonstrated P. *aphanidermatum* causes field resistance to ginger rot (Indrasenan and Paily, 1974). Setty et al. (1995a) experimented on 18 ginger cultivars against rhizome rot (*Pythium* sp.) and concluded that Suprabha and Himachal Pradesh cultivars showed less than 3% disease incidence. Panayanthatta (1997) tested 148 accessions of ginger and seven related taxa for assessing their reaction to rhizome rot caused by *P. aphanidermatum.*

11.4   FUSARIUM ROT DISEASE (YELLOWS)

Fusarium rot is a very serious disease because *Fusarium oxysporum* f.spp *zingiberi* can grow saprophytically in the absence of the host plant. It is more widespread than bacterial wilt and it can devastate the crop field entirely. In South Africa, *F. oxysporum* f.spp *zingiberi* causes ginger wilt. Yellow was first reported in Queensland then in Hawaii followed by India and Asia.

11.4.1   Symptoms

The first symptom that is seen is the yellowing of the margins of the older leaves followed by the younger ones. Plants also show drooping, wilting, drying in patches or wholly, and yellowing. Plants infected by this fungus do not wilt rapidly as in bacterial wilt.

The plants do not fall on the ground. Wilting of ginger leaves in this case is very slow and growth of the plants is stunted. The lower leaves dry out over an extended period. It is not difficult to notice the presences of stunted and yellowed shoots aboveground in between the green shoots.

The fungus invades the entire vascular system of the plant, and hence, the plant dries out eventually. The basal portion of the plant becomes water-soaked and soft and the plants can be pulled out easily from the rhizome. The rhizomes show creamy to brown discoloration of vascular tissues and a very prominent darkening of the cortical tissues. Central portion of the rhizome shows prominent rotting. Later, only the fibrous tissue is seen in the rhizome. In storage, a white, cottony fungal growth may be seen.

11.4.2 Pathogen detection

Trujillo (1963) reported the causal organism to be *F. oxysporum* f. spp *zingiberi*. Sharma and Dohroo (1990) found *F. oxysporum* to be the major cause of yellows, which was present in all the ginger-growing areas of Himachal Pradesh, India. Moreover, 1 out of the 3 isolates of this causal fungus caused 100% mortality of inoculated ginger seedlings while the other to isolates showed 60% and 80% mortality respectively. The other 4 species isolated were *F. solani*, *F. moniliforme* (*Gibbrella fujikuroi*), *F. graminearum* (*G. zeae*), and *F. equiseti*.

Spread of disease and the growth of the pathogen was seen when the moisture of the soil was 25%–30% and the soil temperature was 24°C–25°C. Galled root (galls of *Meloidogyne incognita*) incorporated extract of ginger showed better growth of pathogen.

11.4.3 Pathogen Life

The pathogen survives in the soil and spreads through the infected soil, water, and rhizomes. The pathogen survives as chlamydospores which remain viable for many years. Microconidia, macroconidia, and chlamydospores are 3 types of asexual spores produced by *F. oxysporum*. The plants are invaded either through root tips, wounded roots, or formation point of the lateral roots, and then, either via a sporangial germ tube or mycelium and grows through the root cortex. Gradually the mycelium reaches the vessels and invades the xylem through the pits. Eventually the mycelium grows and produces microconidia. These small characters are pulled upward by the plant's sap stream. The infection can also spread laterally as via the adjacent xylem vessels and pits.[3]

11.4.4 Control Measures

1. *Chemical control*: Good quality rhizome germination acquired by Kothari (1966) when the seeds were drenched in 0.1% HgCl2 followed by thiram (0.5%), ceresan wet (0.5%), and Dithane Z-78 (0.2%). Seed treatment with mercurial fungicides at 2 lb/40 gal of water, trizone as a fumigant, Arasan as a seed treatment, and Benzimidazole-type fungicides is proved to be beneficial in the control of the disease.
2. *Seed selection*: It is obvious to select the healthy non-infected rhizome to protect the future crops from the pathogen and loss of yield.
3. *Intercropping*: Intercropping will help in eradication of pathogen from the soil. Intercropping with capsicum is advised; it is observed that intercropping with capsicum results in 76% control of yellows disease.
4. *Biological control*: In-vitro studies have shown that *T. harzanium* and *T. virens* inhibit the action of *F. oxysporum* f. sp. *Zingiberi*. In addition to this, *Bacillus stubtilis* also showed its antagonistic effect on the pathogen in the pot culture study. When *T. harzanium*, *T. hamatum*, and *T. virens* are used with chemicals like mancozeb and carbendazim as seed treatments and soil application, they proved to reduce the disease. The action of *F. oxysporum* in the presence of 6 different *Streptomyces* sp. (SSC-MB-01 to SSC-MB-06), was studied. The bacteria are successful in suppressing the fungal activity of the pathogen and were found to protect the ginger plant against *F. oxysporum* by both dual culture method as well as agar culture method. A study showed that concentrated cow urine, prepared by dry heating in oven at 50°C for 4 days, and 10% aqueous extract of leaves of *Parthenium hysterophorus* inhibited the growth of the fungi.
5. *Biotechnological approach*: Shoot buds of the rhizome encapsulated in 4% sodium alginate gel (in-vitro and on field) were studied by Prachi et al. (2001). The results were found

to be ginger siblings free of ginger yellows. Yellows is a disease caused by Fusarium oxy-sporum f. sp. zingiberi on Zingiber officinale.

6. *Resistant cultivars*: Using cultured rhizomes, which are resistant or less susceptible to the disease provides good-quality yield. SG 666 is a cultivar, which was proved resistant to the pathogen filtrate.

11.5 PHYLLOSTICTA LEAF SPOT

The disease was seen for the first time at Godavari and Malabar regions of India by Ramakrishnan. Sohi et al. had observed disease in Himachal Pradesh and Kerela. Singh et al. observed this disease in Chhattisgarh. Currently, the disease is widespread in all the ginger growing countries.

11.5.1 THE PATHOGEN

The leaf spot disease, also known as phyllosticta leaf spot is caused by *Phyllosticta zingiberi* T.S. Ramakr.

11.5.2 SYMPTOMS

Initially the symptoms are observed on younger leaves in form of small, oval, elongated spots, usually measuring 1–10 mm×0.5–4 mm. A white papery center and brown margins are visible on spots, further surrounded by yellowish halos. The spots eventually increase in size and coalesce to form larger lesion. The affected leaves become shredded and disfigured and suffer desiccation. The crop acquires gray, disheveled look because of infection (Figure 11.13).

11.5.3 DISEASE CYCLE

The disease appears at the end of June when the plants are at the most susceptible stage, 3–4 leaf stage. During the monsoon season, the disease aggravates and spreads quickly. The rainfall is conductive for disease growth and spread. During this season, the relative humidity is 80%–90%. The disease spread is proportional to number days rains persist. The ginger plant is susceptible to the disease up to the age of 6–7 months. High quantities of rain accompanied by wind spurs the growth of the disease. The diseased leaf infects the field, serving as inoculum for the next season. Pycnidiospores and mycelia remain viable in leaves for about 14 months in laboratory conditions. Pycnidia can survive in the leaf at a temperature range of 30°C–35°C. Pycnidiospores can remain viable up to 6 months at a depth of 25 cm.[4]

FIGURE 11.13   Phyllosticta leaf spot on ginger.

11.5.4 MANAGEMENT

Sanitation and shade: It has been observed that burning of crop debris causes reduction in primary inoculum of the disease. Lesser incidence of disease is observed under the shade. At the Indian Institute of Spices Research, Calicut, India, the germ plasm collection performed under the open conditions became infected, and the same conditions under the shade net showed significantly less infection.

Chemical control: Ramkrishnan observed with the application of the Bordeaux mixture, the disease could be controlled. Sohi et al. demonstrated management of disease utilizing Dithane Z-78 (0.2 %) 6 times at 2-week intervals. They had also stated the use of other fungicides: Flit 406 (0.3%), Dithane M-22 (0.2%), or Bordeaux mixture. Grech and Frean performed comparative experiments for 5 fungicidal treatments and concluded Benomyl (0.1%) + mancozeb (0.2%) + soluble boron (0.1%) and iprodione (0.2%) reduced the average number of lesions per leaf and increased the yield. Cerezine et al. found significant reduction in the disease progress with the use of chlorothanil. Other chemicals shown to have a significant reduction are dithionan, copper oxychloride, folpet, mancozeb, Captan, bavistin (0.15%), and Dithane M-45 (0.25%).

Ruchi Sood and co-worker performed a comparative experiment for 14 fungicides. On a pre-planted rhizome, following were tested Bavistin (carbendazim, 0.05%), Kri-Benomyl (benomyl, 0.05%), Baycor bitertanol, 0.05%), Contaf (hexaconazole, 0.05%), Topsin M (thiophanate-methyl, 0.05%), Antracol (propineb, 0.25%), Blitox-50 (copper oxychloride, 0.25%), Bordeaux mixture (Bordeaux mixture, 1%), Captan (captan, 0.25%), Companion (mancozeb 63% + carbendazim 12%, 0.20%), Indofil M-45 (mancozeb, 0.25%), Kocide-101 (copper hydroxide, 0.20%), Ridomil-MZ (metalaxyl 8% + mancozeb 64%, 0.25%) and Unilax (metalaxyl 8% + mancozeb 64%, (0.20%).

Resistant cultivars: Setty et al. (1995) observed the reaction of 18 cultivars of ginger for 6 months under the coastal climate of Karnataka state (India). All the cultivars were non-resistant to disease. However, the cultivars Narasapatom, Tura, Nadia, Tetraploid, and Thingpuri are grouped as moderately resistant having a disease index less than 5%. Rio de Janeiro, Kunduli Local, Waynad Local, Kurupampady, Suravi, and Karakal are susceptible with a disease index greater than 10%. In Himachal Pradesh, none of the tested material of ginger is rated resistant to *P. zingiberi*.

REFERENCES

1. Nelson, S., 2013. *Bacterial Wilt of Edible Ginger in Hawaii*, University of Hawaii. https://cms.ctahr. hawaii.edu/gingerwilt.
2. Trujillo, E. E., 1964. *Diseases of Ginger*, Hawaii Agricultural Experiment Station, University of Hawaii, Circular 62, 6.
3. Belgrove, A., 2007. *Biological Control of Fusarium oxysporum f.sp. Cubense Using Non-Pathogenic F. oxysporum Endophytes*, University of Pretoria, 132–134.
4. Ravindran, P. N., and Babu, K. N., 2004. *Ginger: The Genus Zingiber*, CRC Press, 321.
5. Kim, C. H., and Yang, S. S., *Effects of Soil Disinfection, Fungicide Application, and Narrow Ridge Cultivation on Development of Ginger Rhizome Rot Caused by Pythium myriotylum in Fields*, Plant Pathology Division, National Institute of Agricultural Science & Technology, Suwon, 129–135.
6. Ram, D., Kusum, M., Lodha, B. C., Webster, J., and Mathur, K., 2000. Evaluation of resident biocontrol agents as seed treatments against ginger rhizome rot. *Indian Phytopath*, 53(4): 450–454.
7. Setty, T. A. S., Guruprasad, T. R., Mohan, E. and Reddy, M. N. N, 1995. Susceptibility of ginger cultivars to rhizome rot at west coast conditions. *Environ. Ecol*, 13: 242–244.
8. Huqa, M. I., and Nowsher Ali Khanb, A. Z. M., 2007. Efficacy in-vivo of different fungicides in controlling stemphylium blight of lentil during 1998–2001. *Bangladesh J. Sci. Ind. Res.* 42(1), 89–96.

9. Yabuuchi, E., Kosako, Y., Yano, I., Hotta, H., and Nishiuchi, Y., 1995. Transfer of two Burkholderia and an Alcaligenes species to Ralstonia gen. nov.: proposal of Ralstonia pickettii (Ralston, Palleroni and Douderoff 1973) comb.nov., Ralstonia solanacearum (Smith 1896) comb. nov. & Ralstonia eutropha (Davis 1969) comb. nov. *Microbiology and Immunology* 39, 897–904.

10. Prachi, T. R., Sharma, T., and Singh, B. M., 2001. In vitro and in vivo phytotoxic effect of culture filtrates of Fusarium oxysporum f. sp. zingiberi on Zingiber officinale. *Advances in Horticultural Sciences* 14(2): 52–58.

11. Sohi, H. S., Sharma, S. L., and Verma, B. R., 1973. Chemical control of Phyllosticta leaf spot of ginger (Zingiber officinale). Pesticides 7: 21–22.

12. Singh, A. K., 2011. Management of rhizome rot caused by Pythium, Fusarium and Ralstonia spp. in ginger (Zingiber officinale) under natural field conditions. *Indian J. Agric. Sci.* 81: 268–270.

13. Setty, T. A. S., Guruprasad, T. R., Mohan, E. and Reddy, M. N. N., 1995. Susceptibility of ginger cultivars to rhizome rot at west coast conditions. *Environ. Ecol.* 13: 242–244.

12 Maize

Maize, commonly also known as corn, is a cereal grain first domesticated by indigenous peoples in Southern Mexico about 10,000 years ago. The leafy stalk of the plant produces separate pollen and ovuliferous inflorescences or ears, which are fruits, yielding kernels or seeds. Over time maize has become a staple food in many parts of the world, with total production surpassing that of wheat or rice. However, not all this maize is consumed directly by humans. Some of the maize production is used for corn ethanol, animal feed, and other maize products, such as corn starch and corn syrup. The word *maize* is derived from the Spanish form of the indigenous Taino word for the plant, *mahiz*. Maize is widely cultivated throughout the world, and a greater weight of maize is produced each year than any other grain. In 2014, total world production was 1.04 billion tons, led by the United States with 35% of the total. China produced 21% of the global total.

12.1 ANTHRACNOSE OF MAIZE

Anthracnose is a relatively common disease but seldom is a cause for major economic losses in corn crops. This fungal disease can be present as leaf blight and/or as stalk rot on corn. 1 phase may be present without the other. It can also occur on all small grains, sorghum, and many grasses. This disease was first detected in Ohio in 1961, causing stalk rot in research plots at Wooster. Later reports of the disease were mainly concerned with the leaf blight phase and slight damage was attributed to the stalk rot phase except in a few localized areas. Since 1972, the evidence of anthracnose has increased greatly.

12.1.1 CAUSAL ORGANISM

Anthracnose is mostly caused by the fungi mentioned below.

Species	Associated Disease Phase	Economic Importance
Colletotrichum graminicola	Leaves and Stalk	High

12.1.2 SYMPTOMS

Anthracnose of corn may appear as a leaf blight, stalk-rot, top-kill of the stalk, and kernel rot. However, most damage results from the stalk rot and leaf blight phases. The anthracnose fungus can attack corn plants at any stage of development. Lesions can be found on leaves of very young plants soon after emergence when the fungus has overwintered in the field. Leaf lesions are generally brown, oval to spindle-shaped, about 1/4-inch-wide by 1/2 inch long. Usually, a yellow or yellow-orange area surrounds the disease portion of the leaf. The actual size and shape of the leaf lesions varies greatly among different hybrids making diagnosis in the field very difficult. The fungus can usually be seen on the leaf surface with the aid of a hand lens. Spore masses within characteristic fruiting bodies are easily identified based on the presence of small, black spines (setae) arising from the leaf surface. Spines can usually be detected within fruiting bodies near the midrib of heavily diseased leaves or within older lesions on lightly diseased leaves.

Symptoms of the stalk rot phase are easy to recognize and usually are not confused with other stalk rot diseases. Late in the season shiny, black, linear streaks and blotches appear on the surface of the lower stalk above the brace roots. Occasionally, the entire stalk becomes blackened. The

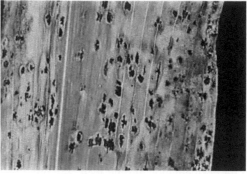

FIGURE 12.1   Typical symptoms of anthracnose leaf blight on a young corn seedling.

internal stalk tissue or pith becomes discolored, turning dark gray to brown and shredded. Severely diseased stalks are weakened and are likely to lodge before harvest. Anthracnose may develop in the upper stalk above the ear, resulting in top dieback. These blighted tops may top-lodge above the ear (Figures 12.1 and 12.2).[1]

12.1.3   CAUSE AND DISEASE DEVELOPMENT

This fungal disease is caused by *C. graminicola* (Ces.) G. W. Wills. The fruiting bodies, called acervuli, are produced as the fungus approaches maturity and can be used as a diagnostic feature. These small, black, oval to elongate acervuli with numerous black setae (spines) develop on the surface of stems and leaf sheaths.

This fungus has many strains, some of which cause various diseases of rye, wheat, barley, forage and weed grasses, as well as corn. The strains appear to have a restricted host range so that only isolates from closely related grasses will cause disease when cross inoculated. For instance, isolates from johnsongrass, sudangrass, and sorghum will attack corn, but are not as pathogenic on corn as those originally isolated from corn. Isolates from small grains do not attack corn.

12.1.4   FAVORABLE CONDITIONS

Anthracnose disease is favored by warm temperatures (21°C–29°C), high relative humidity, and periodic rainfall. Extended periods of high humidity are required for sporulation by the fungus. The spores are dispersed to host tissues in splashing raindrops.

12.1.5   DISEASE CYCLE

The fungus overwinters as a saprophyte on crop residue in the form of conidia or mycelium. It is also seedborne as stroma or hyphae in the endosperm. Infection usually occurs at the base of the plant or on the roots from the soilborne inoculum.

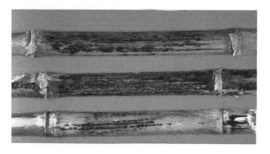

FIGURE 12.2   Anthracnose stalk rot.

There are 3 distinct phases of anthracnose disease: leaf blight, top dieback, and stalk rot. Overwintered corn surface residue is an important source of inoculum for anthracnose leaf blight (thus, it is not surprising that this disease is more common in corn-following-corn fields). Infection of seedling leaves occurs from spores that are produced in fungal fruiting structures (acervuli) and dispersed by splashing and blowing raindrops. Although anthracnose leaf blight can cause significant damage to very young plants and contribute to postemergence stand loss, oftentimes the corn seedlings are growing so quickly, they appear to outgrow the disease. Disease development is less likely on rapidly expanding leaves. The lesions that develop on the lower leaves of the plant provide a source of inoculum (spores) for subsequent infections of leaves higher up on the plant and also the stalk. Stalk wounds allow *C. graminicola* to infect and colonize the vascular system of the corn plant. Under favorable conditions, vascular infections can result in anthracnose top dieback (high temperatures and frequent rains) or stalk rot (high temperatures and plant stress following pollination).

12.1.6 MANAGEMENT

The disease can be managed using the following strategies:

12.1.6.1 Chemical/Biological Control
There are no currently labeled fungicides effective against the anthracnose pathogen.

12.1.6.2 Cultural Practices
Tillage can reduce the risk when the residue is incorporated into the soil and decomposition results. Rotation to crops other than corn for at least 1 year may minimize early season anthracnose but have little impact on late season disease.

Scouting procedure—Examine corn once or twice at 2-week intervals in June, and monthly until the dough stage is reached.

Sow resistant varieties—Hybrids susceptible to the leaf blight stage may be quite resistant to the stalk rot phase and vice versa.

Crop rotation with non-grass crops, e.g. legumes.

Sanitation—Plow under infected residue where feasible. Anthracnose is more of a problem in minimum tillage situations where infected plant residue remains on the soil surface.

Maintain a balanced soil fertility and pH.

Practice effective weed control, especially controlling grass weeds. Selection of well-drained soils is also essential.

12.2 COMMON RUST OF CORN

Common rust of corn occurs every year to some extent, and is caused by the fungus *Puccinia sorghi*. The other rust disease, southern rust, is caused by *Puccinia polysora* and occurs less frequently in the United States. Both diseases can increase in severity rapidly, particularly if infection occurs early in the growing season and favorable environmental conditions persist for extended periods of time.

Timing of disease development and its severity determine the extent of yield loss. Yield loss estimates of up to 45% have been reported in Florida, due to severe southern rust. For each 10% leaf area infected by common rust, approximately 6% of yield loss was reported in Illinois sweet corn.

Common rust and southern rust occur wherever sweet corn and field corn are grown, but cause no real economic damage on dent corn in the Midwest. In commercial sweet corn, yield losses range from 0 to nearly 50% depending on the environment and the type of resistance in the individual hybrid or variety. Quality factors and quantity are also affected, such as ear length, ear diameter, percentage of moisture, and percent soluble solids in the kernel. 1 important difference between the 2 rusts is that southern rust may possibly kill the corn plant while common rust seldom does. Southern rust occurs primarily in relatively warm regions.

12.2.1 CAUSAL ORGANISM

Common rust and southern rust is mostly caused by the fungi mentioned in the table below.

Species	Associated Disease Phase	Economic Importance
P. sorghi	Leaves	High
P. polysora	Leaves	High

12.2.2 SYMPTOMS OF COMMON RUST

Common rust is caused by the fungus *P. sorghi*. Symptoms are circular to elongate (usually 0.2–2 mm long), golden or reddish brown to cinnamon brown pustules that appear on the upper and lower leaf surfaces, and less frequently on other aboveground plant parts. Most of the infection is found on the upper leaf surface and may occur in bands. The pustules break through the epidermis early in their development and become powdery as spores (urediniospores) are produced.

Later in the season they become brownish black as a second spore stage (teliospores) develops which also breaks through the epidermis. If severe, chlorosis and premature drying of the leaves may occur, reducing sweet corn yields. Symptoms may vary slightly depending upon the hybrid being grown. Common rust typically appears much earlier in the growing season than southern rust, often developing on the plant during the vegetative growth stages. The number of pustules (uredinia) on a leaf can vary from 1 to as many as several dozen. Frequently, the pustules appear in a band because of infection that took place when the leaf tissue was still in the whorl (Figure 12.3). The dark-brown reproductive spores (urediniospores) appear about 7 days after infection. Pustules develop on both the upper and lower leaf surfaces. This helps distinguish it from southern rust, which has sparse, if any, pustule development on the lower leaf surface.

12.2.3 SYMPTOMS OF SOUTHERN RUST

Southern rust is caused by the fungus *P. polysora*. Like common rust, it does not overwinter in Kansas, but blows in from southern corn production areas. In Kansas, the disease generally arrives around the first of August, with about a 2-week variance. In years when corn is planted late and the disease arrives early, yield loss can be significant in susceptible hybrids. Temperatures above 27°C with high relative humidity encourage southern rust development. Symptoms of southern rust

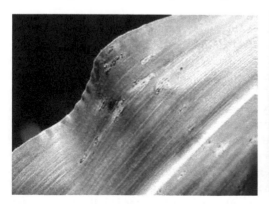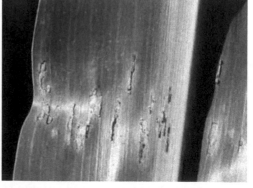

FIGURE 12.3 Common rust pustules in a band on the upper leaf and on the lower leaf surface.

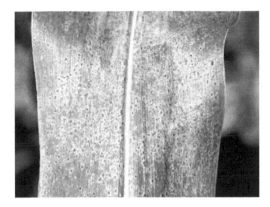

FIGURE 12.4   Densely packed, circular southern rust pustules on the upper leaf surface.

resemble those of common rust, with some subtle differences. The pustules are usually smaller and circular to oval, with a diameter of 0.2–2.0 mms. They typically are densely scattered on the upper leaf surface (Figure 12.4). The light brown to orange spores are lighter in color compared to common rust. Sporulation can be so profuse that the leaf surface becomes covered with a layer of "spore dust" that transfers easily to clothing as a person walks through an infected field. Light-colored clothing will quickly take on an orange-brown color (Figure 12.5).

12.2.4   CAUSE AND DISEASE DEVELOPMENT

P. sorghi is the fungus causing common rust in corn. The reddish-brown color of the pustule is the coloration of the repeating spore or uredospores. These spores are produced throughout the summer and infect new leaf tissue and are responsible for the spread of the disease. Southern rust is caused by the fungus *P. polysora*.

12.2.5   FAVORABLE CONDITIONS

Mild temperatures (15.5°C–23°C) and high relative humidity (near 100%) favor rust development. Temperature plays a critical role in the life cycle of both rusts. The common rust fungus prefers cooler weather (16°C–25°C) for optimal infection. Common rust symptoms may be observed early in the season, depending on how quickly spores arrive from the South. Cool evenings with dew formation can lead to severe common rust on susceptible hybrids and inbreeds. Hot, dry weather will limit common rust disease development and cause pustules to become inactive, leaving small

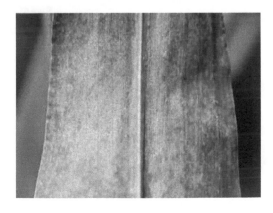

FIGURE 12.5   Lower leaf surface absents of southern rust pustules.

dead areas of leaf tissue that can be confused with Southern rust. Southern rust prefers warmer weather—fungal infection occurs between 25°C–28°C. This disease usually appears in Indiana in late August or September.

12.2.6 Disease Cycle

The complete life cycle of *P. sorghi* includes 5 different spore types and 2 hosts, corn and species of wood sorrel (*Oxalis* spp.). The spore types and the hosts they infect are teliospores, basidiospores, pycniospores, aeciospores, and urediniospores. All spore types occur in Mexico, but those involving the alternate host *Oxalis* spp. are of little importance in the life cycle of the fungus as it occurs in temperate areas of the United States. The aecial stage (called "cluster-cups") appears on the underneath surface of *Oxalis* leaves, producing aeciospores, which are windborne and infect corn leaves. These infections give rise to urediniospores, which are the most important spore type in the northern United States. Urediniospores occur on corn leaves throughout the growing season and continue cyclic infections. The disease cycle for common rust is illustrated in Figure 12.6.

12.2.7 Management

The disease can be managed using the following strategies:

12.2.7.1 Chemical Control

There are numerous fungicides labeled for use on corn that can effectively control rust diseases with timely application. However, a fungicide application typically costs $15–20/acre (including $5/acre cost of application), often making it uneconomical. It should be carefully considered on #2 yellow dent corn. When corn prices are low, fungicide applications are most likely to provide economic returns when applied shortly after infection on a susceptible hybrid with a high yield potential or with higher economic value, such as seed, white, or popcorn. Evaluate weather forecasts before

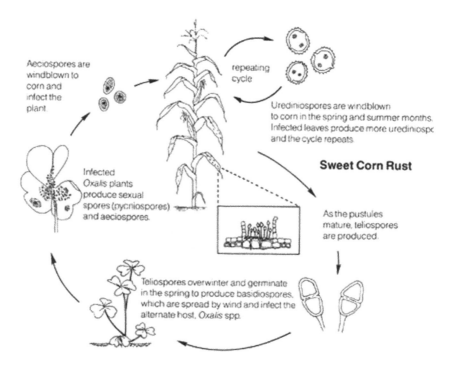

FIGURE 12.6   Disease cycle for common rust.

applying to determine if weather conditions favorable for disease will likely persist, particularly high humidity and warm temperatures.

The infrequency of rust development in Nebraska has prevented the establishment of reliable treatment threshold data under local conditions. Fungicide trials for corn rust control from other states typically show results on seed corn or sweet corn. Data from Illinois trials on sweet corn suggest that applying fungicide when common rust severity over the entire plant was less than 15% provided the most effective control. However, the recommended treatment thresholds for common rust of sweet corn in the state of New York are lower, at only 1%, which is equivalent to only six rust pustules per leaf. Other management complications include the fact that rusts, particularly southern rust, tend to develop late in the season. This limits fungicide options because of pre-harvest interval (PHI) restrictions. Data from North Carolina suggests that applying fungicide for control of rust within two weeks of black layer development (during the dough stage) is unlikely to provide economic returns (Table 12.1).[2]

12.2.7.2 Cultural Control

Younger leaf tissue is more susceptible to fungal infection than older, mature leaves. Delaying disease development until crops are more advanced reduces the likelihood for yield loss. In regions where rusts are more consistently a problem, some producers avoid disease or minimize its effects by not planting late or by using shorter season hybrids. By doing so, they have a more mature and resistant crop in the field when rust spores arrive and sometimes disease can be completely avoided altogether.

Although most of the current popular sweet corn hybrids are susceptible to rust, resistant varieties are becoming available. 2 types of resistance are being used by commercial sweet corn breeders: race-specific resistance and partial rust resistance. A partial list of hybrid reaction to rust severity at harvest, from most to least resistant, includes aRRestor, Excellency, and Prevailer (possess specific resistance with 0% rust severity and fungus unable to sporulate); Sweetie, Miracle, Country Gentleman, Sucro, Sugar Time (partial resistance); Dandy, Gold Dust, Golden Glade, Patriot, Tendertreat EH, Sugar Loaf (moderate resistance); Seneca Horizon, Gold Cup, Seneca Sentry, Kandy Corn EH, Jubilee, Sweet Sal, Commander, Stylepak, Merit, Silver Queen, Florida Staysweet, and Sweet Sue (least resistance). Resistant or moderately resistant varieties should be used for late plantings when fungal spore density in the air is likely to be high because of infections of earlier-planted sweet corn. The varieties listed are examples only, and no endorsement is implied.

12.3   COMMON SMUT OF CORN

Corn smut is an extremely common disease of sweet, pop, and dent corn in Ohio and throughout the world. It is usually not economically important, although in some years yield losses in sweet corn may be as high as 20%. In Mexico, immature smut galls are consumed as an edible delicacy known as cuit-lacoche, and sweet corn smut galls have become a high-value crop for some growers in the northeastern United States who sell them to Mexican restaurants. The fungus attacks only corn-field corn (dent and flint), Indian or ornamental corn, popcorn, and sweet corn and the closely related teosinte (*Zea mays* subsp. *mexicana*) but is most destructive to sweet corn. The smut is most prevalent on young, actively growing plants that have been injured by detasseling in seed fields, hail, blowing soil, or and particles, insects, "buggy-whipping", and by cultivation or spraying equipment. Corn smut differs from other cereal smuts in that any part of the plant aboveground may be attacked, from the seedling stage to maturity.

12.3.1   CAUSAL ORGANISM

Corn smut is majorly caused by the fungi mentioned:

Species	Associated Disease Phase	Economic Importance
Ustilago zeae (synonym *U. maydis*)	Leaves, Fruits	High

TABLE 12.1

Management of Common Rust of Corn

Systemic Products	Active Ingredient	Mode of Action	Rate/A	Pre-Harvest Interval (days)
Headline® (BASF)	Pyraclostrobin	Preventative	6–9 oz	7
Quadris® Flowable (Syngenta)	azoxystrobin	Preventative	6–9 oz	7
Tilt® (Syngenta)	propiconazole	Curative	4 oz	30
PropiMax® EC (Dow AgroSciences)	propiconazole	Curative	4 oz	30
Folicur® 3.6F (Bayer CropScience)	Tebuconazole	Curative	4–6 oz	36
Quilt® (Syngenta)	azoxystrobin + propiconazole	Both	10.5–14.0 oz	30
Stratego® (Bayer CropScience)	trifloxystrobin + propiconazole	Both	7–10 oz	30
Contact (Preventative) Products Containing Chlorothalonil				
Dithane™ DF, Dithane DF Rainshield™, Dithane™ M45 (Dow AgroSciences)			1.5 lb	40
Dithane™ F-45 (Dow AgroSciences)			1.2 qt	40
DuPont™ Manzate® Pro-Stick™			1.5 lb	40
Manzate® 75DF (Griffin L.L.C.)			1.2 qt	40
Manzate® Flowable (Griffin L.L.C.)			1.5 lb	
Penncozeb® 4FL (Cerexagri-Nisso L.L.C.)			0.8–1.2 qt	40
Penncozeb® 75DF (Cerexagri-Nisso L.L.C.)			1.0–1.5 lb	
Penncozeb® 80WP (Cerexagri-Nisso L.L.C.)			1.0–1.5 lb	
Contact (Preventative) Products Containing Chlorothalonil				
Bravo Ultrex® (Syngenta Crop Protection)			0.7–1.8 lb	14
Echo® 720 Agricultural Fungicides (Sipcam Agro USA, Inc.)			1 1/4–1 5/8 lb	14
Echo® 90DF Agricultural Fungicides (Sipcam Agro USA, Inc.)			1.5–2.0 pints	14

12.3.2 Symptoms

Common corn smut is easily recognized and is probably the best-known disease on corn. All actively growing or embryonic corn tissue is susceptible. Galls are commonly found on the tassels, husks, ears and kernels, stalks, leaves, axillary buds and, rarely, on the aerial roots. As the smut galls enlarge, they are covered by a glistening, greenish to silvery-white membrane. Later, the inner tissue darkens because of spore formation (Figure 12.7). Mature galls may reach 6 inches (15 cms) in diameter and are filled with millions of microscopic, dark, olive brown to black, greasy to powdery spores—except for the small, hard, pea-sized galls that form on the leaves. The spores (teliospores, sometimes called chlamydospores) are released when the whitish outer membrane of the gall ruptures at maturity (Figures 12.8 and 12.9).

The fungus infects the plant through wounds caused by cultivation, hail, or insects. It can also infect newly formed silks. Galls are usually larger and more obvious on the ears, but you may also find them on the leaves, stalk, tassels, and aerial roots. The number, size and location of galls depend on the age of plants at the time of infection. Leaf galls differ greatly in size but usually are small when compared to stalk and ear galls. Galls 20-30 cm (8–12 inches) in diameter are common on the stalk. Rudimentary ear shoots below the fertile ear are commonly infected. Galls can replace most of the tassel or individual florets depending on the time and severity of infection. Ear galls usually result from infection of individual ovaries. Commonly, a few kernels at the basal or tip ends of the ear are infected, although nearly every kernel on an ear may be replaced by smut galls if most ovaries are infected.

12.3.3 Cause and Disease Development

Ustilago maydis (Syn. Ustilago zeae) is taxonomically grouped with the heterobasidiomycete fungi. Corn smut is caused by the fungus, *Ustilago zeae,* that survives as a resistant spore in the soil over winter, and possibly for 2–3 years. These spores, called teliospores, can be blown long distances

FIGURE 12.7   Initial symptoms of common smut galls on corn ear and mature common smut galls on corn ear.

FIGURE 12.8   Small, white, firm galls develop 9–10 days after infection.

with soil particles or carried into a new area on unshelled seed corn and in manure from animals that are fed infected corn stalks. The teliospores germinate in moist air and give rise to tiny spores called sporidia. The sporidia bud like yeast, forming new spores that germinate in rainwater that collects in the leaf sheaths. This leads to infections that are visible in 10 days or more. Wounds from various injuries provide points for the fungus to enter the plant.

12.3.4   Favorable Conditions

The smut fungus is sensitive to temperature and moisture changes. In a warm season, the amount of smut is related closely to the amount of soil moisture, especially during June. When temperatures are lower than normal, there may be little smut even though soil moisture remains high. Favorable temperature is between 10°C–35°C (50°F–95°F).

12.3.5   Disease Cycle

The diploid (2N) teliospores, which are very resistant to freezing and drying, may survive in the soil or crop debris for several years. During spring and summer, when the temperature is between 10°C–35°C (50°F–95°F) and moisture is present, the black, globose or subglobose, spiny teliospores which are 8–12 m in diameter, germinate to form a 4-celled basidium (promycelium) that produce from 4 to many oval, haploid (N) sporidia or basidiospores, which are of 2 sexes, (+) and (–). The sporidia are blown about by air currents or are splashed by water to young, developing tissues of corn plants. Infection occurs when the + and - sporidia germinate to form fine hyphae that penetrate the corn tissue through stomata, wounds, or directly through cell walls. The hyphae of 2 compatible sporidia fuse. The resulting hypha enlarges in diameter, becomes binucleate and dikaryotic (N+B), and stimulates an increase in size and number of cells in the corn plant forming a gall. Infection may

FIGURE 12.9   Galls start to have a gray, silvery appearance as streaks of blackened tissues (teliospores) begin to from 14 to 15 days after infection.

also occur from hyphae arising directly from germinating teliospores. Ears of corn are infected by hyphae growing through the silks much like pollen grains.

The smut mycelium in a gall grows between the corn cells until just before the teliospores are formed. The enlarged corn cells are then invaded, collapse, and die. The smut fungus feeds on the cell contents for its further growth and the gall then consists primarily of dikaryotic mycelium and the remains of corn cells. Most of the smut cells develop into teliospores which absorb and utilize the contents of other mycelial cells. Only the membrane covering the smut gall is unaffected by the fungus.[3]

The interval between infection and the formation of mature galls varies from one to several weeks under favorable conditions. Spores formed in the first smut galls may germinate and infect the same or other corn plants, although most spores fall to the ground or remain in corn debris. Galls form and spores are disseminated continuously through the summer growing period.

When animals eat "smutty" stalks, leaves, and ears, the spores may remain alive when passing through the alimentary canal and can be carried in the manure. When infested manure is spread on crop land, sporidia produced by the germinating teliospores may be blown or washed to the surface of a corn plant, germinate, and cause infection. Smut spores are killed by the acids in silage (Figure 12.10).

12.3.6 Management

The disease can be managed using the following strategies:

1. Corn hybrids differ in apparent resistance. Choosing the best adapted, resistant hybrids and varieties available is the best means of controlling common smut. Such hybrids possess generalized or field resistance to the corn-smut fungus. The difference in apparent resistance between corn lines is often based on the protection given by the sheath and husks. Corn breeders generally avoid using very smut-susceptible inbreeds and their hybrids or varieties.
2. Maintain well-balanced soil fertility based on a soil test.
3. Avoid mechanical injuries to plants during cultivation and spraying.

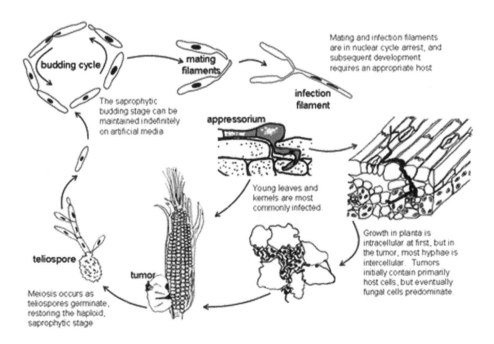

FIGURE 12.10   Disease cycle of smut.

4. Protect corn against insects (such as corn earworms and European corn borers) by timely applications of insecticides recommended by University of Illinois Extension Entomologists. This often decreases the incidence of common smut in sweet corn.

5. In home gardens, cut out and destroy the galls before the smut "boils" rupture and the teliosporses are released.

12.4 DOWNY MILDEW OF CORN (CRAZY TOP OF CORN)

Crazy top of corn, caused by the fungus *Scleroph thora macrospora* (Sclerospora macrospora), has occurred widely but sporadically in Illinois. The disease is seldom prevalent enough to cause much damage, although losses of at least 60% in grain yields have been reported in parts of some fields. Crazy top, when it does occur, is invariably found in localized areas of fields and gardens where the soil becomes waterlogged or flooded between the time the corn kernels germinate and the seedlings are 6–10 inches tall.

The causal fungus attacks all types of corn and more than 140 species of wild and cultivated grasses. The disease hosts include oats, rice, sorghums, wheats, crabgrasses, witchgrasses, foxtails, and barnyardgrass. The fungus is incapable of developing in the absence of a host plant and must re-infect living corn or grass plants each season. In the absence of corn, the crazy top fungus maintains itself on wild grasses.

12.4.1 CAUSAL ORGANISM

Downy mildew is mostly caused by the fungi mentioned below.

Species	Associated Disease Phase	Economic Importance
S. thora macrospora (Sclerospora macrospora)	Leaves, Fruits, and Flowers	High

12.4.2 SYMPTOMS

The symptoms vary greatly according to the time of infection and the degree of host colonization by the fungus. The most conspicuous symptom is the partial or complete replacement of the normal tassel by a large, bushy mass of small leaves (Figure 12.11). These modified leaflike inflorescences are described as crazy top. No pollen is produced, since normal flower parts in the tassel are completely deformed. Ear formation may also be checked, causing the ear shoots to be numerous, elongated, leafy, and barren. Plants affected by crazy top vary greatly in height. Some may be severely stunted, with narrow leaves that are strap-like, leathery, yellow to brown, and streaked, and with 6–10 tillers per plant (Figure 12.12). Generally, excessive tillering, rolling, and twisting of the upper

FIGURE 12.11   Crazy top with abnormal tassel.

FIGURE 12.12   Crazy top and tillering.

leaves appear first. Other plants may be taller than average, with additional nodes and leaves above the ear and in the shank. The principal effect, however, is the development of the tassel and ears into leafy tissues. Common corn smut frequently occurs on these abnormal leafy growths (Figure 12.13).

12.4.3   CAUSE AND DISEASE DEVELOPMENT

Crazy top of corn caused by the fungus *S. thora macrospora* (Sclerospora macrospora). The fungus is like Phytophthora in that it survives as oospores in the soil or in infected plant tissue.

12.4.4   FAVORABLE CONDITIONS

Infection occurs over a wide range of soil temperatures. The optimum for sporangial germination is 12°C–16°C (53°F–63°F).

12.4.5   DISEASE CYCLE

The *Sclerophthora* fungus overseasons in diseased corn or grass residue as microscopic round-ish oospores that are thick-walled and colorless to yellowish. The numerous oospores presumably germinate in soil that is saturated for 24–48 hours, forming a thin-walled tube that bears a lemon-shaped sporangium. The sporangium, in turn, germinates to produce numerous motile zoospores. After swimming about in the soil water for a short time, the zoospores encyst and produce a germ tube that penetrates seedling host tissue sometime during the period from shortly after sowing to

FIGURE 12.13   Field symptoms of downy mildew in corn.

before the plants are in the 4- to 5-leaf stage. Following infection, the fungus develops systemically and invades the entire corn plants, being most abundant in meristematic tissues.

12.4.6 MANAGEMENT

The disease can be managed using the following strategies.

No highly effective control measures can be recommended for crazy top. Very little is known about the level of resistance in corn hybrids to this disease. Proper soil drainage will reduce the risk of flooding and subsequent infection. Avoid planting corn in low wet spots where the disease is known to occur.

12.5 EYESPOT DISEASE OF CORN

Eyespot, caused by the fungus *Aureobasidium zeae* (previously known as *Kabatiella zeae*), has become a common disease in Midwest American corn. It is prevalent in the north central and north-eastern states and in Ontario and Quebec, but sometimes extends much farther south. In Iowa, it is most often found in the northern half of the state. The disease also is found in Europe and South America. The disease is commonly associated with continuous corn culture and reduced tillage practices. The adoption of high-residue farming practices has contributed to the increasing importance of eyespot. Eyespot is favored by long periods of cool, wet weather during the growing season. Therefore, it is more of a problem in the northern regions of the corn belt.

Early and severe leaf blighting from eyespot in no-till and reduced tillage fields have resulted in yield losses when susceptible hybrids were grown. Yield losses can be expected when much of the leaf area is blighted within three to four weeks after silking. Defoliation from leaf blighting also increases the amount of stalk rot, resulting in additional losses from lodged corn.

12.5.1 CAUSAL ORGANISM

Eyespot disease is majorly caused by the fungi mentioned:

Species	Associated Disease Phase	Economic Importance
A. zeae (previously known as *K. zeae*)	Leaves and Stalk	High

12.5.2 SYMPTOMS

The disease usually spreads from the lower leaves upward. When spores are blown in from neighboring fields, lesions can be random or concentrated in the upper leaves. Infections appear initially as small, water-soaked, circular lesions that are about 1/16 inch in diameter. The lesions can enlarge to about 1/8 inch in diameter, and become chlorotic, then necrotic, with a tan center and a darker brown or purple margin (Figure 12.14). The spot is usually surrounded by a larger yellow "halo" which is most visible when light passes through the leaf (Figure 12.15). You can observe the halo by holding a leaf up toward a blue-sky background.

Spots can vary in size and color depending on the hybrid. Spots can coalesce into larger necrotic areas (Figure 12.16). It is common to observe bands of lesions across a leaf, indicating that infection took place in the moist environment of the whorl after a period of spore dispersal. Lesions often are concentrated along the leaf edges and leaf tips. Severely infected leaves can be entirely blighted. The dark margins of the lesions remain visible on dead leaves.

Eyespot symptoms can be confused with physiologic or genetic leaf spots, which are noninfectious, or with insect feeding wounds. Eyespot also can be confused with Curvularia leaf spots and the early symptoms of northern leaf spot or gray leaf spot. Noninfectious leaf spots sometimes do not develop

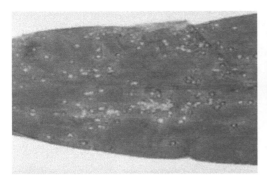 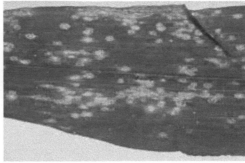

FIGURE 12.14   Earlier chlorotic spot with a tan center and a darker brown or purple margin.

a necrotic center, and if they do, it does not have the darker margin. Northern leaf spot lesions can enlarge to be substantially larger than eyespot lesions, and usually are less circular in shape. Fully developed gray leaf spot lesions are long, rectangular, and easily distinguished from eyespot.

12.5.3   CAUSE AND DISEASE DEVELOPMENT

The eyespot disease is caused by the fungus, *Kabatiella zeae*. Spores produced by this fungus are widely dispersed by the wind. Eyespot appears to be most severe in fields where residues from the previous corn crop are left on the soil surface and in fields of continuous corn for 2 or more years.

12.5.4   FAVORABLE CONDITIONS

Eyespot is favored by long periods of cool, wet weather during the growing season and is more of a problem in the northern regions of the corn belt.

Germination of the stromas occurs when temperature is over 10°C. Conidial germination occurs on the maize leaves only after at least 7 hours continuous leaf wetness. Maize susceptibility is highest at 10°C–12°C., therefore, eyespot mostly spreads during the cold rainy seasons. The highest eyespot intensity occurs in August, September, and early October. *Kabatiella* infection on maize leaves remains visible even when the leaves die.

12.5.5   DISEASE CYCLE

The eyespot disease is caused by the fungus, *Kabatiella zeae*. Spores produced by this fungus are widely dispersed by the wind. Spores settling on a susceptible corn leaf may germinate and initiate infection within a week, especially during cool, wet weather. Older corn leaves appear to be more

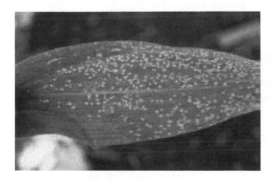 

FIGURE 12.15   Larger yellow "halo" spot which is most visible when light passes through the leaf.

FIGURE 12.16   Spots coalesce into larger necrotic areas.

susceptible to infection. Under favorable conditions new spores are produced and quickly spread to other plants. When infection is severe, essentially all plants within a field may be killed within two weeks by this fungus. Because the eyespot fungus produces spores during cooler weather, the disease is most prevalent during August, September, and early October. However, during abnormally cool seasons it can attack corn much earlier and cause significant losses. The fungus overwinters and survives between corn crops on residue left on the soil surface. In the spring, the fungus again produces spores that are carried to the new corn crop. The fungus may also be seedborne, but this source of fungal inoculum is negligible when compared to the number of spores produced on infested crop residues.

12.5.6   Management

The disease can be managed using the following strategies:

12.5.6.1   Chemical Control

Applying fungicide sprays early in the epidemic can have a significant impact on the disease and yield. Fungicides can be economically beneficial, especially in seed corn production. Fungicides should be considered only when corn grown in the field the previous year was affected by eyespot and reduced tillage practices are being used. Resistant hybrids should be the first choice.

Fungicides registered for use against *K. zeae* include mancozeb, propiconazole, chlorothalonil and benomyl (Gay and Cassini, 1973; Pronczuk et al., 1996). For effective protection against *K. zeae*, seed dressings are recommended, followed by spraying the plants at the early stages of disease development when 1% or less of the leaf area is infected. More than 1 application may be necessary when conditions are favorable to the disease. The use of fungicides against eyespot can be prohibitively expensive, except on seed production fields. Pest control plays an important role in reducing the occurrence of eyespot, particularly the control of Aphididae and Thysanoptera, which feed on maize and can facilitate the penetration of conidia.

12.5.6.2   Biological Control

Resistance to *K. zeae* is important in the control of eyespot and hybrids with some resistance to the disease should be planted. Susceptible hydrids include Julia, Heros, Agio, and Aura; more resistant hybrids include Kosmo and Elsa. Even a known source of resistance such as the line Oh43 can be subject to infection in the case of epiphytosis (Reifschneider and Arny, 1983).

12.5.6.3   Cultural Control

Thorough cultivation and crop rotation can reduce early infection by *K. zeae*. Arny et al. (1971) recommend using a 3–4-year interval for maize cultivation in the same field. The amount of infectious material can be reduced by a suitable crop rotation and thorough plowing and destruction of after-harvest residues, particularly from seriously infected plants. Deep plowing of crop debris prevents sporulation of the stromas and promotes decomposition, thus limiting early season spread.

12.6 GRAY LEAF SPOT OF CORN

Gray leaf spot of corn, caused by the fungus *Cercospora zeae-maydis*, has been known in the United States since 1924 when it was first reported in Illinois. Until the 1970s, the disease was a minor pathogen except for occasional outbreaks. Gray leaf spot of corn, while of little consequence in the United States before 1970, has become a major concern to many corn producers in recent years. Serious outbreaks of the disease first occurred in the early to mid-1970s in low-lying areas in the mountainous regions of Kentucky, North Carolina, Tennessee, and Virginia. Since this time, the disease has spread to most corn-producing areas of western Kentucky and parts of Delaware, Illinois, Indiana, Iowa, Maryland, Missouri, Ohio, Pennsylvania, and west Tennessee. The recent increase in gray leaf spot's distribution and severity has been attributed to the increased use of no-tillage practices and the more frequent monoculture of corn. Both factors favor the survival and increase of the gray leaf spot fungus from 1 year to the next. The disease typically develops after tasseling. However, greatest losses occur when the disease occurs before tasseling. Yield reductions can range from 0–50 bu/A, depending on the time of disease onset, disease severity, and the corn hybrid's susceptibility and yield potential.

12.6.1 Causal Organism

Gray leaf spot is mostly caused by the fungi mentioned below.

Species	Associated Disease Phase	Economic Importance
C. zeae-maydis	Leaves	High

12.6.2 Symptoms

The early lesion produced on the corn leaves by *C. zeae-maydis* are yellow to tan in color and look like those produced by other diseases except they have a faint watery halo which can be seen when held up to the light. At this stage, it can be hard to correctly identify the disease, but as lesions mature, they elongate into narrow, rectangular, brown to gray spots. Lesions expand parallel to leaf veins and may become 1.5–2 inches long. On susceptible hybrids, lesions may also appear on leaf sheaths and husks. The major leaf veins restrict lateral expansion of leaf lesions, giving the lesions a blocky shape (Figure 12.17). Under favorable conditions, lesions can coalesce to form large, irregular areas of dead tissue on the leaves (Figure 12.18).

FIGURE 12.17 Leaf veins limit gray leaf spot lesions. As lesions mature, they expand to form long, rectangular areas of dead tissue.

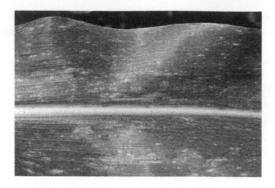

FIGURE 12.18   Severe leaf tissue blighting can occur and result in yield loss.

Symptoms vary by hybrid susceptibility. Hybrids with partial gray leaf spot resistance may not experience the characteristic lesion expansion. These hybrids restrict lesion growth, so they may have lesions that remain small and have a round or jagged shape instead of the long, rectangular shape characteristic of lesions on more susceptible hybrids. Gray leaf spot symptoms may be confused with symptoms of other foliar fungal diseases such as anthracnose leaf blight, eyespot, or common rust. Under periods of prolonged favorable conditions, severe blighting can occur. This blighting may extend to the leaf sheath, which remains on the cut stalk after harvest. Sheath lesions (Figure 12.19) are likely to serve as a source of fungal inoculum the following spring.

12.6.3   CAUSE AND DISEASE DEVELOPMENT

C. zeae-maydis, like many other foliar fungal pathogens of corn, is a poor competitor in the soil and can survive only if infected corn debris is present. Infected corn debris on the soil surface is the primary source of inoculum for the next corn crop. The fungus colonizing this debris produces conidia (spores) as early as May. These airborne spores are how the fungus infects the new corn crop.

FIGURE 12.19   Typical gray leaf spot lesion on leaf sheath.

12.6.4 FAVORABLE CONDITIONS

Gray leaf spot is a highly weather-dependent disease. The pathogen requires long periods of high relative humidity and free moisture (dew) on the leaves for infection to occur. Corn gray leaf spot flourishes under extended periods of high relative humidity (>2 days) and free moisture on leaves due to fog, dew, or light rain. Additionally, heavy rains tend to assist in dispersal of the pathogen. Temperatures between 24°C and 35°C are also required. If temperature drops below 24°C during wet periods or lack 12 hours of wetness, the extent of disease will be greatly diminished. In the Midwest and Mid-Atlantic, these conditions are favorable for spore development during the spring and summer months.

12.6.5 DISEASE CYCLE

C. zeae-maydis survives only as long as infected corn debris is present; however, it is a poor soil competitor. The debris on the soil surface is a cause for primary inoculation that infects the incoming corn crop for the next season. By late spring, conidia (asexual spores) are produced by *C. zeae-maydis* in the debris through wind dispersal or rain. The conidia are disseminated and eventually infect the new corn crop. For the pathogen to infect the host, high relative humidity and moisture (dew) on the leaves are necessary for inoculation. Primary inoculation occurs on lower regions of younger leaves, where conidia germinate across leaf surfaces and penetrate through stomata via a flattened hyphal organ, an appressorium. *C. zeae-maydis* is atypical in that its conidia can grow and survive for days before penetration, unlike most spores that need to penetrate within hours to ensure survival. Once infection occurs, the conidia are produced in these lower leaf regions. Assuming favorable weather conditions, these conidia serve as secondary inoculum for upper leaf regions, as well as husks and sheaths (where it can also overwinter and produce conidia the following season). Additionally, wind and heavy rains tend to disperse the conidia during many secondary cycles to other parts of the field causing more secondary cycles of infection. If conditions are unfavorable for inoculation, the pathogen undergoes a state of dormancy during the winter season and reactivates when conditions favorable to inoculation return (moist, humid) the following season. The fungus overwinters as stromata (mixture of plant tissues and fungal mycelium) in leaf debris, which give rise to conidia causing primary inoculations the following spring and summer (Figure 12.20).[4]

12.6.6 MANAGEMENT

To prevent and manage corn gray leaf spot, the overall approach is to reduce the rate of disease growth and expansion. This is done by limiting the amount of secondary disease cycles and protecting leaf area from damage until after corn grain formation. High risks for corn gray leaf spot are divided into 8 factors, which require specific management strategies.

High risk factors for gray leaf spot in corn are:

1. Susceptible hybrid
2. Continuous corn
3. Late planting date
4. Minimum tillage systems
5. Field history of severe disease
6. Early disease activity (before tasseling)
7. Irrigation
8. Favorable weather forecasts for disease

12.6.6.1 Chemical Control

Fungicides, if sprayed early in season before initial damage, can be effective in reducing disease.

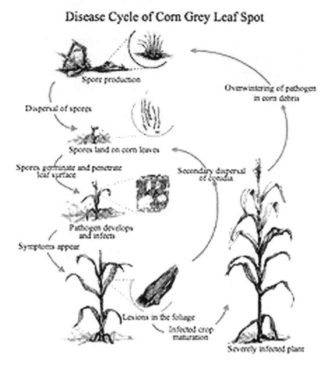

FIGURE 12.20   Life cycle of corn gray leaf spot.

Currently there are 5 known fungicides that treat corn gray leaf spot:

1. Headline EC (active ingredient: pyraclostrobin)
2. Quilt (active ingredient: azoxystrobin + propiconazole)
3. Proline 480 SC (active ingredient: prothioconazole)
4. Tilt 250 E,
5. Bumper 418 EC (active ingredient: propiconazole)

12.6.6.2   Headline EC

Headline is to be applied at 400–600 mL per/hectare (ha). For optimal disease control, begin applications prior to disease development (see disease cycle). This fungicide can only be applied a maximum of 2 applications/year. Ground and aerial application are both acceptable.

12.6.6.3   Quilt

Quilt is to be applied at 0.75–1.0 L per/ha. Application of Quilt is to be made upon first appearance of disease, followed by a second application 14 days after, if environmental conditions are favorable for disease development (see disease cycle). Upon browning of corn sheaths, Quilt is not to be applied. This fungicide can only be applied a maximum 2 applications/yr. Ground and aerial application are both acceptable.

12.6.6.4   Proline 480 SC

Proline 480 SC is to be applied at 420 mL per/ha. This fungicide can only be applied a maximum 1 time/year. It should be note that only ground application is acceptable. A 24-hour re-entry time is required (minimum amount of time that must pass between the time a fungicide is applied to an area or crop and the time that people can go into that area without protective clothing and equipment).

12.6.6.5 Tilt 250 and Bumper 418 EC

Tilt 250 is to be applied at 500 mL per/ha. Bumper 418 EC is to be applied at 300 mL per/ha. Both fungicides are to be applied when rust pustules first appear. If disease is prevalent after primary application, a second application 14 days later may be necessary. Two weeks later, a third application can be made under severe amount of disease. Ground and aerial application are both acceptable.

When spraying fungicides Quilt and Headline EC at 6 oz/A at tassel stage using a tractor-mounted CO_2 powered sprayer using 20 gallons of water/A, average yield was seen to increase. The use of fungicides can be both economically and environmentally costly and should only be applied on susceptible varieties and large-scale corn production. To prevent fungal resistance to fungicides, all fungicides are to be used alternatively, switching fungicides with different modes of action. Pyraclostrobin (Headline EC) is a QoL fungicide, propiconazole is a sterol biosynthesis inhibitor (SBI), azoxystrobin is a quinone outside inhibitor (QoI), and prothioconazole is a deMethylation inhibitor (DMI).

12.6.6.6 Biological Control

The most proficient and economical method to reduce yield losses from corn gray leaf spot is by introducing resistant plant varieties. In places where leaf spot occurs, these crops can ultimately grow and still be resistant to the disease. Although the disease is not eliminated, and resistant varieties show disease symptoms, at the end of the growing season, the disease is not as effective in reducing crop yield. SC 407 have been proven to be common corn variety that are resistant to gray leaf spot. If gray leaf spot infection is high, this variety may require fungicide application to achieve full potential.

12.6.6.7 Cultural Control

12.6.6.7.1 Tillage Practices

Tillage, the turning of corn residues, is beneficial in reducing pathogen survival and inoculum for the succeeding corn crop. The burial of infested debris facilitates rotting and deprives the fungus of a food base. The fungus is unable to survive freely within the soil. It can only overwinter within and on dead corn tissue remaining on or above the soil surface. Disking does not sufficiently bury the infected debris. Mold board plowing does, but it may not be advisable in some fields because of increased erosion potential. Erosion potential can be reduced by fall plowing and seeding to a winter cover crop, followed by no-till planting of corn in the spring. Burial of infected debris, however, may not provide an effective means of reducing gray leaf spot inoculum in regions where widespread use of conservation tillage is practiced because the pathogen may blow into a field from adjacent fields.

12.6.6.7.2 Crop Rotation

Taking a field out of corn production or rotating to a non-host crop for 1 year can reduce gray leaf spot severity. The fungus is unable to survive more than 1 season in infected corn debris. Corn is the only crop this fungus is known to attack. However, the potential for herbicide carryover may restrict the selection of crops in the rotation scheme.

12.6.6.7.3 Corn Silage Production

Growing corn ensilage significantly reduces the amount of inoculum available for infecting the next corn crop in 2 ways. First, silage corn is usually harvested before significant blighting from gray leaf spot occurs, thereby reducing the amount of the pathogen available to survive the winter months. Secondly, removal of corn for ensilage leaves only about 6 inches of stalk in the ground. This practice leaves little, if any, infected debris for overwintering of the fungus.

12.7 MAIZE DWARF MOSAIC VIRUS

Most important viral diseases of corn are maize dwarf mosaic (MDM). Distribution of the diseases generally reflects the geographic distribution of their overwintering host, johnsongrass. In Ohio,

MDM is a serious threat to corn production in the southern half of the state. However, MDM has caused serious economic loss on late-planted corn in northern Ohio, especially in the sweet corn growing region along Lake Erie.

Yield losses in fields where the disease is well established may be severe depending on the susceptibility of the corn hybrid being grown. In general, most loss occurs when the plants become infected at the "knee high" stage of development and is minimal on most hybrids if infection occurs after tasseling.

Diagnosis of virus-infected plants is difficult based on field symptoms. Samples of plants should be tested in the laboratory to confirm the presence of the virus.

12.7.1 Causal Organism

Maize dwarf mosaic is mostly caused by the virus mentioned below.

Species	Associated Disease Phase	Economic Importance
MDMV	Leaves	High

12.7.2 Symptoms

When looking for symptoms of MDMV in corn, one must also be aware of the plant's growth stage as symptoms may affect the plant differently at various stages. Young leaves may experience chlorotic spotting which may eventually turn into a mosaic or mottle pattern. Later in the growing season, the mosaic pattern may bleed into a general yellowing of the leaf and eventually areas of red streaks or blotches may appear if nighttime temperatures are consistently around 15.5°C. Plants affected later in their reproductive cycle may experience a slowing in ear development, while some plants may even become barren. It is not uncommon for plants to have shortened upper internodes or an increase in tiller number. Symptoms of MDMV include narrow, light green to yellow streaks along the veins of leaves, leaf sheaths, and husks. As infected plants continue to grow, and the temperature rises, the mosaic symptoms may disappear while the young leaves become more yellow. Plants may be stunted and have numerous tillers and poor seed set. Infected plants may be predisposed to ear rot and stalk rot (Figures 12.21 and 12.22).

12.7.3 Cause and Disease Development

Mosaic corn virus disease is caused by MDMV is single stranded RNA potyviruses in *Potyviridae*. They are flexuous and rod-shaped, measuring 12 × 750 nm. Antiserum kits are commercially available to confirm virus identity.

FIGURE 12.21   Mosaic corn virus symptom.

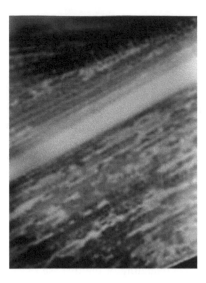

FIGURE 12.22   Chlorotic spots and streaks on green young leaves.

Maize dwarf mosaic occurs where aphid vectors are prevalent and where alternate hosts are cultivated. Diseases can also be transmitted through infected seed and mechanical injury, although these techniques are less common than aphid transmission. Mechanical transmission occurs predominantly in greenhouses and is not considered a major problem in the field. Disease incidence is highest where vector populations are high, a large number of infected plants are present, and susceptible varieties are cultivated.

12.7.4   FAVORABLE CONDITIONS

Average to warm temperatures favors the disease. Nearby johnsongrass infected with MDMV may increase disease.

12.7.5   DISEASE CYCLE

MDMV exists in several strains, the most common being strain A, which infects and overwinters primarily on johnsongrass. Strain B does not infect johnsongrass. Besides corn and johnsongrass, the MDMV may infect over 100 wild and cultivated grasses.[5]

More than 20 species of aphids can transmit MDMV. An aphid can acquire the virus within a few minutes of feeding on an infected corn or johnsongrass plant. The aphid then flies or is carried by the wind to other corn plants and inoculates them with the virus. The aphid retains and is generally able to transmit the virus for 15–30 minutes after acquiring it. The corn leaf aphid and the green peach aphid are common aphid vectors of MDMV.

12.7.6   MANAGEMENT

The disease can be managed using the following strategies:

1. Grow hybrids tolerant or resistant to MDMV. There is good tolerance and resistance to strain A, but only fair tolerance and no resistance to strain B in dent corn. There is no resistance to maize chlorotic dwarf virus and only fair tolerance.
2. Destroy johnsongrass and other grass hosts, including volunteer corn in areas where corn is to be planted. Best control occurs when all farmers in a community cooperate in eradicating johnsongrass.

3. Plant early, since early planted corn will escape damage because aphid populations do not build up until the plants are past the seedling stage.

12.8 NORTHERN CORN LEAF BLIGHT OF MAIZE

Northern corn leaf blight (NCLB), caused by the fungus *Exserohilum turcicum* previously called *Helmithosporium turcicum*, can cause yield losses in humid areas where corn is grown. In Ohio, NCLB can occur throughout the state but usually does not appear in fields before silking. This disease rarely causes significant yield losses during dry weather, but during wet weather it may result in losses of over 30% if established on the upper leaves of the plant by the silking stage of development. If leaf damage is only moderate or is delayed until six weeks after silking, yield losses are minimal. Northern corn leaf blight also predisposes corn to stalk rot by increasing stress on the plants. The disease thrives when relatively cool summer temperatures coincide with high humidity and available moisture. The number of NLB outbreaks has increased considerably over the past 5 years. Corn silage yield and quality losses from this disease can be significant. Therefore, it is important for us to gain a better understanding of the disease cycle, symptoms, and management practices that can be employed to reduce the impact of NLB on the corn crop.

12.8.1 CAUSAL ORGANISM

NCLB is mostly caused by the fungi mentioned below.

Species	Associated Disease Phase	Economic Importance
E. turcicum (Previously Known as *H. turcicum*)	Leaves	Moderate

12.8.2 SYMPTOMS

The telltale sign of NCLB is the 1-to-6-inch-long, cigar-shaped, gray-green to tan-colored lesions on the lower leaves. As the disease develops, the lesions spread to all leafy structures, including the husks. The lesions may become so numerous that the leaves are eventually destroyed causing major reductions in yield due to lack of carbohydrates available to fill the grain. The leaves then become grayish-green and brittle, resembling leaves killed by frost. Yield losses can reach as high as 30–50% if the disease establishes itself before tasseling (Figure 12.23 and 12.24).

The greatest losses from NCLB occur when severe necrosis develops on the upper $^2/_3$ of crop canopy by silking. The reduction in photosynthesis due to the necrosis results in reduced ear fill and when symptoms develop on the husks they appear older and are less marketable.

FIGURE 12.23   Long, narrow lesions that run parallel to the leaf margin are early symptoms of NCLB.

FIGURE 12.24   Oblong lesions develop on leaf tissue after infection by the NCLB fungus.

12.8.3   CAUSE AND DISEASE DEVELOPMENT

NCLB, caused by the fungus *E. turcicum* previously called *H. turcicum*, can cause yield losses in humid areas where corn is grown. The fungal mycelia and conidia can overwinter in plant debris, and the disease is later transferred by wind to new plants. Severe yield loss can occur when leaves become blighted during early grain fill. NCLB will be more severe in fields with corn following corn under reduced tillage. Infection occurs during periods of moderate (18°C–27°C), wet, and humid weather. The fungus requires 6–18 hours of water on the leaf surface to cause infection. Therefore, symptoms are commonly observed following long periods of heavy dew and overcast days, and in bottomlands or fields adjacent to woods where humidity will be higher and dew will persist longer into the morning. In Indiana, symptoms are frequently observed late in the growing season when days become cooler.

12.8.4   FAVORABLE CONDITIONS

Favorable conditions for the pathogens are moderate temperatures (18°C–27°C) and leaf wetness from rain, dew, or fog for at least 6 hours. A cool and humid climate will favor the disease. When there is an extended period of wet weather, the fungus will start multiplying. Continuous growth of only corn might also lead to this corn blight disease. This mainly occurs before the pollination phase or after the pollination phase. Dispersal of the spores will occur with the help of the wind.

12.8.5   DISEASE CYCLE

NCLB is caused by the fungus *E. turcicu*. It overwinters as mycelia and conidia in diseased corn stalks. In the spring and early summer, spores are produced on this crop residue when environmental conditions are favorable. Primary infections occur when spores are spread by rain splash and air currents to the leaves of new crop plants. Infection will occur if free water is present on the leaf surface for 6–18 hours and temperatures are 18°C–27°C. Secondary infections occur readily from plant to plant and even from field to field. Infections generally begin on lower leaves first and then

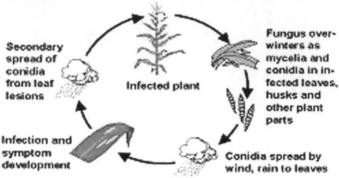

Northern Leaf Blight Disease Cycle
Exserohilum turcicum (Helminthosporium turcicum)

Secondary spread of conidia from leaf lesions

Infected plant

Fungus over-winters as mycelia and conidia in infected leaves, husks and other plant parts

Infection and symptom development

Conidia spread by wind, rain to leaves

FIGURE 12.25   NCLB disease cycle.

progress up the plant. Heavy dews, frequent light showers, high humidity, and moderate temperatures favor the spread of the disease (Figure 12.25).

12.8.6   MANAGEMENT

The disease can be managed using the following strategies:

12.8.6.1   Chemical Control

Fungicide sprays are recommended only for fresh market sweet corn and hybrid seed production fields. The spray schedule should start when the first lesions appear on the leaf below the ear. Several fungicides are available for use on corn for NCLB control.

NCLB specific fungicides include those in FRAC group 11 (strobilurins, e.g. Quadris and Headline) and FRAC group 3 (triazoles, e.g. Tilt). There are also a number of products that contain both FRAC groups (11 + 3, e.g. Quilt and Stratego). Rotate between these FRAC codes and tank mix with a broad-spectrum protectant for resistance management when symptoms are first observed in the field as this will help manage NCLB. PHIs vary between the products so read the labels carefully when the crop is near harvest. Also depending on the label, NCLB might be referred to as Helminthosporium leaf blight which collectively refers to both NCLBand Southern corn leaf blight.

12.8.6.2   Biological Control

Host resistance can still be an effective tool for managing NCLB especially for later sweet corn plantings. There are different types of resistance genes that have been introduced into sweet corn hybrids through traditional breeding (not GMOs). Hybrids can have polygenic (partial) resistance which confers resistance to both races of the pathogen; however, the resistance is not complete for any of the races or monogenic resistance which confers resistance to only specific races of the pathogen. These various resistance genes will limit lesion size, lesion number, and the amount of sporulation within each lesion. When resistance genes are present in a hybrid, lesion size, shape, and color may vary. For example, hybrids that contain 1 of the monogenic resistance genes Ht1, Ht2, and Ht3 will develop chlorotic lesions but sporulation will be limited so the disease does not spread quickly.

Some seed companies indicate the degree of resistance with a numerical rating scale but pay close attention to these scales—individual companies use different values to indicate the level of resistance. In areas where NCLB is a chronic problem, producers should seek out hybrids with race-specific resistance genes (known as *Ht* genes).

12.8.6.3 Cultural Control

A 1-year rotation away from corn, followed by tillage is recommended to prevent disease development in the subsequent corn crop. In no-till or reduced till fields with a history of NCLB, a 2-year rotation out of corn may be needed to reduce the amount of disease in the following corn crop.

12.9 SOUTHERN CORN LEAF BLIGHT

Southern corn leaf blight (SCLB) is a fungal disease of maize caused by the plant pathogen *Bipolaris maydis* (also known as *Cochliobolus heterostrophus* in its teleomorph state). Farming practices and optimal environmental conditions for the propagation of *B. maydis* in the United States led to an epidemic in the year of 1970. In the early 1960s, seed corn companies began to use male sterile cytoplasm so that they could eliminate the previous need for hand detasseling to save both money and time. This seed was eventually bred into hybrid crops until there was an estimated 90% prevalence of Texas male sterile cytoplasm (Tcms) maize, vulnerable to the newly generated Race T. The disease, which first appeared in the United States in 1968, reached epidemic status in 1970 and destroyed about 15% of the corn belt's crop production that year. In 1970, the disease began in the southern United States, and by mid-August, had spread north to Minnesota and Maine. It is estimated that Illinois alone suffered a loss of 250 million bushels of corn to SCLB. The monetary value of the lost corn crop is estimated at 1 billion United States' dollars. In 1971, SCLB losses had basically disappeared. This was due to the return usage of normal cytoplasm corn, not due to conducive weather or residues being buried or planting early. The SCLB epidemic highlighted the issue of genetic uniformity in monoculture crops, which allows for a greater likelihood of new pathogen races and host vulnerability.[6]

In the present day, there are many management methods and better education practices, but the disease can still be an issue in tropical climates, causing devastating yield losses up to 70%.

12.9.1 CAUSAL ORGANISM

Southern corn leaf blight is mostly caused by the fungi mentioned below.

Species	Associated Disease Phase	Economic Importance
B. maydis (Also Known as *C. heterostrophus*)	Leaves	Medium

12.9.2 SYMPTOMS

The symptoms of SCLB are leaf lesions ranging from minute specks to spots of 1/2 inch wide and 11/2 inches in length. They are oblong, parallel-sided, and tan to grayish in color. A purplish to brown border may appear on the lesions depending on the genetic background of the plant. Early and severe infections in susceptible plants predispose them to stalk rots.

In the worst case, the lesions are numerous and can be several cms long and have dark red or purple edges. Ear and cob rot mat occur when kernels become infected with black masses of conidia (asexual spores). Stalks may also be damaged (Figure 12.26 and 12.27).

Because symptoms are a plant response and similar ones can be seen with other plant pathogens, *B. maydis* infection can be confirmed microscopically. The sign (actual pathogen presence) of SCLB is its conidia. They are visible under a microscope and are usually brown and tapered with round edges. NCLB often occurs when SCLB is present and their lesions distinguish between the 2. SCLB lesions are more parallel sided, lighter, and smaller in comparison to NCLB.

FIGURE 12.26   Lesions on corn leaf.

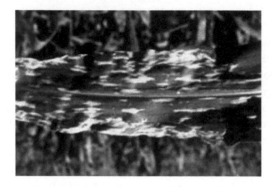

FIGURE 12.27   Severe infection of SCLB on corn leaf.

12.9.3   CAUSE AND DISEASE DEVELOPMENT

SCLB (*B. maydis*) is a member of the ascomycetes, or sac fungi, family. Mycelium and spores can overwinter in soil and crop debris. Spores are blown by wind or in water droplets onto the surface of leaves, and after they have germinated, they enter the plant through the stomata. The fungus produces a toxin that destroys the plants ability to capture energy from metabolism.

3 races known of this pathogen are known race O, race T, and race C. Race T and race C are known to be specifically virulent to corn with cytoplasm male-sterile T and cytoplasm male-sterile C, respectively. Since the switch from cytoplasm male sterile T to normal cytoplasm corn, race T is not considered to be a threat.

12.9.4   FAVORABLE CONDITIONS

SCLB can be found throughout the world—almost everywhere maize is grown. The amount of rainfall, relative humidity, and temperature of the area is critical to the spread and survival of disease. This is because SCLB favors a warm, moist climate. An environment with warm temperatures (20°C–32°C) and a high humidity level is particularly conducive to SCLB. In contrast, long and sunny growing seasons with dry conditions are highly unfavorable.

12.9.5   DISEASE CYCLE

The disease cycle of *Cocholiobolus heterostrophus* is polycyclic and releases either asexual conidia or sexual ascospores to infect corn plants. The asexual cycle is known to occur in nature and is of primary concern. Upon favorable moist and warm conditions, conidia (the primary inoculum) are released from lesions of an infected corn plant and carried to nearby plants via wind or splashing rain. Once conidia have landed on the leaf or sheath of a healthy plant, *B. maydis* will germinate on the tissue by way of polar germ tubes. The germ tubes either penetrate through the leaf or enter through a natural opening such as the stomata. The parenchymatous leaf tissue is invaded by the mycelium of the fungus; cells of the leaf tissue subsequently begin to turn brown and collapse. These lesions give rise to conidiophores which, upon favorable conditions, can either further infect the original host plant (kernels, husks, stalks, leaves) or release conidia to infect other nearby plants. The term "favorable conditions" implies that water is present on the leaf surface and temperature of the environment is between 15.5°C and 27°C. Under these conditions, spores germinate and penetrate the plant in 6 hours. The fungus overwinters in the corn debris as mycelium and spores, waiting once again for these favorable spring conditions. The generation time for new inoculum is only 51 hours.

As previously mentioned, *B. maydis* also has a sexual stage with ascospores, but this has only been observed in laboratory culture. Its ascospores (within asci) are found in the ascocarp *Cochiobolus*, a type of perithecium rare in nature. Thus, the main route of SCLB infection is asexual via conidial infection (Figure 12.28).

12.9.6 MANAGEMENT

The best practice for management of SCLB is breeding for host resistance. Both single gene and polygene resistance sources have been discovered. Normal cytoplasm maize can resist both Race T and Race C, hence, the more widespread presence of Race O. In some resistant hybrids flecking may be found, but is only a reaction to resistance and will not cause loss of economic significance.

Other methods of control can prevent the spread of all races. For example, it is important to manage crop debris between growing seasons, as *B. maydis* overwinters in the leaf and sheath debris. Tillage can be used to help encourage breakdown of any remaining debris. It has been observed that burying residues by plowing has reduced the occurrence of SCLB as opposed to minimal tillage, which can leave residue on soil surface. Another form of cultural control used to limit SCLB is crop rotation with non-host crops.

Additionally, foliar fungicides may be used. Foliar disease control is critical from 14 days before to 21 days after tasseling; this is the most susceptible time for damages from leaf blight to occur. The fungicides should be applied to plants infected by SCLB as soon as lesions become apparent. Depending on the environmental conditions, re-applications may be necessary during the growing season. Common fungicides include Headline, Quadris, Quilt, PropiMax EC, Stratego, and Tilt.

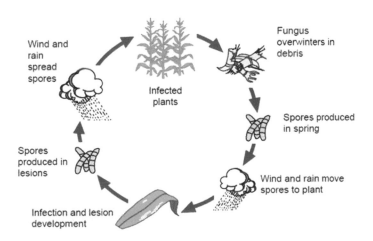

FIGURE 12.28 Disease cycle of *B. maydis*.

12.10 STEWART'S BACTERIAL WILT

Stewart's bacterial wilt of sweet corn was first reported in the United States on Long Island in 1897. This disease remains important in most sweet corn-producing areas of the Northeast, Mid-Atlantic, and Midwestern states. Stewart's wilt is caused by a bacterium *Erwinia stewartii* (syn. *Pantoea stewartii*) that can spread systemically throughout the plant.

Stewart's wilt is also known by other names such as bacterial leaf blight, Stewart's leaf blight, or maize bacteriosis. Although sweet corn and popcorn are more susceptible to Stewart's wilt than field (dent) corn, some very susceptible inbreeds and hybrids are on the market. Stewart's wilt can cause yield reductions directly through stand reductions and the production of fewer and smaller ears, and indirectly through an increased susceptibility of wilt-infected plants to stalk rotting organisms.

12.10.1 CAUSAL ORGANISM

Stewart's wilt is mostly caused by the fungi mentioned below.

Species	Associated Disease Phase	Economic Importance
E. stewartii (syn. *P. stewartii*)	Leaves, Fruits, and Flowers	High

12.10.2 SYMPTOMS

The bacterial wilt organism infects sweet corn plants at any stage of growth. Infected seedlings may die prematurely. The disease is usually most conspicuous and serious in young plants under 2 ft tall. In seedlings, the bacterium often spreads systemically throughout plants of susceptible hybrids. Symptoms are limited to localized areas of leaves in hybrids with moderate levels of resistance. The older leaves of young plants develop narrow yellowish streaks, which later turn brown. Several streaks on a leaf cause it to shrivel, and die. These symptoms may be confused with symptoms of frost damage, drought, nutrient disorders, or insect injury.

Symptoms on more mature plants commonly appear as irregular, pale green to yellowish streaks with wavy margins that sometimes extend the length of the leaf blade. The streaks can often be traced back to flea beetle wounds, usually on the top half of the leaf. The streaks later become dry and brown. On extremely susceptible hybrids, plants are stunted and die prematurely. In older plants, necrotic tissue resulting from Stewart's wilt may resemble severe symptoms caused by multiple infections by the northern leaf blight pathogen, *E. turcicum*.

When a wilted or dying plant with a normal green stalk is cut through and squeezed, small droplets of yellowish bacterial ooze appear on the cut ends of the vascular bundles. Cavities may develop within the lower stalk of a severely infected plant.

The bacteria in such plants are systemic and may pass through the cob into the kernels. On very susceptible hybrids a yellow, slimy ooze infrequently collects on the surface of the inner ear husks or covers the kernels. Other kernels may have grayish spots (lesions) with dark margins or they may be deformed and shrunken (Figures 12.29 and 12.30).

12.10.3 CAUSE AND DISEASE DEVELOPMENT

The bacterium that causes Stewart's wilt survives through the winter in the alimentary tract or "gut" of adult corn flea beetles, and is spread to sweet corn plants in the spring as the beetles feed. Bacteria do not overwinter in soil or plant debris. Beetles deposit the bacteria in the feeding sites, and the bacteria then colonize the leaf tissue. Eventually the bacteria enter the vascular system and then spread throughout the plant (Figure 12.31).[7]

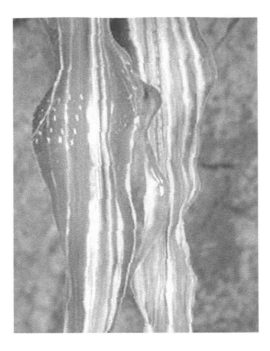

FIGURE 12.29   Foliar symptoms on mature plants.

The severity of Stewart's wilt depends on 3 factors: (1) the winter temperatures prior to plant-ing, (2) the amount of Stewart's wilt the previous season, and (3) the susceptibility of the hybrid to the disease. If cold winter temperatures prevail, then fewer flea beetles survive to emerge in the spring, and consequently, fewer beetles are available for transmitting the disease. Only 10–30% of the emerging beetles carry the bacteria. Generally, the prevalence of the disease in the previous season determines the range of percentage of emerging beetles carrying Stewart's wilt bacterium.

The causal bacteria may live for several months in seed, manure, soil, and old cornstalks; how-ever, the number of plants that become infected from these sources is insignificant. The toothed flea beetle, adult 12-spotted cucumber beetle, and larvae of corn rootworms, seed corn maggot, wheat wireworm, and white grubs also may carry the wilt bacteria from 1 plant to another during the summer.

12.10.4   FAVORABLE CONDITIONS

The number of flea beetles emerging in spring from hibernation depends on the severity of winter temperatures. Warm winter temperatures favor the survival of flea beetle vectors and increase the

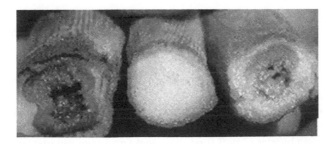

FIGURE 12.30   Water soaking and rotting symptoms in stem tissues.

FIGURE 12.31   The corn flea beetle.

risk of Stewart's disease. Low temperatures are highly unfavorable for beetle survival. The numbers of emerging adults can be estimated by calculating a winter temperature index by averaging the mean temperatures (expressed in °C) for December, January, and February. If the sum of the mean temperatures is 32°C or greater, the beetles will survive in high numbers and the disease risk is high; if the sum is between 29°C and 32°C, the risk is moderate to high; 27°C to 29°C, moderate to low; and a sum less than 27°C represents low risk.

12.10.5   Disease Cycle

P. stewartii overwinters in the alimentary tract of adult corn flea's beetles, not in the soil or in plant debris. The beetles hibernate as adults, and in the spring, feed on corn and other grasses allowing the bacteria to enter the plant. Seedling wilt occurs at or before the 5-leaf stage as the bacterium enters the plant, spreads to the developing stalk, and kills the growing point.

12.10.6   Management

The disease can be managed using the following strategies:

12.10.6.1   Chemical Control

Use insecticides to control flea beetles, particularly on susceptible varieties in the seedling stage. This is not as effective as resistant varieties, but reduces losses where susceptible hybrids must be planted. Gaucho seed treatments provide systemic control of corn flea beetles and reduce the severity of Stewart's wilt.

Several insecticides may be used as foliar sprays for corn flea beetle control; trade names include Sevin, Asana, Pounce/Ambush, and Warrior. Although some of these products persist a little longer than others, rapid growth of leaf tissue means untreated surfaces are available to flea beetles that migrate into fields a few days after treatment. A key for suppressing Stewart's wilt is to scout frequently for flea beetles (2 or 3 times per week) and re-apply insecticides if populations are rebuilding.

TABLE 12.2
Management of Stewart's Bacterial Wilt

Hybrid	Days To Harvest	Source
Yellow, shrunken-2		
Apollo	85	BMM
Flagship	85	BMM
GSS 4606	85	RS
Maxim	82	HM
Midship	75	BMM
Natural Sweet 9000	87	WCI
Punchline	76	ASG
Sch 5005	78	IFS
Sch 5276	84	IFS
Sch 11069	85	IFS
Sch 20777	86	IFS
Sch 30375	84	IFS
Summer Sweet 7620	82	AC
Summer Sweet 7630	85	AC
Summer Sweet 7710	83	AC
Sweet Season	83	SUN
Ultimate	83	HM
Wisc. Natural Sweet	85	WCI
XPH 3082	80	ASG

Days to Harvest = estimated number of days from planting to harvest.

Seed Sources: AC, Abbot & Cobb; AGW, Agway/Seedway; ASG, Asgrow; CR, Crookham; FM, Ferry Morse; HM, Harris Moran; IFS, Illinois Foundation Seeds; LSC, Liberty Seed; BMM, Burpee-Market More; PARK, Park Seeds; RS, Rogers Seeds; ROB, Robson; AGW, Agway-Seedway; STO, Stokes; SUN, Sunseeds; WCI, Wisconsin Crop Improvement

12.10.6.2   Cultural Control

- Grow varieties that are resistant to the disease. Hybrids with greater levels of resistance can tolerate more infection with less yield loss. Resistance restricts the movement of the bacteria in the plant.
- Many dent corn hybrids and inbreeds have resistance to the seedling blight phase.
- Insecticides applied as seed treatments or in-furrow have been effective.
- Avoid excess nitrogen and phosphorous levels as well as excess soil moisture.

12.10.6.3   Biological Control

Grow well-adapted, wilt-resistant sweet corn varieties. Sweet corn hybrids with high levels of resistance to Stewart's wilt are presented in the table below. At present, there are very few early maturing hybrids with high levels of resistance to Stewart's wilt. Consult current seed catalogs and trade publications for additional information on disease resistant hybrids (Table 12.2).

REFERENCES

1. Lipps, P. E., and Mills, D. R., 2001. *Anthracnose Leaf Blight and Stalk Rot of Corn*, The Ohio State University Extension Bulletin 802.

2. Jackson, T. A., 2014. *Rust Diseases of Corn in Nebraska*, University of Nebraska – Lincoln Extension, 4.

3. Report on Plant Diseases, 1990. *Corn Smuts*, Department of Crop Sciences, University of Illinois at Urbana Champaign. http://ipm.illinois.edu/diseases/series200/rpd203.

4. Crous, P. W., Groenewald, J. Z., Groenewald, M., Caldwell, P., Braun, U., and Harrington, T. C., 2006., Species of Cercospora associated with grey leaf spot of maize. *Stud Mycol.* 55: 189–197.

5. Mishra, S. R., 2014, *Virus and Plant Diseases*, Discovery Publishing House, 141–142.

6. Singh, R., and Srivastava, R.P., 2012, *Southern Corn Leaf Blight- An Important Disease of Maize: An Extension Fact Sheet.* Indian Research Journal of Extension Education Special Issue (Volume I), pp. 334–337.

7. Williams, K. M., 2014, *Characterization of an RTX-Like Toxin and an Alpha-2-Macroglobulin in Pantoea stewartii subsp. stewartii, Causal Agent of Stewart's Wilt of Sweet Corn.* UC Riverside Electronic Theses and Dissertations.

8. Gay, J. P., and Cassini, R., 1973. Possibilities of control of maize diseases by fungicide treatments during growth. *Phytiatrie Phytopharmacie*, 22(1): 19–26.

9. Pronczuk, M, Bojanowski, J, and Warzecha, R, 1996. Preliminary evaluation of effectiveness of fungicides in protecting maize plants against diseases. *Biuletyn Instytutu Hodowli i Aklimatyzacji RoSlin*, No. 197:151–155; 6 ref.

10. Arny, D. C., Smallej, E. B., Ullstrup, A. J., Worf, G. L., and Ahrens, R. W., 1971. Eyespot of maize, a disease new to North America. *Phytopathology*, 61: 54–57.

11. Reifschneider, F.J.B., Arny, D.C., 1983. Yield loss of maize caused by Kabatiella zeae. Phytopathology, 73(4):607–609.

12. HYP3, 2005. Eyespot of maize. HYP3 on line. http://www.inra.fr/Internet/Produits/HYP3/pathogene/6kabzea.html.

13 Grape

A grape is a botanically a berry fruit that belongs to the deciduous woody vines of the flowering plant genus *Vitis*. Grapes can be eaten fresh or they can be used for making wine, jam, juice, jelly, grape seed extract, raisins, vinegar, and grape seed oil. Grapes are a non-climacteric type of fruit, generally occurring in clusters. Grapes grow in clusters of 15–300, and can be crimson, black, dark blue, yellow, green, orange, and pink. White grapes are green in color, and are evolutionarily derived from the purple grape. Mutations in 2 regulatory genes of white grapes turn off production of anthocyanins, which are responsible for the color of purple grapes. Grapes are typically an ellipsoid shape resembling a prolate spheroid. According to the Food and Agriculture Organization (FAO), 75,866 km^2 of the world are dedicated to grapes. Approximately 71% of world grape production is used for wine, 27% as fresh fruit, and 2% as dried fruit. The top grape producing countries are China, United States, Italy, France, Spain, and Turkey. This chapter contains information regarding grape fruit diseases and methods for their treatment.

13.1 ANTHRACNOSE OF GRAPES

Anthracnose is a southern disease that occurs in northern regions. Some table grape varieties are particularly susceptible. Symptoms occur on all aboveground parts of the vine, particularly on young tissues. Leaves develop numerous deep brown spots, 1/25–1/5 inch (1–5 mm) in diameter. As the centers fall out, lesions take on a "shot-hole" appearance. Severe infections curl and distort leaves. Lesions on shoots are sunken and dark brown with grayish centers.[1]

Lesions on shoots are sunken and dark brown with grayish centers. On green berries, "bird's-eye" spots are purplish brown or bleached with a dark edge. Berries remain firm, crack, and then shrivel. The fungus overwinters in infected parts of the vine, and spores are dispersed by wind and rain splash in the spring. Anthracnose can be severe in rainy years.

13.1.1 CAUSAL ORGANISM

Species	Associated Disease Phase	Economic Importance
Elsinoe ampelina	On Young Tissues	Moderate

13.1.2 SYMPTOMS

The shoots, berries, and leaves of the anthracnose are all attacked, but symptoms on shoots and berries are easiest to recognize.

Leaves: The fungus will cause small round spots (Figure 13.1) which, as they age, give way to small holes (leaving a "shot-hole" appearance) (Figure 13.2). During severe infections, the leaves shrivel up and drop.

Shoots: Deep elongated cankers, grayish in the center with a black edge (Figure 13.3).

Inflorescences: Inflorescences are highly susceptible. During severe infections, they can turn yellow, brown, then dry out completely (Figure 13.4).

Berries: Deep spots, violet turning grayish in the center, with a black edge (Figure 13.5). Severely infected berries dry up and drop prematurely (Figures 13.6 and 13.7).

FIGURE 13.1   Fungus: (Anthracnose) on berries causing round spots.

FIGURE 13.2   Leaves: The fungus will cause small round spots.

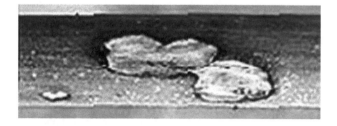

FIGURE 13.3   Leaves: As they age, they give way to small holes (leaving a "shot-hole" appearance).

FIGURE 13.4   Shoots: Deep elongated cankers, grayish in the center with a black edge.

FIGURE 13.5   Inflorescences: During severe infections, they can turn yellow, brown, then dry out completely.

FIGURE 13.6   Berries: Deep spots, violet turning grayish in the center, with a black edge.

FIGURE 13.7   Berries: Severely infected berries dry up and drop prematurely.

13.1.3   Cause and Disease Development

The bird's eye rot disease is common, but seldom is a major problem except in very wet years. When the disease is severe, anthracnose can kill the tips of many new shoots by girdling them. Leaf spots are also whitish in the centers at first, but the centers soon drop out to create a "shot-hole" effect. Young leaves are most susceptible; anthracnose often results in twisting and deformation because it damages part but not the entire expanding leaf blade. Berry spots are the most distinctive—round, purple spots that later turn ashy gray in the centers with dark brown to black margins. Most or all of the berries in a cluster can show symptoms. Berry infection often cracks the fruit skin, which leads to decay of the berry. Anthracnose can infect all green grape tissues.

The disease development first appears early in the year on the first few internodes of new shoots. They are deep lesions with dark margins and a gray center (Figure 13.8) If the disease spreads to

FIGURE 13.8   Disease development.

young tissue, it can distort and kill the shoot tips, giving the shoots a burned appearance. Leaf lesions often cause the leaf to distort and curl. Centers of the spots often fall out, leaving a shot-hole appearance. The disease spreads to developing berries. Berry lesions appear as a dark spot with a gray center, giving the disease its common name, bird's eye rot (Figure 13.9). Bird's eye rot is mostly cosmetic as it does not affect the eating or processing quality of the fruit. Severe infection, however, can reduce vine vigor and yield.

13.1.4   FAVORABLE CONDITIONS OF DISEASE DEVELOPMENT

The causal organism of anthracnose *E. ampelina*, probably overwinters in cankers formed on the canes, in infected berries on the ground, and on berries left in the trellis. In the spring, small fruiting bodies, called acervuli, form and produce spores (conidia). The spores are covered with a mucilaginous substance that enables them to stick to the site of infection. They are dispersed by rain. The longer the leaves and stems remain wet (>12 hours), the more severe the infection. The disease develops at temperatures of 10°C–35°C; the optimum temperatures for disease development are between 20°C and 26°C. Symptoms appear 4–12 days after infection.

13.1.5   DISEASE CYCLE

The following section describes the disease cycle of *E. ampelina*.

Overwintering structures called sclerotia stay on infected shoots and produce many spores, conidia, in the spring when there is a wet period of 24 hours and temperatures above 2°C (36°F). The

FIGURE 13.9   Bird's eye rot.

conidia are spread to other plant tissue by free water or rain over 2 mm or more. These conidia will germinate, causing a primary infection when free water is present for 12 hours and the temperature is between 2°C–32°C (36°F–90°F). The higher the temperature the faster infection will take place. Disease symptoms will develop within 13 days at 2°C and within 4 days at 32°C. Ascospores, spores produced within a sexual fruiting body, also form on infected canes or berries left on the trellis or on the vineyard floor. Asexual fruiting bodies called acervuli, form on necrotic areas once the disease is established. These acervuli produce conidia in wet weather, which are the secondary source of inoculum for the rest of the growing season.

Temperature and moisture are the key components in influencing disease development.

Anthracnose can be very damaging during heavy rainfall and hail.

13.1.6 MANAGEMENT

There are many factors that must be considered in selecting grape varieties. Planting varieties that are less susceptible to disease is a good way to prevent outbreaks. Orienting rows N–S and using topographical gradients will maximize air and soil drainage. Pruning debris may act as a reservoir for inoculum and should be removed and burned. Working the soil in the spring is a form of cultural control. Destroying and burying infected debris will reduce pathogen populations at the start of the season, therefore, reducing the risk of disease.

Canopy management (pruning, shoot positioning, and leaf removal) facilitates air circulation, which promotes drying of the leaves and increased penetration of fungicides. It is also recommended that vineyards be systematically inspected throughout the growing season to monitor the appearance and development of diseases. At harvest and at the end of the season, assess the presence of anthracnose on the leaves to estimate the level of inoculum that will be present the following spring.[2]

13.1.7 CONTROL

There are many fungicide application strategies. Where minimum risk is desired, fungicides can be applied in keeping with an established schedule rather than based on observations made in the vineyard. This approach is not without its consequences on production costs and the environment. It is also possible to develop rational IPM programs for diseases of grapes. In such cases, the decision to spray is made after evaluating the consequences of this decision on production and on the environment. In other words, the decision will be based on factors such as the weather conditions, the stage of development of the vine, the grape variety, parasitic pressure, etc. Prune out damaged shoots and clusters, remove the prunings from the vineyard, and rake the ground under the plants in fall to remove all fallen berries (Table 13.1).

GRAPES: Fungal Diseases

TABLE 13.1
Chemicals and their Applications

Chemical	Application
Liquid lime sulfur or Sulforix (calcium polysulfide)	Lime sulfur will burn tender foliage, so it must be applied just as buds are swelling, but before the leaves are exposed. It effectively kills the developing spores (primary inoculum) at the beginning of the season, and prevents the disease from becoming established. Single application provides nearly complete control
Mancozeb, Captan, Abound or Sovran	After bud break application can provide some control and keep the disease from spreading

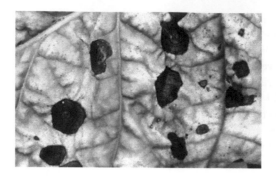

FIGURE 13.10   Alternaria rot on grape leaf.

13.2   ALTERNARIA ROT

Alternaria rot is ubiquitous and distributed world-wide. A number of fruits may be affected. Alternaria rot causes merely marginal losses of berry quality in viticulture and colonizes many ripe berries with leaked sugar. Colonized berries show a black smut on the surface. Only occasionally injured berries are affected. In this, Alternaria rot raises a moldy taste of grapes and wines and produces mycotoxins. Therefore, infected clusters must be sorted at harvest. The agent causing the disease is a fungus called *Alternaria alternata* that affects grapes it is also known as leaf blight and bunch necrosis. (Figures 13.10 and 13.11).

13.2.1   CAUSAL ORGANISM

Species	Associated Disease Phase	Economic Importance
A. alternata	Leaf Blight and Bunch Necrosis	Moderate

13.2.2   TAXONOMY

To the taxon, *Alternaria* belongs to numerous species of which *A. alternate Keissler* is most common on grapevine. The genus *Lewia* is described as a telomorph for *Alternaria* (Pleosporales, Ascomycetes). The organism is known with various synonyms as *Alternaria fasciculata* (1897),

FIGURE 13.11   *A. alternata* spores.

Alternaria rugosa, Alternaria tenuis Nees (1817), *Macrosporium fasciculatum*, and *Torula alternata* Fr. (1832).

13.2.3 SYMPTOMS

Alternaria appears in the month of June and December. The disease attacks both leaves and fruits. Small yellowish spots first appear along the leaf margins, which gradually enlarge and turn into brownish patches with concentric rings. Severe infection leads to drying and defoliation of leaves. Symptoms in the form of dark brown, purplish patches appear on the infected berries, rachis, and bunch stalk just below its attachment to the shoots. *A. alternata* causes bunch rot of export of table grapes, which is characterized by firm, superficial, dark-brown to black lesions on berries near the pedicels, and fluffy gray tufts of fungus growing on rachis and pedicels. The disease develops in some cold stored fruit and its sporadic occurrence in some consignments is a serious challenge to the prolonged storage of table grapes at low temperatures (Figure 13.12).

13.2.4 CAUSE AND DISEASE DEVELOPMENT

A. alternata has been recorded causing leaf spot and other diseases on over 380 host species. It is an opportunistic pathogen on numerous hosts causing leaf spots, rots, and blights on many plant parts.

Studies showed that *A. alternata* colonizes berries, pedicels, and rachises during the entire period of bunch development but that stress factors during cold storage might predispose table grape bunches to *A. alternaria* decay. *A. alternata* was found to be the cause of a decay of cold-stored table grapes in the Cape Province of South Africa during 1986–1989. The disease was characterized by localized dark-brown to black lesions on berries, pedicels, and rachis. The fungus was not considered to be a new pathotype *of A. alternata*, but an opportunist. It was not highly virulent, and caused disease only after prolonged cold storage and grew extensively without lesion formation on parts of mature bunches, especially rachi. It is suggested that sulfur dioxide damage, especially of rachis, might have a predisposing effect. It is concluded that Alternaria rot cannot be considered of great importance to the local table grape industry, but the sporadic occurrence of the disease, and the relatively high incidence of bunches with *A. alternata* growth on rachis reported for some consignments, might pose a threat to individual farmers.

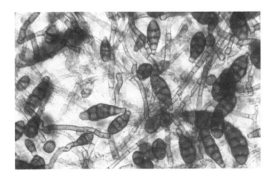

FIGURE 13.12 (a) Initial stage of bunch rot development—Tissues of affected berries shrink and turn brown. Berry flesh changes slightly in consistency but remains firm. Skins of the berries are dull and easily detached from the berry flesh. (b) Grape berries infected with *A. alternate*. (c) Mycelium of *Alternaria* growing in cracks in berry skins—Berry skin ruptures, a narrow strip of mycelium appears along the line of rupture and the mycelium grows rapidly onto other berries. (d) Mycelium of *Alternaria* spreading to other berries on a bunch—Rotting berries have no noticeable odor. With time, the berries become necrotic and fall from bunches. (e) Culture of *A. alternata* on malt agar. (f) Conidia of *A. alternata* (scale bar: 20 mm).

13.2.5   FAVORABLE CONDITIONS OF DISEASE DEVELOPMENT

The conidiophores of *A. alternata* produces pale to medium brown colonies in long and often branched chains. The brown to olive-green conidia have transverse and longitudinal septae and a cylindrical or short conical beak. The fungus has a saprophytic lifestyle and prefers a sugary substrate but occasionally it become parasitic. For setting an infection, high relative humidity is necessary (98–100%). Under these conditions the germinating peg of the conidia is able to penetrate the epidermis directly. Therefore, frequent rain in late summer and autumn is favorable for the infection process (Figure 13.13).

13.2.6   MANAGEMENT

If the disease on the berries is not controlled in the field, it can lead to berry rotting during transit and storage. The following concentration is to be sprayed alternatively at weekly intervals from June–August and again from December until harvest to keep this disease under check. 2–3 sprays of systemic fungicides should be given per season (Table 13.2).

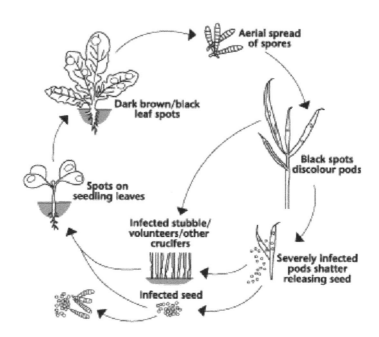

FIGURE 13.13   Life cycle of *A. alternate*.

TABLE 13.2

Chemicals and their Concentrations

Chemicals	Concentration
Bordeaux mixture	1.0%
Mancozeb	0.2%
Topsin M	0.1%
Ziram	0.35%
Captan	0.2%

13.3 GRAY MOLD (*BOTRYTIS CINEREA*)

In viticulture, *B. cinerea* is commonly known as botrytis bunch rot; in horticulture, it is usually called *grey mould* or *gray mold*. Botrytis bunch rot (gray mold) and blight of leaves, shoots, and blossom clusters occurs throughout the viticultural world. The fungus causing the disease grows and reproduces on senescent or dead plant tissue. Botrytis bunch rot is especially severe in grape cultivars with tight, closely packed clusters of fruit. It is 1 of the most important diseases in storage and can grow at low temperatures. In the vineyards, the fungus attacks the shoots and clusters or destroys stalks leading to premature fruit drop. In the early stages of infection, the skin of the affected berries just below the infection become loose. When rubbed with fingers the skin slips from the berry leaving the firm pulp exposed. The infected berries shrivel, rot, and turn dark brown showing the presence of grayish growth of the fungus.

13.3.1 CAUSAL ORGANISM

Species	Associated Disease Phase	Economic Importance
B. cinerea (Asexual Stage) *Botryotinia cinerea* (Sexual Stage)	Bunch Rot	Moderate

13.3.2 SYMPTOMS

Botrytis infection of leaves begins as a dull, green spot, commonly surrounding a vein, which rapidly becomes a brown necrotic lesion. The fungus may also cause blossom blight or a shoot blight, which can result in significant crop losses. Debris, i.e. dead blossom parts, in the cluster may be colonized by the fungus which can then move from berry to berry within the bunch prior to the beginning of ripening, and initiate development of an early season sour rot. However, the most common phase of this disease is the infection and rot of ripening berries (Figure 13.14). This will spread rapidly throughout the cluster. The berries of white cultivars become brown and shriveled and those of purple cultivars develop a reddish color. Under proper weather conditions, the fungus produces a fluffy, gray-brown growth containing spores (Figure 13.15).

13.3.3 CAUSE AND DISEASE DEVELOPMENT

The fungus gives rise to 2 different kinds of infections on grapes. The first, gray rot, is the result of consistently wet or humid conditions, and typically results in the loss of the affected bunch. The second, noble rot, occurs when drier conditions follow wetter, and can result in distinctive sweet dessert wines. The fungus is usually referred to by its anamorph (asexual form) name, because the sexual phase is rarely observed. The teleomorph (sexual form) is an ascomycete *Botryotinia fuckeliana*,

FIGURE 13.14   Botrytis bunch rot of grape.

FIGURE 13.15   Close-up showing botrytis sporulating on infected berries.

also known as *Botryotinia cinerea*. *B. cinerea* is characterized by abundant hyaline conidia (asexual spores) borne on gray, branching tree-like conidiophores. The fungus also produces highly resistant sclerotia as survival structures in older cultures. It overwinters as sclerotia or intact mycelia, both of which germinate in spring to produce conidiophores. The conidia are dispersed by wind and rain-water and cause new infections. A considerable genetic variability has been observed in different *B. cinerea* strains (polyploidy). *Gliocladium roseum* is a fungal parasite of *B. cinerea*.

13.3.4   FAVORABLE CONDITIONS OF DISEASE DEVELOPMENT

Botrytis overwinters on debris in the vineyard floor and on the vine. The fungus produces small, dark, hard, resting structures called sclerotia. Sclerotia are resistant to adverse weather conditions and usually germinate in spring. The fungus then produces conidia, which spread the disease. Sporulation may occur on debris left on the vine during the previous growing season, such as cluster stems remaining after mechanical harvest or mummified fruit, or it may occur on sclerotia on canes. The fungus usually gains a foothold by colonizing dead tissue prior to infection of healthy tissue. Tissue injured by hail, wind, birds, or insects is readily colonized by botrytis. Ripe berries that split because of internal pressure or because of early season infection by powdery mildew are especially susceptible to infection by botrytis. Botrytis conidia are usually present in the vineyard throughout the growing season. Moisture in the form of fog or dew and temperatures of 15°C–25°C are ideal for conidia production and infection. Rainfall is not required for disease development, although periods of rainfall are highly conducive to disease development (Figure 13.16).

13.3.5   MANAGEMENT

Careful handling in the field, precooling, and refrigeration helps in controlling the disease. Pruning and thinning of the vineyard reduces humidity around the clusters. Prophylactic sprays with Captan (0.2%) and Benomyl or Bavistin (Carbendazim) (0.1%) minimize the development of the fungus during transit and storage. Many products are currently available or currently being introduced as "biological control agents" or "biopesticides". These include living microorganisms, "natural chemicals" such as plant extracts, and "plant activators" that induce resistance in plants to disease. For most of these products, independent evaluations are currently being conducted; however, their effectiveness under moderate to high disease pressure is uncertain. Although many of these new products have great potential for use within organic production systems, their effectiveness needs to be determined in field tests. It is important to remember that registration of these materials for control of a specific disease on a crop is no guarantee that they will provide effective control under moderate to heavy disease pressure. In addition, many products may be effective for only 1 or a few

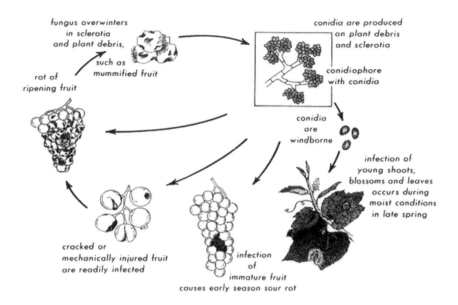

FIGURE 13.16 Botrytis bunch rot disease cycle.

diseases and most have very limited residual activity (they must be applied often). It is also important to remember that these are registered pesticides and growers need to be certain that their use is permitted within their organic certification program (Tables 13.3 and 13.4).

13.4 BLACK ROT (*GUIGNARDIA BIDWELLI*)

Black Rot is a severe disease that is caused by the fungus named *G. bidwelli*. Warm and moist climate with extended periods of rain and cloudy weather favors the development of the disease. The disease attacks the leaves, stem, flowers, and berries. All the new growth on the vine is prone to attack during the growing season.

The symptoms are in the form of irregularly shaped reddish-brown spots on the leaves and a black scab on berries. Occasionally, small elliptical dark colored canker lesions occur on the young stems and tendrils. Leaf, cane, and tendril infection can occur only when the tissue is young, but berries can be infected until almost fully grown if an active fungicide residue is not present. The affected berries shrivel and become hard black mummies. There is a wide variation in susceptibility to this disease among Native American and hybrid cultivars, whereas all common cultivars of *Vitis vinifera* appear to be highly susceptible.

13.4.1 CAUSAL ORGANISM

Species	Associated Disease Phase	Economic Importance
G. bidwelli	Leaves, Stem, Flowers, and Berries	Severe

13.4.2 SYMPTOMS

All young green tissues of the vine are susceptible to infection. Symptoms of black rot first appear as small yellowish spots on leaves. As the spots (lesions) enlarge, a dark border forms around the margins. The centers of the lesions become reddish brown. By the time the lesions reach 1/8–1/4

TABLE 13.3
Chemicals and their Applications

Chemical Control	Rate	Comment
Topsin M WSB *or*	1-1.5 lb	Apply Topsin M at 1-1.5 lb/A at first bloom (no later than 5% bloom), and repeat 14 days later if severe disease conditions persist. Topsin M is also available in 70WDG and 4.5 FL formulations.
Rovral 50WP or	1.5-2 lb	Rovral may be applied at 1.5-2.0 lb/A four times: 1. Early to midbloom; 2. Prior to bunch closing; 3. Beginning of fruit ripening; 4. Prior to harvest if needed. Do not make more than 4 applications of Rovral per season. Do not apply within 7 days of harvest.
Vangard 75WG *or*	10 oz	Vangard is registered for use at 10 oz/A when used alone, or at 5–10 oz/A when used in a tank mix. Timing of application is approximately the same as for Rovral. No more than 20 oz of Vangard may be applied per acre per crop season. Vangard cannot be applied within 7 days of harvest.
Elevate 50WG or	1 lb	Elevate may be applied at 1 lb/A and the timing of application is approximately the same as Rovral and Vangard. No more than 3 lb of Elevate may be applied per acre per season. Elevate can be applied up to, and including, the day of harvest (0-day PHI).
Scala 5SC *or*	18 fl oz	Scala is registered for use at 18 fl oz alone, or at 9 fl oz when used in a tank mix. Timing of application is approximately the same as for Rovral.
Switch 62.5WG	11-14 oz	Switch is also registered for control of sour rot (caused by a complex of organisms). Preharvest applications may be beneficial for control of sour rot. See the label for additional information.

inch in diameter (approximately two weeks after infection), minute black dots appear. Relatively small, brown circular lesions develop on infected leaves (Figure 13.17), and within a few days tiny black spherical fruiting bodies (pycnidia) protrude from them. These fungal fruiting bodies (pycnidia) contain thousands of summer spores (conidia). Pycnidia are often arranged in a ring pattern, just inside the margin of the lesions. (Figure 13.18). Elongated black lesions can be seen on the petiole. Lesions may also appear on young shoots, cluster stems, and tendrils. The lesions are purple to black, oval in outline, and sunken. Pycnidia also form in these lesions (Figure 13.19); they may eventually girdle these organs (Figure 13.20), causing the affected leaves to wilt (Figure 13.21). Shoot infection results in large black elliptical lesions. These lesions may contribute to breakage of shoots by wind, or in severe cases, may girdle and kill young shoots altogether.

TABLE 13.4
Biological Control Methods

Biological Control	Rate	Comment
Serenade (*Bacillus subtilis*)	Applications are recommended on a 7-10-day schedule. No maximum seasonal application rate and 0-day PHI.	Moderate level of control
Trichodex (Trichoderma harzianum)	Sold as wettable powder formulation that is mixed with water and sprayed directly onto the plants.	Primary control

FIGURE 13.17   Small, circular lesions on leaves.

Infection of the fruit is by far the most serious phase of the disease and may result in substantial economic loss. Infected berries first appear light or chocolate brown (Figure 13.22), but quickly turn darker brown, with masses of black pycnidia developing on the surface (Figure 13.23). Finally, infected berries shrivel and turn into hard, black, raisin-like bodies that are called mummies (Figures 13.24 and 13.25).

13.4.3   CAUSE AND DISEASE DEVELOPMENT

G. bidwelli affects grape varieties and cause a black rot disease in grapes. Spreading of disease is dependent on the temperature and the age of tissue. Spring rains trigger release of airborne spores (ascospores) that form within mummies on the ground and in the trellis, and these can be blown for moderate distances by wind. Spores of a second type (conidia) can also form, both within cane

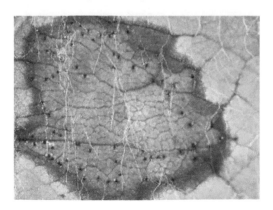

FIGURE 13.18   Tiny, black pycnidia in leaf lesion.

FIGURE 13.19   Elongated lesions on petiole.

lesions or on mummies that have remained within the trellis, and these are dispersed short distances (inches to ft) by splashing rain drops. The period during which these overwintering spores are available to cause infections depends on their source. From mummies on the ground, significant discharge of ascospores begins about two to three weeks after bud break and is virtually complete within one to two weeks after the start of bloom. In contrast, mummies within the trellis can continue to release both conidia and ascospores from the early pre-bloom period through veraison. From overwintering cane lesions, conidia can be dispersed from bud break through mid-summer.

13.4.4   FAVORABLE CONDITIONS FOR THE DISEASE DEVELOPMENT

The black rot fungus overwinters primarily in mummies within the vine and on the ground, although it also can overwinter for at least 2 years within lesions of infected shoots that are retained as canes

FIGURE 13.20   Girdled petioles cause leaves to sag and wilt.

FIGURE 13.21   Shoot and petiole lesions from spores in mummies attached to wire.

or spurs. The period required for the disease development appears after the occurrence of an infection period which depends on both the temperature and the age of the tissue at the time it's infected. Infection occurs when either spore type lands on susceptible green tissue or it remains wet for a sufficient length of time, which depends on the temperature

13.4.5   MANAGEMENT

Black rot should be managed through a combination of cultural and chemical methods. The success of any fungicide program will be greatly enhanced by sanitation practices designed to reduce inoculum of the black rot fungus, and these may be essential for avoiding losses in vineyards where the disease is a perennial problem. It is critical to remove all mummies from the canopy during the dormant pruning process because such mummies produce spores immediately next to susceptible grapevine tissues throughout the season; even relatively few can cause significant damage (Figure 13.27). Cultivating beneath the vines near bud break to bury mummies will also greatly reduce the number of spores that are released from them, which could otherwise cause infection. As with all fungal diseases, control also is improved by canopy management practices that promote air circulation, speed drying of the leaves and fruit, and improve spray penetration.

FIGURE 13.22   Elliptical lesion on shoot.

FIGURE 13.23   Early symptoms of berry infection.

Traditional fungicide recommendations specified regular applications from the early shoot growth stage through veraison.

However, because fruit are most susceptible during the first few weeks after the start of bloom, this is when the fungicidal component of black rot management programs should be focused most strongly, whether additional sprays are applied or not. Sulfur is not effective for black rot control.

FIGURE 13.24   Infected berries with numerous black pycnidia.

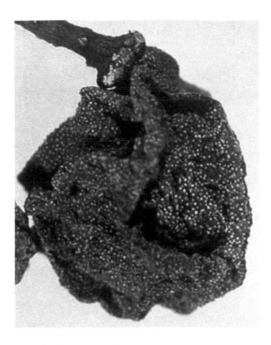

FIGURE 13.25   Infected grape, black rot mummies.

Copper fungicides are not highly effective, but will provide some level of control. The most critical period to control black rot with fungicide is from immediate pre-bloom through two to four weeks after bloom. Common grape fungicides differ greatly in their effectiveness against black rot. Understanding the traits of individual fungicides will improve one's ability to use these tools most efficiently. Mummified berries left on vines should be collected and destroyed. Cultivation practices

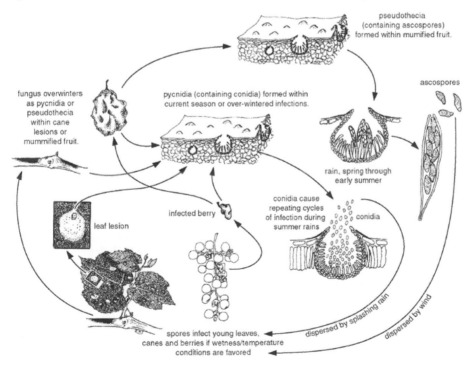

FIGURE 13.26   Disease cycle of *G. bidwelli*.

FIGURE 13.27   Mycelial growth seen beneath the bark of the root.

TABLE 13.5
Common Fungicides

Fungicide	Rate	Comments
Sovran 50WG	3.2-4.8 oz/A	Sovran is excellent for control of black rot
Flint 50WG	2.0 oz/A	registered for the control of black rot
Abound	11–15.4 fl oz/A	Provides good control
Pristine 38WDG	6–10.5 oz/A	Read the label carefully before the
Combination of two active ingredients (pyraclostrobin, 12.8% and boscalid 25.2%)	A maximum of six applications may be made per season	use as not prescribed for certain varieties due to foliar injury.

should ensure free circulation of air. Spraying Bordeaux mixture (4:4:100) once or twice on young bunches prevents the infection. Copper fungicides are preferred for spraying on bunches, as they do not leave any visible deposits on the fruit surface. Other than these, the most common fungicides that prove to be excellent for certain regions of the United States for controlling black rot are Sovran 50WG, Flint 50WG, Abound Flowable (2.08F), and Pristine 38WDG (Table 13.5).

13.5   ARMILLARIA ROOT ROT IN GRAPES

Armillaria root rot in grapes caused by the fungus *Armillaria mellea* infects vine roots, killing the cambium and decaying the underlying xylem (the water-conducting system). Often found on newly cleared land, this root pathogen is native to the Pacific Northwest where it occurs on the roots of many forest tree species including Douglas-fir, madrone, oak, willow, and yellow pine. It also attacks black and red raspberries and trailing berries. The host range includes over 500 species of woody plants, making its common name of "oak root fungus" slightly misleading.

This fungus may form mushrooms at the base of infected vines in fall and winter. Mushrooms produce windblown spores, but these spores are not a significant means of infecting healthy vines. The fungus spreads vegetatively belowground, which leads to the formation of groups of dead and dying plants called "disease centers". The fungus can survive on woody host roots long after the host dies. Its vegetative fungal tissue (mycelium) decomposes root wood for nutrients as it grows. When infected plants are removed, infected roots that remain below ground serve as a source of inoculum for vines planted in the same location.

Infection occurs when grape roots come in direct contact with partially decayed tree roots and are colonized by mycelium. Infection can also occur when grape roots contact rhizomorphs (black, shoestring-like fungal structures) that grow out from partially decayed roots and through the soil. Once vine roots are infected, whether they are living or dead, they serve as a source of inoculum for neighboring vines. The infection process takes months to happen. When spread between neighboring vines, it may take more than 10 years to occur. The disease negatively affects vine mineral—nutrition status and fruit quality.

13.5.1 Causal Organism

Armillaria root rot is mostly caused by the fungi mentioned below.

Species	Associated Disease Phase	Economic Importance
A. mellea	Roots and Stems	Moderate

13.5.2 Symptoms

Mildly symptomatic grapevines have shorter canes than healthy grapevines. Severe symptoms not only include shorter canes, but also include dwarfed and chlorotic leaves.

Diagnostic mycelial fans can be seen beneath the bark of the root crown of infected plants. Mycelial fans are thick, white layers of fungus that adhere to the root bark and/or the wood beneath the bark. These structures can often be observed in symptomatic vines by digging down about a foot below the soil line and using a pocketknife to remove thin layers of bark from the root collar.

The *Armillaria* fungus also makes black, shoestring-like structures called rhizomorphs, which are occasionally found within the bark and/or extending into surrounding soil. Rhizomorphs may look like roots on the outside but are obviously made up of fungal mycelium when cut open in cross-section.

The disease negatively affects vine mineral—nutrition status and fruit quality (Figure 13.28).

13.5.3 Cause and Disease Development

A. mellea is a fungus that infects grapevine roots, killing the cambium, and decaying the underlying xylem. It lives in soil but needs woody tissue to survive. It can live in decaying roots for up to 50 years, depending on their mass. Rhizomorphs, the agents of infection, grow through the soil from an infected root, but they die if they are separated from the roots they feed on.

FIGURE 13.28 After scraping away the dead bark, white mycelial plaques can be seen on the lower portion of vine.

This fungus may form mushrooms at the base of infected vines in fall and winter. Mushrooms produce windblown spores, but these spores are not a significant means of infecting healthy vines. The fungus spreads vegetatively (using a microscopic, threadlike structure called mycelium), belowground, which leads to the formation of groups of dead and dying plants called "disease centers". The fungus can survive on woody host roots long after the host dies. *A. mellea* mycelium decomposes root wood for nutrients as it grows. When infected plants are removed, infected roots that remain below ground serve as a source of inoculum for vines planted in the same location.

13.5.4 FAVORABLE CONDITIONS

Wet soils, excess salt conditions of the soil, pH imbalance, and winter injury favor the development of the disease.

13.5.5 DISEASE CYCLE

Armillaria can attack any woody part of a grapevine's root system. Vines become infected when roots grow into contact with old *Armillaria*-infected root pieces, or when rhizomorphs grow from these inoculum sources and contact vine roots. In either case, the mode of infection is the same; once the fungus contacts a root, it bores through the bark with the aid of lytic enzymes. Below the bark, *Armillaria* kills the cambium and a mycelial fan forms. The mycelial fan expands beneath the root bark, and the fungus decays the wood.

Once a vine is infected, *Armillaria* can move to neighboring vines in two ways: by direct vine and root-to-root contact or via rhizomorphs. Hyphae (strands of fungal tissue) grow from infected roots to healthy roots that are touching them. Rhizomorphs grow from an infected root, through the soil, to the roots of a nearby vine. Vine-to-vine spread of *Armillaria* is usually quite slow. The rate of spread depends on many factors, including soil moisture and temperature, rootstock growth rate and tolerance, amount of inoculum, and vine-spacing. Three things definitely hasten *Armillaria* infection and spread: excessive soil moisture, large quantities of inoculum, and close vine-spacing.

13.5.6 MANAGEMENT

The disease can be managed using the following strategies:

- Chemical control
- Biological control
- Cultural control

13.5.6.1 Chemical Control

Preplant soil fumigation is most effective if the soil has been thoroughly cleared of woody debris. Methyl bromide fumigation has been found to provide the most effective, albeit limited, control. Methyl bromide is more effective if soil is extremely dry. It works better on fine soils with few rocks. Contrary to soil fumigation for nematodes, soil should be as warm and dry as possible. Fumigation in late summer before any rain is best. Apply fumigant as deeply as possible; some spot fumigation may be necessary a few years after planting. Methyl bromide is being phased out of use, so other fumigants might be used.

Sodium tetrathiocarbanate is registered for control of armillaria root rot. This alternative fumigant is a liquid that breaks down into carbon disulfide gas. Make applications one to four weeks before planting when soil moisture is at or near field capacity.

13.5.6.2 Biological Control

There is great promise and potential for use of biological control of armillaria root disease. Most concepts target stumps, using antagonistic fungi to preemptively colonize or to eliminate *Armillaria* species in the wood. However, more research, much of it tailored to local hosts, conditions, and fungal communities, is needed before it can become operational.

13.5.6.3 Cultural Control

Cultural controls are more promising for long-term control of *Armillaria* than chemical controls, especially those that decrease soil moisture at the base of the vine.

When clearing a new site of native forest trees and shrubs or infected plants (disease centers) there are several precautions to take. First, girdle large trees before removal to hasten decay of roots. After removing aboveground vegetation, clear the soil of stumps and large roots. Deep-rip the soil in more than 1 direction to bring large roots to the soil surface. If possible, remove all roots greater than 1 inch in diameter from the soil. Try to burn all woody debris and leave the ground fallow for at least 1 year.

Trenches lined with vertical plastic sheeting may help to prevent infection if inoculum is coming in from an adjacent stand of infected vegetation. If using drip irrigation, move drip-line emitters away from the trunk and place between vines after the first year of planting.

Once vines are infected, there is little that can be done to control armillaria root rot. Remove and destroy severely infected vines, being careful to remove as much root material as possible from soil. Permanently removing soil in a 3-ft radius around the crown and main trunk root area has been effective in California. If practical, do not replant where infected vines have been removed. Be sure to keep root collars free of soil, especially in vineyards with high gopher populations.

13.6 DOWNY MILDEW ON GRAPES

Downy mildew is a major disease of grapes. The fungus causes direct yield losses by rotting inflorescences, berries, clusters, and shoots. Indirect losses can result from premature defoliation of vines due to foliar infections. This premature defoliation is a serious problem because it predisposes the vine to winter injury. It may take a vineyard several years to fully recover after severe winter injury.

The causal fungus *Plasmopara viticola* is an obligate parasite that overwinters as oospores, or sexual spores, in dead leaves. At maturity, the oospores produce new spores (sporangia). The quantity of mature oospores in the spring is determined by the amount of precipitation the preceding fall. During heavy rainfall, the spores are dispersed to the leaves and fruit by water splash where they release another type of spore (zoospores, motile spores) which infects grape tissue. Young grape tissue tends to be more susceptible to downy mildew than older tissue. Once leaves are fully expanded they tend to be less susceptible to infection. Fruit are susceptible only from bloom through four weeks post-bloom. However, since new succulent leaves are produced throughout the growing season, it is important to maintain good fungicide coverage for downy mildew.

13.6.1 CAUSAL ORGANISM

Species	Associated Disease Phase	Economic Importance
P. viticola	The Grape Tissue	High

13.6.2 SYMPTOMS

On leaves, infections can occur throughout the growing season. Young infections are very small, greenish-yellow, translucent spots that are difficult to see. With time the lesions enlarge, appearing on the upper

FIGURE 13.29   Pale yellow leaf spots on upper surface of grape leaf caused by downy mildew.

leaf surface as irregular pale-yellow to greenish-yellow spots up to 1/4 inch or more in diameter (Figure 13.29). On the underside of the leaf, the fungus mycelium (the "downy mildew") can be seen within the border of the lesion as a delicate, dense, and white to grayish, cotton-like growth (Figure 13.30).

Infected tissue gradually becomes dark brown, irregular, and brittle. Severely infected leaves eventually turn brown, wither, curl ("shepherd's crook"), and drop. The disease attacks older leaves in late summer and autumn, producing a mosaic of small, angular, yellow to red-brown spots on the upper surface. Lesions commonly form along veins, and the fungus sporulates in these areas on the lower leaf surface during periods of wet weather and high humidity.

On fruit, most infection occurs during the period from early bloom through to three to four weeks after bloom. By three to four weeks after bloom, fruit are resistant to infection; however, the fruit stems (pedicels) remain susceptible. When infected at this stage, young berries turn light brown and soft, shatter easily, and under humid conditions, are often covered with the downy-like growth of the fungus (Figure 13.31). Generally, little infection occurs during hot summer months. Infected fruit will never mature normally. On shoots and tendrils, early symptoms appear as water-soaked, shiny depressions on which the dense downy mildew growth appears. Young shoots usually are stunted and become thickened and distorted. Severely infected shoots and tendrils usually die.

13.6.3   CAUSE AND DISEASE DEVELOPMENT

Downy mildew is caused by the fungus *P. viticola*. The fungus overwinters in infected leaves on the ground and possibly in diseased shoots. The overwintering spore (oospore) germinates in the spring and produces a different type of spore (sporangium). These sporangia are spread by wind and splashing rain. When plant parts are covered with a film of moisture, the sporangia release small swimming spores, called zoospores. Zoospores, which also are spread by splashing rain, germinate by producing a germ tube that enters the leaf through stomata (tiny pores) on the lower leaf surface.

FIGURE 13.30   White downy fungus on the underside of infected leaves. This downy growth is directly under the pale-yellow yellow spots on the upper surface.

FIGURE 13.31   Grape berries infected with downy mildew. Observe the cottony growth on the berries.

Sporulating lesions on the upper and lower surface of leaves and whitish sporulation on the berries are shown in the following Figures (13.32–13.37).

13.6.4   FAVORABLE CONDITIONS OF DISEASE DEVELOPMENT

Rain (presence of free water) then becomes the principal factor in the development of the disease. The optimum temperature for disease development is 18°C–25°C (64°F–76°F). The disease can tolerate a minimum temperature of 12°C–13°C (54°F–58°F), and a maximum temperature of about 30°C (86°F). Once inside the plant, the fungus grows and spreads through tissues. Infections are usually visible as lesions in about 7–12 days. At night during periods of high humidity and temperatures above 13°C (55°F), the fungus grows out through the stomata of infected tissue and produces microscopic, branched, tree-like structures (sporangiophores) on the lower leaf surface. More spores (sporangia) are produced on the tips of these tree-like structures. The small sporangiophores and sporangia make up the cottony, downy mildew growth. Sporangia cause secondary infections and are spread by rain (Figures 13.38 and 13.39).

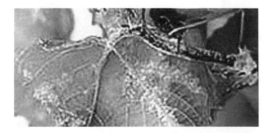

FIGURE 13.32   Sporulating lesions—lower surface (90° angle) downy appearance.

FIGURE 13.33   Sporulating lesions—lower surface (90° angle) downy appearance.

FIGURE 13.34   Whitish sporulation—downy appearance.

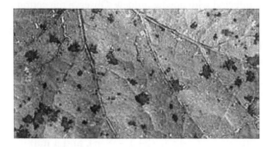

FIGURE 13.35   Lesions after sporulation—upper surface (90° angle).

FIGURE 13.36   Lesions after sporulation—upper and lower surfaces (90° angle).

13.6.5   Management

Any practice that speeds the drying time of leaves and fruit will reduce the potential for infection. Select a planting site where vines are exposed to all-day sun, with good air circulation and soil drainage. Space vines properly in the row, and, if possible, orient the rows to maximize air movement down the row.[4]

Sanitation is important. Remove dead leaves and berries from vines and the ground after leaf drop. It may be beneficial to cultivate the vineyard before bud break to cover old berries and other debris with soil. Cultivation also prevents overwintering spores from reaching developing vines in the spring.[4]

To improve air circulation, control weeds and tall grasses in the vineyard and surrounding areas. When pruning, select only strong, healthy, well-colored canes of the previous year's growth.

FIGURE 13.37 Downy mildew.

Practices such as shoot positioning and leaf removal that help to open the canopy for improved air circulation and spray coverage are also very important.

Grape varieties vary greatly in their susceptibility to downy mildew. In general, vinifera (*V. vinifera*) varieties are much more susceptible than American types, and the French hybrids are somewhat intermediate in susceptibility. Cabernet Franc, Cabernet Sauvignon, Catawba, Chancellor,

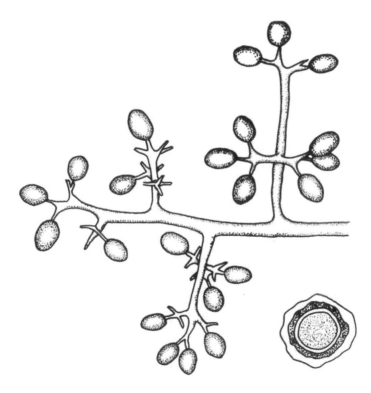

FIGURE 13.38 *Plasmopora viticola*, the grape downy mildew pathogen, seen under a high-power microscope. Thick-walled oospore (lower left). Remainder, branched sporangiophores bearing terminal, lemon-shaped sporangia.

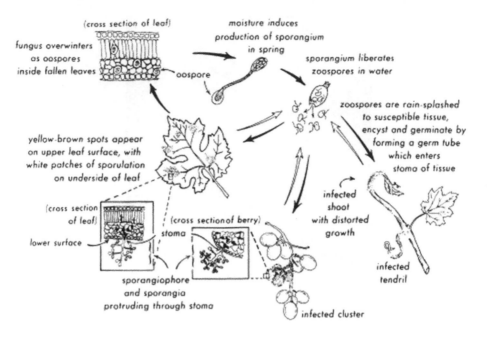

FIGURE 13.39   Disease cycle of downy mildew on grape.

Chardonnay, Delaware, Fredonia, Gewurytraminer, Ives, Merlot, Niagra, Pinot Blanc, Pinot Noir, Riesling, Rougeon, and Sauvignon Blanc are reported to be highly susceptible to downy mildew.[4]

A good fungicide spray program is extremely important. Downy mildew can be effectively controlled by properly timed and effective fungicides (Table 13.6).

13.7 POWDERY MILDEW

Powdery mildew, also known as oidium, is caused by the fungus *Uncinula necator*. This fungus has a narrow host range attacking only grape plants and a few related species.

It is an important disease of grapes worldwide. If uncontrolled, the disease can be devastating on susceptible varieties under the proper environmental conditions. Powdery mildew can result in reduced vine growth, yield, fruit quality, and winter hardiness. Varieties of *V. vinifera* and its hybrids generally are much more susceptible than American varieties.

TABLE 13.6
Control using Chemicals

Chemicals	Application	Comments
Ridomil Gold MZ	Apply 2.5 lb/A of Ridomil Gold MZ. Make up to four applications beginning before bloom.	Do not apply within 66 days of harvest. For late season downy mildew control, apply other registered fungicides and other restrictions also to be followed.
Ridomil Gold Copper:	Apply 2 lb/A of Ridomil Gold Copper. Make up to four applications beginning before Bloom.	Do not make an application within 42 days of harvest. For late season downy mildew control, apply other registered fungicides. NOTE: Always obtain and read the most current label.

The disease develops under warm and dry conditions. Shade or diffused light also helps in the development of this disease. The diseases is characterized by the presence of white powdery (ash-like) coating in patches on both sides of the leaves, young shoots, and immature berries. The affected leaves turn pale and curl up. Affected shoots remain weak and immature. The buds affected during growing season fail to sprout after October pruning. Thus, the productivity of the cane and the number of productive canes are reduced. If blossoms are affected, they fail to set fruit. When young berries are attacked they become corky. Berries attacked at 50% maturity turn dark and become distorted in shape. If severely attacked they are enveloped with a white powdery coating and crack eventually. Loss of yield results from both berry drop and reduced size of berries.

13.7.1 CAUSAL ORGANISM

Species	Associated Disease Phase	Economic Importance
Erysiphe necator; (*U. necator*)	Leaves, Young Shoots, and Immature Berries	Severe

13.7.2 SYMPTOMS

Powdery mildew symptoms can be seen on foliage, fruit, flower parts, and canes. Mildew usually appears first as whitish or greenish-white powdery patches on the undersides of basal leaves. It may cause mottling or distortion of severely infected leaves, as well as leaf curling and withering. Lateral shoots are very susceptible. Infected blossoms may fail to set fruit. Berries are most susceptible to infection during the first three to four weeks after bloom, but shoots, petioles, and other cluster parts are susceptible all season. Infected berries may develop a netlike pattern of russet and may crack open and dry up or never ripen at all. Old infections appear as reddish-brown areas on dormant canes.

Early powdery mildew infections can cause reduced berry size and reduced sugar content. Scarring and cracking of berries may be so severe as to make fruit unsuitable for any purpose.

Leaves: The first powdery mildew lesions are frequently found on the undersides of leaves (a). As the epidemic progresses, lesions become apparent on the upper sides of leaves as well. These lesions will increase in size and number if the disease is left unchecked. Severely infected leaves may become brittle and drop off. Starting as early as late July, very small orange to black spherical structures called cleistothecia develop on the upper and lower surfaces of leaves (b).

Shoots: Brown to black irregular blotches that can measure up to a few cms, follow the gradual degeneration of the fungus over the course of the season (c). The spots have indistinct margins and remain visible even following shoot lignification.

Inflorescences and rachis: Usually seen on rachis, powdery mildew has the appearance of a gray to whitish powder. Severe infections of the rachis can result in clusters being dropped, especially if mechanical harvesting is done. Symptoms on the rachis are similar to those on shoots

Berries: Berries can be infected from immediately after bloom through to four weeks post-bloom. They turn an ash gray color and quickly become covered in spores (d), giving them a floury appearance. At the end of the season, cleistothecia also appear on the berries (e). Affected berries dry out and may drop off (f) (Figures 13.40 and 13.41).

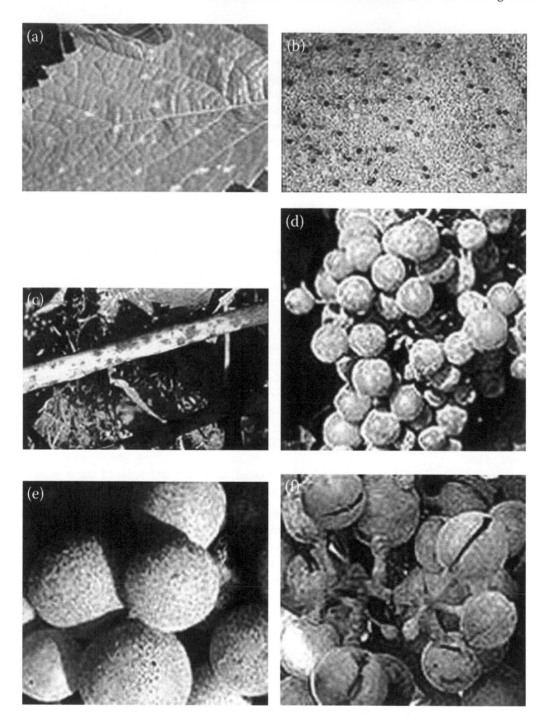

FIGURE 13.40   (a) The first powdery mildew lesions are frequently found on the undersides of leaves. (b) Very small orange to black spherical structures called cleistothecia develop on the upper and lower surfaces of leaves. (c) The gradual degeneration of the fungus over the course of the season. (d) They turn an ash gray color and quickly become covered in spores. (e) Cleistothecia also appear on the berries. (f) Affected berries dry out and may drop off.

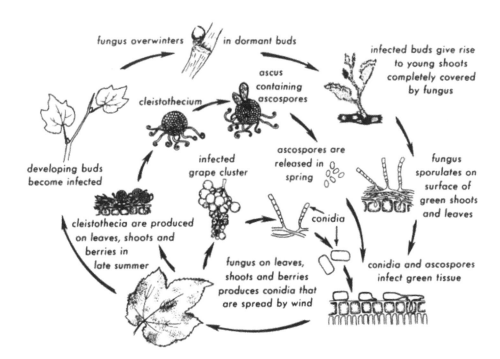

FIGURE 13.41   Disease cycle of grape powdery mildew.

13.7.3   FAVORABLE CONDITIONS OF DISEASE DEVELOPMENT

E. necator (U. necator) is an obligate parasite of grapevines, i.e. it can develop only on living grapevine tissue. The conidia of *E. necator* do not need free water on the tissue to infect it. However, high relative humidity promotes germination of the conidia and therefore infections. Powdery mildew of grape is promoted by hot (optimum temperature of 25°C), dry (but humid) weather since water inhibits germination of the conidia.

The powdery mildew fungus overwinters as cleistothecia (tiny, round, black fruiting bodies), in bark, on canes, in leftover fruit, and on leaves on the ground. Spores (ascospores) from the overwintering cleistothecia are released in the spring after a rainfall of at least 2.5 mm. For primary infection to occur the spores require at least 12–15 hours of continuous wetness at 10°C–15°C to infect developing plant tissue.

Once primary infection has occurred, the disease switches to its secondary phase. Secondary colonies (white mildew patches) form in 7–10 days, although the disease is not noticeable early in the season. The white patches of powdery mildew produce millions of spores (conidia) which are spread by wind to cause more infections. Free moisture is not needed for secondary infection; temperature is the most important environmental factor. The disease spreads quickly in early summer when temperatures are moderate. The incubation time (the time between infection and the production of spores) can be as short as 5–6 days under optimal temperatures. Shaded and sheltered locations favor mildew development. High temperatures and sunlight are inhibitory to powdery mildew. Extended periods of hot weather (>32°C) will slow the reproductive rate of grape powdery mildew, as well as reduce spore germination and infection (Figures 13.42 through 13.47).

13.7.4 Management

1. Manage canopies to increase air drainage and light penetration by removing lateral shoots in dense canopies. If necessary, remove leaves in the fruiting zone. Dense canopies provide low light intensity, which favors powdery mildew development.
2. Use an under-vine irrigation system (drip or micro-jet).
3. Manage irrigation carefully. Excessive irrigation leads to excessive vigor and higher disease potential.
4. Select varieties that are less susceptible to mildew.

13.7.5 Control

Protect grape foliage from primary infection by application of fungicides from early shoot growth until after bloom. Good control early in the season to prevent establishment of the disease is the key to preventing a powdery mildew epidemic later in the summer.

Apply fungicides such as Kumulus (sulfur), Nova, Lance, Pristine, Sovran, Flint, Milstop, or Serenade at the following growth stage.

When new growth is 5–10 cm long

1. Just before or immediately after bloom.
2. Every 10–14 days until grapes begin to soften and red varieties begin the development of color and white varieties change from green to white or yellow. If Kumulus (sulfur) is used, shorten the spray interval to 7–10 days.

FIGURE 13.42   Powdery mildew cleistothecia on grape shoot.

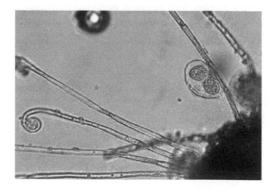

FIGURE 13.43   Magnified view of a cleistothecium and ascus containing ascospores.

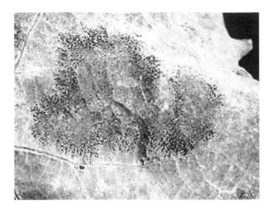

FIGURE 13.44   Severe powdery mildew infection on "Chancellor" grape leaf, with developing cleistothecia.

FIGURE 13.45   Spore (condia) production of powdery mildew on a grape leaf.

FIGURE 13.46   Powdery mildew infection on "Chancellor" grape foliage. Note whitish mildew and purple discoloration.

FIGURE 13.47   Powdery mildew infection on "Pinot Noir" grape cane.

Dormant spray: Lime Sulfur is effective at suppressing the overwintering population of powdery mildew. It should be applied in early spring before bud break to dormant vines to kill powdery mildew cleistothecia (initial inoculum). Good spray coverage of dormant vines is important.

Post-harvest powdery mildew spray: Post-harvest sprays to control powdery mildew are beneficial. Harvest date will determine the need to keep foliage and canes protected. Severe powdery mildew conditions are generally a result of poor control of this organism during the growing season. Additional sprays for powdery mildew under such conditions after harvest will not protect canes (Table 13.7).

13.8   BACTERIAL BLIGHT (BACTERIAL NEROSIS)

Bacterial blight of grapevine is a serious, chronic, and destructive vascular disease of grapevine, affecting commercially important cultivars. It is widespread in the Mediterranean region and South Africa and may occur in other regions. The pathogen survives in the vascular tissues of infected plants.[5]

The disease can be transmitted by propagation material, during grafting and by pruning knives. Bacteria are spread by moisture to plants where infection may take place through wounds, leaf scars, and other sites.

The only known host of bacterial blight is *V. vinifera* (Common grape vine).

13.8.1   CAUSAL ORGANISM

Bacterial blight (necrosis) is mostly caused by the bacteria mentioned below.

Species	Associated Disease Phase	Economic Importance
Xylophilus ampelinus	Stems and Petioles, Leaves, Fruit Cluster	Moderate

TABLE 13.7

Fungicides Registered for Control of Powdery Mildew on Grape

Fungicide	Chemical Group[a]	Rate/ha	Rate/acre	PHI[b] (days)	Notes
Vivando (metrafenone)	U8	750 g	300 g	14	Apply at 14-21-day intervals; use shorter interval for high disease pressure or rapid growth phases. Do not apply more than 2 sequential sprays.
Quintec (quinoxyfen)	13	300 mL/ha	122 mL/acre	14	Excellent mildew fungicide. Apply on a 14 day interval. Do not exceed 5 applications/ season. Alternate with other fungicides.
Flint (trifloxystrobin 50% WG)	11	105-140 g/ha	43–57 g/acre	14	Excellent mildew fungicide. Apply preventively using a 14-21-day interval. Do not use Flint or other group 11 fungicides more than 2 times per season. Alternate with fungicides from other groups.
Sovran (kresoxim-methyl 50% WG)	11	240-300 g/ha	100-122 g/acre	14	Excellent mildew fungicide. Apply at 14-21-day intervals. Do not use Sovran or other group 11 fungicides more than 2 times per season. Alternate with fungicides from other groups.
Pristine (pyraclostrobin + boscalid)	11+7	420–735 g/ha	170–300 g/acre	14	Excellent mildew fungicide; also provides suppression of botrytis. See label for details on rates and spray intervals. Do not use Pristine or other group 11 fungicides more than 2 times per season.
Nova (myclobutanil 40%)	3	200 g/ha	81 g/acre	14	Excellent mildew fungicide. Apply at 21-day intervals. Do not use more than 2 times per season. Alternate with fungicides from other groups.
Kumulus DF (sulfur 80%)	M	4.2 kg/ha	1.7 kg/acre	21/1[c]	Good mildew fungicide. Apply at 10-day intervals.
wettable sulfur (sulfur 92%)	M	2.25 kg/ha pre-bloom 4.5–6.0 kg/ha post-bloom	910 g/acre pre-bloom 1.8–2.4 kg/acre post-bloom	21/1[c]	Use the higher rate when vines are in full leaf. Apply at 10-day intervals. Re-apply after rain.
Lance (boscalid 70% WDG)	7	315 g/ha	128 g/acre	14	Alternate with fungicides from other groups.
Milstop (potassium bicarbonate 85%)	NC	2.8-5.6 kg/ha	1.1–2.3 kg/acre	0	Apply at 7-14 day intervals.

(Continued)

TABLE 13.7 (CONTINUED)
Fungicides Registered for Control of Powdery Mildew on Grape

Fungicide	Chemical Group[a]	Rate/ha	Rate/acre	PHI[b] (days)	Notes
Serenade MAX (*Bacillus subtilis*)	NC	3.0–6.0 kg/ha	1.2–2.4 kg/acre	0	Biofungicide. Disease suppression only. Do not tank mix with other products or fertilizers.
Lime Sulfur (calcium polysulfide 22%)	M	100 L lime sulfur in 1000 L of water.		120 (apply dormant)	Apply 500L of spray mixture per hectare once per season during dormant stage prior to bud swell (early March to early April). Spray to point of runoff, cover completely.

[a] *Chemical group*: Products with the same number belong to the same class of compounds. Alternate products with different chemical groups to help delay or prevent the development of resistance.

[b] *PHI* (pre-harvest interval), or the minimum number of days between the last spray and harvest.

[c] *Sulfur* can be used on table grapes up to the day of harvest (1-day PHI), but the pre-harvest interval for wine grapes is 21 days. Excessive amounts of sulfur are detrimental to winery yeasts. It is suggested that the last application to wine grapes be made no later than 30 days before harvest.

FIGURE 13.48 Cankers on bunch stalks causing partial or total death of a fruit bunch.

13.8.2 Symptoms

Bacterial necrosis of grapevines is characterized by typical symptoms such as cankers on stems and petioles, by necrotic foliar spots and by bud death.

In early spring, buds on infected spurs fail to open or are stunted in growth and eventually die. Affected spurs often appear slightly swollen because of hyperplasia of the cambial tissue. Cracks that appear along such spurs become deeper and longer forming cankers. Young shoots may develop pale yellowish-green spots on the lowest internodes. These expand upward on the shoot, darken, crack, and develop into cankers. Cracks, and later cankers, also form on more woody branches later in spring. In summer, cankers are often seen on the sides of petioles, causing a characteristic 1-sided necrosis of the leaf. They may also appear on main and secondary flower and fruit stalks. Leaf spots and marginal necrosis occur sometimes. Gum formation is not necessarily a symptom.

Infection usually occurs on the lower 2–3 nodes of shoots that are 12–30 cm long, and it spreads slowly upward. Initially, linear reddish-brown streaks appear, extending from the base to the shoot tip. Lens-shaped cankers then develop. Shoots subsequently wilt, droop, and dry up. Discoloration is less common on very young shoots, but the whole shoot dies back. Where there is severe infection, a large number of adventitious buds develop, but these quickly dieback. Infected shoots are shorter, giving the vine a stunted appearance. Tissue browning is revealed in stem cross-sections. Infected grape bunch stalks show symptoms similar to the infected shoots.

Leaves may be penetrated via the petiole and then the veins, in which case the whole leaf dies. Alternatively, leaves are penetrated directly via the stomata, resulting in the development of angular, reddish-brown lesions. Infection through the hydathodes results in reddish-brown discolorations in the leaf tips. Pale yellow bacterial ooze may be seen on infected leaves when humidity is high.

Immature flowers turn black and dieback. Roots may also be attacked resulting in retardation of shoot growth, regardless of grafted or natural rootstock (Figures 13.48 and 13.49).

FIGURE 13.49 Shoots severely infected by bacterial blight.

13.8.3   CAUSE AND DISEASE DEVELOPMENT

X. ampelinus is a strictly aerobic, non-spore forming, gram-positive rod, motile by 1 polar flagellum. Occasionally isolates have some filamentous cells 8–10 times longer than usual. Occurs singly or in pairs.

It produces a yellow insoluble pigment and metabolizes sugars oxidatively. Primary infections occur mainly on shoots 1 or 2 years old, via leaves, blossoms, and grapes.

13.8.4   FAVORABLE CONDITIONS OF DISEASE DEVELOPMENT

Epidemiology of bacterial blight indicates that no insect vector of importance has been found. The major sources of infection are apparently infected propagating material and epiphytic bacteria that enter through wounds.

Bacteria overwinter in the vines, emerge, probably in spring, and are carried to healthy shoots most likely through wind and rain. Wounds may facilitate entry but are not needed for primary infection.

Considerable spread can occur via propagating material, grafting, and pruning. Bleeding sap appears to be an important source of contamination. Illegally imported plants pose the greatest risk and if such material is infected, the disease is likely to become established.

Bleeding sap appears to be an important source of contamination.

The disease can be transmitted by propagation material, during grafting and by pruning knives. Bacteria are spread by moisture to plants where infection may occur through wounds, leaf scars, and other sites. Infection may also occur without wounds

The disease is associated with warm moist conditions and spread is favored by overhead sprinkler irrigation. From initial disease foci, local spread in vineyards tends to occur along rows.

13.8.4.1   Dispersal

Spread can occur via propagating material, grafting, and pruning. Bleeding sap appears to be an important source of contamination. Illegally imported plants pose the greatest risk and if such material is infected, the disease is likely to become established.

13.8.4.2   Survival

The bacterium overwinters in the vines, emerges in spring, and is carried to healthy shoots mainly by rain splash. Shoots are susceptible to infection during autumn and winter and non-susceptible during spring and summer. The bacterium can survive in wood, and thus, may be transmitted between locations in infected cuttings.

The ability of *X. ampelinus* to survive for several years inside plants without inducing symptoms may result in a latency period.

13.8.5   DISEASE CYCLE

X. ampelinus is gram-negative and aerobic bacteria that only survives on alive wood and pruning wood for 5–6 months. Primary infection occurs on shoots that are 1 or 2 years old. Infection usually occurs on the lower 2–3 nodes of shoots that are 12–30 cm long and spreads slowly upward. Infection may be transmitted systemically though the plant or the bacterium can penetrate leaves through open stomates.

Bacterial blight of grapevine is readily transmitted with pruning tools and enters healthy tissues mainly through pruning wounds. Bacterial transmission is greatest in wet and windy weather. Bacteria can spread between shoots in the early summer. This is a disease associated with humid climates. The dissemination is favored by rain or irrigation.

13.8.6 MANAGEMENT

Diseases can be managed using the following strategies:

- Chemical control
- Cultural control

13.8.6.1 Chemical Control

Bordeaux or copper sprays after pruning and up until half leaf expansion can be effective to control this disease (Table 13.8).

13.8.6.2 Cultural Control

Sanitation management is essential.

- Always use disease-free stock.
- Infected branches and canes should be pruned and burned.

13.9 CROWN GALL OF GRAPE

Crown gall of grape is an important disease in all areas where grapes are grown worldwide, but is particularly severe in regions with cold climates. Formerly designated as *Agrobacterium tumefaciens* and *Agrobacterium vitis*, the bacterium that causes of the disease only occurs on grape.

A. vitis (Ophel and Kerr, 1990) survives systemically in grapevines and initiates infections at wound sites, such as those caused by freeze injuries. As a result, vineyards in climates with cold winters are prone to suffer extensive damage from crown gall. In addition to freeze-induced wounds, graft unions are also common sites for infection.

Crown gall can reduce vine vigor and growth, thus, reducing crop yield. Cultivars of *V. vinifera* tend to be highly susceptible to crown gall, although certain French—American hybrids and Native American varieties may also become severely infected. Crown gall can kill young vines to the soil line thereby reducing cropping potential and requiring establishment of new trunks. Vines that are completely killed need to be removed and replaced at significant costs to growers and wine and juice producers.

The causal agent of grape crown gall was first identified in 1897 in Italy. In those investigations, a bacterium was identified as the infectious agent causing disease of the vines. Since this discovery, it has been demonstrated that crown gall of grape is caused predominately by the bacterium

TABLE 13.8

Fungicides Registered to Control Powdery Mildew on Ornamentals Sorted by their Active Ingredient

Trade Name	Active Ingredient	Fungicide Group
Champ DP Dry Prill Champ WG Kentan DF Nu-Cop 50 WP Nu-Cop 3 L	Copper Hydroxide	M1
COC DF, COC WP	copper oxychloride	M1
Nordox, Nordox 75 WG	Copper oxide	M1

A. vitis. However, *A. tumefaciens* (the predominant causal agent of crown gall of other crops) has also been isolated from galls on grape and is associated with the disease at a much lower frequency than *A. vitis.* Reports of grape crown gall have come from many parts of the world including China, Japan, South Africa, several European countries, the Middle East, and North and South America. In Oklahoma, crown gall is probably the second most significant disease of grape after black rot.

13.9.1 CAUSAL ORGANISM

Crown Gall is mostly caused by a single species of *Agrobacterium* mentioned below.

Species	Associated Disease Phase	Economic Importance
A. vitis (Ophel and Kerr, 1990)	Gall (Tumor) Formation at the Lower Trunk, In Canes, at Graft Unions, and at Nodes	High

13.9.2 SYMPTOMS

The disease occurs most frequently by the appearance on the roots and stems.

Swellings or galls are on the roots and stem near the soil line. Galls at first are creamy white to greenish white, soft textured, and have no bark or covering. Young galls frequently look like callus tissue formed in wounds and are often overlooked. As galls age, they become dark brown, woody, rough, and may be bigger than a baseball. The gall surface turns black as it ages. Galls initially develop near or below the ground level. Secondary galls can develop higher up on the trunk and arms. Galls on the canes are less common. Gall formation can disrupt the food- and water-conducting tissues of the vine. Infected young vines may become stunted and grow poorly. On older vines, the galls can also serve as entry points for secondary wood rotting organisms, which further weaken the plants. Foliage on affected Concord vines shows yellow discoloration and resembles the coloration associated with the onset of fall season. The leaves on some wine grape varieties may turn red or yellow. In extreme cases, fruit production decreases (Figures 13.50 and 13.51).

FIGURE 13.50   Tumors at root region.

FIGURE 13.51 Tumors at stem region.

13.9.3 CAUSE AND DISEASE DEVELOPMENT

Crown gall tumors in plants caused by *A. tumefaciens* and *A. vitis* represents a unique disease involving the transfer of DNA from the bacterium to the nucleus of the plant.

Agrobacterium grow aerobically, is a gram-negative, motile, rod-shaped bacterium that is non-sporing and is closely related to the N-fixing rhizobium bacteria which form root nodules on leguminous plants. The bacterium is surrounded by a small number of peritrichous flagella. Virulent bacteria contain 1 or more large plasmids, 1 of which carries the genes for tumor induction and is known as the Ti (tumor inducing) plasmid. The Ti plasmid also contains the genes that determine the host range and the symptoms that the infection will produce. Without this Ti plasmid, the bacterium is described as being non-virulent and will not be able to cause disease in the plant.

Infection process takes place in 4 distinct phases: injury to host plants; bacterial cells attach to the surface of plant cells located on the surface injured; Ti plasmid transfer (which induces tumor formation) in host cells; and Ti plasmid integration into the host cell genome. It follows that the host plant injury is essential in the transformation process and that the bacteria attaching to cells in the injured areas of the plant is necessary for tumor initiation. After the wound serves as a "gateway", probably by removing mechanical barriers and stimulating the metabolic processes of plant cells. Injuries release different chemical signals that are received by the bacteria and resulting installation of virulent genes (vir) in the Ti plasmid. Briefly, amino acids, organic acids, and sugars released from wounded plant cells act as chemoattractants to tumorigenic agrobacteria, which bind to plant cells in a polar orientation upon reaching the wound site.

13.9.4 FAVORABLE CONDITIONS OF DISEASE DEVELOPMENT

The incubation period varies depending on plant age, environmental conditions, etc. At a temperature of 20°C–25°C incubation period is 13–14 days and at 10°C–15°C is 27–28 days. Relative humidity greater than 80–90% favors infection. Light has an inhibitory effect. The disease is favored by, nitrogen fertilizers, the lack of affinity between scion and rootstock, hail, attack of nematodes and frost. It is also favored by wet, compact soils with slightly alkaline pH having humidity of 25 to 50% (of water holding capacity).

13.9.5 DISEASE CYCLE

A. vitis survives in living as well as dead grape tissues that may persist in soil. It is unclear as to whether the root necrosis affects vine and root growth; however, it may provide access points through which the pathogen enters the vine. *A. vitis* is known to colonize vines systemically and can be isolated from grape sap.

Consequently, the pathogen is disseminated in cuttings that are used for propagating new plants. As indicated, crown gall develops at wound sites, primarily during wound healing in the cambium, where dividing cells are susceptible to infection. Infected cells overproduce plant growth hormones leading to gall formation and enlargement.

13.9.6 MANAGEMENT

Diseases can be managed using the following strategies:

- Chemical control
- Biological control
- Cultural control

13.9.6.1 Chemical Control

There are no effective chemical controls available to manage grape crown gall. A petroleum-based product called Gallex can be painted on individual galls, causing them to shrink temporarily. This treatment is costly and needs to be reapplied periodically.

13.9.6.2 Biological Control

Biological control of crown gall on fruit and ornamental plants has been very effective, and commercial preparations of a non-pathogenic *A. radiobacter* (K84) are sold in many regions of the world. Unfortunately, K84 is not effective against *A. vitis* on grape. However, it has been shown that when a non-pathogenic strain of *A. vitis*, F2/5, is applied to wounded grape tissue in advance of gall-forming pathogen, crown gall is prevented. The mechanism by which F2/5 prevents crown gall is unknown. 2 interesting points are that it only inhibits crown gall on grape, and that it must arrive at the wounded grape tissue before the pathogen.

Although F2/5 has been shown to be highly effective for controlling crown gall in greenhouse experiments, its effectiveness has not yet been proven in the field. Several experiments are underway, and in these cases, vines are soaked in suspensions of F2/5 prior to planting. The objective is to allow F2/5 to colonize wounds on the roots and crown and to ideally establish itself in the grapevine. Thus far, variable success in control of crown gall with F2/5 in the field has been noted. F2/5 is not commercially available, and research is being done to determine its efficacy in field trials.

13.9.6.3 Cultural Control

Cultural practices that result in limiting mechanical and freeze injury have proven most useful for managing this disease. Proper site selection is critical for new plantings. Avoid heavy soils in wet areas where frost is likely (low areas). Limiting exposure to the north is also desirable for cold-tender cultivars. Good sanitation practices when removing infected vines is critical. Care should be taken to remove as much of the plant root system as possible. The crown gall pathogen can be present at high levels in the root system of infected plants. Removing and destroying as much of the plant debris as possible will reduce the level of pathogen propagules in the soil. The success of management strategies such as leaving soil fallow for extended periods or planting non-hosts to rid vineyards of the bacterium will have varying success depending on the level of infestation. Care should be taken to limit soilborne nematode damage. Studies have shown that crown gall incidence was positively correlated with root knot nematode damage. Growers should have soil in potential vineyard sites tested for root knot nematode prior to planting. Areas with infestations of root knot nematode should be avoided.

Wounding plants during cultivation or pruning should be avoided. If wounds are caused, they should be treated with a protective coating like "Tree Seal" or "Bordeaux Paste".

Using resistant rootstocks may help in the case of root infection by *A. vitis*. Rootstocks such as Courderc 3309 and Mgt 101–14 are resistant. Rootstocks such as Richter 110 and Teleki 5C are considered susceptible. However, there are drawbacks to using rootstocks. Own-rooted vines allows viticulturists to retrain after severe winter cold damage. However, this may not be possible if vines are grafted. The cost of replanting after every damaging event versus retraining needs to be carefully considered when considering the use of rootstocks for crown gall management.

13.10   EUTYPA DIEBACK

Eutypa dieback is a major disease of grapevines in Australia and worldwide caused by the fungus *Eutypa lata* that infects vines through pruning wounds, colonizes wood tissue and causes dieback of cordons, stunting of green shoots and leaf distortion. Eutypa dieback threatens the sustainability of premium vineyards and is becoming a problem in most cool climate growing regions. This disease reduces growth and yield and if unmanaged eventually kills vines. *E. lata* is widespread in most areas of the world where grapevines or apricots are cultivated.

13.10.1   CAUSAL ORGANISM

Eutypa dieback is mostly caused by the fungi mentioned below.

Species	Associated Disease Phase	Economic Importance
E. lata	Stem (Trunk), Leaves, Fruits	High
Eutypa armeniacae		

13.10.2   SYMPTOMS

Symptoms are best seen in spring when healthy grapevine shoots are 10–15 inches long. Sometimes, symptoms may not appear on diseased vines for more than 3 years after infection.

13.10.2.1   Shoot-Leaves Symptom

Shoot symptoms are most evident during the spring, when healthy shoots are 20–40 cm long. Shoots arising from infected wood are stunted with small, chlorotic, distorted leaves. The leaves become necrotic and tattered as the season progresses. Fruit fails to develop or develops very poorly. No other pathogens are known to cause these shoot symptoms.

Spring shoot growth is weak and stunted above the cankered area. Leaves are at first smaller than normal, cupped, misshapen, and yellowed. Later in the season (mid-July), these leaf and shoot symptoms may disappear from all but the basal leaves of affected shoots. The vines may appear to have recovered. However, the infected trunk and all growth above it will eventually die (Figures 13.52 and 13.53).

13.10.2.2   Wood Symptom

The most conspicuous symptom of this disease is the appearance of dead or weakened canes in the spring. Eutypa dieback shoot symptoms are always accompanied by a canker, which often appears v-shaped in a cross-section of the perennial wood. The fungus infects the trunk or a main branch, resulting in the

FIGURE 13.52   Stunted shoot growth with yellow leaves.

FIGURE 13.53   Cluster development on an affected shoot (right) and a healthy shoot (left) from the same vine.

formation of a canker. Each year the canker enlarges until it eventually girdles the trunk or branch and kills the distal portion. New shoots developing near the canker are weak and stunted. These may result in failure of the vine to leaf-out or in the formation of dwarfed, yellowish, crinkled leaves. Infected canes which survive through the summer often die during the following winter (Figure 13.54).

13.10.3   CAUSE AND DISEASE DEVELOPMENT

Eutypa dieback of grapes is caused by the fungus *E. lata*. The fungus survives in diseased wood and produces perithecia in old, affected host tissue under conditions of high moisture. Eutypa dieback is not generally visible in vines younger than 5–6 years old although vines may still be infected. The disease is most easily seen in vines established for 10 or more years. The fungus survives in diseased wood and produces perithecia in old, infected host tissue under conditions of high moisture. In California several plants in addition to grape serve as reservoirs for the pathogen including almond, apricot, blueberry, cherry, crab apple, *Ceanothus* spp., kiwi, pear, oleander, and native plants including California buckeye, big leaf maple, and willow. Ascopores are discharged from perithecia soon after rainfall. Infection occurs through pruning wounds, which remain susceptible

FIGURE 13.54 V-shaped canker caused by *E. lata* in the xylem of a grapevine cordon.

much longer, early in the dormant season rather than later in the dormant season. Overall susceptibility is about six weeks.

Moderately infected vineyards can lose 19%–50% of yield, severely affected vineyards can lose 62%–94%.

13.10.4 Disease Cycle

Eutypa dieback is caused by the fungus *E. lata*. The fungus survives in infected trunks for long periods of time, whether they remain as part of the in-place vine or as prunings in the vineyard. Eventually, the fungus produces reproductive structures (perithelia) on the surface of infected wood. Spores (ascospore) are produced in these structures and are discharged into the air. Ascospore discharge is initiated by the presence of free water (either rainfall or snow melt). Most spores appear to be released during winter or early spring, with relatively few being released during the summer months. Unfortunately, most spores are released at the same time pruning is being conducted. The ascospore can be carried considerable distances by air currents to recent wounds on the trunk. Pruning wounds are by far the most important points of infection. When the ascospore meet newly cut wood, they germinate, and a new infection is initiated (Figure 13.55).[6]

13.10.5 Management

The disease can be managed using the following strategies:

- Chemical control
- Biological control
- Cultural control

13.10.5.1 Chemical Control

Treatment of fresh, large pruning wounds could offer some disease control, but currently there are no chemicals labeled for control of this disease.

Grapevine pruning wounds have been successfully treated with benzimidazole fungicides such as benomyl and carbendazim to control infection by *E. lata*. Other wound protectants effective against infection by *E. lata* include acrylic paint and pastes, which produce a physical barrier to infection. Boron is an effective alternative for protection of pruning wounds.

13.10.5.2 Biological Control

Biocontrol agents, such as *Fusarium lateritium*, *Cladosporium herbarum*, and *Trichoderma* spp., have also been used to protect wounds from invasion by *E. lata* spores with varying degrees of efficacy (Carter, 1991; Munkvold and Marois, 1993; John et al., 2005).

DISEASE CYCLE OF EUTYPA DIEBACK

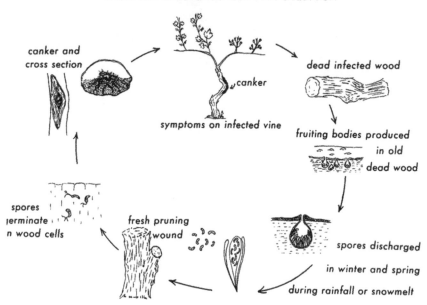

FIGURE 13.55   Disease cycle of Eutypa dieback.

13.10.5.3 Cultural Control

Prune late in the dormant season to promote rapid healing of wounds. Remove and burn infected wood inside the vineyard and dead wood in adjacent vineyards and orchards to reduce the spread of the pathogen. Cut out and remove dead arms and cordons from the vineyard during dormancy. Completely remove all cankers, pruning below the canker on the vine or trunks until no darkened canker tissue remains. Make large cuts directly after rain because the risk for infection is lowest at this time, as the atmospheric spore load has been washed out temporarily. Double pruning cordon-trained vines can help final pruning cuts to be made quickly and late in dormancy, thus, reducing the chance for infection. For additional protection, consider treating pruning wounds.

13.11   GRAPE LEAF ROLL DISEASE

Among the virus and virus-like diseases infecting grapevines worldwide, grapevine leafroll disease (GLD) is the most economically destructive. It accounts for an estimated 60% of yield losses due to virus diseases in grape production worldwide.

GLD can affect all native and *V. vinifera* cultivars, hybrids, and rootstocks, although symptoms are not expressed on all infected vines. The disease was described first in Europe as early as the nineteenth century but its graft-transmissibility was not demonstrated until 1937. In 1979, a specific type of virus (*closterovirus*) was reported in a leafroll-affected vine, and shortly thereafter, in 1983, the capacity of mealybugs to transmit 1 of the viruses associated with this disease was shown.

13.11.1   Causal Organism

Species	Associated Disease Phase	Economic Importance
Grape Leafroll-Associated Viruses (GLRaVs)	Leaf Roll	High

13.11.2 Symptoms

In general, symptoms are more dramatic in red-fruited *V. vinifera* cultivars than in white-fruited cultivars. Infected vines typically exhibit no symptoms until late July or early August when the crop moves toward veraison. 1 of the early visual signs of GLD in red-fruited cultivars is the appearance of red and reddish-purple discolorations in the interveinal areas of mature leaves near the basal part of the shoots. As summer progresses, the symptoms extend upward to other leaves and the foliar discolorations expand and coalesce to form a reddish-purple color within the interveinal areas of the leaf; a narrow strip of leaf tissue remains green on either side of the main veins. So, by the later part of the season (August–October), a typical infection in a red-fruited cultivar will consist of green veins and reddish interveinal areas. In the advanced stages, the margins of infected leaves roll downward, expressing the symptom that gives the disease its common name.

GLD symptoms vary within and among vineyards due to several factors including the variety, age of the vineyard, stage of infection, complex of viruses presents, viticultural practices, and environmental conditions. Symptoms also vary based on the year and the part of the plant. Foliar symptoms tend to be more pronounced during cooler growing seasons and on the shaded side of the vine.

White-fruited cultivars express GLD symptoms differently if at all. In some cultivars like Chardonnay, infected leaves may show general yellowing or chlorotic mottling toward the end of the season, and in some cases, leaf margins may roll downward toward the end of the season (Figures 13.56 and 13.57).

FIGURE 13.56 Curling of leaves in red-fruit variety.

FIGURE 13.57 Curling of leaves in white-fruit variety.

13.11.3 Cause and Disease Development

To date, 10 different filamentous viruses identified as GLRaVs have been isolated and character-ized from leafroll-infected grapevines. These 10 viruses are numbered from 1–10 (GLRaV-1 to GLRaV-10) based on the order of their discovery. They all belong to the family Closteroviridae. The GLRaVs are serologically unrelated, and their particle length ranges between 1,400 and 2,200 nms.

Most of these viruses can be detected by wood or green grafting onto indicator vines of *V. vinif-era* cv. Pinot noir, Cabernet franc, or Gamay. Lab tests including serological assays, such as double antibody sandwich enzyme-linked immunosorbent assay (ELISA), and molecular assays, based on polymerase chain reaction (PCR), can also be used to detect GLRaVs in grapevine tissue.

Among the 10 viruses associated with leafroll disease, GLRaV-1, GLRaV-2, and GLRaV-3 usu-ally prevail in leafroll-affected grapevines.

Grape mealybug (*Pseudococcus maritimus* Ehrhorn) is a documented vector for GLRaV-3 under laboratory conditions. The grape mealybug is the predominant mealybug found in Washington vineyards. Other mealybug species have been documented as vectors of GLRaVs in western United States' vineyards, but these species have not been found in Washington. While grape mealybug has 2 generations per season, other species such as vine mealybug (*Planococcus ficus*) can have up to 9 generations per season, dramatically increasing the number of potential vector insects within a vineyard in a relatively short period of time. Because of this danger, a concerted effort is necessary to prevent the establishment of vine mealybug in Washington State; vine mealybug has been established as a quarantine pest in this state. Other mealybugs implicated in the spread of GLD include *Planococcus citri*, *Pseudococcus longispinus*, *P. affinis*, *P. calceolaria*, *P. comstocki*, *P. viburni*, *Heliococcus bohemicus*, and *Phenacoccus aceris* (Figures 13.58 and 13.59).

13.11.4 Management

The disease can be managed using the following strategies:

- Chemical control
- Biological control
- Cultural control
- Integrated pest management

There are no effective curative measures for eliminating the virus once it is established in a vine—the only recourse is to minimize the impacts of the disease. The focus in this situation is on curbing the spread of the disease and minimizing economic losses.

FIGURE 13.58   Grape mealybug.

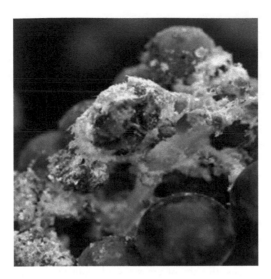

FIGURE 13.59   A waxy, cotton-like residue is common with mealybug infestation.

13.11.4.1   Chemical (Vector) Control

Grape mealybug is a documented vector for the causal agents of GLD. Mealybugs overwinter as eggs or crawlers in the egg sacs, usually in the bark cracks or under the bark scales on the grapevine trunk and in the arms or laterals. In the spring, crawlers move quickly to a new growth to feed. They mature in June, and adults move back to older wood to lay eggs. The second generation of crawlers will move to new growth, including the fruit, where they mature through July and August. In addition to their potential to vector GLD, this second generation may contaminate fruit by production of honeydew, which may further lead to favorable conditions for sooty mold development.

Generally speaking, control procedures for grape mealybug are most effective when the insects are in the crawler stage. Chemigation treatments with chloronicotinyl insecticides are registered for use on grapes. Several products can be effective against mealybugs at any time during the growing season. Irrigation water requirements for adequate distribution of systemic insecticides vary among products. Chemigation of imidacloprid is an effective treatment available for grape mealybug applied mid to late spring when the vineyard soil moisture is being held at or near field capacity. Soil moisture is important in transporting imidacloprid. Chemigation with thiamethoxam and dinotefuran has proven effective in deficit irrigation situations.

If a vineyard is not drip irrigated, foliar treatments can be applied for mealybug control. Foliar sprays of chlorpyrifos are labeled exclusively for dormant or delayed dormant applications and, if utilizing this organophosphate, care should be taken to avoid runoff. Research has demonstrated that foliar sprays of imidacloprid (Provado) are not very effective in controlling grape mealybug infestations. Foliar sprays of thiamethoxam, acetamiprid, and dinotefuran should be directed toward the trunk and main laterals. When applying foliar sprays for this pest, sufficient water and pressure must be used to loosen bark and drive the pesticide into cracks and under loose bark.

Late summer spray applications for grape mealybug control are usually ineffective.

13.11.4.2   Biological and Cultural Control

To date, the only way to manage grapevine leafroll disease and secure a healthy and high-quality crop is to ensure that the planting material originates from virus-tested, virus-free mother plants and that factors contributing to infestation via insect vectors are well controlled. GLRaVs can be eliminated from prospective propagation material by heat treatment and tissue culture, among other techniques. The importance of establishing new blocks with clean planting material cannot be over-emphasized because, once a vine is infected, there is no cure in the vineyard.

TABLE 13.9

List of Insecticides Recommended to Control Mealybug

Insecticide	Dose	Pre-Harvest Interval
Buprofezin 25 SC	1.25 mL/L	40 days
Methomyl 40 SP	1 g/L	61 days
Dichlorvos 76 EC	2 mL/L	15 days
Azadirachtin 1%	2 mL/L	3 days
Chlorpyriphos 20 EC	2 mL/L	40 days

No sources of resistance against any of the GLRaVs have been identified in wild or cultivated grapes. Therefore, conventional breeding is not a viable option to develop GLRaV-resistant material. Research is ongoing to develop resistant material through genetic engineering.

In nurseries and mother blocks, the use of virus-tested material followed by regular and routine monitoring for the disease, its causal agents, and insect vectors is paramount for providing planting material of high phytosanitary standards. Stocks in nurseries and mother blocks should be tested regularly for GLRaVs and such blocks should be isolated from commercial vineyards to avoid infection through vector transmission. Also, mealybugs and soft scales should be intensively surveyed and managed (Table 13.9).

13.12   GRAPEVINE FAN LEAF DISEASE

Grapevine degeneration caused by grapevine fanleaf virus (GFLV) has been documented in many viticultural regions worldwide (Andret-Link et al., 2004). It is 1 of the major economically important virus diseases affecting the longevity of grapevines and reducing the fruit yield and fruit quality. It causes extensive leaf yellowing, stem and leaf deformation, reduced fruit quality, substantial crop loss, and shortened life-span of vineyards.

Infected grapevines show a range of foliar symptoms consisting of leaf deformation, yellow mosaic, vein banding, ring and line patterns, and flecks. GFLV cause yield reductions as high as 80% depending on the cultivar and severity of infection (Martelli and Savino, 1990). Plant-to-plant spread of GFLV is known to occur by dagger nematode (*Xiphinema index*), and hence, the infected grapevines appear in patches in the field. Long-distance spread of the virus, however, occurs by transfer of infected propagation material.

13.12.1   CAUSAL ORGANISM

Fanleaf degeneration/decline disease is caused by several different virus species. Viruses causing fanleaf degeneration/decline are nepoviruses (acronym for ne = nematode-borne; po = polyhedral particle). GFLV is the most well-characterized nepovirus and is by far the most widespread and important cause of the disease worldwide.

Species	Associated Disease Phase	Economic Importance
GFLV	Leaves and Fruits	High
Arabis Mosaic Virus (ArMV)		
Tomato Ringspot Virus (ToRSV)		
Tobacco Ringspot Virus (TRSV)		
Peach Rosette Mosaic Virus (PRMV)		
Blueberry Leaf Mottle Virus (BLMoV)		

13.12.2 SYMPTOMS

Fanleaf degeneration disease gets its name from the fan-like leaf shape that may be exhibited on infected vines and the gradual decline in growth and vigor of infected vines over time. The fan-shaped leaves are caused by abnormally gathered primary veins and widely open petiolar sinuses. In addition to this symptom, which may not be present in all infected vines, leaves may also show yellowing, puckering, deep lobes, bright chrome yellow coloring or mosaics with mottling. Yellow and distorted leaf symptoms often occur in the spring and fade as the summer progresses. Shoots of affected vines may have shortened internodes and abnormal branching. Infected vines may also exhibit a decreased resistance to adverse climatic factors such as drought or freeze events. Infected propagation materials may show reduced ability to root or poor graft take.

Plants that are infected with GFLV may be reduced in size compared to healthy plants. Fruit quality and winter hardiness are often reduced. Fruit losses of up to 80% have been reported in some varieties.

The disease is spread by dagger nematodes; however, presence of the nematode is not required for infection in a vineyard since the virus is frequently spread by movement of infected plant material (Figures 13.60 through 13.62).

13.12.3 CAUSE AND DISEASE DEVELOPMENT

Spread at a site (i.e. within a vineyard or between adjacent vineyards) is mediated by nematodes. 2 longidorid nematode vectors of GFLV are known: *X. index* and *Xiphinema italiae* (Hewitt, Raski and Goheen, 1958; Cohn, Tanne and Nitzany, 1970). The former species is by far the more efficient vector under natural and experimental conditions. Although not all *X. index* populations are equally efficient in transmitting virus isolates, this nematode is to be regarded as the major, if not the only, natural and economically important GFLV vector. It has a limited range of alternate natural hosts (e.g. fig, mulberry, rose), but these hosts are immune to GFLV. No natural virus reservoirs are known other than grapevines. GFLV persists in volunteer plants and in the roots of lifted vines that remain viable in the soil, constituting an important source of inoculum.

FIGURE 13.60   Bright yellow vein banding on leaf.

FIGURE 13.61   Yellow mosaic symptoms caused by GFLV grapevine fanleaf.

Transmission of GFLV through grapevine seeds has been reported, but it has negligible epidemiological significance.

Long-distance spread is passively but efficiently effected through dissemination of infected propagating material.

13.12.3.1 Favorable Conditions

Favorable conditions of GFLV are higher humidity in soils.

FIGURE 13.62   Infected grapevines produced small clusters with poor fruit set, irregular ripening and shot berries.

13.12.4 Management

Once a vine is infected, there is no cure for nepoviruses in a vineyard, and once infected vines and soilborne nematode vectors are established in a vineyard setting, control of fanleaf degeneration/ decline can be extremely challenging.

The disease can be managed using the following strategies:

- Chemical control
- Biological control
- Cultural control
- Integrated pest management

13.12.4.1 Chemical Control

Currently, there are no chemical control options for grapevine fanleaf degeneration disease.

13.12.4.2 Cultural Control

Management would require soil fumigation, deep plowing, covering crops with nematicidal properties, lengthy fallow periods (up to 10 years), and use of nematode-tolerant rootstocks.

However, if the vectors of a given virus are not known to be established in the eastern United States, vine-to-vine spread of these viruses is limited or impossible, meaning that removal of infected vines are sufficient for control of most of the viruses that cause fanleaf degeneration/ decline. Therefore, growers should take appropriate precautions to prevent the introduction of both the virus and its nematode vector.

In the case of ToRSV and TRSV, which have nematode vectors that are already widely distributed in eastern United States' regions, management of these viruses relies on the following:

1. Prior to replanting, perform a soil test to determine the presence of dagger nematodes, which may harbor and spread the viruses of concern.
2. Devise the best preparation strategy for the vineyard replant site based on soil test results.
3. Plant only virus-tested, clean planting material originating from certified, clean mother stocks to ensure a healthy and high-quality crop.
4. Eliminate alternate hosts—especially weed hosts—that can serve as a viral reservoir in vineyard settings.

At present, there are no sources of true resistance in either wild or cultivated grapevines toward most of the viruses that cause fanleaf degeneration/decline, so conventional breeding to develop fanleaf resistant material is not possible in most cases. *Vitis labrusca* is resistant to ToRSV and TRSV, and some interspecific hybrids (DeChaunac, Baco noir, Vidal blanc, Vincent, among others) show some resistance to TRSV but are susceptible to ToRSV. The rootstocks commonly used in the eastern United States (3309 C, SO4, Kober 5BB, St. George, 44–53 Malegue, 110 Richter, 1616 C, among others) show field resistance to ToRSV and rootstocks O39-16, RS-3 and RS-4, among others, show field resistance to GFLV in California, but are not useful in the eastern United States since *X. index* is not currently found here. Rootstocks resistant to the nematode vectors of viruses causing fanleaf degeneration/decline are available for *X. americanum* and *X. index* but they do not prevent infection of scions with ToRSV and GFLV, respectively. Research is ongoing to develop virus-resistant grapevines, in particular, GFLV-resistant rootstocks through genetic engineering.

13.13 PIERCE'S DISEASE

The single greatest threat to the long-term survivability of susceptible cultivars is Pierce's disease. Pierce's disease (PD) is caused by a xylem-limited bacterium that clogs the vascular tissue of susceptible grape cultivars. The causal organism is a gram-negative, rod-shaped bacterium named *Xylella fastidiosa* that is indigenous to the Gulf Coast region of the United States.

The *X. fastidiosa* bacterium resides in the xylem of the grapevine and is transmitted to the grapevine by insects, specifically sharpshooter leafhoppers and spittlebugs. Numerous diseases in a range of crop plants are caused by *X. fastidiosa* including almond leaf scorch, oleander leaf scorch, citrus variegated chlorosis, oak leaf scorch, and alfalfa leaf scorch, among others. PD severely limits areas where vinifera vines can be grown. Within the United States, *X. fastidiosa* diseases are restricted to areas of the country with warmer winter temperatures. PD has been present in southeast Virginia vineyards since at least 1990, when suspicious symptoms on Eastern Shore Chardonnay vines were confirmed as being caused by PD.

Once a grapevine is infected, the bacteria multiply and colonize the xylem. This vascular constriction inhibits the movement of water through the grapevine and often results in first visible symptoms noted during periods of heat or drought stress.

13.13.1 CAUSAL ORGANISM

PD is mostly caused by the bacteria mentioned below.

Species	Associated Disease Phase	Economic Importance
X. fastidiosa	Stems and Petioles, Leaves, Fruit Cluster	Moderate

13.13.2 SYMPTOMS

There are numerous symptoms expressed by susceptible cultivars after infection.

The first symptom is usually uneven marginal leaf necrosis that often appears near the point of infection. Since the disease inhibits water movement in the vine, symptoms often appear during heat stress or near veraison (color change) in the cluster. In disease-susceptible vines, the bacteria can multiply to such levels that they cause vascular blockage, either directly or via defensive gums produced by the vine. The vascular blockage ultimately leads to the characteristic disease symptoms including drying or scorching of leaves in irregular patterns in sections or along margins of the leaf blades.

Additionally, leaf blades abscise leaving petioles attached to the cane (matchsticks), periderm develops irregularly (green-islands), and fruit clusters shrivel (Figures 13.63 through 13.65).

PD may kill the vine within 1–2 years. Symptoms first develop during the latter part of the growing season after the bacteria have colonized the vine. Drought conditions and hot temperatures speed symptom development. Spring symptoms develop if infection occurred the previous year. This is called chronic infection. Spring symptoms appear as delayed bud break, stunted growth, and zigzag internodes on developing shoots. The infected vine may die within a year of infection or vines may persist for 5 or more years.

FIGURE 13.63 A common symptom of this disease is irregular, patchy bark maturity. Note that half the shoots are brown and half are green with islands of green and brown.

FIGURE 13.64   PD causes petioles of leaves to remain attached to canes after leaf fall.

13.13.3   CAUSE, FAVORABLE CONDITIONS, AND DISEASE DEVELOPMENT

The bacterium that causes PD lives in the xylem and is spread from plant to plant by sap-feeding insects that feed on the xylem. Symptoms appear when a significant amount of xylem become blocked by the growth of the bacteria. (This bacterium is also responsible for alfalfa dwarf disease and almond leaf scorch in California.)

Insect vectors for PD belong to the sharpshooter (Cicadellidae) and spittlebug (Cercopidae) families. The blue-green sharpshooter (*Graphocephala atropunctata*) is the most important vector in

FIGURE 13.65   Advanced late summer or fall symptoms of PD on foliage shows concentric rings of drying from the outer edge toward the center.

FIGURE 13.66　Adult blue-green sharpshooter.

FIGURE 13.67　Adult green sharpshooter.

coastal areas. The green sharpshooter (*Draeculacephala minerva*) and the red-headed sharpshooter (*Carneocephala fulgida*) are also present in coastal areas but are more important as vectors of this disease in the Central Valley, California. Other sucking insects, such as grape leafhoppers, are not vectors (Figures 13.66 and 13.67).[7]

A new PD vector, the glassy-winged sharpshooter, has recently become established in California. This vector is a serious threat to California vineyards because it moves faster and fly's greater distances into vineyards than the other species of sharpshooters.

Glassy-winged sharpshooter feeds and reproduces on a wide variety of trees, woody ornamentals, and annuals in its region of origin—the southeastern United States. Crepe myrtle and sumac are especially preferred. It reproduces on eucalyptus, coast live oaks, and a wide range of trees in southern California.

The principal breeding habitat for the blue-green sharpshooter is riparian (riverbank) vegetation, although ornamental landscape plants may also harbor breeding populations. As the season progresses, these insects shift their feeding preference, always preferring to feed on plants with succulent growth. In the Central Valley, irrigated pastures, hay fields, or grasses on ditch backs are the principal breeding and feeding habitats for the green and red-headed sharpshooters. These 2 grass-feeding sharpshooters also occur along ditches, streams, or roadsides where grasses and hedges provide suitable breeding habitat.

Some vines recover from PD the first winter following infection. The probability of recovery depends on the date of infection. Infections that occur until June have the greatest probability of surviving until the following year. Recovery rates also depend on grape variety; recovery is higher in Chenin Blanc, Sylvaner, Ruby Cabernet, and White Riesling, compared to Barbera, Chardonnay, Mission, Fiesta, and Pinot Noir. Thompson Seedless, Cabernet Sauvignon, Gray Riesling, Merlot, Napa Gamay, Petite Sirah, and Sauvignon Blanc are intermediate in their susceptibility to this disease and in their probability of recovery. In tolerant cultivars the bacteria spread more slowly within the plant than in more susceptible cultivars. Once the vine has been infected for over a year (i.e. bacteria survive the first winter) recovery is much less likely.

Young vines are more susceptible than mature vines. Rootstock species and hybrids vary greatly in susceptibility. Many rootstock species are resistant to PD, but the rootstock does not confer resistance to susceptible *Vinifera* varieties grafted on to it. Finally, the date of infection strongly influences the likelihood of recovery. Late infections (after June) by blue-green sharpshooters, green sharpshooters, and red-headed sharpshooters are least likely to persist the following growing season. This may not be the case with the glassy-winged sharpshooter, however, because it feeds on leaves near the base of the cane, as well as on 2-year old dormant wood.

13.13.4 DISEASE CYCLE

PD infection is dependent upon the presence of a susceptible host, a source of the bacteria, and an insect vector to inoculate the susceptible host. In addition to native grapevines, there are other indigenous plant species that harbor the bacteria without visual symptoms. Surveys in California have identified several alternate hosts, but in the area where the disease is endemic, there are undoubtedly many more plant species capable of supporting the causal agent.

13.13.5 MANAGEMENT

The disease can be managed using the following strategies:

- Chemical control
- Cultural control
- Vector management

TABLE 13.10
Insecticides Registered to Control Pierce's Disease and its Vector on Grape Vines.

Trade Name	Active Ingredient	Insecticide Type	Maximum Rate Per Acre
Platinum	thiamethoxam	neonicotinoid	8–17 fl. oz. per acre
Assail 70 WP	acetamiprid	Neonicotinoid	2.5 oz. per acre.

13.13.5.1 Chemical Control

Insecticide treatments aimed at controlling the vector in areas adjacent to the vineyard have reduced the incidence of PD by reducing the numbers of sharpshooters immigrating into the vineyards in early spring. The degree of control, however, is not effective for very susceptible varieties such as Chardonnay and Pinot Noir or for vines less than 3 years old. If a vineyard is near an area with a history of PD, plant varieties that are less susceptible to this disease (Table 13.10).

13.13.5.2 Cultural Control

During the dormant season, remove vines that have had Pierce's symptoms for more than 1 year; they may be chronically infected and are unlikely to recover or continue to produce a significant crop. Also, remove vines with extensive foliar symptoms on most canes and with tip dieback of canes even if it is the first year that symptoms have been evident. From summer through harvest, mark slightly symptomatic vines; reexamine for symptoms the following spring through late summer or fall; and remove vines that have symptoms for a second year. Pruning a few inches above the graft union of vines with moderate foliar symptoms (some canes on entire cordons without symptoms or no symptoms at the bases of most canes) may eliminate PD and allow vigorous regrowth the following year, but symptoms will reappear in many (30–40%) or most of these severely pruned vines the second year.

For table grapes, examine vines for poor bud break in spring. Later in the season, look for pests and damage.

Because the glassy-winged sharpshooter feeds much lower on the cane than other sharpshooters in California, late-season (after May–June) infections and infections occurring during dormancy made by the glassy-winged sharpshooter can survive the winter to cause chronic PD. This enables vine-to-vine to spread of PD, which has not been the case in California. Vine-to-vine spread can be expected to increase the incidence of PD exponentially rather than linearly over time, as has been normal for California vineyards affected by PD. Insecticide treatments of adjacent breeding habitats, such as citrus groves, has been the most effective approach.

Removing diseased vines as soon as possible when PD first appears in a vineyard is also critical to help reduce the infection rate. Early and vigilant disease detection and vine removal is recommended for any vineyards that experience influxes of the glassy-winged sharpshooter.

- The use of tolerant cultivars is the only effective control for PD in areas at high risk for PD development. The muscadines (*Vitis rotundifolia*) are not immune to PD but appear to tolerate the disease. Hybrids and other Native American grapes, such as *Vitis aestivalis* (e.g. "Norton") may have variable tolerance, but this has not been fully evaluated in Virginia.
- Minimize the amount of reservoir host vegetation within and around the vineyard.
- Rouging symptomatic vines may slow disease spread from vine to vine.

13.13.5.3 Vector Management

The difficulty of vector management as a means to manage PD is the inability to identify all potential vectors within and adjacent to the vineyard, so chemical control of vectors is tenuous at best.

Nonetheless, the current thinking in California is that vector transmission occurs primarily from host plants adjacent to the vineyard, so California growers practice vector control in areas adjacent to the vineyard. Growers should use caution when choosing insecticides to ensure that specific pesticide labels permit such use.

Care should be exercised in judiciously using insecticides. Unfortunately, the greater the number of sprays, the more likely secondary pest outbreaks will be created, especially with spider mites.

Because there is limited information as to other species may serve as a source of the PD organism, many growers are utilizing clean cultivation to eliminate any possible inoculum source within the vineyard. Weed growth under the trellis can be controlled with cultivation, or herbicides, but management of the vineyard floor between the rows has become problematic. Clean cultivation can have serious drawbacks such as the potential for serious soil loss due to erosion. The use of cover crops in vineyard row centers has several advantages over cultivation including increased equipment mobility, the preservation of soil structure within the vineyard and erosion control.

Cover crop height can be managed by mowing and is easily controlled during the spring with low rate glyphosate applications. This practice keeps cover crop roots in place to support equipment traffic, helps reduce erosion, and establishes an organic material layer that inhibits the germination of indigenous weed species. When annual rye grass is used for this purpose, additional suppression of weed seed germination may be observed due to the allelopathic properties of rye. Additional applications of glyphosate or glufosinate can be used throughout the growing season to keep developing weed populations in check. Pre-emergence herbicides can also be incorporated into a vineyard floor management program.

13.14 PHOMOPSIS CANE AND LEAF SPOT

Phomopsis cane and leaf spot of grapevine is caused by the fungal pathogen *Phomopsis viticola*. *Phomopsis* infects grapevines grown in most viticulture regions of Australia but has not been reported in WA or Tasmania. Spores are spread by rain splash, and disease symptoms include shoot lesions, leaf spots, and bleached cane. Crop loss is generally through girdling of shoots and weakening and cracking of canes, which consequently lowers productivity of vines. Berry rot is rare in Australia.

Disease incidence of phomopsis cane and leaf spot appears to be increasing in many vineyards throughout the Midwest. Crop losses up to 30% have been reported in some Ohio vineyards in growing seasons with weather conducive to disease development. Phomopsis cane and leaf spot can affect most parts of the grapevine, including canes, leaves, rachises (cluster stems), flowers, tendrils, and berries.

13.14.1 Causal Organism

Phomopsis cane and leaf spot is mostly caused by the fungi mentioned below.

Species	Associated Disease Phase	Economic Importance
P. viticola	Leaves, Fruits, and Stems	High

13.14.2 Symptoms

Symptoms in winter are seen as bleached white areas on dormant canes speckled with small black spots. Where severe infection has taken place, black cracks are also evident. Cane bleaching is not a reliable indicator of *Phomopsis* infection, however, as bleaching can also be caused by a range of factors such as weather extremes and other types of fungi. Leaf and shoot symptoms can be seen in spring and early summer.

13.14.2.1   Leaves

Leaf symptoms first appear in spring on the lower leaves of shoots as small, light-green spots with irregular, occasionally star-shaped margins. Severe symptoms that appear on the leaf look like small dark brown spots, which are usually less than 1 mm, the spot is surrounded by 2–3 mm of yellowish halo. Leaves can distort and be partially killed or stunted. Spots can become necrotic, darken, and drop out. Leaves with badly affected stems can turn yellow and fall (Figures 13.68 and 13.69).

13.14.2.2   Fruit

Lesions on the rachises are sunken and black causing the rachis to become brittle. Clusters may break at brittle points under the weight of the maturing berries or during harvest, leading to reduction of yield. If lesions girdle the rachis, berries below the infection site shrivel and may fall. Phomopsis can also cause fruit rot after berries begin to mature in mid to late summer preharvest

FIGURE 13.68   Symptoms on young leaves early in the season.

FIGURE 13.69   Symptoms on older leaves late in the season.

period. Infected berries turn brown, often beginning where the berry is attached to the pedicel. Once completely rotted, pycnidia (spore producing structures) erupt through the skin, giving berries a rough texture. Berries eventually shrivel and are indistinguishable from the mummies caused by black rot (Figures 13.70 and 13.71).

13.14.2.3 Green Shoots

Small spots with black centers develop, usually on the lower internodes, gradually expanding and elongating to form black crack-like lesions up to 5–6 mm long. Large numbers of merging spots on badly infected shoots may give a "scabby" or "corky" appearance that may crack and scar. Girdled shoots can fail to mature or become stunted and die. Severe infections can lead to dwarfing, deformation, and death of infected shoots, which break off near the base. Weakened older shoots (30–60 cm long) can break in strong winds, usually where lesions are numerous (Figure 13.72).

13.14.2.4 Canes

Infected canes might be bleached white in winter. Bleached areas, particularly those around the nodes, become speckled with small black spots (the resting structures of the fungus). These spots are prominent in the cortex of infected 1-year-old canes, on spurs, bunch and berry stems, and tendrils (Figure 13.73).

FIGURE 13.70 Symptoms of fruit infection.

FIGURE 13.71 Lesions on berries due to infection.

13.14.3   Cause and Disease Development

Phomopsis cane and leaf spot of grapevine is caused by the fungal pathogen, *P. viticola*. The disease was associated with 2 fungi, *Phomopsis* type 1 (now known as *Diaporthe*) and *Phomopsis* type 2 (simply *Phomopsis*). Because *Diaporthe* does not cause injury to grapevine, only *Phomopsis* causes the disease. Spread of *Phomopsis* occurs in wet spring weather as water is required for dispersal of asexual spores and infection. *Phomopsis* can remain dormant in infected canes, spurs, and dead wood for a number of years.

13.14.4   Favorable Conditions

P. viticola overwinters in its vegetative form as well as in pycnidia in the bark or canes. Spores are exuded from the pycnidia during wet periods of spring and splash onto young shoots. The spores infect young green tissues when temperature (4°C–7°C) and moisture conditions (several days of rain) are conducive to infection.

In spring the resting structures of the fungus release threads of jelly-like spore masses if wetted for at least 10 hours at an optimum temperature of 23°C–30°C (suitable temperatures between 1°C and 30°C). Spores on the cane are spread by water and rain-splashed on to young newly developed green shoots. Infections can be localized in the vineyard. Spores infect vines via leaves or stems if conditions remain wet for a further 8 or more hours. Spores need moist conditions to germinate and infect the vine. The risk of *Phomopsis* infection is low if there are few extended rainfall periods in spring.

FIGURE 13.72   Longitudinal lesions on green shoot.

FIGURE 13.73   Bleached cane with black spots (pycnidia).

13.14.5 Disease Cycle

The fungus overwinters in lesions or spots on old canes and rachises infected during previous seasons and requires cool, wet weather for spore release and infection. The fungus produces flask-shaped fruiting bodies called pycnidia in the old diseased wood. These pycnidia release spores in early spring and are spread by splashing rain droplets onto developing shoots, leaves, and clusters. In the presence of free water, the spores germinate and cause infection. Shoot infection is most likely during the period from bud break until shoots are 6–8 inches long. The optimum temperature for leaf and cane infections is between 15.5°C and 20°C, and at least 6 hours wetness duration is required at these temperatures for infection to occur.

As the wetness duration increases, the opportunity for infection greatly increases. Lesions on leaves appear at 7–10 days after infection. Fully expanded leaves become resistant to infection. Lesions on canes require two to four weeks to develop. The fungus does not appear to be active during the warm summer months, but it can become active during cool, wet weather later in the growing season. Pycnidia eventually develop in infected wood and will provide the initial inoculum for infections during the next growing season. Infected canes and rachises do not produce additional inoculum during the same growing season in which they were infected (Figure 13.74).

13.14.6 Management

The disease can be managed using the following strategies:

* Chemical control
* Cultural control

13.14.6.1 Chemical Control
Protectant fungicides will be necessary for effective control where phomopsis is present (2006 VT Pest Management Guide).

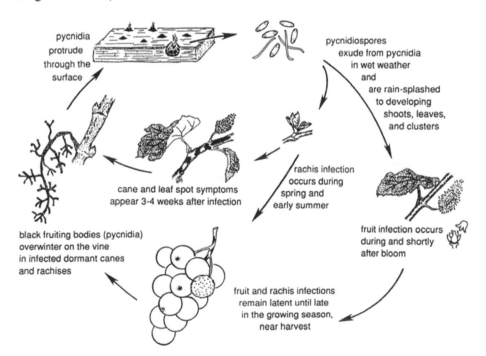

pycnidia protrude through the surface

pycnidiospores exude from pycnidia in wet weather and are rain-splashed to developing shoots, leaves, and clusters

rachis infection occurs during spring and early summer

cane and leaf spot symptoms appear 3-4 weeks after infection

fruit infection occurs during and shortly after bloom

black fruiting bodies (pycnidia) overwinter on the vine in infected dormant canes and rachises

fruit and rachis infections remain latent until late in the growing season, near harvest

FIGURE 13.74   Disease cycle of phomopsis cane and leaf spot.

Currently only 2 fungicides are recommended for phomopsis control on grape in Virginia: Captan and Mancozeb.

1 or the other of these materials should be applied at label rates as soon as possible after bud break and repeated as conditions warrant until after berry set. The most important period for rachis infection control is the first few weeks after cluster emergence. Repeated rains, cool weather, retarded shoot growth, and heavy past phomopsis incidence would all warrant tightening the protectant spray schedule to 5–7 days between sprays.

- Mancozeb (including Manzate, Penncozeb, Dithane, and others) formulations have a 66-day PHI during which time they may not be applied.
- Captan has a 0-day PHI but a 3-day restricted entry interval (REI) with the newest formulation. Therefore, as harvest approaches Captan should be used instead of mancozeb.
- Ziram also does well for phomopsis control and only has a 21-day PHI.
- A dormant spray of lime sulfur may reduce overwintering inoculum of phomopsis. When applying lime sulfur, it is important to thoroughly soak the vines. Therefore, tractor speed and spray volume should be adjusted. If vines are thoroughly soaked, this treatment may also reduce the overwintering spores of powdery mildew.
- Abound (strobilurin fungicide) and sulfur are also registered for phomopsis control. Efficacy of sulfur is unknown in Virginia, although reports from California suggest using 10 lb sulfur per acre provides good phomopsis control. Abound has not been shown to be as effective in controlling phomopsis as mancozeb.

13.14.6.2 Cultural Control

Select planting sites with direct, all-day sunlight (avoid shade). Good soil drainage and air circulation are also very important. Orient rows to take full advantage of sunlight and wind movement. Cultural practices that increase air circulation and light penetration in the vineyard will reduce wetting periods and should be beneficial for control.

2. While dormant pruning, cut out infected or dead canes and destroy them. Remove or destroy all rachises. Select only strong, healthy canes that are uniform in color to produce the next season's crop.

- Use proper pruning techniques and prune out as much apparently infected wood as possible.
- Do not leave dead spurs or pruning stubs on the vine (spur-pruned cultivars have most frequent infection).
- Remove prunings from vineyard and burn or immediately chop prunings with a flail mower.
- Avoid planting in low-lying areas because poor air circulation favors infection.
- Use cultural practices that increase air circulation and improve drying.
- Consider hand-pruning instead of mechanical pruning to remove more old wood.

13.15 RUGOSE WOOD OF GRAPEVINES

The rugose wood (RW) complex is 1 of the major disease complexes affecting grapevines (*Vitis* species) and was reported by Graniti in 1966. RW affects grapevines. On the woody cylinder, RW causes pitting, grooving, and severe aberration of the zone underneath the bark. Today the incidence of this complex is recognized to have a strong economic impact worldwide on the grape industry.

Rugose wood of grapevines is the name given to a group of 5 serious diseases: Kober stem rooving, corky bark, LN33 stem grooving, corky wood, and rupestris stem pitting. These are of major importance to viticulture worldwide.

Many grapevine cultivars affected by rugose wood diseases can have reduced quality and yield of fruit and/or reduced quality and production of wood for propagation.

13.15.1 CAUSAL ORGANISM

Rugose wood is mostly caused by the virus mentioned below.

Species	Associated Disease Phase	Economic Importance
GVA, GVB, GVD, Grapevine Virus E (gvE) and RSPav	Stems and Leaves	High

13.15.2 SYMPTOMS

Individual diseases cannot readily be distinguished in the field because of the absence of differential symptoms on various cultivars. In general, affected vines may be dwarfed and less vigorous than normal and may have delayed bud break in the spring. Some vines decline and die within a few years after planting. Grafted vines often show swelling of the scion above the graft union. With certain cultivars, the bark of the scion above the graft union is exceedingly thick and corky and has a spongy texture and a rough appearance, often marked by pits or grooves. These alterations may occur on the scion, rootstock, or both, according to the cultivar/stock combination and possibly individual susceptibility. In most cases no specific symptoms are seen on the foliage, but bunches may be smaller and fewer than normal. Certain cultivars show symptoms similar to those induced by leafroll, i.e. rolling, yellowing, or reddening of the leaf blades. These symptoms, when they occur, are more severe than those induced by ordinary forms of leafroll.

13.15.2.1 Wood

Removing the bark aids the observation of symptoms on the wood. Specific symptoms associated with GVA, GVB, and RSPa are observed on sensitive indicators during biological indexing. Swelling of the graft union may be observed on other commercial varieties and rootstocks. The diameter of the scion and rootstock may be significantly different from each other and the scion is often larger. The wood around or above the graft union may be rough, pitted, grooved, and/or corky. Necrosis may be observed at the join of the graft union. Graft incompatibilities can occur that may result in decline and death of the grafted scion (Figure 13.75).

FIGURE 13.75 Stem pitting symptoms (vines 1, 3, and 4, arrows) on *Vitis rupestris* (Rupestris St. George) inoculated with a strain of RSPaV from Cabernet Franc C24-1 compared to un-inoculated *V. rupestris* (Rupestris St. George, vine 2).

13.15.2.2   Vine Growth

In rugose wood-affected grapevines, budburst may be delayed. Affected grapevines can have reduced vigor and growth, they may decline and die affecting the life span of the vineyard. A reduction in cane pruning weight associated with a reduction in the circumference and/or the length of the canes can occur.

13.15.2.3   Foliage

In many varieties, no foliar symptoms are observed. However, leafroll-like symptoms have been reported in some varieties in which reddening or yellowing of the entire leaf blade, including the veins, and downward rolling of the leaves was observed.

Symptoms may include necrosis of the veinlets on the underside of the leaf blade, which can extend to the upper surface with time. Symptom expression initially occurs on the basal leaves of a shoot and then progresses to the younger leaves of the shoot. Tendrils may also express necrosis (Figure 13.76).

13.15.2.4   Fruit

Significant yield reduction may be observed in rugose wood. Specially in Shiraz and Syrah disease-affected grapevines due to smaller and fewer bunches. Up to 50% yield losses have been reported in varieties infected with GVA and up to 70% in vines affected by corky bark associated with GVB.

13.15.3   Cause and Disease Development

Rugose wood—associated viruses found worldwide 2 genera of the family Betaflexiviridae are associated with the rugose wood complex.

Vitivirus species include GVA, GVB, GVD, and grapevine virus E (GVE). RSPav is the sole *Foveavirus* species.

There are many strains of each virus species. The role of GVE in grapevine disease is unknown. GVA, GVB, and RSPav occur in Australia, however, corky bark disease, which is associated with strains of GVB, is not known to occur in this country. In a recent survey of grape growing districts of mainland Australia (NSW, SA, QLD, WA, and VIC), RSPav was detected in 214 of GV8 (94%) grapevines.

In the same survey GVA was more frequently detected (82 of 218 vines; 34%) than GV B (2 of 218 vines; 1%).

13.15.3.1   Virus Movement and Disease Development

Virus titer may be low and distribution can be uneven in grapevines at certain times of the year, particularly in the first season after an infection event. It can take more than 12 months for viruses

FIGURE 13.76   Vein necrosis symptoms on grape leaves.

to move from the point of infection to shoots and cordons of the grapevine. This can have important implications for virus detection and disease expression.[8]

13.15.4 MANAGEMENT

The disease can be managed using the following strategies:

- Plant healthy stock. Use planting material certified free of viruses that cause the diseases of rugose wood complex. Clones of most rootstocks and cultivars that have been tested and found to be free of all known viruses are available.
- There is no way to cure an infected vine. Remove and destroy virus-infected vines. Top-grafting is not advisable as rootstocks will be infected.
- Clean pruning equipment with a 1/10 dilution of household bleach between vines.

REFERENCES

1. Report on Integrated Pest Management, Michigan State University. http://www.ipm.msu.edu/grape_diseases/anthracnose
2. Hed, B., 2017. *2017 Summer Disease Management Review*, Penn State Extension. https://psuwineand-grapes.wordpress.com/category/viticulture-2/canopy-management
3. Wayne, F. W., 2003. *Black Rot of Grapes*, Cornell University, NYAES.
4. Ellis, M. A., 2016. *Downy Mildew of Grape*, Ohio State University Extension. https://ohioline.osu.edu/factsheet/plpath-fru-33
5. Borkar, S. G., and Yumlembam, R. A., 2017. *Bacterial Diseases of Crop Plants*, CRC Press.
6. Ellis, M. A., 2016. *Eutypa Dieback Of Grape*, Ohio State University Extension. https://ohioline.osu.edu/factsheet/plpath-fru-11
7. Agrios, G. N., *Transmission of Plant Diseases By Insects*, University of Florida.
8. Constable, F., and Rodoni, B., 2014. *Grapevine Leafroll-Associated Viruses*, Department of Environment and Primary Industries (DEPI), Victoria.
9. Halleen, F., Fourie, P. H., and Lombard, P. J., 2010. *Protection of Grapevine Pruning Wounds against Eutypa lata by Biological and Chemical Methods*, Article in South African Journal for Enology and Viticulture.
10. Tefera, M., Robert, R. M. and Rayapati, A. N., July 9–11, 2008. The Occurrence of Grapevine Fanleaf Virus in Washington State Vineyards, Proceedings of the 2nd Annual National Viticulture Research Conference University of California, Davis.
11. Jones, L. R. & Grout, A. J. 1897. Alternaria fasciculata Bull. Torrey bot. Club 24: 257.
12. Ophel, K., Allen, K., 1990, *Agrobacterium vitis* sp. nov. for Strains of *Agrobacterium* biovar 3 from Grapevines, International Journal of Systematic and Evolutionary Microbiology 40: 236–241.
13. Carter, M. V., 1991. The Status of Eutypa lata as a Pathogen. Kew, UK: International Mycological Institute, Phytopathological Papers no. 32.
14. Munkvold G.P., Marois J.J., 1993b. Efficacy of natural epiphytes and colonizers of grapevine pruning wounds for biological control of eutypa dieback. Phytopathology 83, 624–9.
15. John, S., Wicks, T. J., Hunt, J. S., Lorimer, M. F., Oakey, H., Scott, E. S., 2005. Protection of grapevine pruning wounds from infection by Eutypa lata using Trichoderma harzianum and Fusarium lateritium. Australasian Plant Pathology 34, 569–75.
16. Andret-Link, P., Laporte, C., Valat, L., Ritzenthaler, C., Demangeat, G., Vigne, E., Laval, V., et al., 2004. Grapevine fanleaf virus: Still a major threat to the grapevine industry. J. Plant Pathol. 86(3), 183–195.
17. Martelli, G. P., Savino, V., 1990. Fanleaf degeneration. In: Pearson, R. C and Goheen, A. (eds.) Compendium of grape diseases., St Paul, MN: APS Pres, 48–49.
18. Hewitt, W.B., Raski, D.J., and Goheen, A.C., 1958. Nematode vector of sail-borne fanleaf virus of grapevines. Phytopathology, 48: 586-595.
19. Cohn, E., Tanne, E., and Nitzany, F. E., 1970. Xiphinema italiae, a new vector of grapevine fanleaf virus. Phytopathology, 60: 181–182.
20. Graniti, A. & Martelli, G. P. 1966. Further observations on legno riccio rugose wood, a graft transmissible stem pitting of grapevine. Proc. Int. Conf: Virus Vector Perennial Hosts and Vitis, 1965, p. 168–179. Div. Agric. Sci., Univ. Calif., Davis.

14 Sugarcane

Sugarcane have several species of tall perennial true grasses of the genus *Saccharum*, tribe Andropogoneae that is native to the warm temperate tropical regions of South Asia and Melanesia and that are used for sugar production. It has stout, jointed, fibrous stalks that are rich in the sugar sucrose, which accumulates in the stalk internodes. The plant is 2 to 6 m (6 to 20 ft) tall.

All sugarcane species interbreed, and the major commercial cultivars are complex hybrids. Sugarcane belongs to the grass family Poaceae, an economically important seed plant family that includes maize, wheat, rice, and sorghum, and many forage crops. Sugarcane is a tropical, perennial grass that forms lateral shoots at the base to produce multiple stems typically 3 to 4 m (10 to 13 ft) high and about 5 cm (2 inches) in diameter. The stems grow into cane stalk, which when mature constitutes around 75% of the entire plant.

The average worldwide yield of sugarcane crops in 2013 was 70.77 tons per hectare. The most productive farms in the world were in Peru with a nationwide average sugarcane crop yield of 133.71 tons per hectare. Brazil led the world in sugarcane production in 2013 with a 739,267 TMT harvest. India was the second largest producer with 341,200 TMT tons, and China the third largest producer with 125,536 TMT tons harvest. In the United States, sugarcane is grown commercially in Florida, Hawaii, Louisiana, and Texas. With such a tremendous yield of sugarcane, it is important to detect and manage diseases arising while cultivating sugarcane. This chapter covers the identification and management of sugarcane diseases.

14.1 RED ROT OF SUGARCANE

The red rot of sugarcane was reported first time from Java (Went, 1893) as red-smut and it became a serious cause of decline of several popular varieties of sugarcane in other countries. Butler (1906) reported the disease causing heavy losses in Indian cane varieties and renamed this disease as red rot. This disease is caused by a fungus *Colletotrichum falcatum*. Leaves start losing color and wither. The stalk becomes dry, wrinkled, hollow, and an alcoholic smell is emitted. The red rot fungus may infect any part of the sugarcane plant, but the disease is of principal importance in the stalks of standing cane, in cuttings and seed-cane, and in the leaf midrib. Under favorable conditions, the disease can also attack the rations or stubble pieces (Figures 14.1 through 14.4).

14.1.1 CAUSAL ORGANISM

Species	Associated Disease Phase	Economic Importance
C. falcatum Went (Imperfect State); Perfect/Ascigerous State = *Glomerella tucumanensis* (Speg.)	Upper Leaves, the Entire Crown, Internodes, and Stem	Severe

14.1.2 SYMPTOMS

The disease appears in the field after the rainy season when growth of the plants stops, and formation of sucrose starts. However, the disease is difficult to recognize at its early stage in the field. The disease is initiated as a discoloration and drooping of upper leaves of the shoots which is usually maturing. The margins of the infected plants wither and continue until the whole crown withers and results in plant death. The stem of diseased plants shows a characteristic red color, especially throughout the vascular bundles, and a sour odor. The diagnostic character for this disease

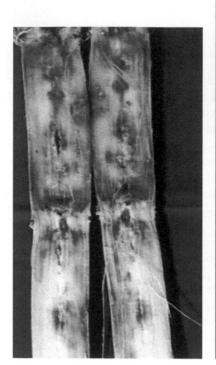

FIGURE 14.1   Red rot internal and external symptoms.

FIGURE 14.2   Drying due to red rot.

FIGURE 14.3   Spreading of red rot through sets.

FIGURE 14.4   Affected midrib lesions.

identification is white patches between the red tissues. As the disease advances, the whole affected stem rots. The internodes of the plants may be shortened, and profuse whitish mycelial growth is seen in the tissues. The fungal pathogen sometimes produces black colored minute velvety bodies representing acervuli. The leaves also show dark blood red lesions on the midrib, which elongates throughout the leaf length. The infected leaves often break at the point of the lesions and hang off (Figures 14.5 and 14.6).

14.1.2.1   Stalk Symptoms

- Drying up of the third and fourth leaf of the crown at margins. Later, the entire crown dries up and drops down.
- Brown or reddish-brown stripes appear externally at nodal region. On splitting, the internal tissue becomes red with white transverse bands.
- Tissues emit an alcoholic sour smell.

FIGURE 14.5   Withering of the crown.

FIGURE 14.6   Red rot.

- Tiny acervuli develop on outer surface of the shrinked upper internodes. Cottony gray fungal mass develops in the pith region of the internodes and sporulates abundantly.

14.1.2.2   Leaf Symptoms
- Tiny reddish lesions on the upper surface of the lamina. These lesions are 2 to 3 mm in length and about 0.5 mm in width.
- Minute red spots are seen on the upper surface of the midrib in both directions. Later, becomes straw colored in the center with the development of black acervuli and dark reddish-brown margins.

14.1.3   CAUSE AND DISEASE DEVELOPMENT

Red rot of sugarcane is caused by the fungus *C. falcatum* and its final stage is *Glomerella tucumanesis*. The septate fungal mycelium is found inter- and intra-cellularly in the parenchyma tissues of the host. The fungal hyphae are hyaline, septate, thin, profusely branched, and have oil droplets. The velvety, minute, black colored acervuli are formed in the midrib, pith and on the rind surface. The setae are long, rigid, bristle-like, septate and dark brown, which are produced by compact hymenial layers. The conidiophores, arising between setae, are linear or club-shaped, hyaline, single celled, and measuring 20 x 8 μm which bears thin-walled, hyaline, falcate, granular or guttulate, measuring 20–48 × 4–7 μm. The pathogen also produced thick-walled intercalary chlamydospores. The characteristic blood red color of the disease is produced due to the interaction of host—pathogen. Primary transmission is through soil and diseased sets, while the secondary transmission is through air, rain splash, and soil (Figure 14.7).

14.1.4   FAVORABLE CONDITIONS OF DISEASE DEVELOPMENT

The first symptom of red rot in the field is discoloration of the young leaves. The margins and tips of the leaves wither and the leaves drop. The discoloration and withering continue from the tip to the

FIGURE 14.7   Appearance of red rot.

leaf base until the whole crown withers and the plant dies within 4 to 8 days. In a single stool, most of the stalks may wither almost simultaneously. As the disease advances the entire stem rots and the central tissues become pithy. The tissues are reddened throughout the basal portion, especially the vascular bundles, which are intensely red; there may be cross-wise, white patches interrupting the reddened tissues. The internodes may shrink, and when such canes are split open, large cavities may be found in the center and the pithy tissues may appear brown.[1]

Often a profuse whitish growth of the fungal mycelium may be found in the brown background of the host tissue. In some cases, black, minute, velvety bodies, representing the acervuli of tire fungus, may also be seen. Since reddening is a common symptom of other diseases of sugarcane, the white patch symptom is an important diagnostic characteristic of red rot.

When a diseased plant is open, a characteristic becomes evident. In the infected plants the leaves may show symptoms in the form of dark red lesions in the mid-rib, which may elongate, turning blood red with dark margins and later with straw-colored centers. In the older lesions, minute black dots representing the acervuli can be seen. Often the infected leaves may break at the lesions and fall down. The fungal mycelium is present inter-cellularly, mainly in the parenchymatous cells of the path. The hyphae are thin, hyaline, septate, profusely branching, and contain oil droplets. The acervuli, which are characteristic of the genus, are formed on the surface of the rind in the leaf mid-rib and sometimes in the pith region. They are minute, black, and velvety. The compact hymeneal layer produces the long, rigid, bristle-like setae, which are septate, dark brown at the base, and lighter toward the tip. Inter-spaced between the setae are numerous club-shaped to linear, hyaline, single-celled conidiophores, measuring approximately 20 x 8 m. The conidia, which contain the conidiophores, are single celled, thin-walled, hyaline, falcate, granular, and guttulate, measuring $16-48 \times 4-8$ m.

The spindle leaves display drying. At a later stage, stalks become discolored and hollow. Acervuli (black fruiting bodies) develop on the rind and nodes. After splitting open the diseased stalk, a sour smell emanates. The internal tissues are reddened with intermingled transverse white spots. In advanced stage of the disease, the color becomes earthy brown with pith cavity in the center showing white cottony hyphae and sometimes fruiting bodies of fungus (acervuli). In rainy season, the disease spreads so fast that whole crop dries and not a single millable cane is obtained.

14.1.5 Disease Cycle

The pathogen of red rot can grow in the soil saprophytically by forming acervuli up to 3–4 months in active stage. This duration of survival as saprophyte is sufficient to carry over from 1 crop season to the next. The chlamydospores produced by the fungal pathogen can survive for a long period in the soil. The planting of infected or diseased sets are the main cause of survival and the spread of disease. When such infected seed sets are sown in the field, they give rise to infected shoots. The diseased shoots in the field produce a huge number of conidia in acervuli and these are transmitted through insects, wind, and water to other healthy plants, which in turn, spread secondary infection. The infection through conidia can be facilitated by mechanical injuries. The conidia germinate and produce mycelium which grows in the parenchyma tissues, but spores can migrate through vascular tissues of the stem and midrib. The disease is favored by high humidity, waterlogging conditions, improper cultural practices, and continuous cultivation (Figures 14.8 and 14.9).

14.1.6 Management

Red rot disease can cause severe economic losses, especially if susceptible varieties can overmature in the field, and if conditions are not conducive to rapid establishment of planted cuttings.

14.1.6.1 Chemical Control

The best way to minimize the red rot disease incidence is the selection of healthy seed sets from disease-free area. The cut ends of seed sets should be dipped in 1% Bordeaux mixture before planting.

FIGURE 14.8   Causal organism of red rot.

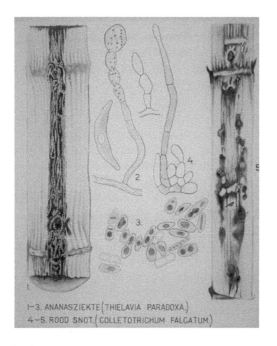

FIGURE 14.9   Spreading of red rot.

TABLE 14.1

Management of Red Spot of Sugarcane

Fungicide	Concentration (%)	Time
Bavistin,	0.1	16–18 mins give complete eradication of red rot infection.
Benomyl,	0.1	
Topsin and	0.1	
Aretan	0.1	
Thiophanate-methyl compound	0.2	Effective to control sett-borne infection

The diseased canes are noticed in the field; they should be collected and burn immediately. The long crop rotation for up to 4–5 years with non-host crops is helpful in destroying disease inoculum. Fresh sowing should be done with seed sets from a resistant variety, dipping these in 0.25% solution of Agallol or Aretan for 2–3 minutes (Table 14.1).

14.1.6.2 Biological Control

In addition to the various sanitary precautions mentioned above, red rot may also be controlled by growing resistant or tolerant varieties. Inter-generic and inter-specific crosses with *Saccharum* spp., in the latter case using *S. spopntaneum* and *S. robustum* with *S. officinarum*, have yielded many economically important cane varieties, some of which are highly resistant, or at least tolerant, to red rot.

14.1.7 CONTROL

- Red rot management in sugarcane has become an important issue in all sugarcane areas. While early stage detection may not be quite easy, during later stages the cane breaks down.
- The canes are to be split open lengthwise to see dull red tissue interrupted by white patches across the stalk. These patches are characteristic of red rot of sugarcane. Monsoon period enables faster disease to spread and drying of the crop.
- The best remedy for avoiding this fungal disease is to cultivate only resistant sugarcane varieties that have been released for cultivation in different sugarcane-growing states. Phytosanitation, being the key to manage this disease, stringent domestic quarantine measures to prevent movement of cane sets from endemic areas to new areas has to be enforced by all concerned agencies including sugar mills.
- In recent times, the disease has spread to many states.
- Following good cultural practices such as clearing fields of excessive trash and ensuring efficient drainage.
- Healthy sets only are to be planted to avoid poor plant stand due to rotting.
- Affected fields should be isolated through bunding to prevent movement of water to adjacent fields.
- Ratooning of infected fields should be strictly avoided.
- Crop rotation in the affected fields could reduce disease inoculum.
- Hot-water treatment of sets before planting at 52°C for 30 minutes is also recommended.
- Planting fungicide treated seed sets is to be followed as a general practice to prevent introduction of the pathogen into sugarcane fields.

14.2 WILT ON SUGARCANE

Wilt is an important fungal disease of sugarcane. The term "wilt" is a misnomer in sugarcane as affected plants dry up gradually without displaying any conventional wilt syndrome as evidenced in other crops (there is no suddenness in death/drying of plant). The causal organism of this disease is believed that it is a "disease complex" produced by different pathogens rather than the act of a single pathogen species. The fungi implicated in wilt disease is, *F. sacchari* (*Cephalosporium sacchari*).[2]

The disease manifests itself mostly in the mature cane. Affected plants appear pale yellowish-green with a marked drying of lower leaves. At a later stage, when the stalk is badly damaged, the crown dries up, resulting in death of the cane. The affected canes are lighter in weight with completely hollow internodes but unaffected nodes and buds. Diseased canes, on splitting longitudinally, display whitish to dark reddish colorations of internal tissues depending on the stage of the disease. In wilt, the nodes and buds remain unaffected till the cane dries out. In wilt the affected canes simply dry out and canes do not break easily.[2]

Causal Organism: *Cephalosporium sacchari*
Class: *Deuteromycetes*
Order: *Moniliales*
Family: Moniliacease

14.2.1 Causal Organism

Species	Associated Disease Phase	Economic Importance
C. sacchari	Infected Sets	Severe

14.2.2 Symptoms

The disease symptoms appear during the monsoon and post monsoon periods, when the plants are 4–5 months old. The affected plants, either singly or in small groups, displays conspicuous stunting and unthrift appearance. This is followed by yellowing and withering of crown leaves. The midrib of all leaves in a crown generally turns yellow while the leaf lamina may remain green. If affected clumps are cut and examined, the diffused purple or muddy red color is seen as conical patches on each node just above the growth rind. The cottony white mycelium is often seen in the pith. Frequently this fungal disease is associated with a saprophytic bacterial growth and often the bacteria are mistaken as causal agents. Disease reduces germination and in severe cases total cane yield losses occur due to drying up of shoots and wilting of the stalks. In severe cases, the spindle-shaped cavities, tapering toward the nodes, develop in each internode because of general recession and rapid desiccation of tissues (Figures 14.10 through 14.13).

14.2.3 Cause and Disease Development

The wilt disease of sugarcane is caused by *C. sacchari*. The fungal pathogen produces abundant mycelium in infected canes. The hyphae are hyaline, thin-walled, septate, and produce a huge number of micro conidia on simple or branched, lateral or terminal hyphae but no macro conidia are produced by the fungus. The single-celled micro conidia are oval to elliptical.

14.2.4 Favorable Conditions of Disease Development

The fungal mycelium is abundant in the infected canes. The hyphae are hyaline, thin-walled and septate. They produce numerous micro conidia on simple or branched, lateral or terminal hyphae, but no macro conidia are produced. The conidia are oval to elliptical, and measure $4–12 \times 2–3\,\mu m$ in size. They are mostly unicellular, but the ones formed later in the advanced growth of the fungus may be septate. Conidia readily germinate to produce single germ tubes. The fungus is transmitted

FIGURE 14.10   Wilt in sugarcane.

FIGURE 14.11   Wilt in sugarcane.

FIGURE 14.12   Wilt in sugarcane.

FIGURE 14.13   Wilt in sugarcane.

from place to place through the infected seed sets. When the diseased sets are planted, the eyes may fail to develop, or often, the shoots arising from the eyes may wilt due to the infection spreading to the shoots. Root formation in such sets may be very poor. The fungus can also survive in soil as a saprophyte for 2–3 years. Near-neutral and alkaline soils are favored by the fungus. The perfect stage is not known.

14.2.5 DISEASE CYCLE

The pathogen is transmitted from place to place by infected seed sets. If the infected seed sets are planted in the field, they fail to develop eyes, or the shoots arising from the eyes may wilt because of the infection spreading to the shoot. The root system is very poorly developed in such sets. The fungal pathogen can survive as a saprophyte in the soil for 2–3 years.

14.2.6 MANAGEMENT AND CONTROL

The seed sets selected for cultivation should be from disease-free areas. The alkali soils favor the fungal growth hence such type of soil should be avoided for cultivation purposes. The seed sets chosen for planting from disease prone areas should be dipped in organomercurials fungicide solutions.

Adopting 1 or more of the following measures can minimize the disease incidence (Table 14.2).

14.3 SMUT DISEASE OF SUGARCANE

The smuts are multicellular fungi, which are characterized by their large numbers of teliospores. It is dark, thick-walled and has dust-like teliospores as the name suggest. The important hosts of this disease include maize, barley, wheat, oats, sugarcane, and forage grasses. It eventually hijacks the plants reproductive systems forming galls, which darken and burst releasing fungal teliospores that infect other plants nearby. Before infection can occur, the smuts need to undergo a successful mating to form dikaryotic.

The smut "whip" is a curved black structure which emerges from the leaf whorl that aids in the spreading of the disease. Sugarcane smut causes significant losses to the economic value of a sugarcane crop.

TABLE 14.2
Measures to Minimize Wilt Disease

Chemical	Comment
MHAT (54°C for 150 minutes), followed by seed dipping for 10 to 15 minutes in 0.1% carbendazim (Bavistin) solution.	
Soil application of quinolphos 5 G @ 1.5 Kg a.i. per ha (30 Kgs per hectare) appears to be very effective against the root borer.	Association of root borer with wilt pathogen has been observed. Cane injury by bores provides means of wilt infection and the pathogen spreads through irrigation water, wind, soil, rain and setts.
(0.1% Bavistin or 0.1% Bayleton – 100 gm of fungicide in 100 Lit. of water, sett dipping for 10 to 15 min.)	Setts are treated with fungi-toxicants before planting.
Dipping the setts in 40 ppm of boron or manganese, or spraying the plants with either of these minor elements reduces the disease intensity.	

FIGURE 14.14 Smut infected clump. Characteristic symptoms of profuse tillering and poor cane formation (left) as compared to healthy canes (right).

For the sugarcane crop to be infected by the disease, large spore concentrations are needed and fungus uses its *smut-whip* to ensure that the disease is spread to other plants, which usually occurs over a period of 3 months. As the inoculum is spread, the younger sugarcane buds just coming out of the soil will be the most susceptible of all the crops. Because water is necessary for spore germination, irrigation has been shown to be a factor in spreading the disease. Therefore, special precautions need to be taken when irrigating the crop to prevent the spread of the smut (Figure 14.14).

14.3.1 CASUAL ORGANISM

Species	Associated Disease Phase	Economic Importance
Ustilago scitaminea	Shoot	Moderate

14.3.2 SYMPTOMS

1. The smut disease of sugarcane is the most easily recognizable disease among other diseases of this crop even from a long distance. The pathogen affected plants produced long, whip-like, dusty-black colored, and several-feet-long curved shoots, which remain stunted as shown in Figure 14.15b.
2. In the early stage of disease development, this structure is covered by a thin silvery membrane and exposes dense black smut spores on rupturing. The affected plants have long, slender, and thin canes compared to the rest of the crops. The production of the long, whip-like structure from the terminal bud of the stalk is black in color and covered by thin silvery membrane. This silvery membrane ruptures releasing millions of reproductive spores of smut fungus, present in the form of a powdery mass.
3. 2 to 4 months after the fungus has infected the plant, black whip-like structures, instead of a spindle leaf, emerge from the meristem, or growing point, of the plant. The developing whip is a mixture of plant tissue and fungal tissue. The whip reaches maturity between the sixth and the seventh month. When spores that are contained inside the whip are released, the core of the whip remains behind and is a straw-like color as shown in Figure 14.15b.
4. Plants infected with the fungus usually appear to have thin stalks and are often stunted. They end up tillering much more than normal, and thus, result in leaves that are slenderer and much weaker. They sometimes appear more grass-like than non-infected plants. Less common symptoms of the disease are stem or leaf galls and proliferating buds.

5. The most obvious symptom of *U. scitaminea* infection is the long, whip-like sorus that emerges from the growing point and frequently extends above the tops of the infected plant. Sori may also be produced from side shoots originating from lateral buds.

6. For much of its life cycle, however, the fungus is systemic in the plant and produces no identifiable symptoms except for a "grassy" appearance in severe cases. This appearance results from the production of numerous weak, spindly stalks in place of the usual vigorous canes. In rare instances, *U. scitaminea* infection has been observed in the flowering panicle, but this is not of economic importance since commercial sugarcane is harvested before flowering.

7. The most recognizable diagnostic feature of sugarcane infected with smut is the emergence of a long, elongated whip. The whip morphology differs from short to long, twisted, multiple whips, etc. (Figure 14.15a). Affected sugarcane plants may tiller profusely with spindly and more erect shoots with small, narrow leaves (i.e. the cane appears "grass-like") with poor cane formation (Figure 14.15b). Other symptoms are leaf and stem galls and bud proliferation (Figure 14.16). The disease can cause significant losses in cane tonnage and juice quality; its development and severity depend on the environmental conditions and the resistance of the sugarcane varieties.

8. Successful management of smut in sugarcane relies more on exploiting host resistance. To enhance smut resistance in commercial hybrids, intensive breeding programs should be formulated by involving exotic clones as sources of resistance from germplasm exchange.

FIGURE 14.15a   Different forms of whip morphology in smut infected sugarcane. (a) Long whip.

FIGURE 14.15b   Different forms of whip morphology in smut infected sugarcane. (b) Closed whip. (c) Twisted whip. (d) Short whip. (e) Multiple whips.

14.3.3   CAUSE OF DISEASE AND DEVELOPMENT

Sugarcane smut infects all sugarcane species unless the species is resistant. The damage caused depends on the susceptibility of the species. Sugarcane fields are planted using vegetative cuttings from mother plants, so they have the same genetic make-up of the parent plant. Seeds are not used in propagation because sugarcane is a multi-species hybrid and therefore is difficult to breed. Sugarcane smuts can also infect some other grass species outside of sugarcane. However, mostly it remains on plants of the genus *Saccharum*. This fungus spreads through wind from plant to plant, hence, 1 infected plant infects its neighboring plants quickly.

14.3.4   FAVORABLE CONDITIONS FOR DISEASE DEVELOPMENT

The spores germinate in wet soil at 25°C–30°C and 100% relative humidity. The fungal pathogen causing whip smut of sugarcane is *U. scitaminea* Sydow. The fungal mycelium is intercellular and found as a dense mass on the surface of spore-bearing shoots, which later produce smut spores or *chlamydospores*. These smut spores are produced in huge number and rupture the silvery membrane to expose themselves.[3]

The spores are globose to sub-globose, light reddish brown, smooth, and measuring 6–10 μm in diameter. The fungal pathogen frequently produced smut spores on potato-dextrose agar (PDA) or malt culture media. The spores germinate in water by producing 2–3 celled promycelia. The sporidia are hyaline, thin-walled, single celled, elliptical to linear, and arise laterally or terminally. The high temperature or dry weather and acute shortage of water leads to high incidence of smut.

14.3.5   DISEASE CYCLE

1. Sugarcane smut is disseminated via teliospores that are produced in the smut-whip. These teliospores, located either in the soil or on the plant, germinate in the presence of water. After germination they produce promycelium and undergo meiosis to create 4 haploid sporidia (Figure 14.17).

FIGURE 14.16 Unusual symptoms due to smut infection. (a) Apical deformity. (b) Floral infection. (c) Malformed spindle. (d) Bud proliferation.

2. Sugarcane smut is bipolar and therefore produces 2 different mating types of sporidia. For infection to occur, 2 sporidia from different mating types must come together and form a dikaryon.

3. This dikaryon then produces hyphae that penetrate the bud scales of the sugarcane plant and infect the meristematic tissue. The fungus grows within the meristematic tissue and induces formation of flowering structures which it colonizes to produce its teliospores.

4. The flowering structures, usually typical grass arrows, are transformed into a whip-like sorus that grows out between the leaf sheaths. At first it is covered by a thin silvery peridium (this is the host tissue), which easily peels back when desiccated to expose the sooty black-brown teliospores.

5. These teliospores are then dispersed via wind and the cycle continues. The spores are reddish brown, round and sub ovoid ,and may be smooth to moderately echinulate. The size varies from 6.5 to 8 µms (Figures 14.18 and through 14.20).

14.3.6 MANAGEMENT

1. The disease whips should be removed from the field and burned.

2. The ratooning should not be practiced in susceptible varieties or such crops which have shown a high degree of the disease.

FIGURE 14.17   Type of smut.

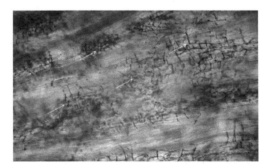

FIGURE 14.18   Trypan blue staining of smut fungus. Arrows indicate proliferation of inter and intracellular mycelial growth of smut fungus in nodal buds of sugarcane.

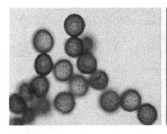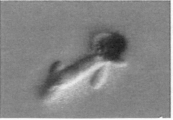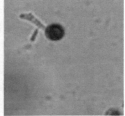

FIGURE 14.19   Smut teliospores and its germination. (a) Teliospores from whip. (b) and (c) germination of teliospores.

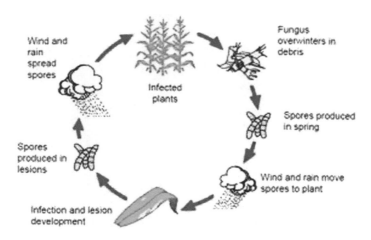

Wind and rain spread spores

Infected plants

Fungus overwinters in debris

Spores produced in spring

Wind and rain move spores to plant

Infection and lesion development

Spores produced in lesions

FIGURE 14.20   Disease cycle of smut.

3. The diseased canes should not be used for planting purposes and obtained from healthy fields.
4. The seed sets should be dipped in a 0.25% water suspension of Agallol or Aretan or Mercuric Chloride (0.1%) or Formalin (1.0 %) for 5 minutes. The systemic fungicides such as Vitavax, Benlate, Dithane M-45, and Bavistin can eradicate internally present dormant mycelium in the seed sets.
5. The hot-water treatment (HWT) is also found to be effective, which involves a 10-minute dip of seed sets at 55°C–60°C before planting.
6. The management of sugarcane smut is done through the use of resistant cultivars, fungicide, and using disease-free planting stock.
7. Control is mainly accomplished using resistant cultivars in areas where the disease is present. Fungicides also are used in the control of this disease but typically resistant cultivars are preferred due to the cost of fungicides.

14.3.7 CONTROL

1. Different ways to prevent the disease from occurring in the sugarcane is to use fungicide, and this can be done by either preplant soaking or post-plant spraying with the specific fungicide.
2. Preplant soaking has been proven to give the best results in preventing the disease, but post-plant spraying is a practical option for large sugarcane cultivations. To control the disease, if affected plants are very few, carefully cut down the apex of the plant in such a way that spores do not spread, by covering with suitable paper-bag and burn it out along with the uprooted affected plant.
3. Or discard and burn the whole field and do not grow sugarcane crop at least for 3 years in the same field. Fresh sowings are done with resistant varieties.
4. Hot-water dip of cane pieces before planting is effective in ensuring clean seed. A short hot-water treatment of 52°C for 30 minutes, or a long hot-water treatment of 50°C for 2 hours, are both adequate in eliminating *U. scitaminea* from sugarcane pieces. This practice is now standard procedure in many plantations.
5. Sets from smutted canes should not be used for planting.
6. Seed sets should be disinfected either in 0.1% mercuric chloride or formalin solution for 5 minutes followed by 2 hours under a moist cloth. The other effective chemicals available in market may also be used.

TABLE 14.3
Control of Smut Disease

Trade Name	Active Ingredients	Rate	Remarks
*Throttle®, *Tyrant®	250 g/L propiconazole	100 mL/100 L water	Apply as a dip treatment to sugarcane setts prior to planting.
*Tyrant®500	500 g/L propiconazole	50 mL/100 L water	Mix with water at ambient temperature and dip setts for 5 Minutes in this mixture.

7. Hot-water treatment of sets at 52°C for 18 minutes can help eliminate the internal infection.
8. Smutted plants should be rouged out and burned before the bursting of the spores.
9. Ratooning of the diseased crop should be undertaken.
10. Suitable rotations with non-host crops should be practiced.
11. Planting should be done in healthy soil.
12. Dry sowing of the crop should be carried out where disease is prevalent.
13. Autumn planting of sugarcane should be avoided.
14. Use of resistant varieties should be encouraged (Table 14.3).

14.4 SUGARCANE GRASSY SHOOT DISEASE (SCGS)

SCGS is caused by small, parasitic bacteria. This bacteria contributes to losses of 5% to 20% in the main crop of sugarcane, and these losses are higher in the ratoon crop. A higher incidence of SCGS has been recorded, resulting in 100% loss in cane yield and sugar production (Figure 14.21).

14.4.1 CAUSAL ORGANISM

Species	Associated Disease Phase	Economic Importance
SCGS Virus (*Candidatus phytoplasma*)	Shoot	Severe

FIGURE 14.21 SCGS.

14.4.2 Symptoms

Phytoplasma-infected sugarcane plants show a proliferation of tillers, which give it a typical grassy appearance, hence the name grassy shoot disease. The leaves of infected plants do not produce chlorophyll, and therefore, appear white or creamy yellow. The leaf veins turn white first as the phytoplasma resides in leaf phloem tissue. Symptoms at the early stage of the plant life cycle include leaf chlorosis, mainly at the central leaf whorl. Infected plants do not have the capacity to produce food in the absence of chlorophyll, which results in no cane formation. These symptoms can be seen prominently in the stubble crop. The eye or lateral buds sprout before the normal time on growing cane.

Symptoms of iron deficiency (interveinal chlorosis) are very similar to those of SCGS. It shows creamy leaves, but no chlorosis occurs in leaf veins, and they remain green. In the case of severe iron deficiency, veins may lose chlorophyll in the absence of iron and appear like SCGS disease. Iron deficiency is caused by a lack of iron nutrients in the soil; therefore, one may observe several plants showing symptoms of iron deficiency in localized patches in a field. Phytoplasma-infected plants, though, may occur anywhere in the field in a more random distribution. Treatment with 0.1% ferrous sulfate, either by spraying or supplying it through fertilizer, cures iron deficiency, but phytoplasma-infected sugarcane does not respond to any treatment. Phytoplasma-infected plants growing in vitro show sensitivity to tetracycline (Figure 14.22).

14.4.3 Cause of Disease and Development

SCGS disease is caused by a phytoplasma (*C. phytoplasma*), which is 1 of the destructive pathogens of sugarcane. Phytoplasmas, formerly called mycoplasma-like organisms (MLOs), are a large group of obligate, intracellular, cell wall-less parasites classified within the class mollicutes. Phytoplasmas are associated with plant diseases and are known to cause more than 600 diseases in several hundred plant species, including gramineous weeds and cereals. The symptoms shown by infected plants include: whitening or yellowing of the leaves, shortening of the internodes (leading to stunted growth), smaller leaves and excessive proliferation of shoots, resulting in a broom phenotype and loss of apical dominance.

14.4.4 Favorable Conditions for Disease Development

Sugarcane is a vegetative propagated crop, so the pathogen is transmitted via seed material and by phloem-feeding leafhopper vectors. *Matsumuratettix hiroglyphicus*, *Deltocephalus vulgaris*, and

FIGURE 14.22   Interveinal chlorosis due to iron deficiency.

FIGURE 14.23   Grassy appearance of phytoplasma-infected sugarcane plant.

Yamatotettix flavovittatus have been confirmed as vectors for phytoplasma transmission in sugarcane (Figure 14.23).

14.4.5   DISEASE CYCLE

Due to low pathogen load in plant crop, the crop suffers less. Once ratooned, the pathogen in the stubbles initiates disease in the newly emerging shoots and such clumps will not have any millable canes. Infected seed-cane serves as a primary source, and insect vectors spread the pathogen from cane to cane in the field.

14.4.6   CONTROL

In SCGS disease, the primary concern is to prevent the disease rather than treat it. Large numbers of phytoplasma-infected seed sets used by the farmers usually cause fast SCGS disease spread. Healthy, certified "disease-free" sugarcane sets are suggested as planting material. If disease symptoms are visible within two weeks after planting, such plants can be replaced by healthy plants. Uprooted infected sugarcane plants need to be disposed of by burning them. Moist hot-air treatment of sets is suggested to control infection before planting. This reduces the percentage of disease incidence, but causes a reduction in the percentage of bud sprouting. Phytoplasma infection also spreads through insect vectors; it is therefore important to control them. Roguing of affected stools and destruction helps. Seed/planting material should be treated with hot water (50°C for 120 minutes) or moist hot air (54°C for 21/2 hours.) that eliminates the pathogen from diseased seed materials. Ratooning of affected crop must be avoided. Crop rotation may be employed to reduce inoculum in the field. Set treatment with 500-ppm solution of ledermycin (an antibiotic) before planting may be incorporated.

14.5   YELLOW LEAF VIRUS OF SUGARCANE

Sugarcane yellow leaf disease was first recognized as yellow leaf syndrome in Hawaii. Later it was reportedly associated with yield losses of 25% or more in the cultivar SP 71-6163 in Brazil. Subsequently, the virus sugarcane yellow leaf virus (SCYLV) was discovered to be associated with the disease. Yellow leaf has been found in numerous countries of the world.

14.5.1   CASUAL ORGANISM

Species	Associated Disease Phase	Economic Importance
Yellow Leaf Virus (SCYLV)	Infected Seed-Canes and Insect Vectors	Moderate to Severe

14.5.2 SYMPTOMS

This disease is caused by sugarcane yellow leaf virus belonging to *Luteovirus*. The symptoms appear initially on third to sixth leaf from the top in a maturing plant or ratoon crop. The symptoms will be very clear after 5 to 6 months of crop growth. On the leaves, the symptoms appear as yellowish midrib on the lower surface. The yellowing may be confined to midrib region or the yellow discoloration may spread laterally to adjoining laminar region parallel to midrib up to 2.0 cm. Reddish discoloration of midrib and laminar region is also noticed in certain varieties. In most susceptible varieties, typical yellowing of midribs and laminar region is noticed on the upper surface of the leaves. Finally, symptoms of necrosis of discolored laminar region from leaf top to bottom and subsequent drying of the entire leaf are noticed in severely affected plants. Cold and nutrient stress appears to intensify the symptoms. In ratoon crop the intensity of the disease will be much higher than in plant crop. The disease incidence in sugarcane is found aggravated by poor maintenance of the crop in the field (Figure 14.24).

14.5.3 CAUSE OF DISEASE AND DEVELOPMENT

Taxonomically, SCYLV is a member of the Luteoviridae family. Its classification within the family has not been fully resolved. The virus is transmitted by aphids, *Melanaphis sacchari* and *Rhopalosiphum maidis,* in a semi-persistent manner. The virus is also spread by planting infected seed-cane. The virus is localized within the phloem cells of the plant and can be diagnosed by either reverse transcription-polymerase chain reaction (RT-PCR) or tissue blot immunoassays using an antibody specific for the virus. There is no thermal treatment effective for eliminating the yellow virus in infected sugarcane (Figure 14.25).[4]

14.5.4 FAVORABLE CONDITIONS FOR DISEASE DEVELOPMENT

There is yellowing of the leaf midrib on the underside of the leaf. The yellowing first appears on leaves 3 to 6 counting down from the top expanding spindle leaf. Yellowing is most prevalent and noticeable in mature cane until the end of harvest. The yellowing expands out from the leaf midrib into the leaf blade as the season progresses until a general yellowing of the leaves can be observed from a distance. Eventually, almost all leaves of the plant turn yellowish. Cold and nutrient stress appears to intensify the factors. In the affected canes, internodal elongation gradually decrease and show bunching of leaves in the apex. In addition to loss in cane yield, sugar recovery is also affected in the infected canes (Figure 14.26).

14.5.5 DISEASE CYCLE

The sugarcane aphid acquires the virus during feeding on an infected plant. The aphid retains the virus for life and can transmit the disease during feeding on healthy plants within the same field or in other fields.

FIGURE 14.24   SCYLV symptoms.

FIGURE 14.25 More general leaf yellowing is shown on mature sugarcane plants.

FIGURE 14.26 Initial symptoms of yellow leaf, with a yellowing of the lower surface of the leaf midrib of leaves 3 to 6 counting from the top expanding spindle leaf.

Under sugarcane plantation conditions, SCYLV appears to be exclusively transmitted by the sugarcane aphid *M. sacchari*. Other common aphids either dislike sugarcane or cannot transmit the virus. Relative specificity of luteoviruses with respect to insect vector and plant group is the norm, although exceptions exist (Miller and Rasochova, 1997). The fact that some sugarcane cultivars appeared resistant to SCYLV did not correlate with aphid preference, since the resistant cultivars showed as much infestation by *M. sacchari* as the susceptible ones in the field and in the experiments. Rassaby et al. (2004) reported that although both *M. sacchari* and *R. maidis* were found on sugarcane, only the former tested positive for SCYLV using RT-PCR.

When a source leaf is infected by an aphid it takes three weeks before the virus has multiplied sufficiently to be detected by immunoassay. The leaf on which the viruliferous aphids had been originally introduced never showed viral infection. It is likely that the fast stream of sieve-tube sap (2.5 m h^{-1}, which corresponds to 10–20 sieve tubes s^{-1}) prevents viral particles from moving sideways into the companion cells, which appear to be the only cell in which they can replicate. Only in the terminal assimilate sink tissue, where sap flow is slowed down and occurs mostly laterally into the adjacent cells, do the viral particles move into the companion cells. Therefore, only the young (sink) leaves are infected after inoculation, not the source leaves. Source leaves are infected only when inoculation occurs when these leaves are still in the sink state. It is not known whether the pre-infection internodes and their buds contain the virus.

The propagation of infection by aphids proceeds slowly and sporadically, in the range of a few meters per year. The infection spreads exclusively within the first 6 months of growth and again after ratooning. Whether this is an indication of "mature plant resistance" or of a "contamination window" for the aphids is unknown. The complete infection of plantation fields as is found today can therefore mostly be attributed to the planting of infected seed pieces from infected seed-cane fields, or probably even from breeding programs. The slow aphid-vectored progression of infection offers a relatively simple, possible way to keep high-yielding but susceptible varieties essentially virus-free in commercial plantation fields. If seed-cane fields are situated in places either remote from commercial fields and/or surrounded by a belt of plants of a resistant cultivar, these seed plants should remain free from SCYLV. Airborne aphid incidence could be monitored by traps in the seed-cane fields. If only SCYLV-free seed pieces were planted in the commercial fields, some airborne infections would possibly occur, but with a growth period of 1.5–2 years per cropping cycle, most of the plantation should remain SCYLV free, even after a first ratoon cycle. The situation may be different in other sugarcane-growing countries, where infected aphids may be moved over greater distances by wind. The result of the study indicates that SCYLV infection, which is a threat for susceptible, high-yielding sugarcane varieties, can be confined relatively easily because of the slow progression of the virus in the field. Thus, with proper seed-field maintenance, it should be possible to plant susceptible cultivars without the danger of YLS in sugarcane-growing areas.

14.5.6 Management and Control

- SCYLV reduces both cane and sugar yield.
- Commercial fields planted with regular seed-cane usually have at least 85% of the plants infected with SCYLV.
- Unfortunately, the heat therapy treatments used to control ratoon stunt do not eliminate SCYLV from sugarcane stalks. Thus, using infected seed-cane transmits the virus to the emerging plants. The only way to eliminate the virus from the plant is to use meristem tip tissue culture techniques. Even this technique is not 100% effective and plants derived using this technique must be tested to ensure the virus is eliminated. The rate at which the plants become infected is cultivar dependent. However, in most cultivars the incidence of SCYLV infection is typically 30–40% after 3 years. Since SCYLV-infected plants have a 4 to 10% yield loss of cane weight and sucrose, yellow leaf is a major economic threat. Actual field losses will depend on the loss per plant and the incidence of yellow leaf in the field.

- The use of the tissue-culture based seed-cane would also be free of the ratoon stunt pathogen and would prevent losses from this disease too. Until resistant cultivars can be developed, only the use of SCYLV-free seed-cane can moderate SCYLV yield losses.

14.6 BANDED DISEASE

The disease is of minor importance in the Philippines. It was observed on the leaf sheath of Phil 8477 in Negros during the wet season. So far, the disease has not reached epiphytotic proportion. It was further observed that during conditions of high moisture and temperature or during the period of high humidity, the disease appears on commercially grown varieties particularly in areas with poor drainage and dense population. The symptoms of the disease are manifested in the older leaves which principally occur in the leaf blade and on the leaf sheath, particularly 3–6 inches aboveground. The causal agent is a soil inhabitant fungus. It attacks grasses, particularly Bermuda grass (*Cynodon dactylon* (L.) Pers.), which are common on the borders of sugarcane fields. Infection is by contact of the sugarcane leaf with the soil or with diseased grass leaf.[5]

14.6.1 SYMPTOMS

The disease on the leaves produces irregularly shaped patches. The first visible symptoms on the leaves and leaf sheaths is the yellowish irregular spots in various sizes, which later are tinged with a reddish color and become irregular, large patches by fusing together (Figure 14.6). In the more advanced stage, the affected parts turn red or reddish brown and eventually become straw colored and dried up. On the infected area, a dirty colored, wood-like sign appears due to the formation of abundant conidiophores and conidia.

14.6.2 CONTROL MEASURES

Improve drainage (Figures 14.27 through 14.29).

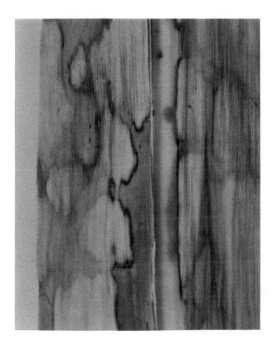

FIGURE 14.27   Symptom of banded sclerotial disease on leaves.

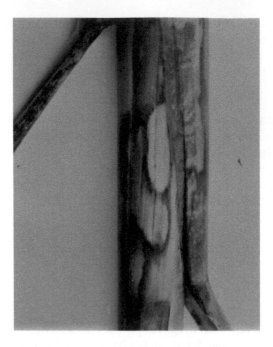

FIGURE 14.28   Symptoms of banded sclerotial disease on the leaf sheath.

FIGURE 14.29   Appearance of the plant affected by banded sclerotial disease on the advanced stage.

14.7   GUMMOSIS OR GUMMING DISEASE

Causal Organism: *Xanthomonas vasculorum*
   Class: Schizomycetes
   Order: Pseudomonadales
   Family: Pseudomonadaceae

- This is 1 of the oldest and most common diseases of sugarcane known in many countries.
- The disease was first reported from Brazil in 1869, since then it has been reported in Puerto Rico, West Indies, Columbia, and Australia.
- In India, Rangaswami reported its occurrence in Tamil Nadu State in 1960, and it is now known in other sugarcane areas of South India (Rangaswami, 1960a,b).

14.7.1   Symptoms

- The disease is characterized by 3 distinct symptoms.

- On the leaves characteristic longitudinal streaks or stripes, 1/8 to 1/4 inch in width and several inches in length are found; these stripes are pale yellow in color, later turning brown.
- The affected canes are stunted with short internodes, giving the plant a bushy appearance. When such canes are split open or cut transversely, a dull yellow bacterial ooze comes from the cut ends.
- In advanced cases, cavities may develop in the pith region.
- The fibro-vascular bundles are deep red, and this is more intense at the nodal regions of the stem.
- The causal bacterium is variously named by different authors, but the above name is universally accepted at present.
- The bacterium is a short rod, 1.0–11.5μm x 0.4–0.5μm, motile by means of a single polar flagellum.
- Gram-negative, non-spore forming, non-capsular, and non-acid fast.
- On beer, extract agar; it produces a yellow spreading and slimy growth.

14.7.2 DISEASE CYCLE

- The disease is primarily spread through the sets taken from diseased plants.
- The secondary spread may be through agricultural implements, including the cutting knife.
- In the field the infection may spread by wind and water, the source of pathogen being the gummy bacterial ooze coming out of diseased tissues.
- Certain insects, particularly flies, play a significant role in transmitting the pathogen from place to place.
- The bacterium can survive in the insect's body for a long time, and in this way, may be transmitted long distances.
- On entry into the host the bacterium reaches the vascular tissues and becomes systemic.
- Besides sugarcane, several plant species falling under *Graminae*, including maize, sorghum, *Panicum* spp., and some other grasses have been found to be infected by this bacterium.
- These hosts may help in many ways the bacterial spread and perpetuation.

14.7.3 CONTROL

- As in other diseases of sugarcane, gummosis can be controlled by selecting sets from disease-free plants and adopting strict field sanitation practices.
- In Fiji Island the disease was completely eradicated by destroying the diseased canes and growing resistant varieties.
- Studies on the availability of resistant stock among the sugarcane varieties in India are needed.

14.8 ROOT KNOT NEMATODE

Casual Organism: *Meloidogyne javanica*
    Class: Nematoda
    Order: Tylenchida
    Family: Heteroderidae

- Over 50 nematode species belonging to 20 different genera are reported to infect sugarcane in various countries of the world.
- At least 10 have been reported in India.
- Among these the *Meloidogyme* spp. are important.
- Sugarcane is among the various hosts affected by nematodes in India.

- The crop is damaged by the root knot nematodes, though several other ectoparasitic nematodes occur on this host.
- The occurrence of root knot and other hosts was first established by Rangaswami (1960a,b) and his associates during 1958–1960.

14.8.1 SYMPTOMS

- The diseased plants are chlorotic and stunted, and yellow stripes show on the young leaves while older leaves appear healthy.
- Crops at all stages of growth are affected and the symptoms are more prominent during summer months.
- When the roots of affected plants are dug out and examined, they are found to be knotted.
- The young, white roots show much less knotting than the older, wiry ones.
- The knots are usually linear and are found more commonly toward the root tips. They are about 5–8 mm in thickness.
- The nematodes are separated from the infected tissues by the Baermann Funnel technique and examined under the microscope.
- The females measure 400–460mm in length and 45–50 mm at the thickest point, oesophagus 40–45 mm and tail 30-35mm.
- The males measure 420–460 mm in length, 20–25 mm in width at the thickest point and oesophagus 55–60 mm and tail 30–35 mm.
- The cyst is pear shaped, measuring 600–650 mm x 380–400mm with a prominent beak.
- The other species of Meloidogyne reported on sugarcane are *M. arenaria* Chitwood and *M. incognita* Chitwood, both of lesser importance than *M. javanica*.

14.8.2 DISEASE CYCLE

- The nematode eggs persist in soil in cyst form and when the susceptible host is planted, the cyst give rise to the larval forms which invade the root and form the galls.
- There are reports on the exudation of specific chemical substances by the various hosts with either attract or repulse the nematode larvae.
- Once the larvae are inside the host tissue, they feed through the vascular bundles and complete the life cycle.
- They produce the cysts, which protrude from the host roots and remain there until harvest.
- The production of growth regulators such as indole acetic acid in the host tissue is reported to be correlated with the growth-promoting and nodule-forming effect of the nematodes, as found by Rangaswami and his associates (Rangaswami, 1960a,b).
- There are over 100 plant species, belonging to diversified families, which are host plants for *M. javanica*. No doubt these host plants, particularly the weeds present in sugarcane fields, and the crops used in rotation with sugarcane, play significant roles in the perpetuation and inoculum build-up of the nematode.

14.8.3 CONTROL

- The root knot nematode can be checked by fumigating the soil with a nematicide such as Nemagon or D.D.T
- The soil must be opened soon after crop harvest.
- The chemical fumigant should be injected into the soil as per the directions given by the manufacturer after which the soil is kept covered with a tarpaulin or plastic sheets for a week or 2 if necessary.
- The fumes permeate the soil, killing the nematode larvae and cysts and other organisms as well.

- The soil is safe for planting in two to three weeks after fumigation.
- In some countries crop rotation with marigold reduces the nematode population.
- The effect of such plants or their plant roots exudates on the nematode population, and the inter-relationships of other soil microbial populations with the nematode, must be investigated in detail to understand this complex phenomenon and to evolve effective control measures.
- Certain nematophagous fungi like *Catenaria vermicola, Arthrobotrys conoides Drech.*, and *Dactylella deodycoides Drech.* are present in soils, and this might help in reducing the nematode population.
- Besides *Meloidogyne javanica* and other species of Meloidogyne, which cause root knot symptoms, *Pratylenchus* sp., *Paratylenchus macrophallus deMan*, *Trichodorus* sp., *Rotylenchus* sp., *Tylenchorhynchus* sp., *Xiphinema* sp., *Rotylenchus* sp., and *Hoplolaimus coronatus* Cobb have been reported by Kishan Singh (Rangaswami, 1960a,b) on sugarcane in some parts of India.
- More detailed studies are needed on the disease symptoms, life cycle of the nematodes, nature, and estimation of damage and control measures.
- A phanerogamous parasite affecting the roots of sugarcane has also been reported from some parts of India.
- It is identified as *Striga euphrasioides Benth.*
- This is a partial root parasite growing up from the roots to form leafy shoots.
- The parasite can synthesize carbohydrates through the green chlorophyll pigments in the leaves but for its other nutrients it depends on the host root.
- It is usually controlled by pulling out the shoots before flowering and seed set.
- Spraying weedicides like 2–4, D will kill the parasite without affecting the sugarcane plant appreciably.

14.9   EYE SPOT IN SUGARCANE

Causal organism: *Helminthosporium sacchari*

- Usually a crop of 6–7 months is more susceptible to the disease.
- Fungus penetrates the host tissue either through stomata, bulliform cells, or directly through the cuticle.
- Cloudy weather, high humidity with drizzle, coupled with low night temperatures, wetting of leaves either through precipitation or dew greatly enhance disease development.
- Waterlogging, high fertility status, and excess nitrogen fertilization also favor the spread of the disease.

14.9.1   SYMPTOMS

Lesions first appear as small water-soaked spots and are darker than the surrounding tissues. The spot becomes more elongated, resembling the shape of an eye and turns straw colored within a few days. Finally, the central portion becomes reddish brown surrounded by straw-colored tissues. Then reddish-brown streaks of a runner develop extending from the lesions toward the leaf tip along the veins. Later the spots and streaks coalesce to form large patches and causes drying of leaves.[6]

14.10   RED STRIPE

The disease is caused by bacteria *Xanthomonas rubrilineans*. It appears in May. Leaves show red streaks. To control this disease, if affected plants are a few in number, then rogue out these and bum them, otherwise discard tile whole field. Fresh sowings are done with resistant variety in well-drained soils.

Causal Organism: *Xanthomonas rubrilineans*
Class: Schizomycetes
Order: Pseudomonadales
Family: Pseudomonadaceae

- This is another bacterial disease of sugarcane which has a world-wide distribution.
- It has been known since 1893 in Java and since 1903 in Australia, but in India only since 1933, when it was reported by Desai and also McRae (Yadav et al., 2016).
- In 1960 Rangaswami made detailed studies on the disease and its causal agent.

14.10.1 SYMPTOMS

- Red stripe is characterized by the appearance on the leaves of chlorotic lesions carrying dark red stripes 0.5–1.0 mm in breadth and several mm in length, either distributed all over the blade, or concentrated in the middle.
- Several of them may coalesce to cover large areas of the leaf blade, and to cause wilting and drying of the leaves.
- Whitish flakes occur on the lower surface of the leaf, corresponding to the red lesions on the upper surface.
- These flakes are the dry bacterial ooze. When young shoots are affected, shoot or top rot may result.
- The growing points of the shoot are yellow and later reddish with dark brown stripes on the shoots.
- The rotting may commence from the tip and spread downward.
- If the affected plants are cut by splitting the shoot downward, dark red dis-coloration of the tissues may be seen.
- In the affected canes cavities may form in the pith region, and the vascular bundles are distinct because of the dark red dis-coloration.
- The diseased and rotting shoots can be easily pulled out and separated from the plant.

14.10.2 DISEASE CYCLE

- The disease spreads in the field by wind and rain, and by cutting, as the basal stem from which the sets are taken is mostly free from the bacterial infection.
- The bacterial ooze forms the infected leaves and shoots dries up to form a thin crust on the surface.
- When dry, the bacterial cells spread freely by wind.
- The cells falling on the host plants, enter them through natural openings or wounds and establish themselves in the various tissues, including the xylem.
- Infected parenchymatous cells may collapse and normal functioning of the plant parts may fail.
- Several grasses, including ragi and bajra, have been reported to be infected by the bacteria.
- These hosts may also play a role in the perpetuation and spread of the pathogen.

14.10.3 CONTROL

- This is a difficult disease of sugarcane to control.
- Whenever the disease is noticed, the affected plants should be removed and burned.
- Such systematic destruction of the affected plants reduces the disease incidence.
- Growing resistant varieties is, however ,the best method of control.

REFERENCES

1. TNAU Agritech Portal, *"Organic Farming: Disease Management of Sugar"*. http://agritech.tnau.ac.in/org_farm/orgfarm_agridiseases.html

2. Majumder, D., 2012, *"Wilt disease of sugarcane"*, IITK Agropedia. http://agropedia.iitk.ac.in/content/wilt-sugarcane

3. Rott, P., and Comstock, J. C., 2002, *"Sugarcane Smut Disease"*, University of Florida, IFAS Extension.

4. Comstock, J. C., Sandhu, H. S., and Odero, D. C., 2015, *"Sugarcane Yellow Leaf Disease"*, University of Florida IFAS Extension.

5. La Granja Agricultural Research and Extension Center (LGAREC), 2012, *"Pellicularia sasakii (Shirai) ITO"*. https://www.bar.gov.ph/index.php/biofuels-home/bioethanol/sugarcane/sugarcane-diseases/1544-banded

6. Nainwal, K., 2009, *"Eye Spot Disease in Sugarcane"*, IITK Agropedia.

7. Lehrer, A. T., Schenck, S., Yan, S.-L., and Komor, E. 23 July 2007, "Movement of aphid-transmitted Sugarcane yellow leaf virus (ScYLV) within and between sugarcane plants", *Plant Pathology* 56, 711–717.

8. Went, F. A. F. C. 1893. Het Rood Snot. Arch. Java Suikerindus. 1: 265-282.

9. Butler, E. J., 1906. Fungus Diseases of Sugar-cane in Bengal. India Dept. Agr. Mem., Bot. Ser. 1 (3): 2–24.

10. Miller, W. A., Rasochova, L., 1997. Barley yellow dwarf virus. *Annual Review of Phytopathology* **35**: 167–90.

11. Rassaby, L., Girard, J. C., Lemaire, O., Costet, L., Irey, M. S., Kodja, H., Lockhart, B. E. L., Rott, P., 2004. Spread of *Sugarcane yellow leaf virus* in sugarcane plants and fields on the island of Réunion. Plant Pathol. 53: 117–125

12. Rangaswami, G. 1960a. Studies on two bacterial diseases of sugarcane. *Current Science* 29: 318–19

13. Rangaswami, G. 1960b. Further studies on bacterial gummosis and red stripe disease of sugarcane *Journal of Annamalai University* 22: 135–50.

14. Yadav, M. K., Dhakad. P. K., Yadav. S. K. Sushreeta, N., and Ram, C., 2016. *A Treatise on Sugarcane Diseases in North India*. Advances in Life Sciences 5(11), 4366–4367.

15 Guava

Guavas are common tropical fruits cultivated and enjoyed in many tropical and subtropical regions. Guava is a small tree in the Myrtle family (Myrtaceae) and it is native to Mexico, Central America, and northern South America. Although related species may also be called guavas, but they may belong to other species or genera, such as the "pineapple guava" Acca sellowiana. The term "guava" appears to be derived from Arawak *guayabo* "guava tree", via the Spanish *guayaba*. It has been adapted in many European and Asian languages, having a similar form. Another term for guavas is *peru*, derived from pear.

Guavas originated from an area thought to extend from Mexico or Central America and were distributed throughout tropical America and the Caribbean region. Guava fruits, usually 4 to 12 cms (1.6 to 4.7 inches) long, are round or oval depending on the species. They have a pronounced and typical fragrance, similar to lemon rind but less sharp. The outer skin may be rough, often with a bitter taste, or soft and sweet. Varying between species, the skin can be any thickness, is usually green before maturity, but becomes yellow, maroon, or green when ripe. The pulp inside may taste sweet or sour. In 2011, India was the world production leader with 17.6 million tons, an amount not exceeded by the accumulative total of the 6 largest guava producers which are China, Thailand Mexico, Pakistan, Brazil, and Bangladesh. Different diseases of Guava are discussed in this chapter.

15.1 ANTHRACNOSE OF GUAVA

Anthracnose in guava is reported to be a serious disease. Anthracnose is the most commonly observed disease that affects both pre- and post-harvest management of guava. This disease can cause considerable postharvest losses and can affect young developing flowers and fruit. It has been reported in all guava-growing areas around the world where high rainfall and humidity are present (Figure 15.1).[1]

15.1.1 CAUSAL ORGANISM

Species	Associated Disease Phase	Economic Importance
Colletotrichum gloeosporiodes (teleomorph: *Glomerella cingulata*)	On Leaves and Fruits	Severe

15.1.2 SYMPTOMS

The disease attacks all parts of the plant except the roots. The growing tips of affected plants turn dark brown and the black necrotic areas extend backward causing dieback of the plant. There is an appearance of small spots of the size of a pin-head on fruits especially during monsoon. Later, several spots coalesce to form bigger lesions. On the tips and margins of the leaves, spots are gray in color.

Also, symptoms of this disease are observed on mature fruits on the tree. The characteristic symptoms consist of sunken, dark colored, necrotic lesions. Under humid conditions, the necrotic lesions become covered with pinkish spore masses. As the disease progresses, the small sunken lesions coalesce to form large necrotic patches affecting the flesh of the fruit (Figure 15.2).

FIGURE 15.1   Guava.

15.1.3   CAUSE AND DISEASE DEVELOPMENT

C. gloeosporioides (teleomorph: *G. cingulata*) is the pathogen responsible for causing anthracnose. The teleomorph stage may or may not play a role in the disease cycle. Colonies of *C. gloeosporioides* on potato-dextrose agar are grayish white to dark gray. Production of aerial mycelia by strains varies, ranging from a thick mat to sparse tufts associated with fructifications. Conidia are hyaline, unicellular, and either cylindrical with obscure ends or ellipsoidal with a rounded apex and a narrow, truncate base. They form on hyaline to faintly brown conidiophores in acervuli that are irregular in shape and approximately 500 μm in diameter. Setae are 1 to 4 septate, brown, slightly swollen at the base, and tapered at the apex (Figures 15.3 through 15.5).[1]

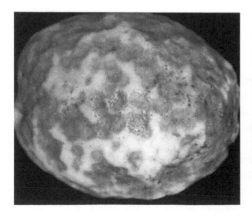 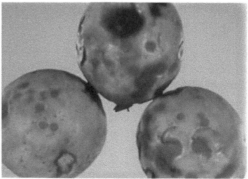

FIGURE 15.2   Anthracnose of guava and symptoms of anthracnose on guava fruit.

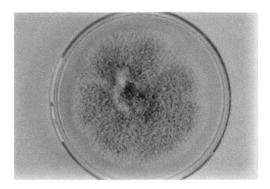

FIGURE 15.3   *C. gloesporeoides* fungal culture.

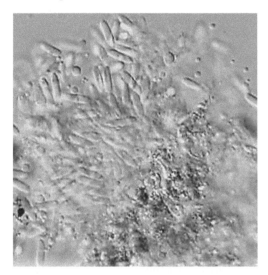

FIGURE 15.4   *C. gloeosporeoides*

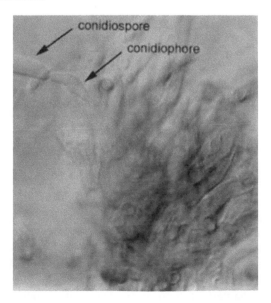

FIGURE 15.5   *C. gloeosporeoides.*

15.1.4 Favorable Conditions of Disease Development

- The first observable symptom of the guava fruit anthracnose on the field appears to be small, slightly sunken, dark or blackened necrotic lesions on immature fruits. The most characteristic symptoms appear during the rainy season as small pin-head sized spots on the unripe fruits.
- They gradually enlarge to form sunken and circular, dark brown to black spots.
- The infected area of the unripe fruits becomes harder and corky.
- Acervuli are formed on fruit stalks.
- The pathogen remains dormant for about 3 months in the young infected fruits.
- In moist weather, acervuli appear as black dots scattered throughout the dead parts of the twigs.
- The conidia are spread by wind or rain.

15.1.5 Disease Cycle

Conidia (asexual spores) are the fungal structures responsible for anthracnose infection. Conidia are produced on dead twigs, necrotic fruit lesions, inflorescences, and leaves. Inflorescences and young fruit are extremely susceptible, and if infected, may cause abortion and abscission. Conidia spread via rain splash and can cause infection; symptoms may develop shortly thereafter on any aboveground host tissue. Latent infections are common with this disease and may remain quiescent for months.[2]

15.1.6 Management

Control measures are needed in commercial guava production. The use of resistant cultivars provides the most efficient tactic in disease management. Additional cultural control tactics used to aid in disease management include disease monitoring and the use of micro-irrigation. Chemical control can be quite effective and several systemic and non-systemic fungicides are available for use on guava. Timely applications shortly before and during flowering and fruit development are crucial for disease management; subsequent applications may alleviate pre-harvest and post-harvest disease on fruit.

15.1.7 Control

Chemicals	Concentration	Use
Mancozeb	0.25% v	Helps to Alleviate Pre-Harvest and Post-Harvest Disease
Bordeaux Mixture or	0.6% v	Spraying the Trees with These Chemicals Before the
Copper Oxychloride	0.2%	Onset of Monsoon Reduces the Disease Incidence

15.2 CANKER [PESTALOTIA PSIDII PAT.]

15.2.1 Symptoms

The disease generally occurs on green fruits and rarely on leaves. The first evidence of infection on fruit is the appearance of minute, brown or rust colored, unbroken, circular, necrotic areas, which in advanced stage of infection, tears open the epidermis in a circinate manner. The margin of the lesion is elevated, and a depressed area is noticeable inside. The crater-like appearance is more noticeable on fruits than on leaves (Figure 15.6). The canker is confined to a very shallow depth and does not penetrate deep into the flesh of the fruit. In older cankers, white mycelium consisting of numerous spores are noticeable. In severe cases, raised, cankerous spots develop in great numbers

FIGURE 15.6   Canker on fruit.

and the fruits break open to expose seeds. The infected fruits remain underdeveloped, become hard, malformed, and mummified and drop. Sometimes, small rusty brown angular spots appear on the leaves. In winter, the cankerous spots are common but in rainy seasons, minute red specks are formed.

15.2.2   Mode of Spread and Reason for Severity

The pathogen is primarily a wound parasite and avoids injury to fruits. Germination of spores is at a maximum of 30°C and does not germinate below 150°C or above 400°C with 98% relative humidity.

15.2.3   Management

The spread of disease (in the early stage of infection) is controlled by 3 to 4 sprays of 1% Bordeaux mixture or lime sulfur at 15-day intervals.

15.3   ALGAL LEAF AND FRUIT SPOT

[*Cephaleuros virescens Kuntze (= C mycoidae Karst.), C. parasiticus*]

15.3.1   Symptoms

Alga infects immature guava leaves during early spring flush. Minute, shallow brown velvety lesions appear on leaves and as the disease progresses, the lesions enlarge to 2–3 mm in diameter. On leaves the spots may vary from specks to big patches. They may be crowded or scattered. Leaf tips, margins or areas near the mid-vein are most often infected (Figure 15.7). On immature fruits the lesions are nearly black. As fruits enlarge, lesions become sunken. Cracks frequently develop on older blemishes as a result of enlargement of fruits; lesions are usually smaller than leaf spots. They are darkish green to brown or black to color. Disease begins to appear to be more serious during later months. The pathogen sporulates readily during periods of high rainfall. In winter, symptoms are not available.

15.3.2   Management

The control of alga can be achieved by sprays of Copper oxychloride (0.3%) 3–4 times at an intervals of 15-days when initial symptoms are noticed.

FIGURE 15.7   Algal spots on leaves.

15.4   CERCOSPORA LEAF SPOT

(*Cercospora sawadae* Yamamoto)

15.4.1   SYMPTOMS

The disease appears as water soaked, brown irregular patches on the lower surface and yellowish color on the upper surface of the leaf. Older leaves are mostly affected and the severely affected leaves curl and subsequently drop off (Figure 15.8).

15.4.2   MANAGEMENT

Spray mancozeb or Dithane-M-45 (0.2%) at monthly intervals.

15.5   SOOTY MOLD

[*Phragmocapnias betle*, *Scorias philippensis*, *Tichomerium grandisporum*, *Limacinula musicola*, *Aithaloderma clavatisporum*, *Tripospermum* sp., *Polychaeton* sp., *Leptoxyphium* sp., and *Conidiocarpus* sp.]

15.5.1   SYMPTOMS

Sooty mold proliferates in abundance on the foliage of guava, subsisting on the honeydew secreted by scale insects, aphids, white flies, and mealy bugs. Symptoms consist of blackish brown, velvety

FIGURE 15.8   Cercospora leaf spot.

FIGURE 15.9   Sooty Mold.

thin, membranous covering on the leaves. In severe cases, the foliage appears black due to heavy infection. The affected leaves curl and shrivel under dry conditions (Figure 15.9).

15.5.2   MANAGEMENT

1. The control of disease consists in removing the cause by destroying the insects. The mold will die out for want of a suitable growth medium if honeydew-secreting insects are killed by suitable insecticides.
2. Foliar spraying of wettasul + chloropyriphos + Gum Accecia (0.2 + 0.1 + 0.3%) at 15-day intervals has been found to be very effective.

15.6   DAMPING OFF OF SEEDLINGS

(*Rhizoctonia solani* Kuhn)

15.6.1   SYMPTOMS

Both pre-emergence and post emergence phases of the disease are observed. In the pre-emergence phase the infected seeds and seedlings show water-soaked discoloration; the seed becomes soft and ultimately rots. The affected young seedlings are killed before they reach the soil surface. In the post emergence phase, hypocotyl at ground level or the upper leaves are discolored a yellowish to brown color, which spreads downward and later turns the plant soft and finally rots and constricts it. The affected seedlings ultimately fall off and die. Strands of mycelium may appear on the surface of the plants under humid conditions. The pathogen is also responsible for causing fruit rot if the fruit are kept on the soil after harvest or if they are touching the ground. The disease occurs in warm wet growing areas during the rainy season. Initially, lesions appear firm, brownish, and water soaked, enlarging into irregular-shaped sunken cankers (Figure 15.10). Rotting is rapid, and infected fruit become covered with a dense grayish-brown mold. *Sclerotial* bodies are seen in due course on fruits.

15.6.2   MANAGEMENT

1. Diseased seedlings and weeds should be removed and burned.
2. Excessive use of water and close planting should be avoided as the organism is moisture loving. Seedbeds should be prepared with proper drainage arrangement.
3. As the fungus survives on several hosts, planting of susceptible hosts should be avoided.
4. 2-minute dipping of guava seeds in Captan/thiram (0.2%) is advocated before seed sowing.
5. Drenching of soil with Copper oxychloride (0.3%) helps in reducing the diseases intensity in the nursery.

FIGURE 15.10   Damping off guava seedlings.

15.7   PHYTOPHTHORA FRUIT ROT

[*Phytophthora parasitica* Dastur/ *P. nicotianae* var. *parasitica, P. citricola*]

15.7.1   SYMPTOMS

The symptom starts at the calyx disc of the fruit during the rainy season. The affected area is covered with a whitish cotton-like growth, which develops very fast as the fruit matures, and pathogen is able to cover almost the entire surface within a period of 3–4 days during humid weather. Under high relative humidity, the fruits near the soil level covered with dense foliage are most severely affected. The fallen fruits are badly affected. The skin of the fruit below the whitish cottony growth becomes a little soft, turns light brown to dark brown and emits a characteristic unpleasant smell. Ultimately, such fruits either remain intact or drop off from the tree (Figure 15.11).

15.7.2   MODE OF SPREAD

1. Rain and wind are important for the spread
2. The pathogen produces a great number of sporangia (spores) on the surface of diseased tissues principally when the temperature is near 25°C, and this is an important source of inoculums in the development of epidemics.

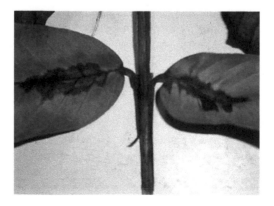

FIGURE 15.11   Phytophthora on foliage.

3. Drops of rain are necessary for the liberation of sporangia from the infected plant material or soil.

4. Infection ceases when temperature is less than 15°C or more than 35°C.

15.7.3 REASON FOR SEVERITY

Cool, wet environmental conditions with high soil moisture favor disease development. High humidity, temperatures from 28°C–32°C, poorly drained soils, and injuries are important for the initiation of the disease. Lack of timely fungicidal sprays and close plantation are the reasons for the severity of disease.

When the disease appears on young and half-grown fruits, they shrink, turn dirty brown to dark brown, turn hard in texture, and either remain intact as mummified fruit or drop off. The disease incidence varies from 8–30%, depending upon the weather and foliage conditions (Figure 15.12).

15.7.4 MANAGEMENT

Dithane Z-78 (0.2%), Ridomil or Aliette (0.2%), or Copper oxychloride (0.3%) are found to be effective in controlling foliar infection. Soil drenching should be completed with Copper oxychloride (0.3%), Ridomil or Aliette (0. 2%). Plant spacing and fertilizer régimes should be managed to avoid unnecessarily dense plant canopy.

15.8 STYLAR END ROT

[*Phomopsis psidii* De camara and *P. destructim*]

15.8.1 SYMPTOMS

The most visible disease symptom is the discoloration in the region lying just below and adjoining the persistent calyx. Such areas gradually increase in size and turn dark brown. Later, the affected area becomes soft. Along with the discoloration of epicarp, the mesocarp tissue also shows discoloration and the diseased area is marked by being pulpy and light brown in color in contrast to the bright white color of the healthy area of the mesocarp (Figure 15.13).

At an advanced stage due to disorganization of the inner affected tissues, size of the fruit shrinks, and concentric wrinkles develop on the skin. Finally, the affected fruit is covered with dark colored pycnidia. Serious losses of up to 10% occur in the orchard if the disease is not properly controlled.

FIGURE 15.12   Phytophthora fruit rot on green fruits.

FIGURE 15.13   Stylar end rot.

15.8.2   MANAGEMENT

- Spray Copper oxychloride (0.3%) or carbendazim or Thiophonate-methyl (0.1%) before the onset of winter fruiting.
- However, care should be taken that there is no spraying 15 days prior to harvesting.
- Fruit injury should be avoided.

15.9   SOFT WATERY ROT

(*Botryodiplodia theobromae Pat*)

15.9.1   SYMPTOMS

The infection starts as a brownish discoloration mostly at the stem end and it gradually proceeds downward in an irregular wavy manner. Finally, the whole fruit may be affected. The decay takes the form of a soft, watery break down, resulting from infection via wound or through the stem end. In advanced cases, numerous small pycnidia are produced over the entire surface of the fruit (Figure 15.14).

15.9.2   MANAGEMENT

1. Ensure careful handling in order to reduce the incidence of wounding.
2. Captan and homeopathic drug arsenic oxide are found to be effective against the fungus.
3. Applications of *Bacillus subtilis* and *Streptosporangium pseudovulgare* on guava fruits have been found to be effective.

FIGURE 15.14   Soft watery rot.

15.10   BOTRYOSPHAERIA ROT

(*Botryosphaeria ribis* Gross. & Duggar)

15.10.1   Symptoms

The disease has been identified as *B. ribis*, which causes serious losses during fall. The infection usually occurs at or near the distal end in the region of persistent calyx. The rot begins with a translucent zone around the distal end that becomes brown in color. With the evolution of the disease, the lesion becomes dark black and wrinkled with dry skin, while maintaining translucent margins (Figure 15.15).[3]

Spots on the skin of the fruit often appear slightly sunken, which tends to increase in size with the formation of Pycnidia. It is embedded in the disease tissues and lesions containing characteristic spores are formed over time. The fungus produces an asexual stage which is responsible for infection in the orchards.

15.10.2   Management

Orchard sprays with Copper oxychloride (0.3%)/Dithane M-45 (0.2%) at 15 day intervals.

15.11   HYALODERMA LEAF SPOT

(*Hyaloderma* sp.)

15.11.1   Symptoms

The fungus preferentially infects mature leaves during wet weather. In the more advanced stage, when the conditions are highly favorable, the disease can cause a severe spot on the leaves around middle lamina. Under humid conditions, it is common to see the brick-red color spots (Figure 15.16).[3]

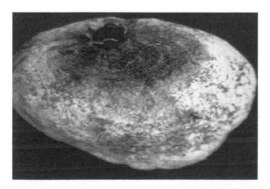

FIGURE 15.15   Botryosphaeria rot.

FIGURE 15.16   Hyaloderma leaf spot (lower surface).

On the lower surface of the leaves, in the areas corresponding to the spot, the growth of the fungus can be observed. Abundant spotting causes defoliation of the leaf. The lesions spread easily to healthy leaves and can coalesce forming a large irregular to semicircular lesions area on the surface of the leaf, up to 4–5 mm in diameter, sharply defined, and occasionally depressed. Initial infections on the underside of the leaf may cause chlorotic patches or spots to occur on the upper side of leaf. Since, it is a new disease, more research is warranted to know its epidemiology, biology, and control, etc. (Figure 15.17).

15.11.2 MANAGEMENT

The diseases could be managed with Copper oxychloride (0.3%) spray during the rainy season.

15.12 PARASITES

(*Loranthus* sp.)

15.12.1 SYMPTOMS

The guava tree is the prey of many parasites and the most interesting of these are so-called phanerogamic parasites and epiphysis. Guava trees are commonly affected with these parasites particularly in the neglected orchards. They are commonly present on the trunk or branches of the tree, thus, making the tree weak. The foliage-infected host plant is sparse, reduced in size, and its bearing capacity and quality of fruit is considerably lowered. Since the appearance of parasitic plants is quite distinct from the guava host, they can be easily distinguished in infected trees. The point at which the guava host is penetrated is usually characterized by swollen growths called "burrs". The burrs help in the identification of sites at which the parasite has entered the host, an important feature while controlling these plants (Figure 15.18).[3]

FIGURE 15.17   Hyaloderma leaf spot (upper surface).

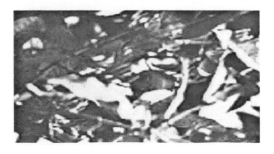

FIGURE 15.18   Parasite (*Loranthus* sp.).

15.12.2 MANAGEMENT

1. The affected branches should be cut sufficiently to eradicate the haustoria. In the early stages, it can be removed easily. If small bit of haustorium is left in the host, it re-grows with renewed vigor and soon starts damaging the host as before.
2. Cutting out affected portions of the to produce enough burrs below surface to remove haustoria and the cut surface is treated with wound dresser viz. Copper oxychloride (0.39%) paste/spray to prevent the secondary pathogens infecting through wounds.
3. Spraying of an emulsion of diesel (30–40%) in soap water is recommended as it was found effective.

REFERENCES

1. Merida, M. and Palmateer, A. J., 2006, *"2013 Florida Plant Disease Management Guide: Guava (Psidium guajava)"*, University of Florida IFAS Extension.
2. Tabua, E., *"Plant Diseases and Their Management", Lecture at Florida National University.*
3. Prakash, O., *"IPM Schedule for Guava Pests"*, National Horticulture Mission Ministry of Agriculture Extension Bulletin No. 2.

Index

Milton Keynes UK
Ingram Content Group UK Ltd.
UKHW052030141024
449569UK00017B/761